A Functional Anatomy
of Invertebrates

A Functional Anatomy
of Invertebrates

V. FRETTER
A. GRAHAM

Department of Zoology.
University of Reading

1976

ACADEMIC PRESS
LONDON   NEW YORK   SAN FRANCISCO
A Subsidiary of Harcourt Brace Jovanovich, Publishers

ACADEMIC PRESS INC. (LONDON) LTD.
24–28 Oval Road
London NW1

US editions published by
ACADEMIC PRESS INC.
111 Fifth Avenue
New York, New York 10003

Library of Congress Catalog Card Number: 75-19636
ISBN: 0-12-267550-9

MADE AND PRINTED IN GREAT BRITAIN BY
THE GARDEN CITY PRESS LIMITED
LETCHWORTH, HERTFORDSHIRE SG6 1JS

Preface

THE LAST HALF century has seen vast changes in the teaching of zoology. The selection of topics by which the student of the nineteen seventies is introduced to the subject is quite unlike that with which his father was presented. In part, this is due to the need to incorporate new ways of studying animals made possible by such techniques as electron microscopy, and to present the results of such new interests as ecology, or those derived from biology's new relations with physical science; in part, it is due to fashion, with its predilection for the novel, which operates in scientific fields with the same apparent inevitability and unpredictability as in dress or hair style. The university or college year, however, has not increased in length to keep pace with the development of new areas of study and they have been accommodated only by the restriction or exclusion of some of its previous content. This has especially affected the courses which dealt with invertebrates and vertebrates, which had, by comparison with others, an old-fashioned look that made them vulnerable to attack: in some cases they have been so severely pruned as to sink below the critical threshold at which intelligibility can be maintained, a situation which inevitably leads to loss of interest on the part of both teacher and taught.

Since Man is a vertebrate along with most of the animals which he has domesticated for work, food, sport or keeping as pets, and since the vertebrates are a zoological unity in a sense that does not apply to invertebrates, it is courses in invertebrate zoology which have come off the worse in these circumstances; indeed, brief study of invertebrates serves only to exaggerate their disunity and to bewilder the student. The average university student, in our experience, finds his courses in invertebrate zoology are those which least capture his affection and enthusiasm, despite the fact that it was a confessed interest in animals at large which attracted him to their scientific study. The reasons for this situation seem to us to be partly the over-brief treatment, partly the emphasis on dead anatomy without relation to function, and partly an over-emphasis of the anatomical bases of the taxonomy of animals about which the student otherwise knows nothing.

It is not necessary to labour the points that animals, because they are animal in nature, have activities which they must carry out to stay alive and reproduce, and that their anatomical organization is the apparatus with which they do this. But to study the anatomy without the approach of the bioengineer, without asking such questions as – What does it do? – How does it do it? – How does the structure limit and define the ways in which the function can be fulfilled? – is dispiriting and sterile, whereas with this point of view the functional study of anatomy is stimulating and rewards the student with fuller insight into each animal's way of life. It is to try to provide the answers to such questions that we have written this book.

The basic activities of animals are identical, from Amoeba to Man – all move, all feed, all respire. It might therefore seem right to describe these activities and their underlying machinery without relation to the systematic position of the animals dealt with, and this is, indeed, what many texts set out to do. Yet there is no doubt that these activities are related to the animal's position in the animal kingdom since its grade of organization determines the anatomical and histological complexity of the apparatus which it can devote to carrying them out. Before the student is in a position to appreciate the similarities and differences between patterns and to recognize their relative advantages and disadvantages, he must have the patterns themselves clear in his mind. For this reason we have, in dealing with most groups (and especially where a phylum contains animals built in diverse ways as in annelids or molluscs), reverted to the type method introduced by Huxley. Where the organization of a group is relatively homogeneous, as in nematodes or brachiopods, we have adopted a more comparative approach.

The book is not intended for the beginner since we have taken elementary ideas and knowledge of many facts and technical terms for granted. Although we hope that it is intelligible to the first year college student it is aimed at those in the later years of their course and at graduate students who need a broad view of the group in which their research interests lie. We hope too that this kind of approach to invertebrate zoology will help the ecologist (to whom, unfortunately, animals are often only so many units in the system which he he is studying) to see more clearly how the activities of each animal contribute to the whole, and will aid the physiologist to relate the particular function in which he is interested to all those which the animal performs.

We have illustrated the text with drawings which we have made as realistic as possible by basing them on animals, whole or dissected. When they have had their starting point in the work of others this fact is acknowledged in the legend of the Figure. Though most of the drawings have been made by one or other of us, we gladly acknowledge much help given by Mrs Margaret Shepherd and Miss Shirley Townend. We are also grateful for valuable textual help given by Dr Elizabeth Andrews, for the photographs in Fig. 9 prepared by Mrs Patricia Hawkins and are glad to thank Mrs Pauline Brown and Mrs Grace Smillie for the care and accuracy with which they typed our manuscript. Finally, we wish especially to thank Mrs Dorothy Sharp of Academic Press for her help in preparing the manuscript for the printer.

July 1975   V. FRETTER

                                                                     A. GRAHAM

Contents

1
Introduction

IF YOU ASK the ordinary man in the street how to tell the difference between an animal and a plant, you are likely to get a variety of answers. One will tell you that animals move whilst plants remain stationary; another that plants are green, whereas green is a relatively uncommon colour in animals; a third may say that animals are usually divisible into two halves along one line only – or, as the biologist would say, are bilaterally symmetrical – whereas plants may be similarly divided along any one of many diameters, or, are radially symmetrical; and still another may mention that whilst animals eat solid food, plants do not seem to eat at all.

All these observations are accurate so far as they go, but it is rather easy to find exceptions to most of them. You have only to mount a drop of pond water on a slide and examine it under a microscope to see many motile organisms, which are also green. Are these animals because of their mobility, or plants because of their green colour? Similarly many animals such as sea anemones exhibit a stationary habit and a radial symmetry, but catch and eat solid food. Are these plants because of their symmetry and immobility, or animals because of their feeding habits? When alleged distinctions between two contrasted groups of objects can be shown to lead to doubt as easily as this it is clear that they are not real differences but merely subsidiary features which accompany an as yet undiscovered point of contrast. And this, as is well known in the case of animals and plants, is a matter of nutrition, that is, the source of the energy by means of which the organism runs its body and carries out the multiplicity of activities which make up its everyday life. Once this fundamental distinction between the two kinds of living organism is admitted, the other differences can be seen to follow from it, though, if an organism adopts a mode of life which is unusual in some way or other, then it may not exhibit the same features as other animals or other plants.

The typical plant possesses the pigment known as chlorophyll and for that reason is green. This allows it to trap the energy in sunlight and so build up molecules of sugar from the simple inorganic substances water and carbon dioxide in a process known as photosynthesis.

$$\text{Water} + \text{carbon dioxide} + \text{energy} \xrightarrow{\text{chlorophyll}} \text{sugar}$$

To do this the plant requires a supply of water, which it obtains from the soil, a supply of carbon dioxide, which it obtains from the atmosphere, and sunlight. Lacking any of these it cannot flourish. Much of the organization of the plant body may be related directly to these requirements. The plant has no need to move about from place to place in search of fresh supplies of carbon dioxide – the continuous movement of the air brings it. Similarly it does not require to move in search of water – rainfall and capillary movement in the soil again bring it within its reach. What the plant does require is a shoot system reaching out into the air to catch the carbon dioxide and light there, and a root system permeating the soil to absorb the water, and since these substances are as likely to be found in one direction as another, the root and shoot systems grow radially from the axis of the plant equally in all directions, branching and rebranching as they go so as to tap the resources of as great a volume of air and soil as possible.

In this way, as a consequence of the mode of nutrition, the characteristic fixed and radially symmetrical body of the plant arises, its structure directly linked to its function.

The sugar manufactured in photosynthesis contains energy obtained from sunlight by virtue of chlorophyll. This may be released by destruction of the sugar molecule in a process known as respiration, which is very obviously the antithesis of photo-synthesis:

$$\text{sugar} \longrightarrow \text{water} + \text{carbon dioxide} + \text{energy}$$

The waste products from this can escape by way of the leaves in gaseous form. The energy may be used by the plant for any activity. One of these will necessarily be the production of new living substance, or protoplasm, either in the process of growth, or in that of producing new plant organisms or reproduction. Both require the manufacture of protein molecules, which contain many elements such as nitrogen, phosphorus, sulphur and so on, not found in the molecules of the substances involved in the chemistry of respiration or photosynthesis in the skeletal form in which it has been represented above. The plant, however, can make proteins provided that it can get these elements as inorganic salts dissolved in the soil water which it absorbs through the root system. The addition of this activity to those already carried out by the organism, therefore, does not call for any modification of the root and shoot system adequate for photosynthesis and respiration. All the substances which a plant requires for its respiratory activities, for growth and reproduction, can be absorbed in gaseous form or in solution, and no uptake of solid food is necessary.

We have therefore been able to relate the pattern on which the body of a plant is built – its sessile habit, its radial symmetry, its uptake of food only in solution, its greenness – to one fundamental attribute of plant life, its particular mode of nutri-tion. Motile green organisms are still plants provided they have this method of nutrition, as are those with bilateral symmetry; but we recognize as plants (though aberrant ones) organisms which are not green and are not capable of photosynthesis (for example, moulds) by comparing their general organization with that of plants of normal structure and function.

If we now turn our attention to typical animals we find that they are biochemically less expert than plants and are incapable of manufacturing for their own use such chemical substances as the sugars which are required as respiratory substrates, or the compounds containing nitrogen (amino acids and the like) which are essential for the manufacture of new protoplasm either in growth or reproduction. If animals require these substances for such activities and yet cannot manufacture them for themselves from an inorganic starting point as plants do, how are they to obtain them? Obviously there is only one way: from the bodies of the plants which have made them, or from the bodies of other animals which have already obtained them from the same source. Thus biochemical inadequacy compels the animal to become a slaughterer of other organisms or a carrion feeder in its search for the materials which it must have for obtaining energy or the manufacture of new protoplasm. And just as its mode of nutrition forced a certain structure on the plant, this different method forces a different structure on the animal.

In the first place, it is clear that animals must move. Imagine an animal anchored to the ground in the way that a plant is. In a very short time it would eat all the food within reach and starvation would become inevitable. Animals, therefore, have locomotor organs: cilia or flagella when they are minute, but appendages of a variety of types when they become larger, worked by muscles which act upon a skeleton. These muscles have to be controlled by the animal so that they may move the body in an appropriate way in relation to the food that the creature is seeking and the environment through which it is moving. The animal must, therefore, be equipped with machinery which will allow it to recognize food, to distinguish the edible from the inedible and to move its body in relation to its external environment, This is provided in a series of sense organs, each detecting some physical or chemical factor in the environment and sending nervous messages to a central coordinating and predicting centre, the brain, which, in the light of all the information presented to it by the sense organs and its memory of what has been presented at times past, "decides" what messages should be sent to the muscles to make the locomotor activity of the animal appropriate to the moment.

The food of a plant is already in solution, ready to diffuse or be actively taken up through the surface of the body; this is not, however, true of the food of an animal, which requires preliminary treatment, digestion, before it can be absorbed. The digestive process renders the food soluble and reduces the size of its component molecules to a level at which they can pass through the membranes of the cells forming the lining of the body. Such a process is not easily carried out on the external surface of the body and the typical animal is therefore provided with an internal space, the gut or alimentary canal, into which the food is taken, often given a preliminary mechanical breaking up, and where it is digested by means of enzymes to produce a solution of substances of small enough molecular size to diffuse into the body. Not all of what is eaten responds to digestive treatment; some parts prove indigestible. These are passed to the opposite end of the alimentary tract from the mouth and escape as faeces from the anus. It is worth noting that this material has never been inside the animal's body in any real sense, but has merely passed through it, and hence has never taken part in any kind of metabolic activity.

In this way animals acquire the sugars necessary for respiration and the production of energy, as well as the nitrogenous and other compounds required for the formation of new protoplasm. These substances, however, are only just within the body of the animal, in the wall of the alimentary tract, whereas respiratory and other vital processes go on in all parts. Some transport system is called for to carry food materials from the gut to these parts, and this is normally supplied by a vascular system, a series of tubes running throughout the body filled with a fluid called blood which can dissolve the food absorbed in the gut and transport it wherever required. The motive power driving the blood along the vessels is provided by a specialized contractile part which is the heart.

Respiration calls not only for a substrate like sugar but for oxygen, which will release the energy contained in the large sugar molecule in an oxidative process. The animal, therefore, requires as regular a supply of oxygen as it does of food in order to obtain a regular supply of energy. Indeed, since most animals appear to have evolved in an environment where food might often be lacking, but where oxygen never was, they have learned to store food against emergencies but have never learned to store oxygen, and many die if their supply is interrupted for more than a very short period of time. Oxygen is taken from the surrounding medium, air or water, by way of special, thin, respiratory surfaces through which diffusion is easy, constituting the animal's lungs or gills; it enters the blood, which thus not only distributes food but also oxygen as it circulates round the body.

Every cell in every part of the body is in this way provided with food and oxygen and can carry out the process of respiration, releasing energy, which can be put to any purpose appropriate to the animal's needs at the time. In this process waste substances will be formed as well as energy released, particularly carbon dioxide and water when the substance respired is carbohydrate, as it most commonly is. These must be evacuated from the body, otherwise it might become waterlogged and too acid, and this involves the animal in the process of excretion. Some of the waste – most of the carbon dioxide and some of the water – may escape to the surrounding medium as the blood passes through the respiratory organs, but a true excretory organ is necessary to expel the rest of the water and a variety of waste products from kinds of metabolic activity other than respiration. This organ is loosely called the kidney, and it expels the waste matter by a duct leading to an excretory opening on the surface of the body, the excretory substances being extracted from the blood as it passes through the kidney. It will be noted that the urine excreted by the kidney contains substances which have been linked with the vital activities of the animal and which have often been intimately concerned in the life of the protoplasm. It is this which distinguishes waste of renal origin from faeces.

The normal animal grows. It may keep on doing so indefinitely, but usually there arrives a time when general bodily growth is replaced by the special type of growth which results in reproduction. New protoplasm is not then used to add to the body of the organism, but to make the starting point of a new individual. Whilst the most primitive organisms may have been able to carry out this reproductive process and survive themselves, it seems that increased complexity of organization brings death of the body as one of its consequences; reproduction then becomes the only way in which a stock of animals can survive. This probably represents a compromise solution to the problems confronting the animal. In order that its protoplasm function properly the surrounding physicochemical environment has to be very carefully controlled and many variable factors kept within narrow limits. As an individual animal ages this becomes more and more difficult. Rather than face the increasing physiological expense of this the individual animal is allowed to die, but meanwhile a new one made of protoplasm which has been kept apart from such contamination is brought into existence in its place, and life continues. The typical animal has, therefore, a reproductive organ which contains the gametes from which the next generation will be formed and a genital duct which leads them to the exterior. For some reason, perhaps of economy of tubes and external openings, this duct becomes joined to the excretory ducts in many kinds of animal so that the excretory and genital system become apparently one.

This rapid review of the organization of the typical animal body shows how, given the fact that an animal has to feed in a certain way, the structure of its body inevitably follows. Because an animal finds its food only in other organisms it *must* be motile, it *must* have sense organs to investigate its environment and control its movements through a brain; an alimentary canal, respiratory organs and a circulatory system are essentials of structure because of the way in which it *must* function; an excretory system and a reproductive system equally inevitably follow as factors in its make-up.

You will realize that we have been talking of the *typical* animal in this argument. If an animal lives in atypical fashion then it may not exhibit all the characteristics of the typical animal: thus sea anemones are radially symmetrical because food may come to them from any direction as it does to a plant, and they reach out equally in all directions to improve their chances of catching prey. Oysters and sea-squirts can be sessile and immobile because they have so modified their way of life as to bring food to themselves instead of chasing it as more ordinary kinds of animal do. Animals which are markedly smaller than their close relatives may often be anatomically simpler because the changed surface to volume ratio may allow diffusion to occur more readily, and so let the creature dispense with organs otherwise essential; adoption of a parasitic mode of life similarly may be reflected in a secondary simplicity of organization. But whether typical in their way of life or not the structure of the organism reflects the functions which its method of living imposes upon it.

When you dissect an animal, whatever its taxonomic position, you ought, therefore, to expect to find within its body a standard equipment of structures allowing it to carry out the standard set of functions which its animal nature imposes upon it. But though an animal must have a gut, a nervous system, an excretory system and so on – though there is this functional need to which it must conform – there is no anatomical pattern to which it must pay equal attention. An animal must have a nervous system and a gut, but it may, so to speak, choose whether the nervous system shall lie above the gut, below the gut or alongside the gut, and there is, therefore, a vast range of anatomical diversity to be found in the animal kingdom superimposed upon the functional uniformity outlined above. Further, there is a great disparity in the physical resources with which different kinds of animals carry out their physiological requirements. A protozoan such as an amoeba may carry out the same kinds of activity as a mammal, but whereas the former has to do them all with the same single piece of protoplasm, the mammal may be able to set aside several million cells for one special activity, with an inevitable increase in the efficiency and delicacy with which it can be performed. The particular pattern on which the body of an animal is constructed, therefore, conditions and may limit the ways in which it carries out its various functions. It is the purpose of the next section of this book to show how these three factors – uniformity of physiological activity, anatomical diversity and the restrictions imposed by the evolutionary grade of organization – interact in the life of a series of different animals.

The anatomical patterns on which the bodies of animals are constructed are extraordinarily varied, particularly if extinct forms are considered alongside living ones, yet similarities exist. It is these likenesses and differences that zoologists have seized upon to erect classifications attempting to categorize the immense number of organisms that constitute the animal kingdom. In pre-Darwinian days there was no very obvious reason why similarities between different kinds of animal should exist, but they are now seen as indicators of evolutionary kinship, just as brothers and sisters, because they have the same ancestry, are more like one another than they are like other members of the population. Similarly, the differences between two kinds of animal are in a general way indicative of their degree of divergence. Using such features the zoologist with an interest in systematics is able to arrange animals into a hierarchy of groups, each called a taxon – species, genera, families, classes, phyla – each member of the series containing more different kinds of animals than the preceding one. The extremes of the series are better defined than its central terms. A species is the population of animals that fruitfully interbreed, or, if limitations of space and time were abolished, could fruitfully interbreed, and it therefore corresponds to what, in nature, is ordinarily understood by a "kind" of animal. Similarly, there is little doubt about what is meant by a phylum – echinoderms are distinct from other kinds of animal and unlikely to be confused with them. Nevertheless there is attached to the word "phylum" the same uncertainty that attaches, though in greater degree, to the words "genus", "family" and the like, in that they stand for ideas in the minds of zoologists and do not correspond to anything in nature: they express a particular systematist's reaction to the animals which he has studied and are a shorthand way of stating his ideas as to their interrelationships and probable evolutionary history. Further, since life is continuous in the ever-rolling stream of time but species change these words represent only ephemeral stages. If our systematist is a good zoologist his ideas, and therefore his classification, may prove acceptable to others, but other, equally good, zoologists can produce different, and apparently equally acceptable classifications. It must be emphasized, therefore, that the classification used in this book is no more "correct" than others which might have been used and is only that classification which seemed to the authors to reflect most adequately the known interrelationships of the animals within each group. It is liable to be upset, or replaced by another, should further study require or should new animals be discovered whose anatomy disclosed new relationships. Zoologists who are systematists tend to be impressed by differences of relatively trivial nature; zoologists with more general interests may be more struck by broad resemblances. These points of view are often revealed by their choice of classification; thus malacologists interested in bivalves may put both protobranchs and lamellibranchs into a single class Bivalvia in order to express their belief in the close relationship of these animals whereas someone who wished to emphasize his belief that they represented distinct evolutionary lines might well abolish the single group Bivalvia and replace it with two taxa of equal rank: Protobranchia and Lamellibranchia.

It was pointed out above that all animals, however classified, were doing fundamentally identical things, though their equipment for this might differ in complexity or topographical layout. In this is concealed a factor which tends to confuse the systematist and may completely confound him. In the evolution of apparatus to carry out any function in which a particular set of mechanical, physical or chemical processes is involved it will frequently come about that similar devices arise, even though the starting points of their evolution may have been unlike. Thus animals which filter their food from a current of water require a mechanism to create the current, a filter and a means of transporting the filtrate. In creatures as unlike as the worm *Sabella*, a bivalved mollusc, a copepod and a sea-squirt, parts of the body are found each carrying out one of these activities, and to that extent these animals, each belonging to a different phylum with its distinctive anatomical pattern, have come to resemble one another. This phenomenon is known as convergence and, though unlikely to mislead the systematist in the example quoted, can on occasion produce structures so similar as to make him believe that the likeness must indicate close relationship and so cause him to propose a false classification. Dr Manton's work on the mandibular arrangements in arthropods (p. 273) is an outstanding illustration of how unrelated animals, in evolving a mechanism to satisfy a particular need, have produced superficially almost identical structures.

The reality of convergence has come to be more readily accepted with a change in our ideas of how the evolution of animals has taken place. It used to be generally believed that a particular group of animals had arisen from a single ancestral stock and, therefore, that there existed a relatively close cousinhood amongst all its members, which had reached their present status by an adaptive radiation. In more technical language the group was said to be monophyletic, and its evolution from its origin expressed diagrammatically as a tree. It is, however, agreed that the evolution of at least some groups may have followed a different pattern; instead of only one ancestral form taking the step which raised it from one level of organization, to another, several may have done so, with the result that the diagrammatic representation of the change would be a herbaceous plant with several uprising stems rather than a tree with one. A group with this origin is said to be polyphyletic and the relationships amongst its members are looser than are those in a group which evolved as a unit. In a polyphyletic group, however, each stock attempts an adaptive radiation and the tendency for different animals to occupy similar niches is thereby exaggerated and the probability of convergence raised. This evolutionary pattern, too, requires us to modify our ideas of what a large taxon may be. It may be less a group of blood relations and more a society of species all of which have attained a given level of organization.

Not all anatomical plans have proved equally successful in their adaptive radiation. This is implicit in the vast number of species which have become extinct, but even amongst living organisms the patterns characteristic of a few phyla and classes seem to have proved better able than others to give rise to what might be described as successful models. This is most marked in the well-known abundance of species of arthropods and, in particular, of insects, of which there are more than of all other kinds of animal put together. In 1962 Mayr put the total number of species of animals which had been described at 1 120 310, of which 923 135 were insects; more have been described since and others certainly await description. Since few people can grasp the meanings of numbers of this order of magnitude and their relationship to smaller numbers, it is worthwhile to approach the matter in a way different from a mere catalogue of the numbers of species in each phylum. Let us suppose that we are the audience in a theatre and that there is about to pass across its stage a procession of animals, one member of every known species, at a rate of one kind of animal per second. We have in prospect an uncomfortably lengthy show, since it will go on continuously for 12 days 23 hours. Of this time no less than 10 days $16\frac{1}{2}$ hours will be occupied by a procession of different arthropods. If we examine this enormous arthropodan series more closely we find that 9 days 20 hours is taken up by a succession of insects and nearly three days of that period – 2 days 21½ hours – by beetles only! The molluscs, the second largest phylum, take 22 hours 13 minutes to pass and the vertebrates 10 hours 57½ minutes, of which time half is occupied by fishes. The next largest group, the protozoans, requires 8 hours 20 minutes, leaving thirteen hours for the whole of the rest of the animal kingdom: coelenterates, flatworms, round worms, annelids, echinoderms and the so-called minor phyla. Obviously, by arthropodan standards, all phyla, even the greatest, are minor.

The classifications given at the end of each chapter are not intended to do more than permit the reader to place the animals dealt with against a taxonomic background. The diagnostic characters of the various taxa are not given. The taxic levels of the groups are also not indicated since this must vary to some extent from authority to authority. Thus whilst Crofton regards the Nematoda as a class in a phylum Aschelminthes, other authorities would raise it to phyletic level; whereas Corliss calls Ciliophora a sub-phylum of Protozoa, others call it a class. We have therefore left the taxa without hierarchical label though the steps in the hierarchy are indicated by degree of indentation.

In the text, partly for brevity, partly from ignorance, not all groups have been treated. We have judged it worthwhile, nevertheless, to give the names of the taxa not described so that the reader may get an indication of how complete the treatment has been. The names enclosed within brackets [] are in this category. They are further marked with an obelus (†) if they are wholly extinct. Whilst all major subdivisions of a phylum have been listed this is not true of the minor ones: in some cases the taxon (at subordinal level or thereabouts) which has been mentioned is named with an indication that other taxa at this level exist. Without this abbreviation the lists would be too cumbersome.

2
Protozoans

SINCE LEEUWENHOEK first demonstrated their existence with his microscope the animals included in the phylum Protozoa have been studied for a number of reasons; partly because of their intrinsic beauty, partly because of their economic and medical importance and partly because their apparent simplicity seemed to suggest that they would reveal the secrets of life more easily than other animals. This impression of more intimate access to living processes is a delusion: the simplicity of protozoans is exclusively in gross anatomy, the electron microscope having revealed an overwhelming complexity in detailed construction, and their fundamental physiological processes are as elaborate as those of any other animal. There is no easy road to the understanding of life through protozoology. Because of the multiple reasons for which they have been studied and because of the chance that one animal has proved useful for the study of a particular topic, our knowledge of the group is unbalanced—much is known of parasites of medical or veterinary importance, much is known of the nutrition of a few ciliates which happen to be rewarding experimental animals, but other areas of protozoological study may hardly have been touched.

The protozoan body usually comprises a mass of protoplasm containing a single nucleus, and to that extent it compares with a single cell extracted from the body of a higher animal. For this reason the Protozoa are often described as unicellular animals or free-living cells, and the origin of the metazoans sought in the association of such to give many-celled units. Alternatively, metazoan origins may be looked for in another type of protozoan, the multinucleate type, in which several or many nuclei lie in a continuous mass of protoplasm; by supposing each nucleus and its surrounding protoplasm to become isolated from its neighbours by cell membranes, a metazoan body is produced. A protozoan of this type is, therefore, distinguished from a true metazoan in not being separated into cells, and, for some zoologists, the protozoans, even those with only a single nucleus, cannot be regarded as "unicellular" animals, but must be called "acellular", i.e. not divided into cells. To some extent arguments of this nature derive from the fact that biologists were long unable to agree as to what precisely constitutes a cell, and the same discussion has gone on in more lively fashion as to whether the word "cell" should be used to describe a bacterium. Modern research with the electron microscope allows a fresh look to be taken at this problem since it has enormously extended detailed information about cellular structure. It has, indeed, shown a near identity in both nuclear and cytoplasmic organization between Protozoa and Metazoa, and, if a line can be drawn to separate different structural patterns it is not between the protozoans and the metazoans, but between the bacteria (which are known as prokaryotic cells) and the rest (eukaryotic cells). Bacteria are cells with only one membrane around them, separating a unity which contains all the constituents of cytoplasmic and nuclear organization from the surrounding medium, whereas in the single unit of the protozoan body and in each of the multiple units of the metazoan, the nucleus lies within its own private, though perforated, membrane and a second, outer, membrane delimits the cytoplasm from the external environment. (According to some, however, the first and second membranes are both parts of the same.) On this basis there is no justification for refusing to apply the same term, cell, to the unit which is the whole protozoan body and to that which helps to form the metazoan body, except, perhaps, the semantic objection put forward by Dobell in 1911 to calling by the same name that which is in one animal a part of its body and in the other the whole body itself. Protozoa, however, are both cells and organisms; as one they parallel the metazoan cell, as the other the whole metazoan.

Classification

Before discussing some of the problems of functional anatomy which arise in the study of protozoans it is necessary to define some of the taxonomic groups to which reference will be made. It is assumed that the reader is already familiar with such protozoans as an amoeba, *Euglena* and the ciliate *Paramecium*. The commonly accepted basis for the division of the group into classes is the type of locomotor organelle which an animal possesses. The most primitive group of the phylum is one known as Mastigophora or Flagellata, which receives its name because the characteristic locomotor structure of its members is the flagellum. One, a few, or many of these may be found on the surface of the organism. The group splits itself into a division Phytomastigina, of more plant-like types, usually with the capacity of photosynthesis, though the pigment responsible may or may not be chlorophyll, and into another, the Zoomastigina, unable to photosynthetize and therefore dependent upon the uptake of organic matter, either in solution or by ingestion of solid food. Some of the commoner or more important members of the Phytomastigina, which are widely accepted as the oldest and most primitive protozoans, are organisms like *Euglena* (Fig. 2A) and *Chlamydomonas*, widely studied as examples of the phylum; the Phytomonadina or Volvocales (Fig. 3), including *Volvox*, not uncommon in fresh water; the dinoflagellates, important members of the freshwater and marine phytoplankton and often provided with a skeleton of cellulose plates, though some are naked. They have two flagella, one which drives them forward, and a second which spins them on their

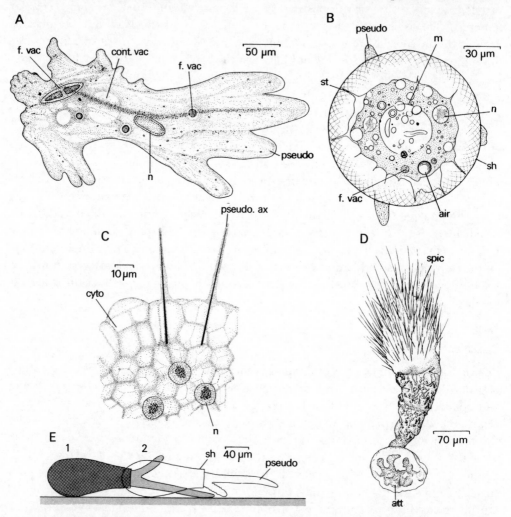

Fig. 1. Rhizopods. A, *Amoeba proteus*, moving to the right with four advancing pseudopodia into which the plasmasol (endoplasm) is streaming. (After Kudo.) B. *Arcella*, a shelled amoeba. (After Bles.) C, *Actinosphaerium*, the sun animalcule, part of the body showing the structure of a pseudopodium. D, *Haliphysema*, an attached foram showing the skeleton of sponge spicules. E, *Difflugia*, a shelled amoeba, creeping to the right. Note the extension of the lower pseudopodium when the animal is in position 1, and its contraction on movement to position 2. (After Grell.) air, air bubble helping in flotation; att, attaching disc; cont.vac, contractile vacuole; cyto, vacuolated ectoplasm; f.vac, food vacuole; m, mouth of shell, seen by transparency; n, nucleus; pseudo, pseudopodium; pseudo.ax, pseudopodium with axial support; sh, shell; spic, sponge spicules projecting from protoplasm; st, strand of protoplasm.

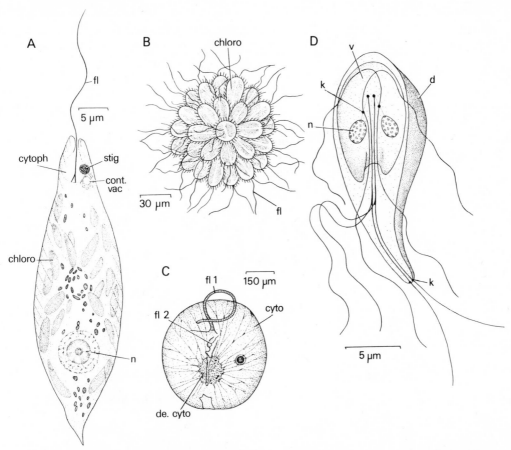

Fig. 2. Flagellates. A, *Euglena gracilis*. (After Grell.) B, *Synura uvella*, a common plant-like colonial form from pond water. (After Jahn.) C, *Noctiluca*, a common marine dinoflagellate which is luminous. (After Pratje.) D, *Giardia intestinalis*, a parasitic flagellate. Note the arched dorsal and concave ventral surfaces and the bilaterally symmetrical equipment of nuclei, kinetosomes and flagella. (After Grell.) chloro, chloroplast; cont.vac, contractile vacuole; cyto, vacuolated cytoplasm; cytoph, cytopharynx; d, dorsal surface; de.cyto, dense cytoplasm round nucleus; fl, flagellum; fl 1, swimming flagellum; fl 2, feeding flagellum; k, kinetosome; n, nucleus; stig, stigma; v, ventral surface.

long axis as they swim. The Zoomastigina includes many minute colourless types common in foul water and collectively known as monads, and some of the serious protozoan parasites of man and his domestic animals such as the trypanosomes responsible for the diseases sleeping sickness, nagana, surra and dourine.

The second major group of protozoans is known as Rhizopoda or Sarcodina. Their distinguishing feature is a dependence upon the type of locomotor organ known as a pseudopodium and the group may be subdivided according to the pattern of pseudium occurring. Some of the subdivisions are: the amoebae (Fig. 1A, B, E), common

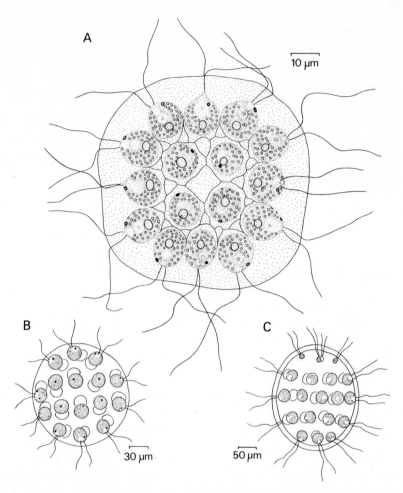

Fig. 3. Volvocales. A, *Gonium pectorale*: colony in surface view showing the sixteen similar zooids, arranged in four groups of three round a central four. All the zooids are alike, with two flagella, an eye-spot, a cup-shaped chloroplast and a nucleus; all are embedded in a common jelly mass. (After Grell.) B, *Eudorina elegans*: globular colony of thirty-two zooids embedded in jelly in five rows of, 4, 8, 8, 8, 4, all morphologically alike. C, *Pleodorina illinoisensis*: small vegetative zooids of the anterior end, each with a stigma, and the large reproductive zooids of the posterior half, each devoid of stigma. (B, C after Jahn.)

in all waters and soil and occasionally parasitic, with blunt, unbranching pseudo-podia; the foraminiferans (colloquially, forams), with bodies often divided into lobes and bearing shells, characterized by a network of fine branching pseudopodia; the heliozoans or sun animalcules (Fig. 1C), common in fresh water, with stiff, unbranched pseudopods radiating from a spherical body; and the radiolarians and acantharians, predominantly marine, with stiff, branching and anastomosing

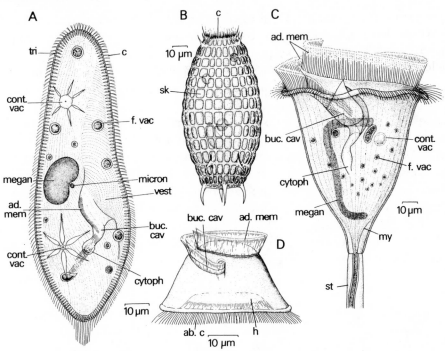

Fig. 4. A, *Paramecium caudatum*. (After Grell.) B, *Coleps hirtus*, a ciliate very common in pond water, showing quadrilateral skeletal plates. (After Kudo.) C, *Vorticella*, a peritrich adapted for sessile life in which the adoral cilia form a circlet spiralling counter-clockwise round the disk before entering the buccal cavity. (Based on Corliss and Grell.) D, *Trichodina urnicola*, a peritrich ciliate with its body modified for creeping over the surface of its host, the freshwater polyp Hydra. At the upper pole lies the spiral wreath of cilia used for feeding whilst at the lower end a circle used for swimming or creeping surrounds a group of hooks by which the animal grips its host. (After Corliss.) ab.c, aboral locomotor cilia; ad.mem, adoral membranelle; buc.cav, buccal cavity; c, cilia; cont.vac, contractile vacuole; cytoph, cytopharynx; f.vac, food vacuole; h, hook; megan, meganucleus; micron, micronucleus; my, myonemes converging on top of stalk; sk, skeletal plate; st, stalk; tri, trichocyst; vest, vestibule, the rows of dots mark the kineties.

pseudopodia extending from a spherical body usually built round an elaborately patterned siliceous shell.

A third division of the phylum has the cilium as its characteristic locomotor structure and is for this reason known as the Ciliophora or Ciliata. Ciliates often show a greater complexity of external shape (Figs 4D and 5A) and internal structure (Fig. 5C) than members of the other protozoan groups. Most swim freely through the water in which they live, but some (known as hypotrichs, Fig. 5B) are also accustomed to movement over surfaces, and others (called peritrichs, Fig. 4C) are commonly attached, either permanently or temporarily. Associated with the ciliates is another group, the Suctoria, devoid of cilia and attached when adult, though their young stages have cilia and swim freely; they have evolved a highly specialized feeding process. All ciliates, except one very primitive genus, *Stephanopogon*, have two types of nucleus, one or more meganuclei responsible for running the body of the animal and a reserve of chromosomes used only in reproductive activity and kept apart in one or more micronuclei.

In addition to these three groups there is a fourth, the Sporozoa, the members of which are parasitic and include animals responsible for such major diseases as malaria. As might be expected, animals in this group have frequently complex life histories involving passage through more than one host. They occur in three groups: the coccidians, the gregarines and the haemosporidians. In the first two groups there is only one host and transmission from host to host is by accidental infection by spores; in the last there are usually two hosts and the parasite passes no part of its life outside the body of one or other of these. The Sporozoa are probably an unnatural grouping of animals flung together because of the absence of such characters

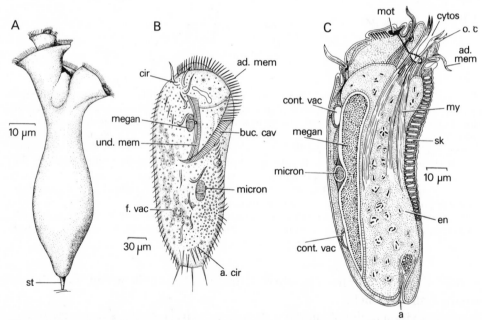

Fig. 5. Spirotrichous ciliates. A, *Spirochona gemmipara*, a sessile form with no general body cilia. (After Corliss.) B, *Stylonychia mytilus*, a hypotrich. (After Grell.) C, *Epidinium ecaudatum*, median section to show complexity of internal structure. a, cell anus; a.cir, anal cirrus; ad.mem, adoral membranelles; buc.cav, buccal cavity; cir, cirrus; cont.vac, contractile vacuole; cytos, cytostome; en, endoplasm; f.vac, food vacuole; megan, meganucleus; micron, micronucleus; mot, motorium, with circumpharyngeal ring; my, myonemes retracting cytopharyngeal region; o.c, oral cilia; sk, skeletal structure in ectoplasm; st, stalk; und.mem, undulating membrane.

as locomotor organs on which a truer classification might be based rather than by the sharing of common structural features.

Patterns of body structure

Within the protozoans several patterns of body structure are found. By far the commonest, and presumably the most primitive, is that in which we find a single nucleus related to a mass of cytoplasm, as in *Amoeba* (Fig. 1A) or *Euglena* (Fig. 2A), a unit sometimes known by Sachs's name "energid". Within this unit must be found all the apparatus necessary for life, since these animals are self-supporting organisms adapted to the ecological niche in which they occur. A second pattern occurs when several energids, by failing to separate after division, remain attached to one another and produce a colony as in the flagellates *Synura* (Fig. 2B) and *Gonium* (Fig. 3A); there is little or no cytoplasmic continuity amongst members of the colony, though they may be interlinked by threads which they have secreted, or lie in chambers of a protective exoskeleton formed in the same way. In the light of this, the advantage of the association to the individual is not clear. A third major plan of construction is the polyenergid, in which, as the name suggests, one finds a number of nuclei, each with cytoplasm, forming a single unit without internal divisions (*Opalina*, a common parasite of the bladder in frogs; the sun animalcule *Actinosphaerium*). This kind of body may undergo considerable modification for carrying out vital functions, and some of the polyenergid protozoans exhibit a complexity of structure which stands comparison with that of metazoan bodies of corresponding size. A similar degree of organization, however, may often also be found in protozoans which are not poly-energid (Fig. 5C). It seems, therefore, that there has been, within the protozoan series, an evolution from the relatively simple to the complex, both in the single energid, and in the polyenergid with multiple nuclei and cytoplasmic organelles, paralleling the organization of the metazoan, and doing so because of the need of all to solve the same functional problems.

This parallelism, indeed, extends further than mere morphological diversification in the well-known case of the Volvocales (Mastigophora Phytomonadina), so suggestive to many biologists of the way in which the evolution of metazoans may have taken place, though it should be emphasized that metazoan origins from protozoan sources are still completely speculative. In Metazoa, by contrast with the simpler Protozoa such as *Amoeba* and *Euglena*, one distinguishes specialized cells, forming a body or soma which perishes at each generation, and a germplasm from which the bodies of the next generation are derived. In the Volvocales is a series of genera of colonial forms the unit of construction in each being a flagellate with almost exactly the same structure as the solitary individuals of the well-known genus *Chlamydomonas* – an ovoid body with a cup-shaped chloroplast within which lies the nucleus, two flagella, two contractile vacuoles and a pigment spot or stigma, all enclosed in a cellulose wall. In the genus *Gonium* (Fig. 3A), 4–16 such zooids lie in the form of a flat plate, embedded in a gelatinous matrix. All are visually alike, but the colony is physiologically polarized and one side of the plate regularly goes first in a clumsy swimming motion. Each individual is capable of asexual multiplication to produce a fresh colony and each may also break away from the colony as a biflagellate gamete, fusing with another to give a zygote from which a new colony grows. The organism thus shows a wholly protozoan character in that all the constituents are equivalent and there is no distinction between soma and germplasm. The same is true of *Pandorina*, which has the form of a globe of up to 16 individuals.

The next members of the series, however, *Eudorina*, *Pleodorina* and *Volvox*, depart significantly from this. *Eudorina* (Fig. 3B) has the form of a sphere containing 32 units, *Pleodorina* (Fig. 3C) 128, whilst in *Volvox* the number averages about 10 000 and may reach twice that. In the first two genera the individuals are isolated from one another within their common bed of jelly, but in *Volvox* they are joined to one another by fine cytoplasmic strands. In all three genera some morphological differentiation of zooids exists: in *Eudorina* the four most anterior are either initially or permanently smaller than the others; in *Pleodorina* the zooids in one hemisphere (the anterior since it travels first) are much smaller than those in the posterior half and possess eye-spots which the latter lack, whilst in *Volvox* the anterior members of the colony have a large eye-spot anteriorly placed whereas it is small and variable in position in the others. This morphological differentiation reflects a more fundamental physiological one in that it is only the larger posterior zooids which can reproduce (in one species of *Eudorina*, *E. elegans*, the anterior cells do reproduce, but at a lower level of activity), either asexually or sexually. In *Eudorina* and *Pleodorina* all posterior units may do this but in *Volvox* the position is more extreme in that at an early stage in the formation of a colony a limited number of posterior zooids are set apart for reproductive activity (a maximum of 8 capable of asexual reproduction, 5 or more for the production of male gametes and 1–15 for female), all other zooids being sterile. When it is added that the female gamete is a large egg-like structure and the male a small biflagellate one resembling a sperm, it will be realized that the genus *Volvox* presents a set of conditions which parallel only those of the Metazoa. Whether this series, *Gonium* to *Volvox*, which exhibits a gradual shift of emphasis from the individual zooid to the second grade which is the colony, is an artificial series, or whether it represents a real phylogeny is unknown; but it is at least useful in illustrating one way in which a metazoan body and reproductive system could have arisen, and in suggesting that the multicellular state may have evolved more than once and perhaps in more than one way in the course of time.

Protozoa occur in a multiplicity of shapes and sizes. The smallest are very small, only 1–2 μm in length, the largest (extinct forams) 5–6 cm, whilst the great majority fall between these extremes and measure a fraction of a millimetre. Many protozoans secrete a shell round their body, which imparts a precise shape to them. When naked their shape may be quite indefinite, as in amoebae, or the outermost layer of material under the bounding cell membrane, the pellicle, may be sufficiently rigid to maintain a definite shape except when temporarily deformed by the contraction of the fibrils known as myonemes. In euglenoid flagellates the pellicle consists of numerous arched strips fitted into one another sideways by a tongued and grooved articulation like so many floorboards. They are movable on one another and allow the organism to change shape, sending waves of contraction and expansion down the body, a type of movement called euglenoid or metabolic. Its functional value is in doubt. Under the pellicle the outermost layers of the body are stiff enough to preserve some degree of rigidity of structure, though the inner layers are fluid. The outer layer is known as the ectoplasm or plasmagel, the inner as the endoplasm or plasmasol. In ciliates the pellicle may be strengthened by incorporation of skeletal materials of a variety of sorts so as to constitute an armour, as in *Coleps* (Fig. 4B, sk, calcareous plates) or in the ciliates (Entodiniomorphida) which live in the stomach of ruminants (Fig. 5C, sk, polysaccharide plates). The same may happen in flagellates, many coloured ones being enclosed in a more or less rigid cellulose envelope, sometimes made of articulating plates as in the so-called armoured dinoflagellates (*Peridinium, Ceratium*), whilst the silicoflagellates have siliceous spicules and some, like *Synura* (Fig. 2B), have the surface covered with scales of silica. The most familiar and conspicuous shelled protozoans, however, are the rhizopods, nearly all of which possess shells of one sort or another. The common amoebae are naked, it is true, but many shelled amoebae occur (common sorts are *Arcella*, Fig. 1B; *Difflugia*, Fig. 1E; *Euglypha*) whilst some heliozoans and practically all foraminiferans, acantharians and radiolarians are provided with a shell, which may range from a modest array of spicules to an object of elaborate build. The usual material for the manufacture of the shell is calcium carbonate, though radiolarians have shells of silica and acantharians use strontium sulphate. Occasionally other substances such as the polysaccharide tectin may be used, and some organisms, like *Difflugia* and many forams, make a shell by embedding in a soft secretion foreign bodies from the environment, producing what is known as an arenaceous shell. The particles used are often grains of sand or other detritus, but forams may select the material from which they make their shell with the same nice choice as do caddis larvae building their cases, and must sort it out of vast quantities of apparently unusable material. Thus, among the foraminiferans, *Haliphysema* uses only sponge spicules (Fig. 1D, spic), and *Verneiulina polymorpha* grains of garnet, magnetite and, perhaps, topaz. That these feats can be accomplished by animals with so little apparent morphological or cytological specialization has always fascinated and aroused the interest of the zoologist. From another point of view, too, foram shells are significant when it is realized that practically the whole of the floor of the oceans at a depth of 2 500–4 500 m is covered by a deposit containing many shells of dead foraminiferans, mainly of the genus *Globigerina*, whence the name "Globigerina ooze" for the deposit. The area amounts to about 50×10^6

square miles. At greater depths calcium carbonate dissolves whereas silica remains insoluble; in the deeper parts of the ocean, therefore, the floor is largely made of "radiolarian ooze" containing a significant number of the siliceous shells of dead radiolarians. The extent to which the accumulation of such shells has gone on in the course of geological time at a rate of about 1 cm 10^{-3} years will be appreciated when it is realized that the bulk of the Cretaceous limestone or chalk rocks has originated in this way.

The shape of protozoans may be related in a general way to the kind of life which the organism leads, and shapes correlated with (1) floating, (2) swimming, (3) creeping over surfaces, and (4) attached life, are distinguishable.

The commonest planktonic protozoans are the heliozoan, foraminiferan, radiolarian and acantharian rhizopods, which float freely in the water for most of their lifetime, moving vertically in relation to light, weather, season, and, in some instances, the stage of their life cycle. The typical member of each of these groups (say the heliozoan *Actinosphaerium*, the foraminiferan *Globigerina* or *Orbulina*, a radiolarian like *Thalassicola*, the acantharian *Acanthometra*) is globular in shape. Few show the perfect spherical symmetry (that is, a condition in which any plane passing through the centre of the sphere subdivides the organism into two equal halves) which might be expected of animals living in such a way. This is understandable since it presupposes little differentiation of the protoplasm; it may, however, be seen in some of the simpler radiolarians such as the thalassicolids. In many pelagic types some of the radii of the sphere end in such special structures as spicular skeletal supports. The symmetry is then no longer spherical but radial, that is, the organism may be subdivided into equal halves only along a limited number of planes passing through the centre of the sphere. Most pelagic protozoans float passively and active movement is restricted to vertical migration, though acantharians and heliozoans are alleged to progress slowly by beating their pseudopodia, which were therefore once believed to be highly modified flagella. Recent electron microscope work has not borne this out. Flotation is facilitated by the development of a number of radiating structures which help to increase surface area. These may be: (1) spicules in the form of rods, which may all meet at the centre of the spherical body (many Acantharia and some radiolarians) or be superficial projections from a complete or fenestrated shell (most radiolarians and the pelagic forams of the family Globigerinidae); (2) radiating pseudopods stiffened by an internal skeletal rod or fibre, known as axopodia (Fig. 1C, pseudo. ax, found in acantharians, the pelagic forams and Heliozoa), sometimes, as in *Actinophrys*, related to a central nucleus, sometimes not; or (3) a network of fine and anastomosing pseudopodia (radiolarians and forams).

The specific gravity of floating protozoans is under limited control by manipulation of the ectoplasm, the highly vacuolated outer layer of the body which extends as a foamy mass beyond the surface of any shell which may occur; by altering its volume the animals are able to affect their specific gravity so that they float higher or lower. The details of how this is done differ from group to group. In the radio-larians and heliozoans (see Fig. 1C, cyto) the outer protoplasmic layer, known as the calymma, has vacuoles filled with fluid with specific gravity less than that of the surrounding water: by collapsing or forming more of these the animal may sink or rise. The same kind of arrangement is found in the dominant pelagic forams of Recent times, those placed in the families Globigerinidae and Globorotaliidae.

Forams often have a body formed of a series of lobes spirally arranged; as an adaptation for pelagic life in these two families the last formed chamber of the shell is globular and encloses the spirally arranged older ones, giving a radial symmetry. The shell is perforated by numerous pores through which the ectoplasm extends as a frothy mass of which the depth, and so the flotation power, may be varied. A more elaborate apparatus occurs in acantharians. In these animals a radiating series of spicules runs through the ectoplasm, which is not so frothy as in the groups dealt with above. In it may be distinguished an outer pellicular layer at the surface and sometimes also an inner one, roughly parallel to the outer, but lying some way below the surface. Both these layers are pulled up into conical extensions around the base of each spicule where it extends into the surrounding water. Between them, embedded in the ectoplasm, run fibres giving stiffness to that layer. In the conical extension of ectoplasm up each spicule lies a bundle of contractile fibres or myonemes (some-times, unnecessarily, known as myophrisks), attached distally to the outer surface of the spicule and, proximally, to one or other of the pellicular membranes. When these contract the ectoplasm is pulled up the spicules, the effective radius of the sphere is increased, and the animal floats higher.

The only other pelagic protozoan which deserves mention at this point is the flagellate, *Noctiluca miliaris* (Fig. 2C), one of the commonest of luminescent marine animals. Like the creatures dealt with in the previous paragraph, *Noctiluca* has an approximately spherical shape but in fact departs much further from truly spherical symmetry, not unexpectedly indeed, since it is a motile rather than a passively floating animal. Its body, however, is swollen with vacuoles which are said to be filled with an aqueous solution isotonic with sea water but of lower specific gravity since it contains ammonium instead of sodium chloride. Many other free-swimming protozoans lighten their bodies on the same principle, usually, however, by means of oil drops which have the advantage of being without osmotic complications.

Free-swimming protozoans are normally more complex in shape than *Noctiluca* and the more adapted they are for this mode of life the more likely they are to be asymmetrical. This may strike the reader as odd, in view of the regular streamlining of the metazoan body for free-swimming life; it must be remembered, however, that the physical conditions which have led to this shape in larger animals do not neces-sarily operate when objects the size of protozoans are concerned and that the modes of locomotion differ. The fundamental departure from spherical symmetry is towards an elongated cigar shape in association with forward movement, but complicated by the fact that in protozoans this, whether it be achieved by flagella or cilia, involves a rotation around a longitudinal axis to give a spiral screwlike motion. In the simplest case the longitudinal axis around which the body spins is that of the animal itself; in another the axis passes through the body but is not coincident with its long axis; whilst a third situation arises when the axis round which the creature is spiralling does not pass through the body at all; in this the path which the animal follows is like a spiral traced out on the surface of a cylinder, and the protozoan turns one of its sides constantly towards the axis around which it is rotating and the other towards the outside. Not surprisingly, these two surfaces are apt to reflect in their organization their different contacts with the environment, and may be recognized as inner and outer, or ventral and dorsal. The animal often becomes flattened between them with the mouth opening on the leading edge. This is exhibited by ciliates such as *Stylonychia* (Fig. 5B) or *Euplotes*. The common arrangement amongst free-swimming flagellates and ciliates is the second, and whilst some degree of flattening may occur, this is not marked, nor, indeed, is any modification of shape apart from a general elongation; many animals are asymmetrical. In animals which swim forward rotating on their own long axis, the mouth is advantageously placed for feeding if terminal; when they follow a screw course the mouth is more effectively placed at the side of the body and, since this is continuously in rotation, the result is that fresh volumes of water are sampled as the organism swims. The surface of these creatures is frequently affected by the spiral course which they follow and presents spiral grooves and ridges, whilst other features such as the gullet, cilia and myonemes may also be arranged along spiral lines. These may follow the curve along which the animals swim, but are more commonly set at right angles to it.

The third mode of life which has provoked a characteristic shape in protozoans is that of the animal which creeps over a surface. Some adaptation towards this is already indicated by the flattened body of the ciliates described above (see Fig. 5B) which swim in a screw path with a lower, ventral side facing the axis around which they spiral. These are the hypotrichous ciliates, which can not only swim thus but are able to move over surfaces. Part of the differentiation of the two sides is concerned with this: on the ventral side the cilia are grouped into compound structures known as cirri (cir), by means of which the animals swim and can, in addition, crawl over surfaces with great agility; on the dorsal side single cilia are sparsely scattered and are alleged to be sensory. The cirri are under control. Whilst most hypotrich ciliates are free-living (*Stylonychia*, *Euplotes* and *Uroleptus* are well known), one genus at least has become adapted for creeping over the surfaces of other animals, using materials which they secrete or the living substance itself as food. This is *Kerona*, which is common on the freshwater polyp Hydra.

A few ciliate protozoans of other groups which show similar adaptations may be mentioned here. These are mainly, like *Kerona*, found skating over the surface of the body of other animals in search of food and are characterized by a flattening of the body to form a locomotor and sometimes adhesive underside. They may be relatively harmless, but some damage the body of their host. Two better-known genera with this habit are *Trichodina* (Fig. 4D), which also creeps over Hydra, and *Urceolaria*, species of which may be found on the surface of the flatworm *Polycelis*, in the mantle cavity of the terrestrial prosobranch mollusc *Pomatias elegans* and on the gills of the limpet *Patella*. Outside the ciliates the parasitic flagellate *Giardia* (Fig. 2D) creeps over the lining of the gut of the animals within which it lives, the under surface (v) arched just as in the hypotrichous ciliates.

The animals dealt with in the preceding paragraphs are those which can creep over surfaces but primarily swim or float freely in fluid; in addition, there is a vast series of protozoans which predominantly use a creeping mode of life. These are rhizopods, which have pseudopodia as their characteristic locomotor organs. The animals are usually larger than the free-swimming protozoans and are often readily visible to the naked eye. They include the amoebae (Fig. 1A) and the bulk of the forams, the former apparently the simpler, and often putting out only a single pseudopodium at a time. This is a broad lobe (lobopodium), blunt at the tip, and containing both gel and fluid cytoplasm, the latter streaming up the centre of a more gelated tube as the pseudopodium extends. At the apex the fluid cytoplasm (plasmasol) gelates and converts itself into plasmagel, whilst at the hinder end of the animal the reverse occurs. The transformation has been linked with the behaviour of the protein molecules of the cytoplasm, which are said to form a layered system of straight chains in the plasmagel and be highly folded in the plasmasol, passing from one to the other at the pseudopodial tip and at the back end of the amoeba. This change by itself would not cause the streaming of the plasmasol, which must be due to the application of an external force. This has been supposed by some workers to be a sucking forwards by the changes which are taking place in the conversion of plasmasol to plasmagel at the tip of the pseudopodium, and by others, to be due to a pressure exerted maximally at the posterior end of the amoeba pushing plasmasol forwards. A third proposal is that there is an interaction between sol and gel where they come in contact so that the one slides on the other, an idea which would link amoeboid movement with the sliding of muscle fibrillae on one another. There is a shortage of evidence to allow any decision to be reached as to which is the most likely.

Sporozoans of the kind called gregarines (which are extracellular parasites when full grown) can also move over surfaces with an apparently effortless gliding. They produce a trail of mucus, rather like a slug, and were once thought to progress by rocket-like reaction to this stream. Now it is believed that some contractions of the surface layers must be occurring to propel them, again in much the same way as happens in a slug.

In some shelled amoebae (*Arcella*, Fig. 1B; *Difflugia*, Fig. 1E) the pseudopodia (pseudo) are broad and blunt; in others (*Euglypha, Gromia*) they are fine, branching and anastomosing outgrowths known as filopodia. A few shelled amoebae (*Allogromia*) and all the forams have fine, branching and anastomosing pseudopodia (reticulopodia) along which a persistent circulation of cytoplasm up one side and down the other is maintained, even to the finest tips, which seem to be enlarged by secretion of mucus. Thickenings occur at the nodes of the network. The pseudopodia are rigid enough to support themselves in water without internal skeletal rods.

With neither axial rod nor, apparently, central plasmasol in these pseudopodia no theory involving forces acting on the plasmasol can explain the circulation of their cytoplasm. It seems likely that the stream of cytoplasm moving in one direction on one side of the pseudopodium is sliding or pushing against the cytoplasm moving in the opposite direction on the other side, a mechanism reminiscent of the sliding of actin and myosin fibrils in the contraction of striped muscle and, like it, apparently linked to ATP. If this is so, it makes a similar explanation of the relative motion of plasmasol and plasmagel in lobopodia that much the more likely.

In *Difflugia* (Fig. 1E) the animal uses its pseudopodia to pull itself along the ground like a looper caterpillar and forams gradually creep over the surface to which they are attached, though how this is managed is not wholly understood. Though mobile, the creatures move too slowly for their shape to be affected by their movement and it seems to be determined by other factors. Some forams after passing through a juvenile motile phase become lethargic and finally wholly sedentary, after which their growth may become arborescent, especially if they occur in quiet situations.

The last type of body pattern which may be mentioned is that which occurs in the sedentary or sessile protozoans. Since food and other substances are likely to come to the animal from all directions, at least an approach to a radial symmetry is as characteristic of these animals as it is of other sessile organisms. The symmetry is rarely perfect, however, and a superficial radial symmetry often masks a truly bilateral one, or even an absence of any. This is more probable when the animal has a free-swimming ancestry and the adoption of the sessile mode of life is relatively recent. In many sessile protozoans, too, colony formation is frequent, as it is among metazoans of like habit, though here its physiological advantage is obscure since the individual members of the colony are often unconnected and cannot share food; perhaps the stronger currents set up by numerous contiguous animals are mutually advantageous when they are ciliary feeders. That the sessile animal is at some disadvantage when it comes to feeding is further suggested by the frequent occurrence of colonies as ectocommensals of other animals, by whose movements they may be carried from place to place and from whose feeding or respiratory currents they may incidentally profit. These protozoans are mainly either flagellates or ciliates, though a few sessile forams and heliozoans are known, and all are attached to the substratum by a stalk. In the flagellates, rhizopods and simpler ciliates (*Spirochona*, Fig. 5A) this is a protein secretion from the basal part of the cell and therefore immobile; in the more advanced ciliates (peritrichs, Fig. 4C, and suctorians) the stalk (st) is an outgrowth of a special basal area of cytoplasm known as the scopula which contains a large number of kinetosomes (granules from which flagella and cilia are produced). These bodies give rise to a bundle of modified cilia which run down the stalk within an enveloping membrane arising from the margin of the scopula. Towards the substratum the cilia lose various elements of their structure and degenerate into a bundle of membranous tubes or fibres. In species in which the stalk is not contractile (*Epistylis*, suctorians) this is all that is in it, but in species in which it is (*Vorticella*, Fig. 4C, *Carchesium*, *Zoothamnium*) the fibres are striated and elastic and their recoil antagonizes the action of a central contractile myoneme which splays out into a cone of fibrils in the main part of the body. Most of these sessile animals have retained their locomotor organs, flagella, pseudopodia or cilia, for feeding and also because some of them are not permanently fixed but move, on occasion, from place to place; even when cilia appear to be absent their kinetosomes are present and are capable of producing cilia at particular phases in the life cycle.

A possible method of functioning of the pseudopodium of an amoeba has been indicated above. Motile flagella and cilia appear to be identical in structure, not only in protozoans but throughout the animal and plant kingdoms, as was predicted in 1952 by the botanist Irene Manton, though variation in detail occurs. Each consists of a sheath enclosing microtubules embedded in a matrix and arranged as an outer ring of nine paired tubules and a central two which are single: the functional significance of this, which one would assume to be great since the pattern is universal, has so far escaped explanation. The central two end at the surface of the cell, the others unite basally and end on a disk resting on a cylindrical body or form the cylindrical body themselves. From this one or more fibrous extensions, known as striated root fibrils, may run further into the cell; sometimes these are present as threads separate from those in the cilium. In flagellates they may end on the nuclear membrane and often relate spatially to the Golgi apparatus; in ciliates this connexion is not made, a difference perhaps to be associated with the use of the basal body as a centriole during division of some flagellate nuclei, something which never happens in ciliates. Cilia further differ from flagella in being simple outgrowths: flagella often bear rows of lateral hair-like processes known as mastigonemes. The flagellum of *Euglena*, for example, bears a single row of such processes which may well make it a more efficient locomotor structure. In metazoan cells the body at the base of a cilium is known as a basal granule and the fibres extending into the cell as the ciliary rootlets; in flagellates, because protozoologists have elaborated a technical language of their own, the granule tends to be known as a blepharoplast and the fibre as the rhizoplast. In ciliates a rather more complicated arrangement has been described by the French protozoologists Chatton and Lwoff and since this has proved of value in understanding the relationships of these animals, it is worth while saying a little more about it. These authors, and others, also endowed this system with some of the properties of a nervous system and made it responsible for the coordination of the ciliary beat. This assumed "neuroid" function is now, however, denied by most protozoologists.

According to Chatton and Lwoff the apparatus of granules and fibres related to cilia is known as the infraciliature or subpellicular fibrillar system (Fig. 6). At the base of each cilium is a granule known as a kinetosome (k) and a fibre extending from it known as a kinetodesmal fibril (kd.f). The cilia (c) are arranged in rows and the kinetodesmal fibrils from adjacent cilia in a row converge and run alongside or become spirally twisted round one another to form a rope of fibrils known as a kinetodesmos (kd) (plural, kinetodesma) though each fibril extends only a short way. Other fibrils may run at right angles to the line of kinetosomes and still others obliquely, but there is no direct link, longitudinally or transversely, between one

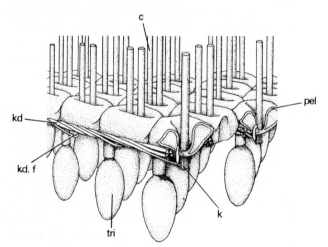

Fig. 6. *Paramecium*, small part of the surface layers showing the pellicle, the bases of some cilia and the infraciliature. (After Ehret and Powers.) c, cilium; k, kinetosome; kd, kinetodesmos; kd.f, kinetodesmal fibre; pel, pellicular corpuscle; tri, trichocyst.

kinetosome and its neighbours. A row of kinetosomes constitutes a unit known as a kinety. The body of the animal is covered with a series of kineties usually running from pole to pole, in a straight course in primitive types, in spiral or still more complicated patterns in more advanced ones. Each kinetodesmal fibril may run only a short distance along the kinetodesmos and then end, or it may run to the anterior end of the kinetodesmos depending on the species. During elongation of a ciliate (after binary fission, for example) the number of kinetosomes increases as the ciliary row lengthens; this is achieved by division of the kinetosomes and the formation by them of new cilia. It is generally believed that new kinetosomes usually, though not inevitably, arise from pre-existing ones and that they react to changes in environment, external or internal, by growing or withdrawing cilia. Kinetosomes have also been linked with the production of trichocysts (structures which can be discharged through the pellicle when the animal is irritated) (Fig. 4A and Fig. 6, tri), trichites (p. 31), and a number of other constituents of the protozoan body.

Kineties, or, more precisely, their constituent kinetosomes, appear to have an inherent topographical value which associates a particular one with a particular part of the body. Thus at transverse binary fission in a ciliate such as *Tetrahymena pyriformis* it is always on one particular kinety that the new mouth of the posterior daughter arises. Other kineties have similar associations with other organelles so that when reorganization occurs after division the inheritance of half of each kinety by each daughter ensures that it is ultimately organized on the same pattern as the parent. The kineties, therefore, form a kind of organizer and, on this view, the organization of the surface layer of the ciliate – where its most characteristic features of structure and function occur – is ultimately their responsibility, each kinetosome being capable of organizing its own territory. They are not, however, so limited in potency that accidental loss of one kinety deprives the organism of the structure associated with it; all are, indeed, totipotent, but they are arranged in an order of dominance, the kinety which normally produces the mouth having the highest rank, and the others following in clockwise order round the animal. If the stomatogenous kinety is lost, that on its right (as seen from the anterior end) replaces it as the source of a mouth. Similarly, a gradient of dominance extends in an anteroposterior direction along each single kinety. This is shown if, for example, the stomatogenous kinety is transected; the anterior end of the posterior section, freed from the dominance of the front part, then produces a new mouth. This is, indeed, what happens normally when binary fission occurs. With the system of transverse and longitudinal dominance in its kineties the surface of the ciliate is marked out by a kind of latitudinal–longitudinal grid which is responsible for the accurate reproduction, generation after generation, of the precise pattern of surface organization characteristic of the species. What maintains this system upon which the morphogenesis of the animals depends, and how it is related to the gene system are subjects of importance under active investigation. It is of interest to note than DNA has been shown to occur in the kinetosomes of both *Tetrahymena* and *Paramecium*; its function could relate to their ability to organize the pellicle.

In simpler genera morphogenesis may amount to little more than the production of an orderly array of cilia, with special groupings round the mouth and the proper siting of cell anus and the openings of contractile vacuoles. In higher types the whole surface may be much more elaborate in its architecture. Recent work on the familiar ciliate *Paramecium* (Fig. 6) – and further investigation is showing this to be generally true – suggests that the kinetosomes (k) organize the surface of this animal into producing a series of closely fitting hexagonal pellicular corpuscles (pel) each with a central hollow from which protrudes a single or double cilium (c); between these open trichocysts (tri) (said to be derived from kinetosomes budded from the ciliary kinetosomes). Each corpuscle contains a cavity or alveolus lined by unit membrane; they seem to be always inflated with fluid in *Paramecium*, but may be nearly collapsed in such ciliates as *Tetrahymena*. In every case, however, the whole surface of the animal, cilia, corpuscles, openings of trichocysts, are all covered by a further unit membrane which delimits the cell. Kinetodesmal fibrils (kd.f) from the cilia pass to the right of the cilia (i.e. in a clockwise direction if the animal is viewed from the anterior pole), intertwine with those from other cilia and form the kinetodesmos (kd) running longitudinally, each individual fibril soon coming to an end, but the kinetodesmos running the whole length of the body. The kinetodesma have been said to be cross-linked, especially in the walls of the gullet, and connected to a special centre there known as the motorium. This has been alleged to act as a kind of central control of the ciliary beat, responsible for the metachronal beat of the cilia and, on occasion, for their reversal, the kinetodesma (sometimes called neuronemes) acting like nerves and the contacts between them as microsynapses. There are many reasons why this proposal about the functioning of the fibrillar system should be approached with great caution, and probably rejected: the connexion between ciliary basal body and fibrillae has been denied; cilia isolated from basal bodies may still beat; cuts or isolation of parts of the body do not disturb metachronal beating of the cilia; there is nothing in the fine structure of the fibrils comparable to that of nerve, and the cross fibres can be seen in electron micrographs not to link kinetosome with kinetosome. However attractive the idea of a subpellicular nerve net may be, most workers are willing to find the basis for metachronal beating of cilia perhaps in some mechanical effect of each cilium on its neighbours, but mainly in the passage over the cell membrane of an electrotonically propagated wave of depolarization comparable to that marking the passage of a nerve impulse or the contraction of a muscle cell. The infraciliature may be no more than a mechanical strengthening of the superficial part of the body against the stresses and strains of locomotion, as is the likely function of swollen pellicular corpuscles.

The mode of contraction of cilia and flagella has been observed and is known to include a rigid stroke in one direction, the effective beat, followed by a more flexible one in the opposite direction, the recovery stroke, with the possibility of more complicated spiral movements in the case of the flagellum. The effective beat is the propulsive stroke since it is faster than the recovery stroke, 2–3 times in *Paramecium caudatum* and 5 times faster in *Urocentrum turbo*, though in *Opalina* effective and recovery strokes occupy equal times. How the strokes may be related to the minute structure of the organ is still speculation The beating has been stated to be due to the contractility of the nine double peripheral microtubules, and various ways in which stimulation might be passed from the central to the peripheral fibrils invoked in explanation of the differences between the beat of a cilium in one plane and that of a flagellum in three. All assume that the basal body initiates the beat and that the contractions are due to the presence of something similar to the ATP-activated actin–myosin system of striated muscle, with fibres perhaps sliding on one another as bending occurs. The ATP can be shown to occur in the fibrils of the cilium. Like other effectors they are sensitive to the charge on the cell membrane. In *Paramecium* straightforward swimming may lead to the animal hitting an obstacle; it reacts to this with a standard response which involves swimming backwards from the obstacle by reversal of beat, rotating slightly to one side and then swimming forwards again on a new path which, hopefully, clears the obstacle. If the animal is touched anteriorly this sequence is followed; if it is prodded posteriorly, however, its response is to swim forward more rapidly. It has been shown that ciliary reversal depends on Ca^{2+} and that it is accompanied by a fall in the membrane potential and an increased entry of calcium ions; this also occurs when the anterior part is stimulated mechanically. Increased rate of forward beating is similarly due to an increased polarization of the membrane and rise in the rate of entry of K^+; again these events also accompany mechanical stimulation of the posterior half.

Feeding and nutrition

When one compares one of the higher animals with one of the higher plants the differences between the feeding of the two are plain to see. The plant is able to synthetize from inorganic sources the organic nitrogen and carbon compounds which it requires, using energy obtained from sunlight, whereas the animal requires both in organic form. A further distinction is that the plant takes its food material in the form of salts dissolved in soil water or as gas from the atmosphere, whilst the animal ingests solid food. This clear-cut difference, however, is not necessarily found in protozoans. Many are obviously animal-like from this point of view – most ciliates and rhizopods, for example, which are micropredators and ingest solid food for digestion. But there are others which seem to partake of the nature of both plant and animal. This fact was one of the main reasons why Haeckel invented the name Protista for the unicellular organisms, to suggest that they were a group with both plant and animal affinities, out of which, by emphasis on one kind of nutritional ability the plants evolved, and, by emphasis on another, the animals.

Protozoa may first be distinguished as osmotrophs or phagotrophs; the terms saprozoic and holozoic are alternatives to these. The former take food in the dissolved state, the latter eat particles. Some are compulsory osmotrophs (*Euglena*, *Volvox*), others may supplement a phagotrophic existence with facultative osmotrophy (chrysomonad flagellates, dinoflagellates) or, alternatively, may be regarded as supplementing an osmotrophic existence with facultative phagotrophy. Another group, the compulsory phagotrophs, includes most ciliates and rhizopods, which depend entirely on ingested material. Even the compulsory phagotrophs are all capable of pinocytosis – cell drinking or the uptake of water with solutes in small vacuoles which ultimately merge with the cytoplasm – and so can perhaps survive without uptake of solid food.

Protozoa may also be catalogued according to the type of carbon and nitrogen compounds which they can use. Phototrophs are those which use carbon dioxide as a source of carbon and obtain energy from sunlight; haplotrophs cannot do this, but obtain energy by breaking down organic compounds of carbon, making use of the same compound (usually an amino acid) as that from which they obtain nitrogen; oxytrophs require two separate sources, one for carbon, the other for nitrogen. If the nitrogen source used by a protozoan is a nitrate the organism is called an autotroph; if an ammonium salt, a mesotroph, whilst the name metatroph is used to indicate that it requires a molecule of the complexity of an amino acid. The two series of names (one referring to the type of carbon compound used, the other to the kind of nitrogen compound) may be combined and Table 1 gives examples of protozoans falling into various categories.

TABLE 1

	Phototroph	Haplotroph	Oxytroph
Autotroph	Green phytomonads	"non-acetate"	"acetate" flagellates
	Euglena gracilis ⎫	flagellates (in dark	(i.e. green phytomonads
	E. stellata ⎪	if green)	in dark using lower fatty
	E. klebsi ⎬ in light		acids as C source) and
	E. pisciformis ⎭		others, colourless
	Plants generally		
Mesotroph	*E. anabaena* in light		
Metatroph	*E. deses* in light	rhizopods ciliates	*E. quartana*
		animals generally	*E. gracilis* in dark

Some protozoans are therefore photoautotropic in the way that plants are. If strictly so, like *Euglena pisciformis*, they cannot be kept in continuous darkness any more than a true green plant can; most, however, are capable of other kinds of nutrition which can be brought into action in the dark as the sole instead of merely an accessory source of food. This clearly affects the ecological conditions within which the organisms are capable of survival, extending them to include environments where light is minimal and permitting the modification of behaviour so as to reduce photic responses.

If organisms with this ability are kept in darkness the chloroplasts which they contain lose chlorophyll and become colourless; they lose their normal internal structure, become very small ($1~\mu$m) and are known as proplastids. They multiply, however, and are shared when daughter cells arise so that the power to become green on re-exposure to light is not lost, and the protozoan may be green again in a few hours, even, it has been recorded, after three years in continuous darkness. It is possible to produce by experimental means – heat at $35°$C, a variety of antibiotics, ultraviolet light – mutant forms in which all chloroplasts and the ability to produce them have been lost, so that the animals are permanently bleached. In organisms like the photoautotrophic *E. pisciformis*, even in the light, this must inevitably cause death by starvation, but in species which are not obligate photoautotrophs such a result does not follow and the creature survives as a colourless flagellate with a morphology like that of its green kindred except in respect of chloroplasts. When this experiment is carried out on *Euglena gracilis*, as has been done by the Austrian botanist Pringsheim who first produced organisms of this kind, the resulting colourless flagellate proves to be an organism very similar to one long known as *Astasia longa*; indeed, Pringsheim thought the two organisms were identical, though more recent work has revealed subtle morphological and biochemical distinctions. It is probable, therefore, that there are two series of flagellates in nature: (1) the ancestral green (and necessarily not obligatory photoautotrophic) types; (2) the colourless forms derived from them. Though morphologically all but identical originally, these will almost certainly with time acquire features which make them distinct. Many of these organisms are the kind known as monads – small, colourless flagellates whose ability to live in darkness on dissolved material has led to their pullulation in liquids in which products of decay and putrefaction are abundant.

It may well be that the source of the animal and plant kingdoms is to be looked for in a primitive flagellate of this nature, the green form of which lost its dependence on the saprophytic (osmotrophic) uptake of dissolved organic matter and became fully photoautotrophic to give rise to true plants. Its colourless derivative, on the other hand, went on from a dependence on the absorption of dissolved food to phagotrophy, and so gave rise to a truly animal stock.

In phagotrophic protozoans the organs seizing the food are often those used in locomotion. This is wholly true of the rhizopods in which pseudopodia engulf particles in food vacuoles arising at any position, though where several kinds of pseudopods occur simultaneously, as in Acantharia and Heliozoa, it is the small ones, often anastomosing, which catch food, the others being concerned with flotation and locomotion. Pseudopodia of this type may also be used by certain flagellates of the group Dinoflagellata to capture prey for ingestion and digestion. Flagellates, however, are often provided with a depressed area known as the cytopharynx, opening to the surface at the cytostome, into which particles are driven by the beating of the flagellum and at the inner end of which food vacuoles form directly in endoplasm. Sometimes the wall of the cytopharynx is armed with trichocysts (*Chilomonas*, *Cryptomonas*) or stiff, rod-like structures known as trichites (*Peranema*) acting like mobile teeth, by means of which the prey is presumably more rapidly caught or is actively grasped. In other flagellates food appears to be ingested at less definite but still reasonably localized positions on the body, though no permanent cytopharynx occurs. In trichomonads (*Trichomonas* from man and many other animals; *Devescovina* from termites) a depression on the ventral side of the body is the site of formation of food vacuoles (not the so-called cytostome which is found in some genera at the base of the flagella), whereas in trichonymphids it is the posterior pole at which this occurs. In the group of sessile animals known as choanoflagellates or craspedomonads, in which the flagellum has a protoplasmic sheath or collar around its basal half, particles are drawn towards the collar by the flagellar beat and caught on its sticky cytoplasm. This seems to be in continual movement so that the particles are carried to the base, where they are taken into food vacuoles, with or without the help of pseudopodia, depending on the species. The collar appears to be formed basally from a fold of cytoplasm, but distally from parallel microvilli, perhaps ensheathed in mucus, which grow from the summit of the fold.

Ciliates often show more elaborate feeding arrangements. In the simplest, conditions are not very different from those in flagellates: in the group known as the gymnostomes (Gymnostomatida) a cytopharynx lies either at the anterior pole or on the ventral side, and is usually armed with trichites though it has no special cilia. Its mouth can often be greatly expanded so as to permit the capture of prey of large size; this is particularly true of those forms in which the mouth is anterior. This group (Rhabdophorina) includes some of the commonest ciliates: *Coleps hirtus* (Fig. 4B), whose barrel shape is maintained by calcareous plates in the ectoplasm (sk), eats ciliates and algae; *Prorodon teres* eats dead animal or vegetable matter; *Didinium nasutum* feeds on other ciliates, particularly *Paramecium*, using a special adhesive organ which is part of the cytopharynx; *Loxodes* eats algae. In those with the mouth placed ventrally (Cyrtophorina) food is smaller and largely bacterial.

In the evolution of the ciliate buccal apparatus the first step towards complexity is taken when the cytostome no longer lies on the general surface of the body as it does in the gymnostomes but at the base of a funnel-shaped depression. This is known as the vestibule, and since it is derived from the general body surface it bears the same kind of cilia as occur there. With further advance, however, and presumably because of increased efficiency in the selection and handling of food particles, the ciliation of this area is modified so that there is formed on the right wall of the cavity a continuous sheet of fused cilia, the undulating membranes, and on the left, a number of plates (primitively three) also formed from fused cilia and known as the adoral zone of membranelles. Other parts may be bare. When this change has occurred the cavity is known as a buccal cavity. Its mouth is flush with the general body surface in the more primitive types (such as *Tetrahymena*), or may come to be at the bottom of a depressed area of the body surface which is still known as the vestibule, though presumably it is a secondary structure, not strictly homologous with the original; into the vestibule may run a groove, the oral groove. This is the arrangement found in *Paramecium* (Fig. 4A), though in this animal more complexity is apparent. The three adoral membranelles are each composed of four rows of cilia, whilst the right wall is corrugated with oblique longitudinal ribs. How all this complicated apparatus works is not at all clear. Current views on the classification of the ciliate protozoans suggest that from some primitive group with a buccal cavity like that of *Tetrahymena* may have been derived the peritrichs (Fig. 4C) by breakdown of the undulating membrane and membranelles and their extension out of the buccal cavity to form a spiral wreath (ad. mem) around the disk.

The ciliates mentioned above, along with many others not named, fall into a major group called the Holotricha in which, if a buccal cavity occurs at all, it is not very prominent and in which, broadly speaking, locomotor cilia are spread evenly over the surface of the animal. The main departure from this occurs in the groups Peritrichida, Chonotrichida and Suctorida, in relation to their predominantly attached mode of life: they tend to have no cilia over the general surface of the body. A second major grouping of ciliates is the Spirotricha, in which runs a trend towards reduction of the general body ciliation, perhaps compensated to some extent by the development of myonemes, contractile fibrils lying under the pellicle and facilitating change of shape. The main feature of the Spirotricha, however, is an increase in the complex ciliation of the buccal cavity, especially in the membranelles, which, instead of numbering three as in holotrichs, become extremely numerous and may spread out of the buccal cavity on to neighbouring parts of the body surface as in *Stentor*, *Halteria*, and most hypotrichs (Fig. 5B).

This evolution has allowed ciliates to adopt two main kinds of feeding habit, the predacious and the microphagous. For the former animals are adapted by a mouth placed at the anterior end of the body and a cytopharynx armed with trichites used for seizing prey, which may be large (as in the holotrichs *Coleps* and *Didinium*); for the latter they are adapted by a multiplication of ciliary structures in the buccal cavity which increase the volume of water sampled for food, and sieve it. In view of the way in which ciliates swim, too, a lateral or ventral position of the mouth increases the amount of food caught and the animals' chances of finding it, since it is continually being rotated to face in different directions. Examples of this are given by innumerable ciliates such as *Paramecium*, *Spirostomum* and the hypotrichs.

One group of Protozoa deserves mention in connexion with their unusual method of feeding: the Suctorida. Once regarded as standing apart from all other ciliates these animals are now believed to be holotrichs highly modified for a sessile life. They live mainly on the bodies of other animals or on weeds, attached by a stalk which is produced by secretion from a scopula (p. 25) but which is not contractile. The body bears a number of tentacles, sometimes grouped, sometimes scattered, sometimes sharp-tipped, sometimes knobbed. They are covered by a pellicle continuous with that over the rest of the body except at the tip, where the plasma membrane is naked. Through it protrude the points of a number of missile-like structures which penetrate the surface of the prey (other protozoans) and probably inject a toxic and digestive substance into them which rapidly paralyses them and starts a liquefaction of their protoplasm. Each tentacle contains internally a tubular arrangement of microtubules which marks off a central canal from a marginal region. The latter is used to transport from their site of manufacture in the body the granules which are stored in the tentacle tip and used to immobilize the prey; the former is used as a channel along which the protoplasm of the prey is sucked into the body of the suctorian. The suction appears to be brought about by an active expansion of the surface of the suctorian.

In phagotrophic protozoans food is taken into vacuoles within which digestion occurs. These lie in the endoplasm, movement of which (cyclosis) circulates them around the body on a usually fixed path. At the end of the digestive period, when the products have entered the cytoplasm, egestion of the indigestible residue takes place. In amoebae, where formation of food vacuoles may occur at any point, so, too, may egestion; in flagellates and ciliates, where their formation occurs at the base of a cytopharynx, egestion is correspondingly restricted to a cell anus or cytopyge. This may be due to the lack of a pellicle with any structural complexity in the first group and its presence in the two latter, or simply to the impossibility of predicting from moment to moment which part of a rhizopod is going to be at the back end and so conveniently placed for emptying a vacuole.

Osmoregulation

Other vacuoles occurring in the cytoplasm of protozoans are the contractile vacuoles which are certainly concerned with the control of the water content of the body and may also excrete nitrogenous material. They are most obvious in freshwater protozoans, which are those most liable to flooding of the tissues by osmotic uptake of water. As with food vacuoles they tend to move about the body and discharge at any point in rhizopods but to be precisely located in flagellates and ciliates. Electron microscopy has shown (in the few animals investigated, but probably as a general feature) that the visible principal vacuole is surrounded by a series of vesicles or canals within which the real separation of water from cytoplasm must take place. As this occurs the vesicles swell, may fuse, and ultimately empty to the main channels of the vacuolar system. The canals appear to be part of, or to link with, the endoplasmic reticulum. Around the vesicles the cytoplasm is extraordinarily rich in mitochondria which presumably provide the energy required for the secretory process. The mechanism of discharge is not yet properly known; in the suctorian *Tokophrya* fibrils run from the wall of the contractile vacuole to the pellicle which may aid in this, and this animal probably has (and others certainly have) fibres running round the duct of the vacuole which may act as a sphincter. It is possible that hydrostatic pressure set up by pellicle or cuticle could empty the vacuole on relaxation of these myonemes, which might happen when pressure within the vacuole reaches a critical level. In *Paramecium* (Fig. 7) conditions are more complex. Each tributary canal (trib.can) has a large number of apparently permanent channels (exc.tub) leading into it which are connected peripherally to endoplasmic reticulum (end). The tributary canals are contractile and discharge their contents to the main vesicle (cont.vac) which also has contractile fibrils (cont.fib) in its wall by means of which it can empty to the exterior (ext.can).

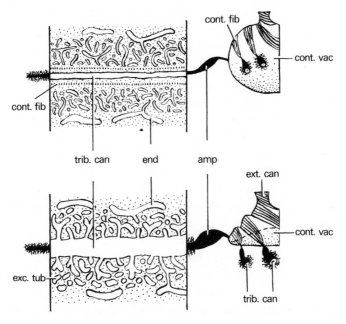

Fig. 7. Diagram of the contractile vacuole and associated structures in *Paramecium*. The upper drawing shows the tributary canal in systole, main vacuole in diastole; lower drawing shows canal in diastole, main vacuole contracted. The sections between the two vertical lines are very highly magnified. (After Schneider.) amp, ampulla of tributary canal; cont.fib, contractile fibrillae; cont.vac, contractile vacuole; end, endoplasmic reticulum; exc.tub, excretory tubules; ext.can, canal to exterior; trib.can, tributary canal.

Reproduction and life history

Protozoa normally divide by binary fission, the nucleus dividing by one or other of a series of processes which produce the same effect as mitosis, though often differing in many details from that process as it occurs in metazoan cells. Thus centrosomes may or may not be present, the spindle may lie within or without the nuclear membrane and the membrane may or may not break down. Since the result seems in all cases to be the same as that achieved by metazoan mitoses, these ingredients of the process – which strike us as necessary parts of it because of their regular occurrence – turn out to be relatively unimportant facets of the underlying essential activity. This seems to be the major lesson to be learnt from a study of the varieties of protozoan nuclear division so that they will not be described here, although they were once upon a time regarded as likely to explain the origin of mitosis. Many protozoan nuclei appear very large, and this is particularly true of the meganucleus of many ciliates. In some forms with two, like *Loxodes*, they are small and diploid, but in other species they are polyploid and contain a very large number of sets of chromosomes, ranging from 76 in *Paramecium bursaria* to over 13 000 in *Spirostomum ambiguum*. The rate of

multiplication of the chromosomes in these nuclei is not at first always tied rigor-
ously to mitotic activity as it is in the typical nucleus, but more usually continues
from the time that the meganucleus is first apparent as a daughter of the fusion
nucleus formed at conjugation (see below). Later the same rhythm of chromosome
replication and division is established as in metazoans. The meganucleus divides at
binary fission by a process which looks amitotic but involves chromosome replication
in the usual way. There is an inherent possibility of disaster since there is no obvious
machinery for ensuring that a chromosome of each sort (and so a complete set of
genes) enters each daughter nucleus, but the degree of ploidy (i.e. the number of
sets of chromosomes) seems to be so great as to reduce the risk of this happening to
negligible proportions; or there may be as yet some undiscovered mechanism for
ensuring that the nuclei always contain balanced sets of chromosomes, such as
subnuclei acting as autonomous units so that each daughter gets at least one.

Fission of the cytoplasm may take place in apparently any plane, as in amoebae,
in what is usually described as a longitudinal or symmetrogenic plane in flagellates,
passing between rows of basal bodies and in a transverse or homothetogenic plane in
ciliates, cutting across kineties. It is a process much more complex than the usual
ideas about it would suggest. When an organism such as *Euglena* or *Peranema* divides
some of the organelles such as chloroplasts may be shared between the daughters,
each receiving half, and since the volume of the organism is simultaneously halved
this may be an appropriate plastid content; other organelles, however, like the eye-
spot or stigma, will be received by one daughter only and the other will have to
develop them *de novo*. In a ciliate such as *Paramecium* transverse division may allocate
half the cytoplasm to each daughter, but such organelles as might be inherited by the
daughters in such a process would clearly be of the wrong size and in the wrong
place. Fission, therefore, calls for dedifferentiation and redifferentiation and it is
this which appears to be under the control of the kineties in ciliates, though what
replaces them in other groups of protozoans is not clear.

Fission may also be complicated in species where the animal is enclosed in a shell.
In some foraminiferans, its presence seems to have made binary fission too difficult
and a schizogony – a subdivision of the protoplasm into multiple naked units – has
replaced it, each daughter growing its own shell thereafter; in radiolarians with
elaborate shells, too, binary fission results in one daughter retaining the parental
shell and the other, from a naked start, secreting its own. This may also happen in
some shelled amoebae like *Arcella*, though in others, like *Euglypha*, a secretion of
reserve skeletal plates precedes fission and these enter the daughter which does not
inherit the original shell. In Acantharia something similar occurs, whereas in some
dinoflagellates each daughter may receive half of the parental set of cellulose plates
and have to reconstitute the whole shell by growing the remainder. Other armoured
dinoflagellates may shed their shell, divide and each daughter grow a new one.

The life history of protozoans may consist of nothing but an endless series of binary fissions without any trace of sexual process as in some amoebae. This is almost certainly a simplification of an originally more complex state. In parasitic amoebae, encystment, permitting transference from one host to another, may be combined with a phase of multiple fission so that from one successfully transferred cyst there emerges a small population of parasites. In most groups a period of asexual multiplication is followed by a sexual phase, as in foraminiferans and many flagellates, when either the whole animal transforms to a gamete (hologamete) or subdivides to form several (merogametes), which fuse with those from another source. In ciliates, sexual processes take the form known as conjugation, in which two individuals pair, lose their original meganuclei and exchange micronuclear material. Each makes a new zygote or fusion nucleus and the animals then part, to reconstitute their nuclear apparatus and resume the business of feeding and multiplication by binary fission. From this point the life of the ex-conjugants is divisible into a period of immaturity during which they are incapable of conjugation, followed by a period of maturity during which they are capable of it and normally carry it out, usually in response to some stress such as starvation; if, for any reason, they are prevented from conjugation they then enter a period of senility during which they may conjugate, but have less and less chance of producing viable ex-conjugants. During each of these periods binary fission continues so that the whole cycle may be spread over a large number of generations. Conjugation is clearly not primarily a way of getting more individuals into existence; this is more easily achieved by fission. Its advantages derive from the new genetic combinations which it makes possible, that is, it is fundamentally sexual.

Classical work, carried out initially by Sonneborn on *Paramecium aurelia*, has revealed an elaborate pattern of rules underlying conjugation in this species, and probably more or less similar ones will be proved to govern the process in other ciliates. *P. aurelia* has been shown to occur in a series of classificatory groups known originally as varieties, now as syngens, which are strictly comparable to closely related metazoan species. They are distinguishable among other things by size, fission rate and geographical distribution. Within a syngen individual animals belong to one or other of physiologically distinct strains known as mating types, and effective conjugation occurs only when the two conjugants belong to different mating types within one syngen. It will, sometimes, occur between animals not so qualified, but the ex-conjugants are then frequently inviable. If syngens thus appear to correspond to what one ordinarily understands as species then mating types correspond to what one calls sexes in metazoans, though unaccompanied by morphological differentiation. The basic characteristic of the mating type seems to be the production, under the control of nuclear genes, of different chemical substances, known as mating type substances, to which conjugating animals react. Furthermore, syngens fall into two assemblages, within one of which animals are outbreeders (preferentially conjugating with animals from another clone or related group of asexually produced animals), whereas in the other they are inbreeders (preferentially mating with fellows from the same clone). There are thus four conditions to be satisfied before conjugation can occur with success – maturity, syngen, mating type, clone. The length of the period of immaturity can be correlated with the inbreeding or outbreeding of the animals, as can also the production of mating types at conjugation. With inbreeding types the period of immaturity is short or effectively absent and the ex-conjugants invariably represent two mating types: this permits conjugation to occur soon, whilst the animals are still within a restricted volume of water. In out-breeding types the period of immaturity is prolonged and the ex-conjugants may all be of the same mating type; this is, however, unimportant, since the animals are immature and can disperse from their fellows into new areas where they may meet those members of another clone with which alone conjugation may occur.

Senility sets in if conjugation is prevented. In certain stocks (inbreeders) this may be cured by a process of self-fertilization or autogamy, in which, in one animal, a new meganucleus and micronucleus are made from the old micronucleus; this appears to have the same success in preventing senility as conjugation. Outbreeders seem to overcome the risk of senility by producing animals of the type appropriate for con-jugation by differentiation of that already in existence, followed by conjugation. Both these devices seem to secure an ultimate escape from senility and extinction; how-ever, external factors also must be appropriate for pairing to occur and although in natural conditions senility may not be the normal fate of a given stock it must in-evitably happen to some.

Parasitic protozoans

Nearly all the references to protozoans made in previous pages are to free-living animals; many, however, have adopted a parasitic mode of life. This is an evolu-tionary change apt to occur in osmotrophic stocks since the dissolved matter on which they rely for food is readily available on the surface or in the bodies of other animals. Phagotrophic protozoans have also become parasitic, nevertheless, and may ingest cells, or portions of cells from the body of their host, like the malarial parasite (*Plasmodium*) which takes part of the red blood cell in which it lives through a cytostome into food vacuoles for subsequent digestion. In most respects the life of such parasitic species does not differ significantly from that of their free-living rela-tives, though motility and osmoregulatory activity are often reduced, but it was with some surprise that early electron microscopists found all the structures associated with the ability to live independently within their bodies. They do, however, face one problem calling for adaptive modification of their life history, in connexion with transference from one host to another. In this they are no different from other para-sites and, in general, have arrived at the same two main solutions as metazoan parasites: either they produce a vast number of resistant infective stages which may, by accident, find their way into the body of the next host; or they use a vector, the feeding habits of which are such as to make transfer of the protozoan from host to host reasonably likely.

The former method is that used by most protozoan parasites – amoebae, the flagellates *Trichomonas* and *Giardia*, *Monocystis* and other gregarines, *Eimeria* and other coccidians – all have a resistant stage in the life history, which escapes from the body of the host, lies passively in the soil and is accidentally taken into the body of another host. This kind of life history is obviously most appropriate for gut parasites since a regular intake of food and output of faeces occurs there, and is less so for parasites of the vascular system, which is harder to enter and leave; and its success also depends upon the production of enormous numbers of the stages capable of existing outside the host. It therefore often includes a rapid multiplicative phase.

Vectors are associated primarily with parasites of systems other than the alimentary canal, and the most notorious of these are the trypanosomes, causing sleeping sickness in man and a variety of diseases in his domestic animals, and the malarial parasite. In both of these groups the life history shows an insect vector transferring the protozoan from one mammalian host to another. They therefore seem comparable. Nevertheless it appears that whilst trypanosomes were originally gut parasites of blood-sucking arthropods, the malarial parasites were originally gut parasites of vertebrates and the life histories have come to resemble one another by convergence. It is interesting to examine examples of these to show how such a complex life history as that of a trypanosome or malarial parasite may have evolved.

The most primitive members of the trypanosome group are the genera *Leptomonas* and *Crithidia*, species of which occur as parasites in the gut of a great variety of insects such as the aquatic bugs *Gerris*, *Velia*, *Nepa* and the house-fly *Musca*, or its relatives *Calliphora* and *Lucilia*. These parasites, at the appropriate stage in the life cycle, pass towards the anus of the host and surround themselves with a cyst wall. They escape with the faeces and a new infection depends upon the chance ingestion of a cyst by a new host. There is thus a straightforward life history and no vector. However, many of the invertebrates in which these protozoans live are not simple carnivores or carrion feeders like the examples mentioned above; some are bloodsucking flies such as *Tabanus*, *Melophagus* or *Haematopota*. The protozoan parasites must, therefore, be bathed in vertebrate blood every time that their host feeds, and although the food is altered as digestion occurs there will be periods when they are living in blood almost exactly as it courses through the vessels of the mammal from which it was drawn. Such trypanosomes, were they somehow to obtain entry to the vascular system of the vertebrate, might, because of their previous conditioning to an environment of blood, be able not only to survive but to flourish; though as they are living in a host to which they are less adapted than the invertebrate, their presence is more likely to upset the mammal than the fly that bit it.

This theory of the origin of the two-host life cycle of trypanosomes has much to commend it, but it is not to be expected that evidence to support it explicitly will be forthcoming. Nevertheless, there are one or two suggestive details in the life history of some trypanosomes. Thus in the case of *Trypanosoma cruzi*, the cause of Chagas' disease in South America, the parasite lives in blood-sucking arthropods, mainly bugs such as *Triatoma* or *Rhodnius;* when these animals bite to suck blood they defecate, as many blood-suckers do, and the infective stages of the trypanosome escape, just like the *Leptomonas* and *Crithidia* referred to above, in the faeces, only naked and not enclosed in a cyst. They are infective and gain entry to the host by being rubbed into the bite made by the bug, or into scratches in the skin made to alleviate irritation due to the bite, or by burrowing through sufficiently delicate surface membranes with which they have chance contact. The next step in the evolution of the characteristic trypanosome life history is an abrupt and rather extensive one, of which it is improbable that evidence could exist – the step by which the trypanosome relinquished the anus as its exit from the invertebrate host and substituted for it the salivary glands. This has the important sequel of direct injection into the vascular system of the vertebrate host, eliminates many of the elements of chance in the life history and produces the alternation from vertebrate to blood-sucking insect vector characteristic of many species of *Trypanosoma* such as *T. gambiense* and *T. rhodesiense* (which cause sleeping sickness in man and are transmitted by the fly *Glossina*), *T. brucei* (causing nagana and aina in horses and cattle, also transmitted by *Glossina*), *T. evansi* (causing surra in horses and cattle, transmitted by tabanid and culicid flies) and *T. equinum* (causing mal de caderas in horses, transmitted by the flies *Tabanus* and *Stomoxys*). The final evolutionary stage in this sequence is represented by *T. equiperdum* (causing the disease dourine in horses), in which the insect host has been completely eliminated and the protozoan is transferred from horse to horse as a venereal organism, in coition. Trypanosomes, therefore, form a group of animals, initially parasitic on invertebrates, which in the course of evolution have become parasites of vertebrates as well.

The malarial parasites and their relatives, on the other hand, seem to have been primarily parasites of vertebrates which have added an invertebrate host to their original life history. The general path which this evolution may have followed is again suggested by several genera of parasites, but not so fully as in the trypanosomes and more of its possible course remains conjectural. The most primitive types of animal from which the malarial parasites may be derived are the eimeriid coccidians, of which the most familiar example is *Eimeria stiedae*, found in rabbits. The asexual stage or trophozoite lives in the cells lining the bile ducts and liver, feeding and undergoing repeated division (schizogony) to produce schizozoites which invade other epithelial cells and so build up a population of parasites. Ultimately, presumably in response to some signal given when the population reaches a certain level, sexual stages are produced, syngamy occurs with a biflagellate microgamete swimming to unite with a stationary megagamete, and a cyst is secreted round the zygote. Within the cyst multiple division (sporogony) takes place as the parasite moves down the gut to escape with the faeces, and its decay liberates many resistant spores each containing two sporozoites. Infection of a new host occurs when food is contaminated by such spores.

The first life cycle which may be noted as showing a significant departure from this is that of *Schellackia bolivari*, which lives in lizards and tortoises. The trophozoites are found in the epithelial cells of the intestine, where they undergo a schizogonic multiplication, the schizozoites entering other intestinal epithelial cells. When gamonts (the stage from which gametes are derived) are produced, however, they pass into the submucous layers of the gut wall and gametogenesis and syngamy occur in that situation. The zygote, enclosed in a thin cyst, divides thrice to give eight sporozoites; these enter blood vessels and penetrate red blood cells within which they will be transported round the body. Some may be sucked into the gut of a tick (*Liponyssus saurarum*) commonly found on such vertebrates; if so, the red blood cell will be ingested by the cells lining the intestine of the tick and digested. The sporozoites resist digestion and are freed to the cell, within which they undergo a limited second sporogony. Infection occurs when an infected tick is eaten by a lizard or tortoise.

The life history of a second protozoan parasite, *Haemoproteus columbae*, does not involve the gut of the vertebrate at all, shows a little increase in the time spent in the vascular system, but demonstrates an increased importance of the invertebrate host. The animal parasitizes pigeons which are infected, not by eating infected food, but by the bite of an infected hippoboscid dipteran fly (*Pseudolynchia canariensis*). The sporozoites injected into the blood penetrate the endothelial cells of the vessels and undergo schizogony there. When gamonts appear, these differ in behaviour from schizozoites in entering erythrocytes rather than endothelial cells; they remain there until ingested by the fly when it sucks blood and the remainder of the sexual cycle and sporogony occur in the invertebrate. Since there is no exposed stage, cyst production is minimized and limited to a delicate membrane secreted around the zygote after it has been formed in the stomach of the fly and bored through the stomach wall. The sporozoites make their way to the salivary glands and enter the bird as the fly feeds. A parallel evolution is found in the life history of *Hepatozoon muris*, though this belongs to a different group of sporozoans and it must have taken place independently.

The final evolutionary step is that taken by *Plasmodium* (as a development from *Haemoproteus*) and by *Haemogregarina* (as a development from *Hepatozoon*). In *Plasmodium*, the sporozoite, injected into the blood stream of a vertebrate by the bite of a mosquito, is treated as any other foreign particle would be and phagocytosed by cells which are found lining blood spaces in the liver, spleen, bone marrow and lymph glands and collectively known as the reticulo-endothelial system, the function of which normally appears to be the elimination of such matter entering blood vessels. The ingested sporozoites undergo a limited schizogony in these cells (the exo-erythrocytic cycle, perhaps representing the older site of infection) and then enter red blood corpuscles, an evolutionarily newer but functionally more important situation, within which a second schizogonic cycle occurs. Ultimately gamonts are produced which invade other erythrocytes: they undergo no further developments until eaten by a mosquito, in the stomach of which gametogenesis and syngamy occur. The zygotes bore through the stomach wall and encyst as in *Haemoproteus*, giving vast numbers of naked sporozoites which migrate to the salivary glands, ready for injection when the mosquito next feeds.

This series thus demonstrates how a parasitic stock, originally living in the gut cells of a vertebrate and infecting new hosts by contamination, can evolve into a parasite predominantly of the blood and transmitted from host to host by a vector, normally a blood-sucking arthropod. It is interesting to note how the evolution seems to have occurred twice, once in the series Coccidia Eimeridea – Haemosporidia (the examples being *Eimeria, Schellackia, Haemoproteus, Plasmodium*) and once in the group Coccidia Adeleidea, as represented by the series *Adelea* (with a life cycle comparable to that of *Eimeria*), *Hepatozoon, Haemogregarina*. Taken in conjunction with the series culminating in the trypanosomes, it offers a striking example of convergent life histories and of adaptation to a parasitic mode of life.

Classification of Protozoa

In this and subsequent schemes the groups enclosed within brackets are not discussed in the text; those marked with an obelus (†) are wholly extinct.

Protozoa
    Mastigophora (= Flagellata)
        Phytomastigina
            Chrysomonadina
            Cryptomonadina
            Phytomonadina
            Euglenoidina
            Dinoflagellata
        Zoomastigina
            Protomonadina
            Diplomonadina
            Polymastigina
            Opalinina
    Rhizopoda (= Sarcodina)
        Amoebina
        Foraminifera
        Heliozoa
        Acantharia
        Radiolaria

Ciliata (= Ciliophora)
  Holotricha
      Gymnostomatida < Rhabdophorina
                       Cyrtophorina
      Trichostomatida
      Chonotrichida
      Suctorida
      [Apostomatida]
      [Astomatida]
      Hymenostomata
      [Thigmotrichida]
      Peritrichida
  Spirotricha
      Heterotrichida
      Oligotrichida
      [Tintinnida]
      Entodiniomorphida
      [Odontostomatida]
      Hypotrichida
Sporozoa
  Telosporidia
    Gregarinida
        Schizogregarina
        Eugregarina
    Piroplasmidea
    Coccidiomorpha
        Coccidia < Adeleidea
                   Eimeridea
        Haemosporidia
  [Neosporidia]

3

Sponges

IT IS NOW ACCEPTED that the animal kingdom exhibits a series of grades or levels of organization and that distinct groups may evolve to attain the same level; membership of a grade does not, therefore, imply genetic relationship. One such grade is the metazoan, contrasting with the protozoan, the multicellular contrasting with the unicellular. In the latter, single cells are capable of continuous life; in the former, single cells may survive for periods of varying length but have an extended life only when associated with other cells in a multicellular organization.

The metazoan grade has been attained certainly on two occasions, probably on three, and possibly by many other early groups which have become extinct without leaving any fossil trace, so that we can never learn of their existence. The two certain groups are the coelenterates and sponges, the third an extinct group, Archaeocyathida, which possibly belongs to the same branch as the sponges. To make it plain that there are these two kinds of metazoan, each with a separate evolution from a protozoan ancestry, the sponges and archaeocyathids are placed in a group Parazoa, and the coelenterates and all other metazoans in the Eumetazoa. The separateness of the sponges has been deduced from the fact that they are built on a pattern, and function in ways which are fundamentally different from those of eumetazoans and which point to them as coordinated aggregations of cells rather than as individuals. This is illustrated by the well-known experiment of dissociating a sponge into its constituent cells which can then re-group themselves into a series of small sponge bodies, each type of cell assuming its correct situation within the new body. Despite this, sponges are truly metazoan in that their cells, when isolated, are incapable of indefinite survival. The idea of individuality is extraordinarily difficult to apply to sponges in that (a) a sponge may be chopped into quite minute pieces each of which then behaves as if it were a complete animal; and (b) separate sponges of the same species which happen to grow against one another fuse to give what appears to be a single individual.

Organization of a simple sponge

A simple sponge like *Leucosolenia variabilis* may be described to give an idea of the organization. It consists of a series of pale, chimney-like tubules arising from a branching and anastomosing feltwork of basal tubes attached to a rock or weed substratum at a low level on the shore. Each uprising tubule may be 5–10 mm high and carries an apical opening almost equal in diameter to that of the tube. This aperture is known as the osculum and it can easily be shown that a current of water emerges from it: the main opening on the body of a sponge is, therefore, an exhalant one and so unlike that of a eumetazoan which is a mouth. It is more difficult to discover where the water emerging from the osculum enters the sponge, but this can be shown to be by way of a large number of microscopic apertures scattered over its external surface. Each opening is called a pore and from their presence derives the name Porifera for the phylum in which sponges are placed. Investigation at the microscopic level shows that each pore (p. 46 Fig. 8F) is a channel running through a single cell, or porocyte (poroc), from the external environment to the space within the sponge from which the osculum leads, known as the gastral cavity or spongocoel (g.cav). The size of the pore is under control and it can be shut; it is always small and may measure no more than two microns – sponges are therefore necessarily microphagous.

The body of the sponge consists of two layers of cells resting on a jelly-like mesogloea in which occurs a variety of isolated cells. The outer surface of the sponge, the oscular rim and the most apical part of the spongocoel are lined by a sheet of flattened cells known as pinacocytes (pin); the layer is known as the pinacoderm. The rest of the spongocoel in *Leucosolenia* is covered by the choanoderm, composed of choanocytes (ch), each like the kind of protozoan known as a choanoflagellate (p. 31) which must enter into the ancestry of sponges, though not into that of eumetazoans. It is the beating of their flagella which sets up the water current whose course has already been traced. The cells of the pinacoderm make contact with their neighbours, though electron microscopy shows only a loose union, but the choanocytes do not necessarily touch. The two layers are not, therefore, true epithelia but rather sheets of cells adhering to the two surfaces of the mesogloea; it is their common adherence to that structure which holds them together.

A sponge shows slow contraction in adverse circumstances and the osculum can close. The latter event appears to be due to a syncytial group of cells, myocytes, containing fibrils which may act like those of true muscle cells, though the proteins actin and myosin have not been demonstrated in them, which run around the opening. Contractility of the rest of the body is due to change in shape of the pinacocytes which can stretch to cover a greater area, or contract to cover a smaller one, extending deeper into the mesogloea as they do so. Choanocytes do not change shape but may be crowded together or separated by movements of the mesogloea.

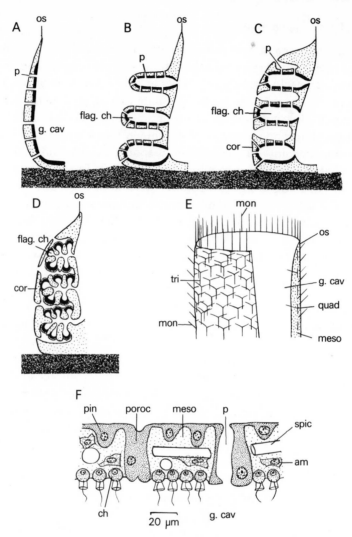

Fig. 8. Sponge structure. A–D, diagrams of the canal system. Each represents a vertical section through part of the body of a sponge, A, of asconoid grade, as in *Leucosolenia*; B, of lower syconoid grade, as in *Sycon*, C, of higher syconoid grade with cortex, as in *Grantia*; D, of leuconoid grade, as in most sponges. The thin continuous line represents pinacoderm, the thick one choanoderm; mesogloea is stippled. E, part of the tip of a piece of *Leucosolenia*, showing arrangement of spicules. F, part of a section through the body wall of *Leucosolenia*, showing arrangement of cells. am, amoebocyte; ch, choanocyte; cor, cortex; flag.ch, flagellated chamber; g.cav, gastral cavity; meso, mesogloea; mon, monaxon spicule; os, rim of osculum; p, pore; pin, pinacocyte; poroc, porocyte; quad, quadriradiate spicule with fourth (gastral) ray projecting into gastral cavity; spic, spicule; tri, triradiate spicule.

Many searches have been made for undoubted nerve and sensory cells in sponges but without success, and tests for such common accompaniments of true nervous tissue as acetylcholine and cholinesterase have been unsuccessful. The pinacoderm and myocytes must therefore be responding directly to internal factors or environmental changes. They are what are known as independent effectors. Nevertheless, numerous areas can be demonstrated with the electron microscope where contacts are made between cells and passage of messenger substances occurs.

In *Leucosolenia* the shape of the sponge is maintained by a skeleton of spicules, made of calcium carbonate, which lie predominantly in the mesogloea (Fig. 8E). Most are triradiate (tri) and consist of three rays spreading from a point; occasional one-rayed (monaxon) or four-rayed (quadriradiate) spicules also occur. The spicules do not lie at random but have each their proper station, often intermeshing to produce what is almost a spicular network on which the flesh is hung. The walls contain many triradiates, all lying with one ray pointing towards the base and the other two diagonally upwards towards the osculum; also embedded in the wall are quadriradiates (quad), three rays lying as before and the fourth projecting into the central gastral cavity, for which reason it is called the gastral ray. Monaxons (mon) project obliquely upwards from the pinacoderm, like so many pins in a pin-cushion, and a collar of larger ones is set on the oscular rim; these are obviously protective. It has been shown that if stresses are applied to the body of the sponge the mesogloea is slowly deformed and some rearrangement of spicules occurs in response to the new set of forces; it is probably in response to such forces, therefore, that the spicules become orientated in the first place.

The flagella of the choanocytes beat so as to push water away from themselves and from the mesogloea on which they rest. There will therefore tend to be a higher pressure in the centre of the gastral cavity and a lower one near the walls. This is the motive power which sucks water in by the pores and expels it by the osculum. Since the beat of the choanocytes is not coordinated the force they exert is trivial – the pressure at the osculum has been measured as equal to 2–4 mm of water (not in *Leucosolenia*, but in a sponge with a more complex arrangement). Particles fine enough to enter the pores may be caught on the collars of the choanocytes and ingested and liquid may also enter by pinocytosis. These substances are not, apparently digested within the choanocytes but are passed to amoebocytes in the mesogloea; it is within their vacuoles that digestion occurs and it is their wandering which is responsible for carrying food round the body.

It will be realized from this description that, in given circumstances, the amount of food available to a sponge of known size is a function of the number of choanocytes creating the feeding current. It is therefore not surprising to find that, from a starting point like *Leucosolenia* (Fig. 8A) which represents what is called the asconoid grade of organization, an evolutionary trend leads to an increased complexity of the choanoderm. The first step in this process is exhibited by such a sponge as *Sycon* (Fig. 8B), where the main cavity leading to the osculum is no longer lined by choanocytes: by a process of folding of the walls these are now restricted to lateral outpouchings, known as flagellated chambers (flag.ch), which draw their water from spaces lying between their walls and those of their neighbours. In *Sycon* the outer wall of the sponge is irregular because it is formed of the tips of these thimble-shaped outgrowths; in the purse sponge *Grantia* (Fig. 8C), however, the tips expand to form a superficially smooth cortex, perforated by a series of apertures leading to the spaces between flagellated chambers.

In most sponges (perhaps 90 per cent of species) a further complexity results from the application of the same principle of outward folding to the walls of the flagellated chambers of an animal like *Sycon* as produced it from an animal like *Leucosolenia*. The result is the leuconoid grade of organization with a thick fleshy wall replacing the thin one of simpler sponges and an extremely large number of small chambers (Fig. 8D). The advantages of these changes are not just an increased area of choanoderm and so of volume of water filtered. Choanocytes are not well adapted for creating a water current except in the line of the long axis of the cell, yet in *Leucosolenia* the bulk of the movement of water which they have to produce is parallel to the choanoderm. In the syconoid grade, and still more in the leuconoid, the volume of the flagellated chamber becomes steadily smaller (ascon chambers average 50 μm $\times$ several mm; sycon chambers, 50 $\times$ 100–500 μm; leucon chambers 20–50 $\times$ 30–100 μm) and more open at the exhalant end. This means that the choanocytes in each chamber come to lie more and more as a curved grid of cells stretched across the direction of flow of the water, the situation in which they work best. A well-known calculation by Bidder shows some of the rates of water movement in a sponge of leuconoid organization. He investigated the sponge *Leucandra*, using a specimen about 10 cm long and 1 cm in diameter with a single terminal osculum. In this sponge there were $2\frac{1}{4}$ million flagellated chambers, each 54 μm in diameter; the osculum of area 0.031 cm² passed a volume of water 0.26 cm³ s⁻¹ travelling at a rate of 8.5 cm s⁻¹; translated into terms of a single chamber this implies flow at a mean rate of 50 μm s⁻¹, but at only 12 μm s⁻¹ past a choanocyte, an event which occupies one second. This means that at the surface of the choanocyte where food is ingested, the water is moving at an extremely slow pace – about three quarters of a millimetre in a minute – and upon this the success of the mechanism depends. The force of the oscular jet is important in that it delimits a torus (a more or less doughnut-shaped space) around the sponge, the upper surface of which is occupied by water evacuated from the animal, the lower by water coming towards it. The greater the force of the oscular jet the greater the diameter of the cross section of the torus and so the less the chance of the immediate reingestion of exhaled water. Though not of significance to littoral sponges washed by restless tidal waters it is important to deep-water species living in quiet surroundings. The torus is limited below by the substratum in sponges of encrusting habit: it can be enlarged by raising the body off the ground on a stalk. This is probably the functional explanation of why so many sponges, particularly those living in deep and quiet waters, have a globular body elevated on a stalk.

Sponges can all reproduce asexually in that fragments transform into new sponges. All sponges become larger, too, so that a sponge covering a given area of substratum at one moment covers a greater area later. In view of the impossibility of deciding what an individual sponge is, it is not clear whether this is growth or asexual reproduction. Sexual reproduction also occurs, though there are gaps in our knowledge of its details. In *Sycon* and *Grantia* eggs are derived from amoebocytes lying in the mesogloea under flagellated chambers; sperm are perhaps also amoebocytic in origin but have been said to arise from choanocytes. Their presence is not apparently essential and larvae identical to those produced from the fertilization of eggs can also arise from agglomerations of amoebocytes. If formed, sperm are liberated and eventually enter the flagellated chambers of another sponge where they are trapped by choanocytes, usually those directly over a developing egg, as if they were chemically attracted to the underlying egg. The choanocyte with the ingested spermatozoon then withdraws from the choanoderm and transfers the sperm to the egg. In other words it treats the sperm as any choanocyte treats any ingested particle but hands this particularly special particle to a special amoebocyte. This unique method of fertilization appears likely to occur in all sponges though verified in only a few species. The development of the embryo includes a process, which parallels one occurring in the development of the protozoan *Volvox*, but nowhere else in the animal kingdom, by which cells initially within the embryo come to lie externally. The typical spherical larva of calcareous sponges, the amphiblastula, and that of other sponges, the parenchymella (or parenchymula) show two types of cell, one flagellated occupying one hemisphere, the other not, though rich in yolk granules, and occupying the other. At settlement and metamorphosis it is the flagellated cells, which have been the locomotor cells of the larva, which are infolded to form the choanoderm, and the yolky ones which remain externally to produce the pinacoderm. The two layers do not therefore match the ectoderm and and endoderm of eumetazoans and these terms are not used in describing sponges.

Asexual reproduction occurs in many sponges by means of gemmules or reduction bodies. These are clumps of amoebocytes, rich in reserve food, protected by membranes and sometimes by special spicules, which are formed when environmental conditions become unsuitable. When circumstances improve the gemmules hatch. In marine sponges, a free-swimming larva emerges; in freshwater ones (*Spongilla* and *Ephydatia*) the sponge often dies at winter, leaving a skeleton full of gemmules which give rise to miniature sponges in spring and reclothe the old skeleton with new flesh.

Other sponges

The phylum Porifera contains none but sessile animals. So far as its living representatives go, it may be split into three classes, Calcarea (or Calcispongiae), Hexactinellida (or Hyalospongiae) and Demospongiae, most species falling into the last. A few species amongst the Calcarea exhibit asconoid or syconoid organization

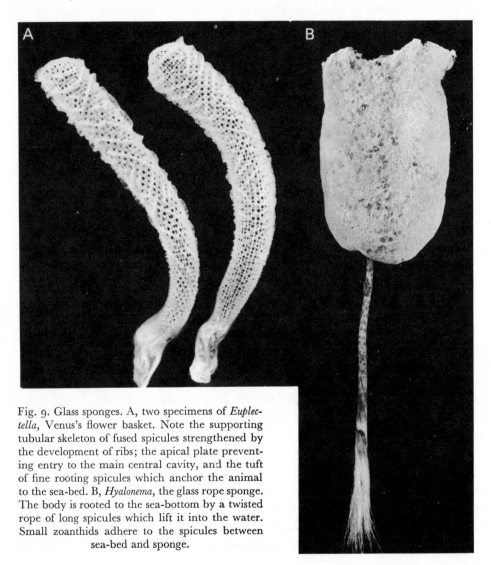

Fig. 9. Glass sponges. A, two specimens of *Euplectella*, Venus's flower basket. Note the supporting tubular skeleton of fused spicules strengthened by the development of ribs; the apical plate preventing entry to the main central cavity, and the tuft of fine rooting spicules which anchor the animal to the sea-bed. B, *Hyalonema*, the glass rope sponge. The body is rooted to the sea-bottom by a twisted rope of long spicules which lift it into the water. Small zoanthids adhere to the spicules between sea-bed and sponge.

but all other sponges are leuconoid. In Calcarea the spicules are made of calcareous material; in the other classes they are siliceous. In Demospongiae an extra-skeletal element may also occur, the keratin-like protein known as spongin. In genera like *Reniera* it occurs as blobs at the nodes of a spicular network of monaxons; in sponges like *Chalina* the network is of spongin with embedded spicules; in some, like *Hircinia*, detritus which falls on the surface of the sponge becomes incorporated in the spongin network as the sponge grows and may lead to the odd position of one species containing spicules derived from another. In sponges in the family Spongiidae the spongin network is the sole skeletal support, spicules and detritus both being absent, features which have allowed their use as the bath sponges of commerce. Finally, in a small group of Demospongiae, the Myxospongida (common genera, *Oscarella*, *Halisarca*), neither spongin nor spicules occur, an arrangement which may be correlated with the flat, film-like body which these slime sponges possess. Whether the lack of skeleton is primitive, or advanced and reached by degeneration, is still debated, but the body of evidence inclines to the former view. Demospongiae also differ from calcareous sponges in that their pores are gaps between cells of the pinacoderm rather than intracellular channels. Some demosponge species (the crumb-of-bread sponge, *Halichondria panicea*, and its red relative *Hymeniacidon sanguinea*, both common littoral sponges found on rocky substrata) have been said to be capable of a kind of slow, gliding movement; how much their change of position could be due to this and how much to growth is not certain.

Hexactinellids or hyalosponges, known as glass sponges, differ from all others in having their spicules six-rayed or based on a six-rayed pattern. The spicules tend to fuse with one another so as to produce a firm skeleton, as in Venus's flower basket (*Euplectella*, Fig. 9A), though isolated spicules are also scattered through the sponge. The flesh consists of a network of strands of cells to which flagellated chambers cling. There appears to be no pinacoderm so that the spaces of the network are widely open to the surrounding sea water; support and strengthening, however, are obtained from the lattice of fused spicules and the tendency of the others to form sheets on the outer surface. All hexactinellids are marine and live in deep water; since the substratum is often soft, many have tufts of long spicules to root them and hold them upright in a situation where there are no currents strong enough to dislodge them (*Hyalonema*, the glass rope sponge, Fig. 9B).

Classification of sponges

Parazoa
  [†Archaeocyathida]
  Porifera
    Calcarea (= Calcispongiae)

Demospongiae
    Myxospongida (= Homosclerophorida)
    Tetractinellida (= Tetractinomorpha)
    Monaxonida (= Ceratinomorpha)
[Sclerospongiae]
Hyalospongiae (= Hexactinellida)

4

Coelenterates

COELENTERATES OR CNIDARIANS are the animals which we call jelly-fish (or medusae) and polyps. Both of these may be regarded as exhibiting fundamentally the same body structure, adapted in the one case for a pelagic existence and in the other for a more or less sedentary attached one and they therefore provide a valuable opportunity of seeing how the same basic anatomical pattern can meet the different demands of contrasting modes of life. The essential feature of coelenterate organization, and the one in which they differ from other metazoans is that their body is built entirely of sheets of cells which have become folded in a variety of ways. Their component cells are capable of only limited histological differentiation. When specialization of function is called for, therefore, it can sometimes not be met by different kinds of *cell*: it can, however, be supplied by specialization of a whole *individual* or zooid if this is a member of a colony. It is probably because it permits this kind of differentiation that the formation of colonies with their accompanying polymorphism has become widespread among coelenterates.

Classification

The phylum is divisible into three classes.

1. *Hydrozoa*, with six orders. The Trachylina are jelly-fish when adult, though they have a relatively trivial polyp-like stage in the early part of the life history. Three orders include the animals usually called hydroids: Gymnoblastea or Athecata, in which there is no special development of the exoskeleton into which the polyp head can be withdrawn; Calyptoblastea or Thecata, in which such a protective structure is present, both orders normally with a medusoid stage in their life history; and Hydrida, the freshwater polyps, with neither protective skeleton nor medusoid phase. The fifth order is a small group, Hydrocorallina, the millepores, related to hydroids and important in the formation of coral reefs. Lastly, the large order Siphonophora contains elaborately constructed colonial and polymorphic animals with a pelagic mode of life, of which the Portuguese man-of-war is perhaps best known.

2. *Scyphozoa*, the common large jelly-fish. They can be arranged in four orders. (a) Cubomedusae, a small group with deep, more or less cubical bells and a long tentacle springing from each lower corner. (b) Coronatae, another small group of animals, often from deep water, in which the bell is divided by a horizontal furrow into higher central and lower marginal parts. (c) Semaeostomeae, the largest order, containing the common jelly-fish of coastal waters like *Aurelia* and *Chrysaora*, often stranded on summer beaches. In these animals the bell tends to be flattened whilst the edges of the mouth are elongated into four oral arms. (d) Rhizostomeae, jelly-fish of warmer waters with usually high bells, the central underpart of which extends downwards as four columns embracing the manubrium (the projection at the apex of which the mouth lies). The edges of the mouth are enlarged and penetrated by many fine coelenteric canals leading to minute openings on the surface, whilst the mouth proper tends to shut by fusion of its lips; the animals are therefore obligatorily microphagous. A fifth order in the class is the Stauromedusae, a small group, not of free swimming jelly-fish, but of attached animals called lucernarians, the most primitive in organization, neither strictly polypoid nor medusoid but sharing some of the features of both.

3. *Anthozoa* (or *Actinozoa*). Animals in this class are invariably polyps and there is no medusoid stage at any point in the life history. It contains two main orders. (a) Alcyonaria (or Octocorallia), the "soft" corals without rigid, massive, calcareous skeletons, such as Gorgonacea (sea-fans and the pink coral of commerce), Pennatulacea (sea-pens) and Alcyonacea (animals like dead men's fingers). The polyps in this order never have more than a single siphonoglyph, eight tentacles, which branch pinnately, nor eight mesenteries projecting into the coelenteron; they are always colonial and often polymorphic. (b) Zoantharia (or Hexacorallia), the sea-anemones (Actiniaria) and the reef-building or hermatypic corals (Madreporaria). The polyps in this order may be solitary or colonial, have two siphonoglyphs (see p. 61), unbranched tentacles and both tentacles and mesenteries are usually numerous and occur in some multiple of six. They are not polymorphic.

To these three classes is sometimes added a fourth, the Ctenophora, the sea gooseberries, but these are now more usually accepted as a separate group at about the same evolutionary level.

As an introduction to a more general discussion of the coelenterates a description of the more important aspects of the organization of a sea-anemone follows.

The organization of a sea-anemone

The body of a sea-anemone (see Fig. 10) conforms to the type known as a polyp. It is a hollow cylinder; the walls of the cylinder form what is called the column (col), and the cylinder is closed below by a basal disk attached to the substratum (base), and, above, by an oral disk (or.d.). At the line of junction of column and oral disk

Fig. 10. *Metridium dianthus*: dissected to show internal anatomy. The animal has been bisected vertically, but half of the base of the half removed has been left. The section passes through a primary septum on the right, but between primary septa on the left. (After Batham and Pantin.) ap.s, aperture in septum; base, basal disk; cn, coelenteron containing mesenteric filaments and acontia; col, column; m, mouth; mes.fil, mesenteric filaments; or.d, oral disk; ret.mu, retractor muscle in septum; sept 1, primary septum; sept 2, secondary septum; sept.att, pedal attachment of septum; siph, siphonoglyph; st, stomodaeum; t, tentacle.

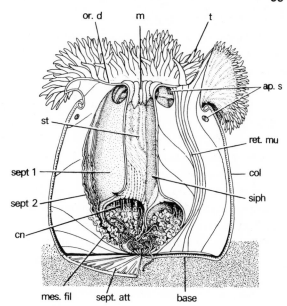

are placed one or more rows of finger-shaped tentacles (t) which reach outwards into the surrounding water; they are hollow and contain extensions of the main cavity of the anemone, the coelenteron (cn). Into this there leads the mouth (m), an opening placed at the centre of the oral disk; it is not in direct communication with the coelenteron but reaches it by way of an inwardly directed tube, the stomodaeum (st). The lining of the coelenteron is flung into a series of radial folds (sept 1, sept 2) which are attached to the column externally, to the oral and basal disk above and below, and which reach for varying distances towards the central axis of the cylinder. The number of these septa (or mesenteries) increases with age, young anemones having at first six pairs, to which others are added as it grows. The six primary pairs (sept 1) join the outer wall of the stomodaeum at their oral end but those formed later (sept 2) do not do this. The free central edges of the septa are thickened into filaments (mes.fil) which form the incomplete outer wall of a central part of the coelenteron into which the stomodaeum leads.

The wall of the cylinder (Fig. 11A) is made of a dead layer of mesogloea (meso) clothed on both sides by an epithelium. The outer layer is called the ectoderm (ect) and the inner the endoderm (end): these terms were invented for the layers by Allman in 1871 and, though they have been appropriated for, and given other meanings in, embryological usage, they will be used in their original sense here.

Cells and fibres occur in the mesogloea, the former occasionally, the latter invariably; the fibres are not secreted by the cells but seem to be merely condensations of mesogloeal material. Like the fibres of tendons or the struts of bone in vertebrates they orientate themselves along the lines of force to which they are exposed and so strengthen the matrix in which they lie. The mesogloea is elastic.

Changes of shape in sea-anemones have been most carefully investigated by Batham and Pantin and are achieved by the action of antagonistic sets of muscle on a hydrostatic skeleton provided by the fluid contents of the coelenteron. Although the coelenteric cavity connects with the external medium by way of the mouth (and, in some anemones, by pores known as cinclides (sing.cinclis) in the body wall as well), it has been shown that it acts physically as if it were a closed space. The anemone may therefore be regarded as of constant volume, but the shape which the volume presents may be changed. By bringing the oral and basal disks closer the anemone becomes shorter but, necessarily, stouter; when it elongates it must also become more slender. In a generalized coelenterate the antagonistic sets of muscle primitively constitute what are known as the muscle fields of the cells of the ectodermal and endodermal sheets. The cells are of the type known as musculo-epithelial, each epithelial cell tapering to a basal tail containing contractile fibrils set along the surface of the mesogloea, the tails of the ectodermal cells running longitudinally on its outer surface, those of the endoderm running in a circular direction on its inner face. Contraction of the ectodermal fibres (the endodermal ones relaxing) tends to shorten the polyp, contraction of the endodermal ones (the ectodermal fibres relaxing) tends to elongate it, the volume remaining constant. Gross change in shape is accompanied by corresponding changes in the ectodermal and endodermal layers and in the mesogloea (see Fig. 11A, B). The endodermal muscle layer crumples and folds horizontally as the body grows shorter, the wrinkles smoothing again as it elongates, extension stopping when the mesogloeal–endodermal boundary has completely straightened, a point reached when the length is about four times normal. The musculo-epithelial cells of the ectoderm are wine-glass shaped, with a narrow stem attached basally to the muscular tail; the endodermal cells are similar but show the shape less clearly. When a polyp is extended the cells have to cover a larger area than when it is contracted and they change from an approximately cubical to a columnar shape on shortening. This does not, however, absorb the full effects of the change in shape and it has been shown that a layer of liquid lies in the basal part of the epithelial layer around the narrow connections between the expanded distal parts of the cells and the muscle tails and shares in the change of shape as well as helping to force the cells into the altered configuration by providing a local hydrostatic skeleton.

In anemones (see Figs 11C and 12A) the arrangement of the muscles is not so simple as that of the typical coelenterate described; they are developed in localities where they can best resist the normal stresses to which the animals are exposed. The longitudinal muscles (long.mu) lie in the mesenteries and the circular muscles (circ.mu) belong to the cells lining the coelenteron between the attachment of the pairs of mesenteries, though forming a complete encirclement of the column. At the summit of the column, below the level of the tentacles, the circular muscles enlarge

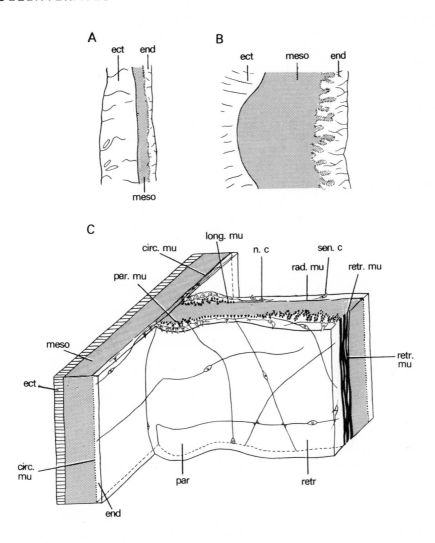

Fig. 11. Sea anemone: A,B, vertical sections of body wall showing circular muscle layer un-buckled in complete expansion of the anemone (A) and highly folded on retraction (B). C, stereogram of part of a sea anemone to show arrangement of muscles and nerve cells. The diagram shows a piece of column wall on the left with the origin of a piece of mesentery to the right. The ectodermal layer of the column is hatched, mesogloea is stippled and endoderm left white. Muscles are shown in black. circ.mu, circular muscle of column; ect, ectoderm; end, endoderm; long.mu, longitudinal muscle; meso, mesogloea; n.c, nerve cell; par, parietal region of mesentery; par.mu, parietal muscle; rad.mu, radial muscle of mesentery; retr, retrac-tor region of mesentery; retr.mu, vertical retractor muscle of mesentery; sen.c, sense cell.

to a sphincter (sph.mu), contraction of which pulls the column over retracted tentacles and disk and protects the animal completely. Both circular and longitudinal muscle fields are endodermal, and the column has no ectodermal musculature. The mesenteric muscles form a general longitudinal field, thickened near the origin of the mesentery from the column to form a special bundle, the parieto-basilar muscle (par.mu), and again half-way along the length of the mesentery to form another, the retractor muscle (retr.mu). The thick mesogloea imparts considerable rigidity to the column wall and shares with the contained muscle resistance to any force that would deform it.

The ways in which the actinian muscular system departs from the general coelenterate pattern rest on a clear functional basis. In a large polyp like an anemone, with many septa, the longitudinal muscles form an extensive series of ties between the two ends of the cylindrical body and thus spread the downward pull over much of the disk instead of concentrating it at the margin as would have occurred had they remained in the ectoderm. Similarly when the circular muscles contract and the coelenteric fluid is pressed against the oral disk these muscles help to resist the pressure at a large number of places. This allows the wall of the disk to be thinner and so more flexible and better adapted for other functions than it would otherwise have to be. The developmental pattern of the mesenteries, too, ensures that, while small anemones have relatively few of these supports, they rapidly increase in number with increase in size. The presence of an inturned stomodaeal tube the walls of which will be forced together by the slightest rise in coelenteric pressure is one of the ways in which the animal can operate, when required, as if it were a closed system whilst retaining connection with the external environment.

In addition to circular and longitudinal muscles the column also contains muscles which run obliquely from the outer ends of the mesenteries near the base on to the central parts of the pedal disk. They are known as parieto-basilar muscles (par.mu) and aid in fastening the animal to the substratum, their contraction tending to raise the middle of the disk from the ground and produce a sucker-like effect.

Some of the endodermal muscles appear superficially to depart from the usual coelenterate arrangement of muscle fibres underlying sheets of epithelium and to constitute solid blocks. This arises from extensive folding of the mesogloea and, sometimes, to the isolation of folded-off parts; within these folds, however, the muscles still lie in sheets on the mesogloeal surface. In this way the retractor muscles of the septa have become so powerful that, in the anemone *Metridium senile*, they can generate pressures of up to 100 mm of water. The column wall of a resting anemone of this species normally exerts a centrally directed pressure of only 2–3 mm water and is, therefore, a low-pressure system. When circumstances provoke a contraction of the retractor muscles, therefore, their greater strength means that the anemone must inevitably shorten.

So far we have been concerned only with the column, where all the musculature lies in the endoderm and the ectodermal cells have no muscle tails, as they have in other kinds of coelenterates. In one part of the body of a sea-anemone, however, the original distribution of the musculature is retained – in the oral disk and tentacles. In the former the ectodermal cells contain a series of radial fibres and in the latter a series of longitudinal ones. This superficial placing of the longitudinal muscles may be correlated with the fact that it makes more effective the most important movement executed by the tentacles, their bending in feeding.

The movements of anemones are controlled by a nervous system in the form of a network of nerve cells and fibres without ganglia, though some variation in the size of the mesh of the nerve net in different parts of the body may reflect varying importance, or fineness of control. The net seems not to enter the mesogloea.

The column wall and disk of *Calliactis* (the anemone first investigated by Pantin) react differently to stimulation; when a moderate stimulus is given to a point on the column the response occurs all round the column, but when applied to a point on the disk the reaction is localized near the point of stimulation. Functionally this is related to the greater appropriateness of responses by individual tentacles for feeding in the latter case and by the whole body for protection in the former. Physiologically it appears to be associated with the fact that the whole net in the column is flung into activity by a series of stimuli given at any one point whereas stimuli entering the nerve net of the disk at one point may die out as they pass across it and so produce a local effect. This is dependent upon a process of interneuronal facilitation and upon the time relations of that to the impulses evoked by the stimulus.

Pantin has shown that the type of response (both of column wall and disk) varies with the mechanical strength of the stimulus. In terms of nervous impulses this means with their frequency, since a weak mechanical stimulus evokes a slow discharge of nerve impulses and a strong stimulus a more rapid one. The arrival of an impulse at a synapse in the disk (or at a neuromuscular junction in the column) does not necessarily stimulate the cell on the farther side but reduces the resistance of the intercellular gap to the passage of the stimulus. The reduction lasts for a certain time but fades to the resting state, and probably depends upon the production of a chemical transmitter substance which is destroyed with the passage of time, though it must be added that chemical polarized synapses have not yet been demonstrated in coelenterates. If a second impulse does not arrive until the resting state has been reached then its effect is no different from the first and no impulse will ever cross to the nerve or muscle. If the second impulse arrives before the effects of the first have decayed then it may succeed in crossing, though it may only prepare the way for the crossing of the third impulse. The number and effect of the impulses which do succeed in crossing the synapse or neuromuscular junction will thus depend upon the frequency with which they reach it. Some muscles (e.g. the parietal muscles of the column of an anemone) react to an extraordinarily low frequency of stimulation, 1 shock per 5 s, and there may be a further latent period of 10 s before contraction begins; others, such as the longitudinal muscles of the septa, require a more rapid one, as would arise should a strong stimulus be given to the animal. These muscles normally contract more speedily than the former so that their effects may have been produced before the slower muscles have even started to contract. In this way the responses of the anemone can be varied to suit the kind of stimuli it receives.

Coelenterates are carnivores and are regarded as incapable of digesting substances characteristic of vegetable food, even though many contain symbiotic algal cells.

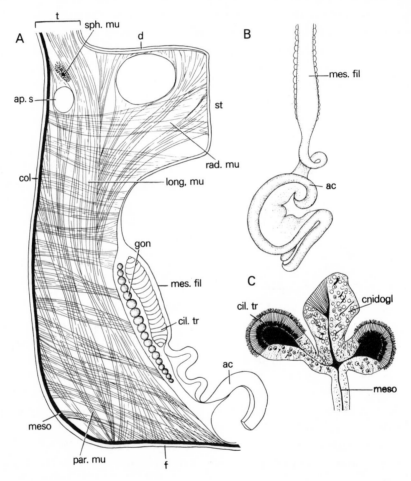

Fig. 12. *Metridium dianthus*: A, thick slice of half sea anemone to show the organization of a mesentery. B, basal half of a mesentery viewed from the coelenteric cavity to show filament. C, transverse section of mesenteric filament. (After Pantin and Stephenson.) ac, acontium; ap.s, aperture in septum; cil.tr, ciliated tract of filament; cnidogl, cnidoglandular tract of filament; col, column; d, disk; f, foot; gon, gonad on mesentery; long.mu, longitudinal muscle of mesentery; mes.fil, mesenteric filament; meso, mesogloea; par.mu, parieto-basilar muscle; rad.mu, radial muscle of mesentery; sph.mu, sphincter muscle; st, stomodaeum; t, tentacle.

Their food is usually caught by means of nematocysts, capsules lying within and manufactured by a cell known as a cnidoblast and liable to occur anywhere on the body, though particularly abundant on the tentacles and the thickened, free edges of the septa, which are called mesenteric filaments. They hold and paralyse the prey (see p. 71.) Normally the food is caught on the tentacles which extend radially from the edge of the oral disk, those of the outermost rows spreading outwards, those of the inner rows tending to stand more and more erect and so increase the volume of water fished. If an animal makes contact with a tentacle it is held by nematocysts whilst other tentacles bend towards it. Any struggling by the prey tends to bring it into contact with more tentacles and so hasten its destruction. When paralysed the prey is carried by the bending of the tentacles to the mouth, which opens by contraction of radial muscles in the disk and primary septa (Fig. 12A, rad.mu), so that the food can pass along the stomodaeum to the coelenteron. In most anemones two grooves run along the stomodaeum at opposite ends of its long diameter. These are the siphonoglyphs, lined by powerful cilia beating inwards, by virtue of which coelenteric turgor is maintained. The rest of the stomodaeum is covered by a more weakly ciliated epithelium and provides an outlet for excess coelenteric fluid and an exhalant pathway for the circulation within the coelenteron set up by the water entering along the siphonoglyphs. Normally, in association with this function, the cilia here beat outwards; when food (and certain constituents of normal food such as glycogen) are brought into contact with them, however, they reverse their beat and so aid transport into the coelenteron.

Once within the coelenteric cavity the process of digestion is begun by the secretion of enzymes, primarily proteinases, from endodermal gland cells, which rapidly break the food masses into microscopic crumbs which are phagocytosed by other endodermal cells. Within these the process of digestion is completed more leisurely in food vacuoles by enzymes such as polypeptidases and dipeptidases. The glandular and absorptive cells are restricted to the cnidoglandular tracts of the mesenteric filaments (Fig. 12C, cnidogl). The septa may also carry free threads called acontia (Fig. 12A, B, ac), which can wrap round the food in the coelenteron and help to kill it, since they carry many nematocysts. The acontia, the filaments and the general surface of the coelenteron in anemones are all ciliated (cil.tr) and this maintains a brisk internal circulation which, in addition to ensuring mixing of food and enzymes, brings the products of the preliminary extracellular phase of digestion within the reach of the phagocytic cells. Since fresh water is always being led into the coelenteron by way of the siphonoglyphs and some of its contents are always escaping by way of the compensation current over the remainder of the stomodaeal surface, this circulation also aids in distributing oxygen to the deeper parts of the body of the anemone and in removing excretory matter, and, in season, reproductive cells from the gonads (Fig. 12A, gon).

Polyp and medusa

As one might expect in animals which are either pelagic or attached most coelenterates exhibit an apparent radial symmetry. Only occasionally, however, does it prove genuine on close examination; often a cryptic bilateral symmetry is present. In some animals this arises from trivial anatomical points, as when the mode of attachment of the polyps of the hydroids *Aglaophenia* and *Sertularella* to the main stem of the colony forces them into a bilaterally symmetrical shape, but in others it has a functional basis, as in anthozoan polyps, where its flattened shape allows the stomodaeal tube to act as a valve, collapsing on pressure and so shutting the coelenteron against the outside.

Considerable argument goes on as to which of the three classes of cnidarians may be regarded as the most basic and eminent students of the group can be found who support the claims of each class: thus Pantin regarded the Anthozoa as the most primitive, whereas Hyman gave this position to the trachyline hydrozoans. Most recently Werner has suggested that Scyphozoa, in particular Cubomedusae, should be thought of as ancestral. The one point on which all agree is that the simplicity of Hydra, the "typical coelenterate" of the elementary student, is deceptive and secondary. Some of the reasons for believing in the primitiveness of the anthozoan polyp as compared with those of the other two classes are the simplicity of their nematocysts, especially the presence of the kind known as spirocysts (p. 71), their lack of sense *organs* as distinct from sense *cells* and the fact that their nervous system appears to take the form of a single nerve net lying in their epithelia. If this is true it is interesting to speculate on the relationship between polyp and medusa since the latter stage is absent from the anthozoan life history.

In Hydrozoa and Scyphozoa, life histories normally exhibit what is called an alternation of generations, part being spent in the polyp state, part as a medusa, the first reproducing asexually, the latter carrying gonads. In hydrozoans the polyp generation is commonly the longer lived and the more conspicuous, whereas in scyphozoans the reverse is true. There are two main ways in which this alternation may be explained. The first may be called the hydroid theory and is due to Leuckart (1851). This supposes that just as different types of polyp have come into being to subserve different functions within a hydroid colony (see p. 79) so have medusae, to act as carriers of the gonads and thus spread the species. On this theory medusae would be later evolutionary developments than polyps and have been added to a previously simpler life history based on the polyp. This theory is in line with the deduction made above from anthozoan structure in accepting the primacy of the polyp: there are, however, good reasons for believing that those hydrozoans in which the polyp is the most obvious component are the most advanced rather than the most primitive.

The second theory is the actinula theory, originally proposed by Böhm but usually

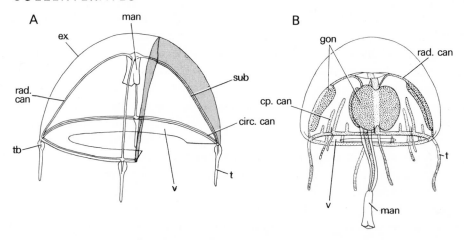

Fig. 13. Hydrozoan medusae. A, hydrozoan medusa to show its parts; the near quadrant on the right has been removed, and the cut surface is hatched. B, *Liriope tetraphylla*. circ.can, circular canal; cp.can, centripetal canal; ex, exumbrella; gon, gonad; man, manubrium; rad.can, radial canal; sub, subumbral surface; t, tentacle; tb, tentacular bulb; v, velum.

associated with the name of Brookes, who expounded it rather fully in 1886. For many reasons, more particularly the primitive nature of their nematocysts, it seemed to Brookes that the most primitive hydrozoans are those classified as Trachylina, such as *Liriope* (Fig. 13B) and *Geryonia*. In these the egg, on fertilization, gives a type of larva known as a planula (Fig. 24B), which may either change directly into a medusa or settle for a while as an insignificant attached stage, which may bud off medusae, but which ultimately becomes a medusa itself. (This fixed stage is often parasitic, but this is presumably secondary and therefore a complication not important from the present point of view.) Brookes supposed that from a life history of this sort there has been evolved that exhibited by *Obelia*, for example, by gradual elaboration of the transitory fixed stage into the complex polyp colony and its adaptation for a sessile mode of life, with the production of medusa buds as before, but with loss of the final transformation of the first settled polyp (and, now, many of its daughters as well) into medusae. On this theory the polyp and medusa are of equally ancient ancestry and the question which came first is a "chicken and egg" type question which cannot be given a straightforward answer.

The same kind of life history occurs in Scyphozoa, as in the common jelly-fish *Aurelia*. Here the planula settles and transforms to a polyp-like stage (hydra-tuba) which may produce a few more polyps, but all ultimately bud off larvae known as ephyrae (Fig. 24A) which grow into adult jelly-fish. Hydrozoans and scyphozoans, on this theory, therefore, are both fundamentally stocks in which the adult stage has a medusiform organization, but in many hydrozoans the life history has been profoundly affected by a neotenic evolution which has converted the original transient, juvenile polyp phase into its main element. In many hydroids the medusae are well formed and live an independent life lasting weeks or months in the plankton. In some, however, the free life of a medusa is to be measured in hours (*Podocoryne carnea*) and there is a large number of genera in which no free-swimming medusae are produced and distribution depends entirely on larvae. This evolutionary loss of the medusa parallels the suppression of distributive stages in other groups of animals when the adult has become specialized in its habitat or when, for any other reason, free-swimming stages are more likely to die at metamorphosis by failing to find an appropriate environment to settle on than to found new colonies.

In both classes the starting point of this evolution is the type of organism which arises at the metamorphosis of the free-swimming planula larva. It was originally, in all probability, still free-swimming but had tentacles and may later have converted itself into a true medusa. In many groups, however, a sessile phase was introduced, perhaps as a transitory device for over-wintering and exploiting the more abundant food on the bottom at that time of year. It is therefore perhaps best regarded as something which predates polyp and medusa and from which, by further adaptation for a fixed life, came the true polyp, and, by further adaptation for pelagic life, the true medusa. This type of animal, called an actinula, appears as a phase in the life history of many coelenterates and is not unlike the larva of *Tubularia* (Fig. 14). As we have just seen it occurs in trachyline hydrozoans and in scyphozoans; the stauro-medusan scyphozoans are not much more than actinulae which have become adult and we can explain the absence of anything corresponding to a medusoid phase in

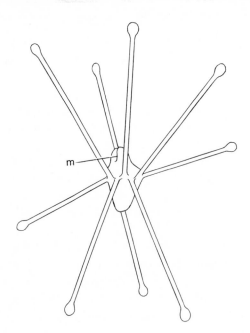

Fig. 14. *Tubularia larynx*: actinula larva, morphologically equivalent to a polyp head with a single circlet of tentacles held alternately forwards and backwards. m, mouth.

anthozoans by supposing that they branched off from the evolutionary line leading to the hydrozoan-scyphozoan group before the free-swimming medusoid phase in the life history had been invented. In the hydrozoans the neotenic colony produced by asexual multiplication of the actinula has become the dominant part of the life history and its proper historical ending in medusae sometimes forgotten; in the scyphozoans – unless it has occurred in stauromedusans – this never takes place and the medusoid phase has remained prominent.

Muscles and movement

The primary pattern on which coelenterates – polyp and medusa – are built is that of a dead layer of mesogloea clothed on both sides by an epithelium and shaped into a hollow vessel. In polyps the mesogloea is either a very thin layer (hydroids) or is moderately thick (actinians); in medusae (see, for example, Figs 13 and 15) it becomes voluminous, mainly by the inclusion of water-filled spaces. The fluid is probably hypertonic to the environment so that the body possesses a certain turgor, and the spaces act as a hydrostatic skeleton. They may be necessary because the coelenteric spaces are of inconsiderable volume in medusae. In polyps, on the other hand, the mesogloea is not voluminous and the main space which acts in this way is the coelenteron. In anthozoans it is topped up with water by the inward beat of the cilia which line the siphonoglyphs. In a resting anemone (*Metridium*) this has been measured as having a pressure equal to 2–3 mm of water. Since the anemone may remain of constant volume for long periods this pressure must counterbalance that exerted by the column wall, which may be due either to muscular contraction or to elasticity of the mesogloeal component, or both. In animals where tentacular extension is important (sea-anemones and Hydra) the coelenteric cavity and circular endodermal muscle fibres are retained in the tentacles to provide a local hydrostatic mechanism, but where it is not important (as in most hydrozoan polyps) the endoderm forms a solid core to the tentacle, which can be pulled in any direction by contraction of the appropriate longitudinal fibres and which returns to a resting position through its own elasticity on removal of the deforming force.

We are accustomed, in dealing with the muscular organization of such animals as vertebrates and arthropods, to think of prime movers which alter the orientation of one part of the body in relation to others, of antagonists which must relax in order that the primary motion can be effected with speed and economy of effort, and of synergists which aid or modify the action of the prime mover; but the interrelationships of muscles in these animals rarely extend further than this. In coelenterates, every muscle in the body is linked to every other by the coelenteric fluid, which is compressed when any muscle contracts and transmits some pressure effect all over the body. The action of a muscle is therefore double in that it has a local effect, brought about by the shortening of the fibres of which it is composed, and a general one due to its action through the hydrostatic skeleton. In this respect the muscles of the body wall of a coelenterate are unique, though comparable to those of the gut wall of one of the higher animals. This effect does not mean that contraction of one muscular area would make others incapable of contraction; it does, however, imply that contraction might call for more energy. Other complications arise from the fact that, unlike arthropod or vertebrate muscle, there are no clear maximal or minimal distances, dictated by a skeleton, beyond which it cannot be extended or shortened. After feeding, or after artificial inflation of the coelenteron, for example, the longitudinal muscles of an anemone may become so stretched as to be capable of only isometric contraction, i.e. without any shortening taking place. The only escape from this condition is to open the mouth by the radial muscles of the primary septa and allow water to spill out. To minimize disturbance of other parts of the body due to the transmission of the effects of local contraction through the coelenteric fluid, movement takes place very slowly.

Patterns of movement in coelenterates relate primarily to locomotion and to feeding. The first of these are inevitably different in polyps and medusae since the one is sedentary and the other not. In both the tempo of feeding will be faster than that of locomotion since a carnivore has to catch its food despite avoiding movements of its prey; this difference is more pronounced in polyps than medusae. Further differences arise from the fact that locomotor movements are cyclic, with a pacemaker, whereas feeding movements are irregular and different from occasion to occasion. To the casual observer sea-anemones seem immobile unless obviously stimulated into feeding, protective or other activity. This, however, is not a true picture but results from the very slow rate at which their movements occur. If steps are taken (by time-lapse cinematography, for example) to speed up the apparent rate of movement, it is discovered that they are as active as many other invertebrates. Batham and Pantin, indeed, observed an apparently regular cycle of movements in the column of *Metridium*, contraction of the parietal muscles being followed by contraction of the sphincter and that by a wave of contraction of the circular muscle. The cycle lasts for several minutes and may be repeated after a pause.

Such observations as have been made on movement in polypoid coelenterates refer to large sea-anemones; their smaller size has prevented comparable studies being made on alcyonarians and hydroids. In jelly-fish, a different anatomical musculature underlies their different patterns of muscular activity. Here, radial and circular muscles, sometimes in overlapping sheets (Hydrozoa), sometimes in distinct bundles (Scyphozoa), run below the ectoderm of the under surface (subumbrella), and there are no muscles in the upper surface (exumbrella). The two sets are clearly not antagonistic but synergic since both reduce the diameter of the bell on contraction. Restoration of shape must therefore depend primarily upon elastic recovery by the mesogloea after it has been deformed. This rests partly on the fact that it contains lacunae filled with water and partly on the occurrence of elastic

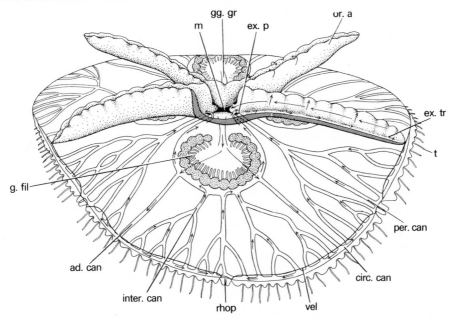

Fig. 15. *Aurelia aurita*: animal seen obliquely from the oral side to show ciliary currents (arrows) involved in feeding, circulating fluid in the coelenteron and rejection of waste. One half of one oral arm has been removed to expose the mouth and the cut surface is hatched. (After Southward.) ad.can, adradial canal; circ.can, circular canal; ex.p, excretory pore; ex.tr, exhalant tract removing waste; g.fil, gastric filament; gg.gr, gastrogenital groove; inter.can, interradial canal; m, mouth; or.a, oral arm; per.can, perradial canal; rhop, marginal sense organ or rhopalium; t, tentacle; vel, velarium.

fibrils, some at least of which run from exumbrella to subumbrella surface and so are stretched by the contraction of the muscles. Contraction of radial and circular muscles results in a decreased radius of the bell and hence in a reduced volume of the subumbrellar cavity and expulsion of water from that space. This is the motor impulse which propels the animals through the water and the more sudden and jet-like the expulsion the more effective the swimming movement. In hydrozoan jellyfish it is made jet-like by the presence of the velum (Fig. 13A, v), a fold of ectoderm supported by mesogloea which projects centrally across the mouth of the subumbrellar cavity from the inner edge of the bell, forces the expelled water into a jet and so propels the animal more vigorously. Scyphozoan medusae have no velum, but compensation is provided by the facts that (1) the circular muscle of the subumbrella has been elaborated into a powerful ring muscle (the coronal muscle), whilst (2) the initial contraction of the radial fibres, which are more marginally placed, pulls the edge of the bell in before the coronal muscle contracts and so narrows the mouth and makes the outflow more jet-like. The structure which occurs in some scyphozoan medusae called a velarium (Fig. 15, vel) is a series of lobes belonging to the inturned margin of the bell and therefore containing endoderm as well as ectoderm. In most scyphozoans it is not functionally equivalent to a velum; in such forms as *Charybdea*, where all the lobes fuse to give rise to an inturned edge, it may be.

In jelly-fish, which are swimming organisms, a regular locomotor rhythm and so a pacemaker, quite uncharacteristic of polyps, is required to keep the animals afloat. In correlation with this active life sense organs in the form of statocysts or eyes are of common occurrence; they contrast with the sensory cells which are all that is found in polyps. In scyphozoan medusae the swimming movements originate either in the elaborate sense organs known as rhopalia, which are modified tentacles, found at the edge of the bell (Fig. 15, rhop), or in adjacent concentrations of nerve cells known as ganglia. As shown by Romanes, provided at least one of these is active all the swimming muscles are flung into rhythmical contraction and movement results. Romanes also showed that the bell could be slit in a variety of ways but so long as a narrow bridge of tissue connected one part with another nervous conduction between the two was possible and swimming occurred. This indicated the presence of a nerve net. Recent investigations have shown that there are, in fact, two nerve nets: (1) the system just described, which carries impulses rapidly from rhopalia to muscles all round the bell and exhibits cyclic activity, known as the giant fibre nerve net; and (2) a more diffuse net, not narrowly related to sense organs and muscles, though connecting with them, which is responsible for feeding and other irregularly occurring activities distinct from swimming – it also modulates the frequency of discharge in the first system so as to relate it to activities governed by the second. Rhythmicity apart, there is a clear resemblance between the first system of scyphozoan jelly-fish and the nerve net of the column and mesenteries of anthozoan polyps in that both give rise to a response affecting the musculature all round the animal even when the applied stimulus is at a point – a symmetrical response to an asymmetrical stimulus; indeed, the effectiveness of the swimming movement of a jelly-fish and of the protective shortening of the column of a sea-anemone depends upon the nearly simultaneous contraction of muscles all round the body. The neural arrangements responsible for these movements are reminiscent of giant fibre systems in other groups of invertebrates (arthropods, annelids, cephalopods) which act in a similar way, sending all the muscles of a particular sort into almost simultaneous contraction. The second system of the jelly-fish parallels in its working the nerve net of the disk of the sea-anemone in which local responses, adjusted to local needs, are called for – an asymmetrical stimulus provoking an asymmetrical reaction.

The anatomical distinction which can be made in sea-anemones between locomotor and feeding muscles is blurred in scyphozoan jelly-fish where the same muscles are involved in feeding as in swimming. In an ephyra larva, for example, a twitch-like contraction causes swimming but a sustained local contraction is necessary for feeding. The one nerve net causes the first type of response, the diffuse nerve net the second. In adult jelly-fish such as *Cassiopeia* and *Cyanea* the same double innervation persists, though primarily to modulate swimming in such a way as to effect righting movements when the animal becomes disequilibriated. In *Aurelia*, which gives up feeding on large food masses on becoming adult (p. 77), the two nets persist but the second acts on the sense organs or ganglia and not directly on the muscle. In hydrozoan jelly-fish matters are more complex in that, in addition to the two nerve nets, epithelial cells provide a third conducting system over the bell. Each nerve net, further, has its own pacemakers and communication round the margin of the bell is accelerated by the presence of a double nerve ring close to the marginal coelenteric canal. This is the basis of a more complex behaviour. Hydrozoan jelly-fish, however, are physiologically simpler in that of the two nerve nets one is related to the radial muscles used in feeding and the other to the circular muscles used in swimming so that no double innervation occurs. In polyps, only one net occurs, though, as indicated above, it shows regional physiological differentiation which allows the same kind of separate feeding and locomotor (protective) activity as is achieved by the two nets of medusae. The polyp is thus equivalent to a medusa deprived of the nervous machinery required for swimming.

In colonies there may sometimes be integrated activity of the various members, so that there must be a continuity of nervous conduction between polyps by way of the intervening common tissue.

Nematocysts

Coelenterates are carnivores and their food is usually caught by means of nematocysts, structures characteristic of the phylum; a few have become microphagous. Nematocysts are largely automatic in action and explode when appropriately stimulated; because of this coelenterates are often destructive of other animals, killing them on contact, without relation to their need for food. This is particularly true of the larger medusae and siphonophores, which may trail tentacles many feet in length through the water, fishing for food. The details of how a nematocyst explodes are not fully known but it seems that the threshold of stimulation may be raised or lowered by nervous control, or by appropriate chemical substances. Those that facilitate explosion are of animal rather than of vegetable origin, a fact presumably related to the carnivorous habit. The discharge almost certainly depends upon uptake of water from the surroundings, though it is said that a basketwork of contractile fibres lies around some nematocysts, and starts the process by mechanical pressure, to which all nematocysts are sensitive. The area of the surface of the discharged nematocyst is the same as that of the undischarged one, so that, if explosion is due to swelling, it is swelling of the capsular contents rather than of the bounding membrane which must be responsible. They have been shown to have an osmotic pressure equal to about twice that of sea water.

A nematocyst such as a large stenotele of a Hydra (Fig. 16) is a capsule (cap), made of material allied to collagen, lying within and manufactured by a cell known as a cnidoblast (cyto). It contains a thread, spirally coiled and armed with spiral rows of projections (barb), and it is provided, at its distal end with a lid, the operculum (op). In other kinds of nematocyst there may be no operculum. The cnidoblast, in the more primitive coelenterates such as the anthozoans, carries cilia, but, in hydrozoans, at least one of these has become modified into a stiff, stationary structure connected at its base to the nematocyst and known as the cnidocil (cnido), and the rest have been lost. The cnidocil or ciliary cone triggers off the explosion

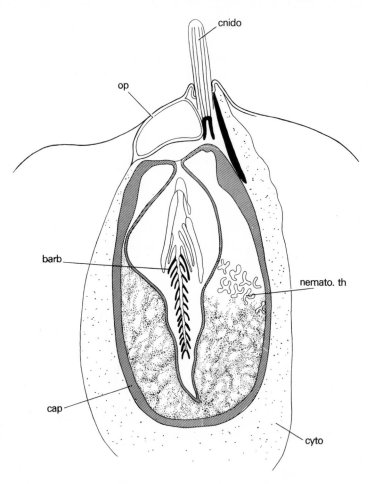

Fig. 16. Section through stenotele nematocyst of Hydra, based on electron microscope photograph. (Based on Loomis.) barb, barbs on base of nematocyst thread invaginated within nematocyst capsule; cap, capsule of nematocyst; cnido, cnidocil with structure comparable to that of a cilium; cyto, cytoplasm of cnidoblast; nemato.th, distal part of invaginated nematocyst thread; op, operculum.

of the nematocyst, either by altering the permeability of the capsule – a rather unlikely event in view of its chemical composition – or of the cell around it; the operculum, if present, is opened and the thread everts, turning inside out as it does so to reveal any projections it may carry as barbed or straight spines. The thread may be hollow and, if the process of swelling which fired the nematocyst continues after discharge, the contents of the capsule pass down the thread and ooze in droplets from the tip; if this has penetrated the body of the prey the nematocyst acts as a microscopic hypodermic needle and may lead toxic material into the prey, paralysing it and facilitating ingestion.

Other types of nematocyst are not open at the tip of the thread and do not penetrate. These are used for entangling prey and may do this by coiling round projections on its body such as setae or appendages. In addition to nematocysts zoantharians contain a similar but simpler structure known as a spirocyst, which lacks a cnidocil, probably does not discharge a thread but acts as an adhesive organ.

A large variety of nematocysts, unfortunately labelled with formidable names (which will be sparingly used), occurs throughout the coelenterates, but their role in the animal's life has been worked out in detail only in the case of the freshwater hydra. These animals possess four kinds of nematocyst, known as penetrants or stenoteles, volvents or desmonemes, and two sorts of glutinants with and without spines on the surface of their threads (the holotrichous and atrichous isorhizas respectively), which differ in their location on the body of the animal, in the use to which they are put and in their reactions to stimuli. The first two types, stenoteles and desmonemes, are primarily concerned with the capture of food and are most abundant on the tentacles. They have their threshold of stimulation lowered by the chemical stimulus of food, the stenoteles more than the desmonemes, so that they explode easily when prey brushes momentarily against the cnidocils of the cells within which they lie. The stenoteles are large nematocysts, exploding within 50–60 ms of stimulation and capable of penetrating the chitinous exoskeleton of small crustaceans; they are the kind with open threads which cause poisoning. The desmonemes, which are not set off so readily, shoot out threads closed at the tip which wrap round setae. The two sorts cooperate to grasp and immobilize prey prior to its being ingested. This act involves the separation of the prey from the tentacles, however, and this is facilitated by the ease with which the two kinds of nematocyst come out of the tentacular epithelium. This, indeed, happens spontaneously if an inedible object provokes their discharge: it is at first held to the tentacles by the nematocyst threads but soon drops off as the nematocysts are loosed from the ectoderm.

The smooth artrichous isorhizas also occur in the tentacles but these do not seem to be used in capturing prey. Instead, they are involved in grasping the substratum during locomotion, for which the tentacles are often used in a looping movement. The adaptations required of a nematocyst put to this use are clearly different from those of one for catching moving animals. The latter must react rapidly to momentary contact whilst this might be inappropriate in nematocysts used in locomotion. The mechanical sensitivity of the atrichous isorhizas is therefore low. On the other hand, once this type of nematocyst has gripped an object it must remain embedded in the ectoderm and not rapidly detach itself as do the stenoteles and desmonemes. This behaviour of the atrichous isorhizas would complicate feeding, were these nematocysts to discharge during that process; it is therefore interesting to find them adapted for this in that the chemical stimulus of food raises their threshold of discharge and makes it less likely to occur. The last type of nematocyst found in freshwater polyps is the spiny holotrichous isorhiza, which observation shows is not discharged either when the polyp feeds or moves, from which it may be deduced that they are not sensitive to mechanical stimulation and that their threshold is not altered by food substances. They appear to be defensive.

Whilst largely independent, nematocysts sometimes seem to be under nervous control. The anemone *Calliactis parasitica* lives on the shells of molluscs occupied by hermit crabs. The original attachment of anemone to shell is due to the discharge of nematocysts on tentacles which touched the shell. The nematocysts on the tentacles of an anemone already established on a shell, however, do not discharge when brought into contact with another shell. This has been explained as due to information fed into the nerve net of the anemone at the pedal disk, where contact with shell is made, which lowers the sensitivity of the nematocysts on the tentacles and effectively inhibits their explosion.

Throughout the coelenterates nematocysts occur most abundantly on tentacles, where there must be continuous replacement, especially of such sorts as are shed after discharge. It is not usual for development of nematocysts to occur at the place where they are used; instead, this takes place in other parts of the body such as the bulbs at the base of the tentacles in many medusae (Fig. 13A, tb), and in the ectoderm of the column wall in polyps. Replacement is rapid. Migration of cnidoblasts containing mature nematocysts must be continually in progress and seems in most cases to be brought about by amoeboid movement of the cnidoblast through the ectoderm, parallel to the mesogloea; occasionally, however, movement into the endoderm or even the coelenteron occurs. In the last case the nematocysts are picked up by amoeboid endoderm cells and passed to their destination in the ectoderm. Effete nematocysts not discharged may be digested by the endodermal cells. In Hydra the distribution of the nematocysts produced on the upper part of the column wall on to the tentacles is largely due to the fact that the tentacles are in continuous growth, new cells being added to their base and, about a week later, dying at their free tips. Nematocysts are caught up among the migrating cells and are carried over the tentacles. It is not known how widespread this method may be.

Feeding and digestion

Normally prey is caught on tentacles which extend out from the body of a coelenterate in radial fashion. There is a tendency for tentacles to be specially widespread in hungry animals, particularly noticeable in Hydra, and in some jelly-fish which fall through the water with all their tentacles outspread. When paralysed, prey is carried by the bending of the tentacles to the mouth, which, in hydroids, is opened by the contraction of radial muscles in the disk, and, in anthozoans, by these and the radial muscles of the primary septa, so that the food can pass to the coelenteron. Opening is apparently facilitated by the leakage of material from wounds in the prey made by nematocyst action; thus reduced glutathione escaping from a damaged *Daphnia* is the trigger which releases mouth opening in Hydra.

In hydrozoan polyps the coelenteric cavity is normally simple, without folded walls, though there may be a marked constriction at the point where a polyp head joins its stalk (Fig. 18A). When the polyp reaches a certain size, however, the coelenteron can no longer remain simple, not, it would seem, because the area devoted to digestive and absorptive activity has failed to keep pace with the needs of the polyp, but partly for mechanical reasons (see below) and partly because, in a large coelenteron, the prey might fail to make the contact with its lining which is an essential part of the digestive process. Subdivision of the coelenteron occurs in all scyphozoan and anthozoan polyps and, though the former are small, it is one of the factors which has permitted the large size of the latter. Here the endoderm is flung into folds (septa) projecting towards the centre of the coelenteron, reducing its effective size so much that the prey enters a cavity of rather limited extent with the walls of which it can hardly fail to make contact (Fig. 10). In Hydrozoa, gigantism occurs only in a small number of genera (*Arum*, Fig. 17B; *Tubularia*, Fig. 20D; *Corymorpha*, Fig. 17A, and, in extreme form, *Branchiocerianthus*, the polyp of which may be 3–6 ft long and inhabits the still water found at depths of several thousand metres); in most of these the need for mechanical support has resulted in the coelenteric cavity being partly filled by parenchymatous tissue. In *Corymorpha*, for example, the mouth leads into a spacious coelenteron in the polyp head, divided into oral and aboral sections by a ring-shaped pad of connective tissue which provides a skeletal base for the ring of long tentacles (ab.t), but in the main stem it is represented only by a series of anastomosing longitudinal canals (end.can), the rest of the stem being filled with parenchyma. In *Arum* the endodermal lining is flung into folds and the coelenteron is not obliterated; the first point seems to indicate that this is one of the few polyps in which increased absorptive area has had to be provided because of increased size, the second may be linked with the fact that *Arum* tends to lie prostrate whereas *Corymorpha* stands erect.

In jelly-fish, both hydrozoan and scyphozoan, hypertrophy of the mesogloea for locomotion has transformed the coelenteron from a single central space into a series of narrow canals. The mouth at the end of a proboscis known as the manubrium (Fig. 13, man) leads into a central stomach from which four, sometimes more, radial canals (rad.can) run to a circular canal (circ.can) at the margin of the umbrella. In some hydrozoans (Fig. 13B, cp.can) a number of blind centripetal canals originate in the marginal canal; in the larger scyphozoans (Fig. 15) these may become

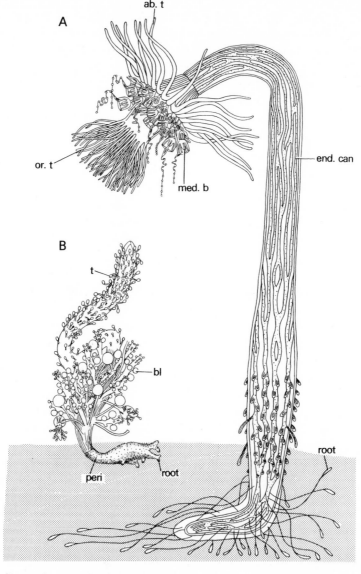

Fig. 17. A, *Corymorpha nutans*, B, *Arum cocksi*. (A, after Allman.) ab.t, aboral tentacles; bl, blastostyle; end.can, endodermal canal; med.b, medusa bud; or.t, oral tentacle; peri, perisarc; root, rooting process; t, tentacle.

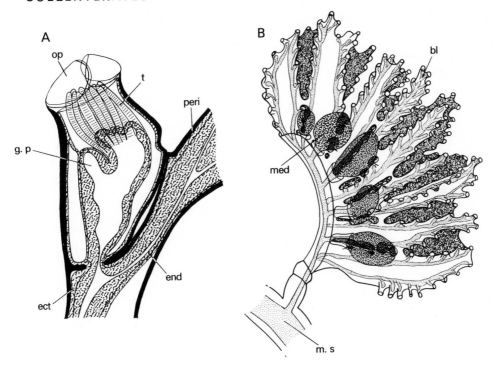

Fig. 18. Some details of hydroid structure: A, *Sertularella polyzonias* to show the gastric pouch; B, *Aglaophenia pluma* to show corbula. bl, blastostyle; ect, ectoderm; end, endoderm; g.p, gastric pouch; med, medusa bud; m.s, main stem; op, operculum; peri, perisarc; t, tentacle.

very numerous and branch and rejoin to create an elaborate reticulation of spaces between stomach and bell edge. In jelly-fish, feeding is accomplished in fundamentally the same fashion as in polyps: the prey is grasped by a tentacle and pushed or pulled towards the mouth. If tentacles are short they are often also numerous (*Obelia* medusa, *Gonionemus*, *Craspedacusta*) and several may be involved in the capture of a small copepod; many medusae, however, have only a few tentacles, which then are often long (*Sarsia*, *Liriope*, Fig. 13B, *Charybdea*) and succeed in capturing prey by coiling round it. In both cases not only do the tentacles convey food to the mouth by bending towards it but the manubrium helps by reaching in the direction of the food. This tendency towards a smaller number of longer tentacles may perhaps be regarded as reaching its climax in some of the siphonophores (*Physalia*, the Portuguese man-of-war) where each large individual of the food-catching type has but one very long tentacle (Fig. 21D, t); it must be remembered, however, that these are colonial animals and from the functional point of view the organism parallels a single medusa with a large number of elongated tentacles trailing below it and acting as a drift net in which fish and other organisms can be snared. On capturing prey the tentacles contract, hauling it up to the mouths of the gastrozooids for ingestion.

Jelly-fish have the surface of the body washed clean by movement through the water. Polyps, however, if capable of motion over a substratum at all, are so slow that they might easily become covered with detrital particles were not some mechanism operating to remove them. Some polyps, those of the madreporarian corals for example, are so sensitive to the presence of particulate matter on their disk and tentacles that they are unable to live in situations where it is abundant and demand clear, agitated water for survival. Other polypoid coelenterates like *Alcyonium* (dead men's fingers) and *Tealia* (the dahlia anemone) must also rely on currents in the water to remove their own faecal material and any particles which settle on them, as they seem to have no mechanism of their own. Most hydroids are too small to be affected by silt (unless it is so abundant as to smother them altogether) and are also cleaned by water scour. Sea-anemones, on the other hand, are usually provided with cilia on the disk and tentacles, and corals have them also on the column and the flesh, known as coenosarc, which links the individual members of a colony, to keep these parts clean. The usual arrangement is that ciliary currents beat up the tentacles from the disk; thus anything falling there, or egested from the mouth, will be carried to the tips of the tentacles and washed off by the surrounding water. The currents, however, are so arranged that, when an object has been seized for food, they will help to transport it, even if only slightly, to the tip of the tentacle which will have bent over towards the mouth. They may therefore be regarded as part of the feeding mechanism and the arrangement is susceptible of evolutionary complication so that it becomes the main element, instead of a minor component, of that process. This is illustrated by the sea-anemone *Metridium senile*, small specimens of which readily accept pieces of appropriate food and swallow them as other anemones do; as they get larger they do this less and some well-grown specimens may be successfully kept without large food ever being given to them. They apparently collect minute particles from the surrounding water by means of their tentacles which have become extremely numerous and are borne on the expanded and lobed margin of the disk, so offering a greater collecting surface; the particles collected are carried to the mouth by the ciliary currents already described. In most hermatypic corals, too, particles – including zooplankton – which fall anywhere on the surface of the colony may be killed by nematocysts, trapped in mucus and transported to the mouth by ciliary currents. In one group of genera (*Seriatopora, Pocillopora, Porites* and some others) this method of feeding is auxiliary to the main trapping of food by the tentacles; in another group (the meandrine and fungiid corals) it has entirely replaced ordinary tentacular feeding and, as might be expected, the tentacles are much reduced in number and size. Ciliary reversal on the disk occurs in animals feeding in this way, but apparently not in corals still provided with long tentacles.

Though active life allows jelly-fish to be washed by water currents most scyphozoan medusae have a ciliated epidermis with centrifugal currents bearing any particulate matter which settles on their surface to the edge of the bell, where it drops off the marginal tentacles. A few have transformed this cleansing mechanism into a food-collecting one, notably *Aurelia* (Fig. 15), a common shallow water form around the British Isles, often left stranded on beaches. Here ciliary currents on both surfaces of the bell stream to the edge where circular currents collect food particles (and the mucus in which they are trapped and transported) at eight points known as food pouches, placed at the ends of the adradial (unbranched) canals (ad.can). At intervals the material accumulated there is picked up by the tips of the oral arms (or.a). These are deeply cleft on their underside and a further series of ciliary currents lying on the lateral walls of the cleft transports the food-mucus material to the mouth. The normal food of *Aurelia* seems to be small copepods, but many other planktonic organisms are also taken. If detrital material of no food value is collected it may be rejected; this is achieved by a change in the direction of the ciliary beat on the oral arms which carries the material out of the groove so that it is washed away. It is not known whether this is due to a reversal of direction of the ciliary beat or to local muscular activity which brings into prominence tracts of cilia which are concealed and inoperative during feeding.

So far as is known, all coelenterates digest their food by the double digestive process, an initial extracellular phase followed by an intracellular one, described above in relation to a sea-anemone. In hydroid polyps the digestive processes tend to be limited to the part of the coelenteron near the mouth and in sea-anemones to the small part lying centrally between the free edges of the mesenteries. Certainly it is to those areas that gland and absorptive cells are restricted. In alcyonarian polyps there is a modification of the arrangement found in anemones for circulating coelenteric fluid: only one siphonoglyph occurs, leading water into the coelenteron, and only eight septa are found no matter how old the individual (Fig. 22C). Furthermore, only six septa have edges capable of secreting enzymes and absorbing food; the other two, on the side opposite that on which the siphonoglyph lies, are thickened at their free edge, but carry only ciliated cells which beat outwards, i.e. in the opposite direction to those of the siphonoglyph, so as to complete the machinery for maintaining a circulation of coelenteric fluid. The small number of mesenteries and the special arrangement of ciliary streams on them are probably correlated with the fact that alcyonarian polyps never grow to a large size. Since madreporarian and other small zoantharian polyps also lack ciliary tracts on the septa, their presence in sea-anemones is presumably another adaptation to large size exhibited by these animals.

Jelly-fish agree with polyps in the broad outline of their digestive activities, and in scyphozoan types have gastric filaments in the main chamber of the coelenteron which are comparable to mesenteric filaments and acontia. However, since the coelenteron of medusae usually forms a rather complex canal system, there are elaborate arrangements for circulating partially digested food or oxygen, or both. In hydrozoan medusae, for example, digestion occurs in the "stomach", the central part of the coelenteron lying at the base of the manubrium (Fig. 13, man), but much of the ingestion and subsequent intracellular digestion of particles takes place in endodermal cells at the base of the tentacles, in what are known as tentacular bulbs (tb). This is probably to be linked with the fact that most of the nematocysts used on the tentacles are made in the tentacular bulbs and migrate thence to their destination, so that a higher metabolic activity is maintained here than in most other parts of the medusa. The circulation in the canal system of jelly-fish has been carefully studied in *Aurelia* (see Fig. 15). Food from the lateral tracts on the oral arms enters the mouth and ciliary currents direct it along the roof of four grooves, the gastro-genital grooves (gg.gr), to the roof of the four main lobes of the coelenteron (gastric pouches). Over these it passes to the gastric filaments (g.fil) which help to break food masses mechanically, and (on the assumption that *Aurelia* behaves like other scyphozoan medusae) secrete enzymes over them. The food may then escape into the unbranched adradial canals (ad.can), which lead it to the marginal canal (circ.can); a return current drains this along collecting canals into the main per- and interradial canals (per.can, inter.can). In the latter case they lead to the floor of the gastric pouches and thence to a tract (ex.tr) along the base of the cleft in the oral arms, which conducts waste to their tips. Each perradial canal, on the other hand, opens directly to the exhalant tract at the base of the oral arm by an aperture near, but independent of, the mouth (ex.p). To what extent this system is responsible for distributing food, either in particulate form for ingestion or in dissolved form for absorption, and to what extent it subserves respiratory and perhaps excretory functions is not known. In the rhizostome genus *Cassiopeia*, which tends to spend its life lying on its exumbrellar surface in tidal pools, catching prey by means of the oral arms, it is known that secretion of enzymes and uptake of food occur primarily on the gastric filaments, whence they are distributed by amoebocytes. The elaborate reticulation of canals which that animal, like all rhizostomes, possesses seems therefore to be primarily respiratory. No comparable work has been done on *Aurelia*, but conditions may well be as in *Cassiopeia*. Presumably the centripetal canals of some hydrozoan medusae have a similar function (Fig. 13B, cp.can).

In colonial forms the coelenteric cavities of the different individuals are usually interconnected. In hydroid colonies the polyp heads arise from a common stem or shared stolon; in siphonophores they all arise from a single stolon, elongated or compressed according to the type; in gorgonians, alcyonaceans and other corals tubes called solenia (Fig. 22C, E, sol) link the main coelenteric cavities of the polyps. Presumably the passage of material from zooid to zooid which all these arrangements permit is one of the advantages which colonial life confers. In hybroid colonies transport of material from one part of the colony to the other is achieved by muscular contractions of the polyp head, parts of which are sometimes (Sertulariidae, Fig. 18A, g.p) specialized for this and maintain a surging of fluid to and fro along the colony. Flagella on the endodermal cells also create currents running along polyp stems and stolons. In madreporarian colonies they, and incidental movement of the polyps, are the effective agents of distribution. In alcyonarian colonies they are also responsible for circulating material along the tubes of the colony but, in addition, some genera (the red coral *Corallium*, the gorgonians and all sea-pens) have a type of polyp adapted for this activity which has lost its tentacles, muscles and mesenteric filaments so that it is dependent on other zooids for food. Its sole activity is to augment the current of water which is passing through the solenial network and so distribute food, oxygen and the like through the colony. Such a zooid is known as a siphonozooid (Fig. 22E, G, siph).

Colonial coelenterates and polymorphism

Coelenterates occur commonly in colonies of mutually dependent individuals or zooids. Though the component individuals are frequently all alike (true corals) this circumstance permits the evolution of polymorphism with specialization of different zooids for different purposes. In hydroid and soft coral colonies (alcyonarians, sea-fans, sea-pens) the types of zooid which occur are modifications of the polyp type; in siphonophores, however, modifications of both polyp and medusa are found, giving a complexity of structure which may become elaborate and bewildering.

Specialization of zooids in coelenterate colonies usually relates to a limited number of functions, feeding, protection and reproduction being the commonest. In hydrozoan polyp colonies it is normally exclusive, so that a feeding polyp does not reproduce and vice versa. Such feeding polyps, known as gastrozooids, otherwise retain all features of the typical polyp; in anthozoan colonies, however, reproductive powers are retained and the polyp is then called an autozooid. Protective zooids, known as dactylozooids or tentaculozooids, lose everything except tentacles and often look like a single enlarged tentacle. Their function is to protect the colony with their nematocysts and, often, to help in the capture of prey. Siphonozooids, aiding circulation within the colony, are limited to anthozoans. Reproductive zooids – gonozooids or blastostyles – also tend to lose tentacles and mouth.

Polymorphism in siphonophores is more complex than that of other hydrozoans or of anthozoans because polymorphs based on the medusa occur along with a polypoid series. The most important types based on the polyp are: (1) gastrozooids (Fig. 21, g), with a single, usually branched tentacle (t); (2) dactylozooids, without a mouth and with an unbranched tentacle (td); (3) gonozooids (gon), often like gastrozooids which have no tentacle, more frequently branching and bearing numerous medusa buds. The medusoid polymorphs are predominantly: (1) swimming bells (sb), nectophores or nectocalyces, in primitive genera medusae of normal structure save that they have lost the mouth (Fig. 21A, C), but often modified into odd and complicated shapes in more advanced ones; (2) bracts (br), sometimes called hydrophyllia, medusae which have become modified into helmet-shaped, or simple leaflike protective covers arching over groups of other zooids; (3) the medusa buds (mb) or gonophores borne on the gonozooids; (4) a single medusoid individual of which the apical exumbrellar part has invaginated into the bell to give a cavity filled with gas from a gland on its floor – this is the pneumatophore (pn), which gives buoyancy to the colony. All the individuals of all sorts are interlinked by a stolon (st) or coenosarc along which food travels from zooid to zooid. The gas in the pneumatophore of two genera *Physalia* and *Nanomia* has been shown to contain 5–20 per cent and 90 per cent respectively of carbon monoxide which is manufactured in the gas gland where the ectoderm of the floor of the invaginated float

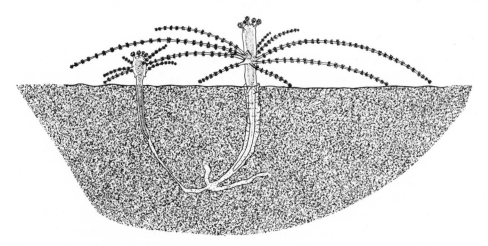

Fig. 19. *Corymorpha aurata* embedded in mud. Note the gelatinous sheath in which the polyp is embedded and the moniliform tentacles spread over the surface of the mud. (After Rees.)

comes intimately in contact with endodermal cells of the coelenteric cavity. The carbon monoxide appears to originate in salts of folic acid but how the organism has been able to accommodate to such proportions of such an unusual substance is not known. This ability is the more remarkable when it has been shown that *Nanomia* has its float filled with carbon monoxide in equilibrium with pressures at a depth of 300 m, that is about 30 atm.

In the most primitive genera like *Muggiaea* (Fig. 21A, B) the stolon is a linear structure springing from the subumbrellar surface of an apical swimming bell whose pulsations keep the colony afloat. Along the stolon are other zooids set in groups called cormidia; each cormidium has a bract, a gastrozooid and a gonozooid, the gonophores on the gonozooid helping in the movement of the colony and providing locomotive power for the cormidium if, as often happens, the stolon fragments and the cormidia become free. An evolution from this pattern can be observed within the siphonophores; in more advanced groups locomotor ability is increased by raising the number of nectocalyces at the upper end of the stolon, so allowing more cormidia to be carried by its lower half, as in *Agalma* and *Nanomia* ("*Halistemma*").

A second evolutionary line is associated with a stolon which forms a disk, perhaps derived by shortening a linear one – a half-way stage may be represented by *Nectalia* (Fig. 21C) – with the zooids set on its sides and edge. This is seen, for example, in *Stephalia*, where a pneumatophore and a ring of swimming bells arise from the upper side of the disk, whilst a single large gastrozooid surrounded by dactylozooids and gonozooids springs from the other. In the Portuguese man-of-war, *Physalia* (Fig. 21D), the swimming bells are lost, but enlargement of the float above the surface of the water produces a sail by which the animal is passively drifted from place to place. The lower face of the stolon gives rise to bunches of gastrozooids, dactylozooids of more than one kind with fishing tentacles that may trail for many metres through the water, and gonozooids.

At one time it would have been accepted that the climax of this evolutionary line was to be found in such genera as *Porpita* (*Porpema*, Fig. 21E) and *Velella* (Fig. 21F), which, like *Physalia*, are surface-dwelling forms drifted by wind and water. Both have a disk-like body with gas-filled spaces which keep them at the surface of the sea; it is raised into a sail in *Velella*. On the underside is a central mouth around which is an area carrying medusa buds whilst a ring of tentacles is set at the margin. These animals were once interpreted as a colony of polymorphic individuals – gastrozooid flanked by gonozooids and ringed by feeding and protective dactylozooids – but it is now believed that their relationship is not at all with siphonophores but with gymnoblastic hydroids, particularly with those like *Corymorpha* (Fig. 17A), which have lost their attachment stalk and now float upside down at the surface of the sea.

The most primitive hydroids are supposed to have been solitary forms like the modern genera *Hypolytus* and *Euphysa*, with the stem encased in a loose gelatinous sheath and embedded in some soft substratum such as mud (Fig. 19). At the base of the stem arise processes the tips of which anchor themselves to particles in the surrounding soil. The polyp head carries a short oral and a longer aboral ring of tentacles, both with a moniliform arrangement of batteries of nematocysts, a type of tentacle said to be more extensile than any other and therefore capable of fishing a wider area. The polyps lie with the aboral tentacles radiating outwards over the surface of the mud so as to form a trap for small animals moving over the surface or alighting on it. Evolution from such a starting point has led in the first place to attachment to a firmer substratum (sand, rock) by means of the rooting processes, with the body more or less erect in the water. The need for extensile tentacles with moniliform nematocysts is less in such a situation since they are exposed in the water and may be moved so as to sample a greater volume, and an even distribution of nematocysts along the tentacle develops, making it filiform. The habit of feeding off the bottom, however, has persisted in some hydroids at this stage of evolution and is often accompanied by gigantism as in *Corymorpha* (Fig. 17A) and *Arum* (formerly known as *Myriothela*) (Fig. 17B). The former of these (though it may grow to a height of 10 cm in the far north) feeds by bending over so as to let mouth and tentacles fall on to the substratum; it then straightens and any particles which may have stuck to the tentacles are eaten. *Arum* does not attempt to hold itself upright but lies prostrate on the rock on which it lives, its tentacles moving over it and reaching up into the water in search of food.

A second evolutionary line from the primitive polyp has led to the adoption of a colonial habit with the stolon and polyp stems, and, in some cases, even the polyp head ensheathed in a tough, protective and supporting chitinous exoskeleton, the

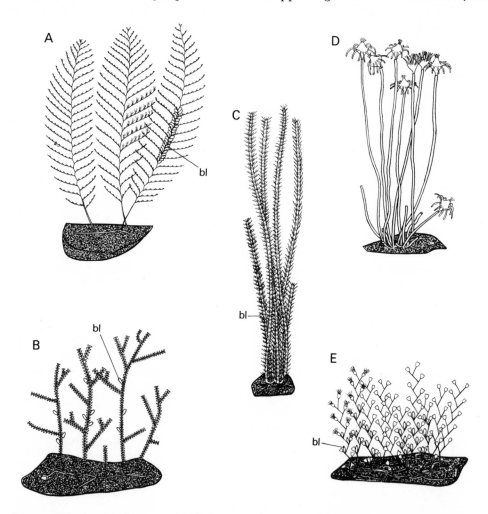

Fig. 20. Colonies of hydrozoans: A, *Kirchenpaueria* (= *Plumularia*) *pinnata*; B, *Dynamena pumila*; C, *Nemertesia* (= *Antennularia*) *antennina*; D, *Tubularia indivisa*; E, *Obelia geniculata*; bl, blasto-styles.

perisarc. The organism in a real sense is now the colony rather than the polyp, since food and other materials may be carried from one part to another by way of the interlinking stems and stolon. There need therefore be no attempt to improve food catching powers by gigantism: individual polyps tend, instead, to be small and arranged in spatial patterns which make the colony an efficient food-catching apparatus (Fig. 20). The reduction mainly affects the longer tentacles of the polyp; had these been retained, they would have been inefficient in the close quarters in which polyps in a colony work. Most polyps, therefore, especially in the elaborately constructed colonies of the advanced calyptoblasts, have but a single row of short ones. The colonies may grow to large size, when strength is imparted to the main stem by fasciation of many stems.

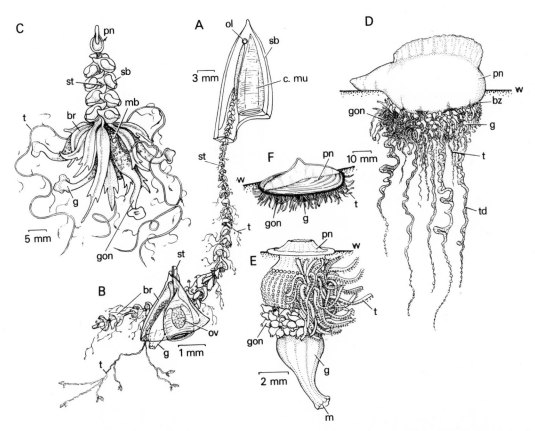

Fig. 21. Organization of siphonophores. A, *Muggiaea*; B, a cormidium enlarged; C, *Nectalia*; D, *Physalia*, the Portuguese man-of-war; E, *Porpema*, like *Porpita*; F, *Velella*. br, bract; bz, budding zone; c.mu, circular muscles; g, gastrozooid; gon, gonozooid; m, mouth; mb, medusa bud; ol, oleocyst; ov, ovary; pn, pneumatophore; sb, swimming bell or nectocalyx; st, stolon; t, tentacle; td, dactylozooid, tentacle-like in form; w, water level.

A further consequence of the adoption of the colonial habit is functional specialization of individual polyps. Some remain the food catchers of the colony (gastrozooids); some are specialized to aid this by supplying extra nematocysts but have lost the mouth and coelenteric cavity which would permit ingestion and digestion (dactylozooids or nematophores). These may also be simply protective. The third main type of individual is the reproductive polyp, on which the medusa buds form. In solitary polyps buds usually grow on the oral cone between the two rings of tentacles (Fig. 17A, mb). This situation, it has been suggested, is most suitable because of its proximity to the place where food is ingested and absorbed. In the simpler types of colony (*Clava*, *Tubularia*, Fig. 20D), which consist largely of groups of single polyps arising from a narrow basal stolon, this arrangement persists, but in more elaborate colonies, in which there is a more effective spatial arrangement of the gastrozooids (*Obelia*, Fig. 20E; *Kirchenpaueria*, Fig. 20A; most higher calyptoblasts), this is no longer necessary because of (a) the greater and more evenly spread food-catching ability, and (b) the more efficient circulation of food. As a result the reproductive polyps, known as blastostyles, tend to be located in situations which are better protected rather than convenient for uptake of food (Fig. 20, bl). They are enclosed within expansions of the perisarc called gonothecae, the blastostyle and gonotheca together forming a gonangium. In some genera even more elaborate protective devices have been evolved; thus the gonangia on the main stem may be arched over by a basketwork of special curved branches bearing numerous dactylozooids and forming a structure known as a corbula (*Aglaophenia*, Fig. 18B), or they may be limited to a length where dactylozooids are particularly well developed giving a coppinia (*Lafoea*).

A similar evolution has attended the formation of colonies in the Antipatharia or black corals, and in the alcyonarian branch of the Anthozoa. Here the primitive members, animals like *Sarcodictyon catenata* (Fig. 22A), are small colonies forming straggling growths on hard substrata with the polyps arising from a meandering stolon and the common flesh strengthened by spicules. Larger colonies occur in genera like *Alcyonium* (dead men's fingers, Fig. 22B), *Corallium* (the red coral of commerce, Fig. 22D) and, generally, in gorgonians (Fig. 22F) and pennatulaceans (Fig. 22G). In these a special spatial arrangement of the zooids frequently occurs, particularly marked in the higher members of the last two groups, giving the animals known as sea-fans and sea-pens respectively. In the characteristic sea-fans of shallow water the colony grows upright at right angles to the prevailing currents and vertical branches are united by cross pieces to give rise to a reticulated colony with all branches lying in one plane. The polyps usually project into the meshes of the net and so are well placed for catching small animals carried through by the current. In deep water, where currents are less or absent, branching is irregular and growth more tufted. The pattern of sea-pen colonies is different, with the polyps (au) usually

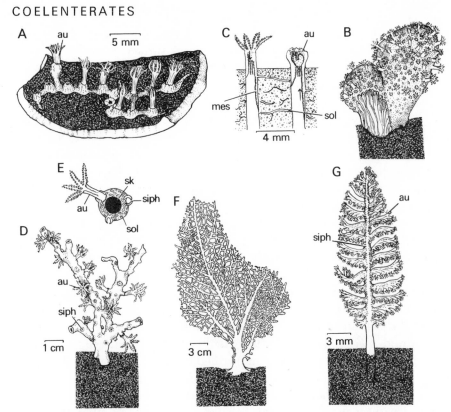

Fig. 22. Alcyonarians. A, *Sarcodictyon*; B, *Alcyonium*, the lower left part of the colony cut to show the extensive coelenteric cavities of the polyps; C, diagrammatic longitudinal section of two polyps of *Alcyonium*; arrows show direction of ciliary currents; D, *Corallium*; E, transverse section of branch of *Corallium*; F, gorgonian; G, pennatulid. au, autozooid; mes, long mesentery on side of polyp opposite to siphonoglyph; siph, siphonozooid; sk, skeleton; sol, solenial tube.

lying on lateral branches from a main vertical stem, the base of which is buried for a certain distance in a soft substratum.

Sea-anemones differ from hydroids and from the other anthozoans in being rarely colonial. The anemones usually accepted as the most primitive are placed in a group called Protantheae, though it ought to be added that they are sometimes regarded as stages in the development of more normal anemones arrested and made sexually mature. The genera *Gonactinia* and *Protanthea* are delicate animals measuring only a few millimetres in length when expanded, having only one imperfect siphonoglyph, with the longitudinal muscles lying along the whole depth, from column wall to free edge of the mesenteries (of which there are only eight primaries and eight secondaries) instead of forming a localized retractor, and they retain longitudinal muscles in the ectoderm. They are much more mobile than the ordinary sea-anemone, with the result that the base is hardly differentiated from the column and has no anchoring (parieto-basal) muscles. From such a starting point two evolutionary lines lead, one to the familiar anemones adherent to a rock surface, the other to the sand-burrowing ones. The second group has retained more primitive features because its mode of life is still like that of its ancestors, and, though a genus such as *Peachia* may be advanced in its large size, most sand-dwelling anemones (*Edwardsia, Halcampa*) are small elongated, worm-like creatures. They have no base for attachment, and only a limited number of mesenteries, though these carry well-developed retractor muscles which cause abrupt, even violent, retraction of the animal when it is disturbed, as is characteristic of all burrowers and tube-dwellers. Since this abrupt movement may raise the coelenteric pressure rapidly, there are pores in the column (cinclides) which permit coelenteric fluid to escape into the sand. The pull of the retractor muscles, too, tends to invaginate the base and thus lets it exert a sucker action anchoring the lower end of the animal and making the effective movement resulting from the contraction a withdrawal of the tentacles and disk from the surface.

Anemones which do not burrow have all tended to become large, and much of their adaptation consists in adjustments to permit large size – particularly of the diameter of the column – to be reached. This, like the parallel development in the large hydroid *Arum*, has called for increased endodermal surface achieved by the continual production of new mesenteries throughout life and the correlated growth of new sets of tentacles. In larger sorts (*Tealia, Bolocera*) the animal may reach up to 30 cm across the disk. The animals are normally attached to a hard substratum by a base provided with anchoring muscles. Because of the great number of mesenteries the muscles on each tend to be reduced in size and closure is never as abrupt as in a burrowing anemone. With increased size and slower closure, cinclides tend to be lost and the column may be roughened and, as in *Tealia* living in exposed situations, artificially protected by the attachment of pebbles and similar detritus to adhesive papillae on its surface. As pointed out above, the limitation of longitudinal muscle to the mesenteries and the resultant scatter over the diameter of the animal, have been important factors in permitting large size.

It is usually assumed that the reef-forming corals arose as offshoots from the line of sea-anemones, with the ability to secrete around the polyp an external supporting skeleton of calcium carbonate. The skeleton forms a cup-like support for the polyp, but also carries on the base of the cup a series of radiating septa which, in general, alternate in position with the mesenteries in the coelenteron of the polyp. When the retractor muscles of the polyp contract, therefore, and pull the oral disk towards the basal disk, the coelenteric cavity is practically obliterated because of the way in which the skeletal septa lift up the tissues of the pedal disk. This, it has been sug-gested, is a point of survival value since it destroys areas where water could stagnate when the polyp is contracted, as it is, for example, in daytime. If this explains why mesenteries are paired – a question to which no valid answer can be given when it is asked about sea-anemones – it may imply that pairing of mesenteries was invented by corals and inherited by anemones, a statement which implies that corals are more likely to be the ancestral rather than the derived group.

In British latitudes only solitary corals such as *Balanophyllia* occur. In tropical conditions, however, colonies are commonly found, the polyps multiplying by a process of division and the skeleton of the daughter joining that of the parent. In this way large masses of calcareous matter are formed, with a living film of conjoined polyps over their surface. Often neighbouring masses join or are cemented together by other coral-like animals such as hydrocorallines to give rise to extensive reefs, so characteristic of appropriate situations in the tropics. The precise shape of a coral colony is partly related to the kind of coral involved and partly to the nature of the environment, rough conditions leading to more solid growth, sheltered to a more delicate arrangement. Many complications of structure appear: the coelenteron of one polyp communicates with that of its neighbours by solenia; sometimes, as a polyp grows, its mouth may subdivide so that many openings on the disk, each with its own stomodaeum, and sometimes with its own circlet of tentacles, lead into a wide coelenteron. This is characteristic of the brain corals of the genus *Meandrina* and of the giant solitary corals in the family Fungiidae which have been recorded up to 55 cm in diameter, with 1500 stomodaeal openings and as many pairs of mesenteries. These animals are largely microphagous.

Symbiosis

It is appropriate here to say a little about symbiotic algal cells which are found within the tissues of many coelenterates and of members of some other phyla. On the assumption that the plants can photosynthetize all such animals must be restricted to habitats where this can occur. The algal cells may be green, when they are known as zoochlorellae, or brown when they are called zooxanthellae. Zoochlorellae are found in protozoans like *Paramecium bursaria*, in green Hydras, in the freshwater sponge *Spongilla* and in such turbellarians as *Convoluta roscoffensis*; they belong to more than one group of algae. Zooxanthellae are the common symbionts of coelenterates and all appear to be peridinians. Whether green or brown, the algae lie, in ordinary circumstances, in vacuoles within cells of the animal host, usually in those belonging to the wall of the gut or coelenteron, though in the scyphozoan jelly-fish *Cassiopeia* they are located within the mesogloea.

One of the most successful early investigations of symbiotic algal cells in animal tissues was that of the acoel turbellarian *Convoluta roscoffensis*, carried out by Gamble and Keeble, the results of which have influenced other workers for a long time. *Convoluta roscoffensis* is a small intertidal flatworm which is green in colour because its inner cells (representing the greatly altered alimentary system) contain vast numbers of zoochlorellae. The habits of the animal are adjusted to the photosynthetic needs of its algal cells and it emerges from interstitial spaces in the sandy beaches on which it lives to bask in the light and allow its symbionts to photosynthetize. The algae lie most densely close under the epidermis and have lost their normal cellulose wall so that their shape is often irregular, with processes extending towards the over-lying epidermis; they have also lost flagella and eyespot.

Gamble and Keeble discovered that whereas young specimens of *Convoluta* ate freely, the worms became more and more abstemious and finally ingested no food at all. In these circumstances they had become nutritionally dependent on their algal cells and the supposition was that they treated these as a kind of internally grown plant crop, digesting a fraction for food, but cultivating the rest by supplying them with the light and the carbon (from their metabolic carbon dioxide) which they needed for photosynthesis, and with the nitrogen (from their metabolic nitrogen) which was required for their manufacture of protein. Indeed so far as these sub-stances were concerned each worm could be described with fair precision as a unit isolated from its environment, with little carbon or nitrogen exchange. The early feeding stage in the life history of *Convoluta* is seen to be necessary when it is realized that the egg has no algae in it: each worm has therefore to infect itself at a very early stage in the life history, a process facilitated by the fact that various stages of the alga are attracted to the egg case and that this is usually eaten by the young worm as soon as it emerges.

It was with this kind of situation in mind that most animal-alga relationships were long explained. Only recently has convincing evidence of the way in which it happens in groups other than acoel turbellarians come to light with the development of tech-niques such as the use of radioactive isotopes which allow accurate tracing of the movement of particular molecules. It can then be shown how the metabolic activities of the two types of organism, heterotroph and autotroph, enmesh one with the other. If radioactively labelled carbon compounds are made available to Hydras or corals it is found that the rate of uptake is greatest when a green or brown animal is kept in the light; it is markedly lower when in the dark, and is very low in an aposymbiotic individual, that is, one deprived of algal cells. These results implicate the algae very clearly in the use of the material. If the time relations of the process are considered and the amount of radioactive carbon in the algae compared with that in the animal it is seen that the carbon moves first into the alga and then back into the tissues of the animal. The amount of carbon which does this and the rate at which it moves are both surprisingly high: it has been shown that in *Zoanthus*, a zoantharian, 25–30 per cent of the photosynthetically fixed carbon is back in the coelenterate 3 h after the start of the experiment, whilst in the anemone *Anthopleura elegantissima* about 75 per cent of the carbon fixed by the algal cells is returned to the coelenterate.

The chemical nature of the material exported by the alga to its host varies from group to group: in coelenterates where zooxanthellae are concerned, it seems to be predominantly glycerol; in animals with zoochlorellae (*Paramecium bursaria*, Hydra) maltose is the form in which it is transferred. Neither substance appears to be the usual end product of photosynthetic activity in these algal cells, or their close relatives, when living outside an animal host. The preparation of a different end product, usually one marked by its soluble nature allowing easy translocation, appears to be part of the adaptive change which the algae undergo in becoming symbiotic. Another change is in the amount of photosynthetically produced material which they allow to escape from their body. This process occurs normally in free-living algae but to a limited extent – only 3–6 per cent of the total made; the zoochlorellae living in green Hydra, however, normally release 85 per cent or more. This is much to the advantage of the animal in which they lodge and it is apparently able to influence the alga in some way in order to bring this about. Zooxanthellae removed from an anemone or coral and allowed to photosynthetize in substances labelled with ^{14}C in sea water release much less labelled photosynthate than if the experiment is carried out in a homogenate of the host animal. How this "animal factor" works to accelerate the slow rate of diffusion from the free-living alga is unknown. There is some evidence, too, that the factor is specific, which would imply that each kind of host had its own physiological strain of symbiont. The production of an excess of carbohydrate material by the symbiotic alga by comparison with its free-living relative implies either a much higher level of metabolism or a diversion towards the host of material used for other purposes when the alga is free. Probably both these changes take place since there is little doubt that the algal cell has richer sources of nitrogen, and probably of carbon, than free-living algae because of its intimate association with an animal. The tissues of aposymbiotic corals, for example, come to contain amoebocytes laden with brown granules, possibly of an excretory nature, which are absent in polyps with zooxanthellae. Symbiotic algae, too, differ from free-living ones in having a much lower rate of reproduction and sometimes in having thinner cell walls – all changes which would allow material to be made available for the use of the host.

Symbiotic algal cells are particularly characteristic of the reef-building or hermatypic corals and other coelenterates of low latitudes, and of many of the protozoans which belong to the plankton of open waters. Both of these are sites where nutrients tend to be in poor supply, yet the population is often staggeringly high in richness and biomass. Many zoologists, indeed, have been impressed firstly by the relatively small amount of zooplanktonic food available to reef corals in relation to their numbers and, secondly, by the fact that the corals do not seem to eat even as much of this as might be expected. If, however, they are able, to greater or lesser extent according to circumstances, to avail themselves of photosynthetic material escaping from their symbiotic algae, then their ability to feed less, to survive and thrive where food is not over-abundant, is explained. The zooxanthellae can also contribute to the nitrogen metabolism of the animal since they are able to make amino acids from ammonia provided by the host. The animal–plant unit is largely self-contained and self-sufficient, recycling carbon and nitrogen with little loss to and perhaps little gain from the environment – like men in a space capsule. If zooplanktonic organisms are captured their carbon and nitrogen enrich the total resources of the unit since little is lost back to the environment. Though platyhelminth worms like *Convoluta*, descended from a stock able to digest plant as well as animal material, may survive by eating their algal cells the carnivorous specialization of coelenterates, acquired before the association with algae was begun, renders them enzymatically incapable of a direct attack on the algae living within their tissues. A very small number of anthozoan species (*Xenia hicksoni, Clavularia hamra* and *Zoanthus sociatus*) stop catching food and so become completely parasitic on their algae.

Similar advantages are conferred on Hydras by the algae within their endodermal

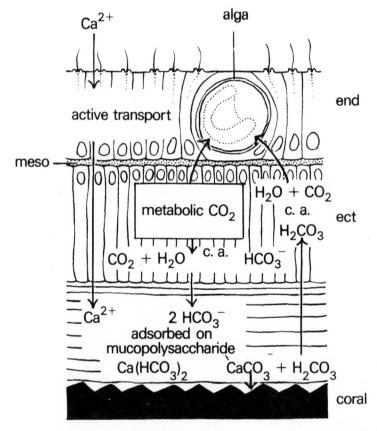

Fig. 23. Diagram to show the possible pathways involving coelenterate and symbiotic alga in the secretion of the skeleton of a coral polyp. The diagram represents a section through part of the body wall of a polyp containing an algal cell and the external skeletal material separated from the ectoderm by a space. (Based on Goreau.) alga, symbiotic algal cell; c.a., carbonic anhydrase; coral, skeleton of calcium carbonate; ect, ectoderm of polyp; end, endoderm of polyp; meso, mesogloea.

cells so that they grow faster than Hydras which have none. In laboratory stocks which are kept well fed, green and colourless Hydras can successfully coexist; when fed little, however, the green survive whilst the colourless ones die. This situation is much the more natural one, since starvation is something wild Hydras must often experience. In all these animals, therefore, natural selection would favour the symbiotic stock as fitter for habitats where conditions are difficult.

The presence of zooxanthellae is also important in the process of calcification by means of which corals build up the massive exoskeletons that, joined to those of neighbouring colonies, produce that most outstanding marine structure of animal origin, the coral reef. The process of secretion of calcium carbonate by a coral is summarized in Fig. 23. Calcium ions are actively transported from the coelenteric fluid through endoderm and ectoderm to an area immediately external to the animal where (as in nearly every animal known to secrete a calcareous skeleton) they are adsorbed on to layers of mucopolysaccharide material derived from epidermal gland cells. Bicarbonate ions diffusing outwards from the polyp and deriving from its respiration are also adsorbed on to this and the two react to give molecules of calcium bicarbonate which break down, liberating calcium carbonate and carbonic acid. The former is precipitated on to the skeleton, while the latter can diffuse back to the polyp. The zooxanthellae can regulate this process because the carbon compounds are involved in their carbon cycle. The production of respiratory carbon dioxide by the coral, its use in the photosynthesis in the plant and the availability of bicarbonate ions for the formation of skeletal material are all tied in interrelated equilibria.

Classification of coelenterates

Hydrozoa
   Trachylina
   Gymnoblastea (= Athecata)
   Calyptoblastea (= Thecata)
   Hydrida
   Hydrocorallina
   Siphonophora
Scyphozoa
   Cubomedusae
   Coronatae
   Semaeostomeae
   Rhizostomeae
   Stauromedusae

Anthozoa (= Actinozoa)
   Alcyonaria (= Octocorallia)
      Alcyonacea
      Gorgonacea
      Pennatulacea
   Zoantharia (= Hexacorallia)
      Actiniaria
      Madreporaria
      [Zoanthidea
      [Antipatharia]
      [Ceriantharia]

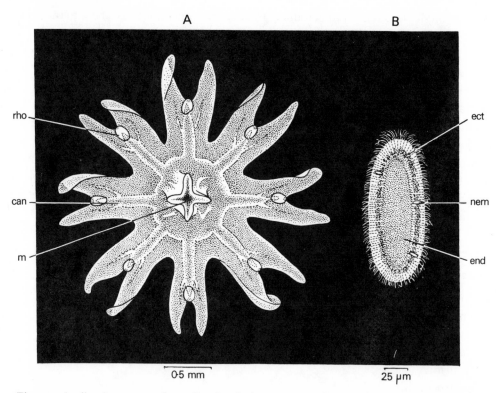

Fig. 24. *Aurelia*: A, young ephyra; B, planula larva. can, radial canal; ect, ectoderm; end, endoderm, seen by transparency; m, mouth at end of short manubrium; nem, nematocyst; rho, rhopalium, marginal sense organ.

5

Ctenophores

THE ANIMALS which have been described in the previous pages as coelenterates are sometimes given the more restricted designation of cnidarians, a name which reflects their possession of stinging cells, in order to distinguish them from another group of animals at about the same level of organization, the ctenophores, sometimes included in the same phylum, sometimes placed in a separate phylum, Ctenophora. The commonest kinds are popularly known as sea-gooseberries, because of their size and shape, or comb jellies because of their way of swimming and consistency. Like cnidarians, ctenophores have a body composed of an outer ectoderm, an inner endoderm and an interposed voluminous, jelly-like mesogloea; they have a superficial radial symmetry which proves on examination to be a hidden bilateral one, and, though nematocysts are absent, they capture their prey by means of thread-like adhesive cells on a pair of tentacles or on their lips; they are strict carnivores. Like jelly-fishes they almost all swim freely amongst planktonic organisms which they capture for food, only a few highly specialized genera having become attached or adopted a creeping mode of life like that of a flatworm. The general conclusion which may perhaps be drawn from these facts is that the ctenophores represent an evolutionary line roughly parallel to that followed by cnidarians, separating from it at a point where some traits common to both groups had already come into existence and offering, therefore, similar adaptations for similar problems.

Organization of a cydippid

Most ctenophores resemble the very common animals placed in the genera *Pleurobrachia* (Fig. 25, top) and *Hormiphora*, which are often caught in tow nettings, or may be found left by the tide in rock pools, or stranded on sandy shores; they are classified in an order Cydippida. The cydippid ctenophore has an egg-shaped body about an inch long and is almost completely transparent when alive. Though superficially radially symmetrical it can be divided into equal halves along only two perradial planes. One of these, the tentacular plane, bisects the bases of the two tentacles by means of which the animal feeds; the other plane, known as the sagittal, lies at right angles to it. Sagittal halves are not identical with tentacular halves. At one end, which goes first in swimming, lies the mouth. Swimming is brought about by the beating of swimming plates, each formed of a transverse row of fused cilia, arranged in eight, evenly spaced, meridional lines. Each plate is called a ctene and the row of ctenes a costa. According to coelenterate terminology the costae occupy adradial positions. They end orally some distance from the mouth and also short of the aboral pole, but from the last ctene of each costa there can be traced a line of ciliated cells running to the aboral pole where they originate in an elaborate structure, the apical organ. The most conspicuous element of this is a statolith containing several hundred cohering grains of calcareous material, each grain 5–10 μm in diameter. It is perched on the summits of four groups of long balancer cilia partially fused with one another; there may be up to two hundred cilia in each group and they are immobile. The cilia arise from tall cells which form the most aboral members of the rows of ciliated cells mentioned above as extensions of the costae and which, in some genera, run, linking ctene with ctene, the whole length of the costae. To complete the description of the apical organ mention must be made of the dome, a hemispherical cover formed of fused cilia which arches over the statolith and its supporting groups of cilia, and of the polar fields, two elongated ciliated grooves lying in the sagittal plane, which extend a short distance from the base of the dome and are presumably sensory in function.

In normal swimming a wave of movement is initiated at the apical organ, and is transmitted from the balancer cells along the ciliated rows to the ctenes which beat in succession from the aboral to the oral end of the costae. The effective beat of cilia and ctenes alike is towards the aboral pole so as to send the ctenophore through the water mouth first. Since an impulse passes simultaneously along each costa the ctenophore swims in a straight line, normally vertically upwards or downwards. The equal frequency of the impulses running along the eight costae depends on the equal loading of the four groups of balancer cilia by the weight of the statolith. Should the ctenophore come to lie at an angle to the vertical the statolith bears more heavily on some of the cilia than on others and this is reflected in an increased number of impulses emanating from the overloaded cells and so in an increased rate of beating of the ctenes to which they are connected, leading to a righting of the animal in the water. It has been shown that the cells of the ciliated tracts act as conductors of the impulses sent out from the apical organ: each cell is elongated in the line of the costa and behaves as a bipolar axone. The apical organ therefore acts as a pace-maker and is functionally equivalent to the rhopalia or ganglia of a scyphozoan jelly-fish.

The mouth leads into a complex array of canals which lie embedded in the gelatinous mesogloea that occupies most of the volume of the animal. The coelenteric cavity of the ctenophore has thus a tubular form like that of the cnidarian jelly-fish, and presumably for the same reason: the hypertrophy of the mesogloea as part of the adaptation for planktonic life. The node of the system is the so-called stomach which lies about two-thirds of the body length away from the mouth. The mouth is connected to the stomach by an elongated stomodaeum the walls of which, flattened in the sagittal plane, are flung into folds secreting enzymes which effect a rapid breakdown of the food to a finely divided state that allows it to enter the stomach and be circulated by cilia along the length of the other canals. Here it is phago-cytosed by the endodermal cells and digestion completed intracellularly, a process identical with digestion in cnidarians. From the stomach two canals run peripherally in the tentacular perradii. Each branches to form two interradial canals from each of which, in turn, two adradial canals arise. The adradial canals end in costal canals which run meridionally, one underneath each costa. In addition to the radial canals a number of others emerge from the stomach: two stomodaeal canals run towards the mouth, one on either side of the stomodaeum, in the tentacular plane; two tentacular canals run along the base of the invaginatious from which the tentacles spring and one aboral canal runs to the aboral pole. Here it divides into four short branches that lie one on either side of each polar field; of these branches two, diagonally placed, are blind, but the other two open by small pores through which the indigestible remains of food have been seen to pass. With this elaborate system of canals and the final stages of absorption taking place within the cells that form their walls, ctenophores clearly do not require any other kind of circulatory system. It will be noticed, too, that special canals run to sites where energy requirements are high – to the stomodaeum in whose walls digestive enzymes are elaborated, to the tentacles which undergo active muscular movements and to the gonads which are developed along the length of the costal canals, an ovary and a testis set on each.

Food is caught by the tentacles of a cydippid ctenophore whilst it hovers, motion-less, in the water. These arise from the base of sheaths (lying one on either side of the stomodaeum) into which they may be retracted. Each tentacle is a solid, muscular structure which can extend to great lengths, up to 50 cm. It carries one (*Pleuro-brachia*) or two (*Hormiphora*) series of lateral filaments up to 4 cm long. These and the main axis of the tentacles are covered with an epithelium containing vast numbers of special adhesive cells, the colloblasts, sometimes called lasso cells. Each colloblast has a hemispherical head projecting above the general epithclial surface and laden with granules which transform to a sticky substance in contact with food. The head is attached to the connective tissue core of the filament by two fibrils, one of which spirals round the other and is contractile so that it can pull on prey even if dragged out of its position in the epithelium. The straight filament arises from the nucleus of an ordinary epidermal cell during its transformation into a colloblast. Like nemato-cysts colloblasts have to be made continuously during the lifetime of a ctenophore since they are destroyed in capturing prey. The tentacles, with their filaments out-spread, hang in the water like a drift net into which the prey swims whilst catching its

Fig. 25. Ctenophores. Upper left, *Pleurobrachia pileus*, 12 mm in diameter. The animal is motionless in the water with the mouth uppermost; the tentacles and their filaments are extended to form a fishing net. (After Greve.) Upper right, cydippid in apical view; the pole is occupied by the apical organ from which eight rows of ciliated cells run adradially to the costae. The circles on either side of the polar fields are the aboral canals two of which (upper left, lower right) open, the other two being blind. The bases of the tentacles lie in the tentacle sheaths at nine o'clock and three o'clock. Perradial canals pass to them from the stomach and divide to form interradial and adradial canals related to the costae. Below, *Cestum veneris* or Venus's girdle, about 1 m long. The body is greatly elongated in the sagittal plane and the four costae which border it share in the elongation, terminating at the tip of the strap-shaped body. The four costae which border the tentacular plane are reduced to four tufts near the apical organ. The tentacles are reduced to small threads lying alongside the stomodaeum and are not involved in the capture of food; this is done by a double row of small tentacles set along the oral edges of the body.

own food. When they catch prey it is held by colloblasts and the tentacles then contract over the mouth, wiping the food off on to the lips, where mucus and ciliary streams entangle it and carry it into the stomodaeum.

Since locomotion in ctenophores is brought about by cilia rather than the musculature of a swimming bell as in cnidarian jelly-fishes, the muscular system is less well developed and, in particular, there is no need for its organization into antagonistic sets. Thin layers of longitudinal and latitudinal muscle fibres, however, lie under the general epidermis and stomodaeal walls and radial fibres run across the mesogloea to link body wall with gut wall. The musculature is largely under the control of a nerve net located under the epidermis and extending into the tentacles, although there is evidence of direct transmission of stimuli from muscle cell to muscle cell. The nerve net stimulates the muscles but inhibits the beating of the ctenes; this happens when any part of the body is stimulated. When the animal hits an obstacle the ctenes stop beating; when it is resumed the direction of beating is reversed so that the animal clears the obstruction. There are thus two nervous systems in ctenophores: (1) the superficial epidermal system which causes movement of body wall, lips and tentacles and so is involved in feeding; and (2) the system formed by the ciliated conducting cells of the costae which is therefore involved in locomotion. This recalls the double nerve nets of both hydrozoam and scyphozoan medusae.

In all swimming coelenterates, cnidarian or ctenophore, the nervous system responsible for locomotion is one which flings the whole locomotor system into simultaneous activity – the muscles of the whole column wall of anemones (the swimming anemone *Stomphia*, for example), the whole musculature of the bell of jelly-fishes and all the costae of a ctenophore. Feeding is controlled by a second neuromuscular mechanism which normally deals with localized activity in tentacles and lips. The swimming coelenterate is, from the neuromuscular point of view, a polyp-like animal on which has been superimposed a mechanism for dealing with the problems of swimming; one may perhaps see in this a persistent actinula-like organization basic to all coelenterates whether they belong to the cnidarian or to the ctenophore evolutionary line.

Other ctenophores

The common ctenophores of temperate waters are cydippiform. A modified shape occurs in the other orders of the phylum, of which there are four: Lobata, Cestida, Beroida and Platyctenea. Although the lobate ctenophores are mainly found in tropical waters the genera *Bolinopsis* and *Mnemiopsis* are often abundant in North Atlantic coastal areas. The body is no longer approximately spherical but markedly elongated in the sagittal plane; its most marked features, however, are six lobes projecting from the equatorial level so as nearly to cover the oral pole and mouth. Four of the lobes are delicate, ciliated structures known as auricles, but the other two are stout and muscular and concave on their oral face. The tentacles of these animals have no sheaths and have come to lie close to the mouth; they are small and play little part in food capture, a process carried out by the oral lobes. Since these lack the great extensibility and adhesive properties of cydippid tentacles *Bolinopsis* and *Mnemiopsis* include a greater proportion of passively drifting and slow swimming members of the zooplankton such as eggs and small larvae in their diet than do the cydippids which catch actively swimming prey, mainly copepod crustaceans.

The trends exhibited by Lobata are carried to greater lengths by the two genera which constitute the Cestida; these are *Cestum* (Fig. 25, bottom) and *Velamen*, differing from one another only in size and trivial anatomical details. They are commonly known as Venus's girdle and are found in the Mediterranean and throughout tropical seas. Elongation of the body in the sagittal plane and its compression in the tentacular plane are extreme – *Cestum* may be over a metre long, six or seven centimetres deep and one or less in thickness. The four costae flanking the sagittal plane are much elongated and set along the aboral edges of the ribbon-shaped body, whereas the four which flank the abbreviated tentacular plane are minute. The tentacles are so small as to have no significance in food capture which is due to many small tentacles set in grooves along the oral edges of the body. The animals can swim by the beating of the ctenes but move more effectively by contractions of a sheet of longitudinal muscle fibres set in each half of the ribbon; a wave of contraction passes outwards from the central regions along each half of the body, sending it into curves which propel it through the water.

The order Beroida contains ctenophores, mostly in the genus *Beroe* (found in all seas), with a thimble-shaped body, mainly occupied by a greatly enlarged stomodaeum opening by a slit-like mouth. The body is almost as compressed as that of a cestid but much less elongated. The size of the stomodaeum limits the canal system to a thin skin of flesh and the canals are further marked by having innumerable short branches which may unite to form a network. There are no tentacles and food is caught by the mobile lips. Within them lies a strip of glandular cells and, internal to that, a ring of cells with giant cilia, known as macrocilia, each of which consists of 2 500–3 500 ciliary shafts all interconnected and bound together so as to behave as a single functional unit. *Beroe gracilis*, the common North Sea species, lives exclusively on the cydippid *Pleurobrachia pileus*, refusing other ctenophores: *B. cucumis*, commoner on western British shores, will take both *Pleurobrachia* and *Bolinopsis*. When about to feed the lips curl outwards exposing the glandular and ciliated rings, the macrocilia start to beat, radial muscles in the mesogloea contract so as to dilate the stomodaeum and, by a combination of manipulation by the lips and suction from the expanding stomodaeal cavity, the prey is engulfed.

The last order, Platyctenea, contains a number of rather extraordinary animals which were hailed on their discovery as bridging the gap between the more typical ctenophores (and so coelenterates in general) and the polyclad turbellarians. Better knowledge of the animals has shown, however, that the platyctenean ctenophores are possessed of all the typical ctenophoran characteristics but that these are modified for a creeping mode of life comparable to that followed by flatworms. As a result the two stocks have acquired many similar features – but as a result of convergence rather than of genetic relationship. There are four genera in the order, *Ctenoplana*, *Coeloplana*, *Tjalfiella* and *Gastrodes*, of which only *Coeloplana*, found in Japan, is common. The first is a planktonic animal, the next two creep on shores, though *Tjalfiella* is all but sessile, whilst *Gastrodes* is parasitic on pelagic tunicates.

Classification of ctenophores

Ctenophora
    Cydippida
    Lobata
    Cestida
    Beroida
    Platyctenea

6

Platyhelminths

ANIMALS in the phylum Platyhelminths are known as flatworms or planarians if they are free-living, flukes or tapeworms if they are parasitic. All names emphasize the fact that they possess a flattened shape, fluke being an old name for a flounder or flatfish. The differences between the free and parasitic types are no more than would be expected of creatures living in such different ways and do not constitute significant departures from a type of organization which lies clearly between the diploblastic structure of the coelenterate, with its emphasis on the folding of sheets of cells, and that of the annelid worms, with solid masses of tissue related to a coelomic body cavity.

Though platyhelminths all exhibit this type of organization there is some hesitation amongst zoologists in admitting all within the confines of a single phylum, though this is usually done. It may well be that the group is polyphyletic. Thus while one group of zoologists regard the polyclad flatworms as derived from simpler turbellarians others would regard them as descendants of sedentary ctenophores, the organization of which has reached a platyhelminth level and therefore converged with that of other flatworms. Similarly, the simplest turbellarians have been interpreted by some as cellularized ciliate protozoans which have taken to a creeping life and by others as coming from an organism like the planula larva of coelenterates modified for the same mode of life.

Organization of a planarian

The common freshwater planarians will be described as a basis for the understanding of platyhelminth organization and functioning. Species of three genera (*Polycelis*, *Dugesia* and *Dendrocoelum*) are common in the more eutrophic fresh waters.

As the popular name suggests, these flatworms have a very depressed body, with the proportions of a rather narrow leaf. They are carnivores, though will eat carrion if fresh, and they search throughout the ponds or streams within which they live for their prey. They thus differ from the ordinary coelenterate in being actively motile.

This mode of life is most effective if one end of the animal normally goes first, since the sense organs by means of which it informs itself about the new territory it is entering can be grouped there, just as the "sense organs" of a motor car (driver, headlight and the like) are placed at the end which normally goes first. This end is an animal's head. Along with the formation of a head the habit of forward movement also involves an alteration in symmetry. Radial symmetry is an adaptation for a mode of life in which there is no reason for supposing that one direction is either more or less important than another, and the equal development of the parts of the body which is its main characteristic reflects this. In an animal which always moves with its head going forward this is no longer true: one direction *is* more important than any other and the body is rearranged to a bilateral symmetry to adapt to this fact.

The planarian body is bounded by a syncytial epidermis or by a cellular one, (Fig. 26A, ep) one layer thick, often with liquid-filled spaces between the bases of the cells which act as a local hydrostatic skeleton as in coelenterates. The other systems of the body do not lie in any cavity but are embedded in a loose network of branching and anastomosing cells with numerous interconnecting intercellular spaces, forming a tissue known as parenchyma (par). Its real nature has not yet been fully explored. In *Polycelis*, and perhaps in all flatworms, the epidermis seems to be derived from parenchymatous cells which have migrated outwards. Though doubtless some movement of intercellular fluid occurs as flatworms glide over the substratum and change shape, there is no regular circulation of the fluid in the parenchymatous spaces, and it seems likely that the flatness of the platyhelminth body is basically a way of increasing surface area, and of keeping all regions of the body close to an external surface from which they can obtain oxygen by diffusion, and through which easily diffusible metabolites may escape. Lacking a blood system, too, flatworms may also find that their highly branched alimentary system (int) brings digested food close to all parts and so either abolishes the need for its transport from the gut by wandering cells, or reduces the distance they have to travel.

A further correlation of the flattened shape may be with the method of locomotion. This is partly due to cilia on the cells of the ventral epidermis beating against a layer of mucus secreted by gland cells which lie in the parenchyma but have been derived from parenchymatous cells, and which discharge by long narrow necks between the epidermal cells. This device (which is met in many other groups) allows an increased amount of secretion to be produced over a given area by permitting gland cells with necks of varying length to be banked in layers one under the other, instead of the single row which is all that could be accommodated were the cells simple goblets. The flatworm, therefore, glides over a mucous trail like a snail. There is also a muscular component in the locomotor effort and the two cooperate. Waves of muscular contraction start at the head and pass back along the body. Each wave comprises an area in which a transverse strip of the body is lifted off the

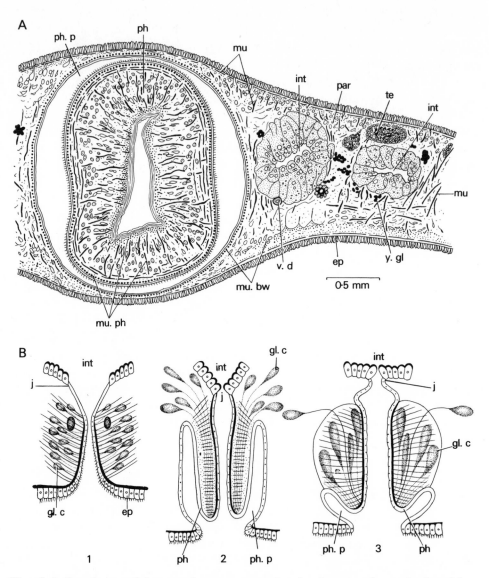

Fig. 26. A, *Dugesia gonocephala*: part of a transverse section showing general organization of the turbellarian body. (After de Beauchamp.) B, diagrams to show the different types of pharynx encountered in turbellarians; 1, simple pharynx; 2, compound pharynx; 3, adherent pharynx. (After Beklemischev.) ep, epidermis; gl.c, gland cell; int, parts of intestine; j, junction of ectoderm and endoderm; mu, dorsoventral muscles; mu.bw, muscles of body wall; mu.ph, muscles of pharynx wall; par, parenchyma; ph, pharynx; ph.p, pharynx pouch; te, testis; v.d, vas deferens; y.gl, parts of yolk gland.

substratum by contraction of the dorsoventral muscles. Contraction of the local longitudinal muscles pulls the posterior part of the body involved in the wave forward to a point where it can make a new attachment to the substratum. In some circumstances (natural and experimental) either the cilia or the muscles may be paralysed and locomotion then becomes less efficient because it is either purely muscular or ciliary. The surface of the body is also rich in receptor cells, particularly at the anterior end on the dorsal epidermis and at the lateral margins. These are apparently chemoreceptors and allow the animal to detect and to orientate itself to food and perhaps other objects in its environment. They are frequently placed on anterior (*Dendrocoelum*) or lateral (*Dugesia, Polycelis*) projections of the head. On this part of the body, too, are eyes, of which a pair (*Dendrocoelum, Dugesia*) or a large number (*Polycelis*) may occur.

The musculature (Fig. 26A, mu.bw) is subepidermal and lies in several layers of fibres running around and along the body, or lying diagonally so as to form a double spiral. In addition dorsoventral muscles (mu) act as ties between the upper and lower surfaces, helping to maintain the flat shape. These muscles act on the fluid which fills the interstitial spaces of the parenchyma, and this therefore acts as a hydrostatic skeleton. Change of shape can be effected by the antagonistic action of the circular and longitudinal muscles as in a coelenterate or worm, with the diagonal fibres acting synergically. There is, however, another factor involved in the flatworm, which is also important in such animals as nemertines and nematodes and which may appropriately be described here. The epidermis of all these animals, it has been shown (mainly by Cowey), rests on a thick basement membrane through which run fibrils, arranged in two sets, disposed diagonally and crossing one another so as to produce a trellis-like pattern. In nemertine worms (see p. 140), but not in the others, a second similarly arranged set of fibrils lies deeper among the muscles of the body wall. Wherever they lie the fibres are inelastic, so that change of shape in the worm is not due to their elastic recovery from deformation; but, since they strengthen and stiffen the basement membrane, their presence does condition the kinds of shape which a worm can assume. In a trellis of this sort the fibres form a double spiral wrapped round the body of the worm and crossing one another at an angle θ. As the worm elongates and becomes thinner θ decreases towards $0°$, whereas when it shortens and thickens θ increases towards $180°$. It is possible to relate the volume of the worm to the value of θ. Clearly, at the limits when θ equals $180°$ or $0°$ the volume of the body must be zero: in the former case it has the shape of a plane disc of infinite area and in the latter of a thread of infinite length and zero cross-section. At intermediate values of θ the volume is measurable. The relationship between the two is given by the curve in Fig. 27, which represents the maximal volume which the worm can attain at a given angle of intersection. At these volumes the body is circular in transverse section, since a cylindrical shape gives the greatest volume for a given surface area, and the worm is, after all,

A

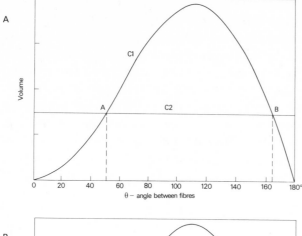

B

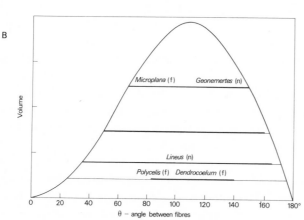

Fig. 27. A, Graph showing the relationship between θ, the angle of intersection of the fibrillae in the body wall of a turbellarian or nemertine, and the volume of the animal. For further explanation see the text. B, Graph showing the relationship between θ and the volume of the worm as in A. Superimposed on this are horizontal lines giving the actual volumes of some common flat worms (f) and nemertines (n). These are thickened to show the range over which changes in length (indicated by change in θ) actually take place by comparison with the range which the fibre system would permit in theory. (After Clark and Cowey.)

bounded by an epidermis of given and definite area. At any moment in its life history the worm has also a particular volume and for one individual animal this is represented by the horizontal line AB on the graph. The area bounded by the line AB, the curve and the x-axis represents volumes which that particular worm cannot achieve because they would involve it in a reduction in size. Similarly the area above the curve represents volumes which the worm cannot reach because they would make it burst. The area of the graph between the line AB and the curve represents the possibilities open to the animal as regards size and shape and within which it must accommodate such alterations in volume as may be brought about by feeding, defecation, the growth and the discharge of gametes. At the left intersection of curve and line (A), the animal has its greatest elongation, and at this length its transverse section is circular; at the right intersection (B) the worm is at its shortest, and again has a circular cross-section. No increase in volume is possible at either of these lengths because the circular cross-section implies maximal volume for that length; the animal cannot therefore feed. However, at lengths intermediate between these (say at C_1) the volume of the animal is less than can be contained within a cylinder of the same length and this allows the animal to become flattened in transverse section. Since it has not then the maximal volume for such a length feeding or increase in volume by maturation of eggs can occur, a limit to swelling being reached when the cross-section once again becomes circular as at C_2. Any increase beyond this might lead to rupture of the body wall.

As will be seen below, there is also a relationship between the kinds of shape which a flatworm can assume and its mode of life.

Freshwater flatworms feed on worms, small crustaceans and insect larvae, dead or alive. They capture these animals by wrapping themselves around them, at the same time secreting much sticky mucus, particularly if the prey struggles. The first part of the gut, a proboscis known as the pharynx (Fig. 26A, ph) and normally kept withdrawn into a pouch (ph.p) is then protruded and explores the surface of the prey, elongating considerably if necessary. It ultimately pushes into the prey and its contents are sucked into the gut, a much branched system of blind tubes extending into all parts of the body. Uptake of food seems to be accomplished by mechanical force since there is no evidence of the secretion of digestive juice at this stage. The pharynx is extremely muscular (mu.ph) and inwardly directed peristaltic waves pass along it throughout the whole feeding process. At the end of a meal the entire gut may be filled with food in a rather fine state of disintegration achieved by purely muscular breakdown. This is phagocytosed by the more abundant of the two types of cell found in the lining of the alimentary tract and digestion appears to be limited to an intracellular process within vacuoles in these cells. This conclusion rests on observations that vacuoles abound just after feeding and get fewer and less conspicuous with the passage of time. A second type of cell in the gut epithelium does not seem to be involved in digestive activity but to act as a store of reserve protein in the form of spherules. The number of these cells depends upon the diet and the degree of starvation (Table 2).

TABLE 2

Diet	Ratio of storage to digestive cells
High protein	1 : 4
Normal	1 : 6.3
Starved 14 days	1 : 6.7
Starved 28 days	1 : 9.5
Starved 2 months	1 : 15.5
Starved 3 months	no storage cells

Flatworms appear to be able to starve for the three-month period indicated in Table 2 without serious effect beyond the depletion of their glycogen reserves (stored in parenchymatous cells), which disappear after a fortnight, and of the protein in the reserve cells of the gut. Their stored fat is still, however, present. Further starvation does not necessarily involve death since planarians are able to survive by consuming their own tissue to obtain energy. This is not just a simple slimming and decrease in size but involves the reversion of many of their systems to a juvenile condition, a process known as dedifferentiation. When food once more becomes available they return to the fully differentiated state.

Because of the extensive surface which planarians offer to the environment and because of their lack of exoskeletal structures, it is likely that diffusion of oxygen, carbon dioxide and perhaps of some nitrogenous waste occurs freely over it and no special respiratory or excretory system need be developed. This permeability, however, lands the freshwater planarian in the position of absorbing water osmotically from the environment and requires the presence of an osmoregulatory system (Fig. 30E), provided by a series of flame cells lying in the parenchyma. These are cells (f.c) with a tuft of languidly beating cilia projecting into the innermost end of a tubule which connects with its neighbours so as to give rise to a network ramifying throughout the body. Main longitudinal trunks (dorsal and ventral in *Dugesia*, dorsal only in *Dendrocoelum* and *Polycelis*) open to the exterior by several minute pores (pore). The flame cells seem to be responsible for the excretion of a watery filtrate of the body fluid and so control the degree of hydration of the body; the cells which line the tubules can resorb salts; they also have the ability to excrete vital stains and contain yellow granules and may, therefore, have some excretory as distinct from osmoregulatory importance. Each channel of this system ends blindly internally in a slight swelling known as a flame bulb (Fig. 28A, B, fl.b) which consists of a single cell with one end drawn out into a hollow tubular extension which fits on to the cells of the rest of the channel (tub.c). The neck of the bulb is made of a series of rod-like pieces (r) interconnected by thin membrane. Within this circlet lies another of long microvilli (mv), and, centrally, the cilia (c) forming the flame. The beating of the cilia driving fluid down the tubule lowers the pressure in the bulb and fluid is sucked through the membrane between the rods from surrounding tissues. It is believed that the microvilli can act as valves, coming away from the membranes when pressure in the bulb is low and allowing fluid to enter, falling back and blocking the membrane when internal pressure is high.

Planarians are hermaphrodite. As in molluscs and earthworms, this is to be correlated with their rather slow rate of movement so that when two animals do encounter one another at the breeding period it is more advantageous for the eggs of both to be fertilized by the sperm of the other than for this to happen to one animal only. Elaborate mechanisms, anatomical and physiological, have been evolved to prevent self-fertilization, which is effectively impossible. The testes (Fig. 26A, te) are scattered throughout the parenchyma and their ducts (v.d) lead to a penis (Fig. 28C, pe), which is a muscular organ often with associated glands (pe.gl) and a central

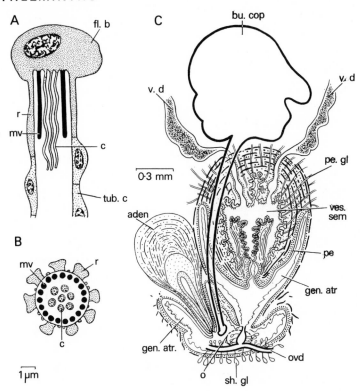

Fig. 28. A, longitudinal section of innermost part of excretory tubule to show details of structure of flame cell; B, flame cell in transverse section at level indicated by arrow; C, *Dendrocoelum lacteum*, diagram of the terminal parts of the reproductive system. (A, B, based on
McKanna, C, on de Beauchamp.) aden, adenodactyl, muscular or pyriform organ; bu.cop,
bursa copulatrix, connected to genital atrium by long stalk; c, cilia of flame cell; fl.b, flame
bulb; gen.atr, parts of the genital atrium; mv, microvillus; o, opening of genital atrium to
outside; ovd, oviduct; pe, penis; pe.gl, ducts of glands; r, riblike thickening of neck of flame
bulb; sh.gl, shell gland; tub.c, tubular cell of excretory canal; v.d, vas deferens, its basal coil
acting as seminal vesicle; ves.sem, so-called seminal vesicle, the cavity of the penis.

space known as a seminal vesicle (ves.sem), though storage of spermatozoa usually
takes place in the basal coils of the vasa deferentia (v.d). When not in use the penis
lies withdrawn into a pouch called the genital atrium (gen.atr) to which the oviducts
(ovd) and some other structures discharge, the atrium in its turn opening by a single
pore (o) to the ventral surface of the worm. One of the organs connected to the
atrium is usually known as the "muscular organ" or adenodactyl (aden): it is a
hollow, pear-shaped structure, with a pore at the narrow end. The walls are muscular and many glands open to it. It appears to discharge its secretion during copulation and so to be comparable to a prostate gland, though not directly connected to
the male duct. At copulation the penis is extended into a lengthy flagellum which is

passed into the genital atrium of the copulating partner. Here it enters still another structure opening to the atrium, a blind pouch known as the bursa copulatrix (bu.cop), and the spermatozoa (with accompanying secretion from both penis and adenodactyl) are discharged into this pouch, well away from the female organs of the animal which produced them. The secretion of the penial glands sometimes forms a spermatophore; that of the bursa activates the sperm. Sperm exchange is reciprocal and simultaneous. Later, after copulation is finished, the now active sperm migrate to the oviducts and swim along them to a point near the ovaries, of which there are two in freshwater flatworms, where they lie in a kind of receptaculum seminis awaiting the discharge of the eggs.

Planarians (and some other members of the phylum) are unusual in that the eggs are small structures with little included yolk for the nourishment of the developing young. This is provided by secretion from a separate yolk gland (Fig. 26A, y.gl) which is composed of strands of cells closely associated with the branches of the alimentary tract, presumably because they obtain the material which they require for the elaboration of yolk most readily in such a situation. The yolk glands produce cells containing at least three kinds of granule, some of which are reserve foods, which enter the oviduct and group themselves around the eggs as they descend, after being fertilized, towards the genital atrium. At the point where the oviducts open to the atrium there also discharge a large number of glands, the so-called shell glands (Fig. 28C, sh.gl). Flatworms lay their eggs in a cocoon, usually fastened to the underside of a stone or other suitable substratum. Recent work suggests that the "shell" glands certainly produce the substance which anchors the cocoon but that its walls, which are made of the tanned protein sclerotin, are derived from granules from another source, probably the yolk glands, and are made available or allowed to fuse to a membrane by the secretion of the shell glands. The polyphenols involved in the tanning process are certainly contained in one kind of granule secreted by the yolk glands. The cocoon is secreted around the eggs as they lie in the atrium, and later is pushed out and attached to the substratum.

Classification of Turbellaria

In most classifications of the animal kingdom the free-living flatworms are all included within a single class, Turbellaria, of the phylum Platyhelminthes, though various authors have expressed doubt as to whether the class is not polyphyletic, that is, comprises animals derived from distinct ancestral stocks which have all reached a similar grade of organization and may exhibit similarities of structure because of adopting similar ways of life. It can be subdivided in several ways, the relative values of the different schemes being beyond discussion here. The following are the most important subgroups, some small ones being omitted.

1. A large group of predominantly marine flatworms, the Polyclada, growing some-

times to large size, the commonest British form *Leptoplana tremellaris* reaching 1.5–2.0 cm and occurring under stones at mid-tide levels. The proboscis leads to a large number of gut branches, whence the name, and the main other characteristic is that there are no separate yolk glands, the ova containing the yolk within themselves. Male and female ducts open by separate pores.

2. The second major group is the Triclada to which belong the common freshwater flatworms described above. Triclads, however, are also marine and some are terrestrial. The name derives from the fact that the proboscis opens to three main gut branches, one going forwards in the mid-line, the others passing back, one on each side. Yolk for eggs is produced in separate yolk glands and the male and female pores open to the genital atrium so that there is only the one genital opening on the surface of the body.

3. Eulecithophora, a group more or less the same as one previously called Rhabdocoela because of the unbranched, finger-like shape of the gut (Fig. 29D). The yolk is produced in cells separate from the eggs and there is a single external genital pore. Most animals in this group are small, only a millimetre or so in length and occur equally in the sea and in fresh waters.

4. The last taxon of any size in the class is known as Archoophora, the most important constituent of which is the group called Acoela. The animals are again small, predominantly marine and are characterized by the fact that the eggs are yolky and there are no separate yolk glands. In many the gut has no cavity and forms a solid mass of cells which can ingest food particles for intracellular digestion (Fig. 29A, end).

The evolutionary significance of these groups, and in particular, their possible relationships with coelenterates and protozoans, have been the subject of much argument since the statement by Hadži that one of them might contain the most primitive surviving multicellular organisms. These are the minute worms placed in the group Acoela (part of the larger group Archoophora) and characterized principally by the fact that the mouth (m) leads into a central mass of endodermal cells which directly phagocytose such food as the animal captures; there is no gut lumen into which prey is trapped. According to Hadži this might be the result of a direct evolution from an ancestor such as a ciliated protozoan by a process of cellularization accompanying increase in size. On other views of the origin of metazoans the peculiar features of these animals are interpreted as due to the simplification of organization which often goes with reduction in size, and this seems to be more in line with current thinking.

Although all free-living flatworms are depressed in shape, some are more so than others. In all, flatness must be deemed to be linked with the need for a large surface area for respiratory and other exchange, but its degree is determined by the habitat in which the animal lives. The differences between worms may be expressed in terms of the curve in Fig. 27B. From this it will be seen that animals of the genus *Microplana*

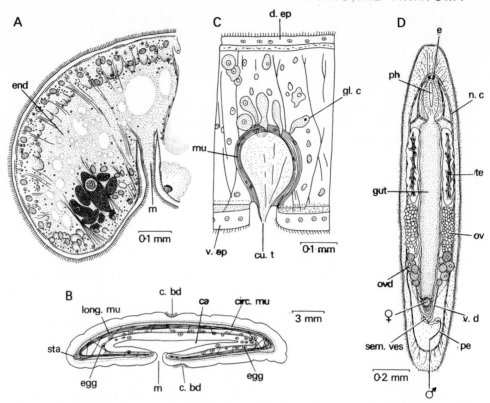

Fig. 29. A, *Convoluta*: part of a transverse section. B, *Xenoturbella bocki*: diagrammatic sagittal section. (After Westblad.) C, *Apidioplana mira*: part of a vertical section through the body of the polyclad turbellarian to show adenodactyl unconnected with reproductive system. (After Block.) D, *Macrostomun gigas*. (After Hyman.) ca, cavity of gut; c.bd, equatorial ciliated band; circ.mu, circular muscle layer; cu.t, cuticularized tip; d.ep, dorsal epidermis; e, eye; egg, egg; end, endodermal syncytium; gen.o, genital opening; gl.c, gland cell; gut, simple, unbranched gut; long.mu, longitudinal muscle layer; m, mouth; mu, muscle coat; n.c, nerve cord; ov, ovary; ovd, oviduct containing ripe eggs; pe, copulatory apparatus; ph, simple pharynx; sem.ves, seminal vesicle; sta, statocyst; te, testis; v.d, male duct; v.ep, ventral epidermis; ♂, male pore; ♀, female pore leading to antrum, occupied by an egg to which the shell is being applied.

(a terrestrial triclad) have a very low extensibility, theoretical and actual, and the limited changes in length which are possible for them to carry out confine them to the part of the graph near the peak of the curve. In transverse section they therefore remain nearly circular whatever their length. This – which gives the animal a minimal surface for the contained volume – is a limitation on their mobility, but is advantageous when correlated with their terrestrial habit, where control of water content becomes a matter of extreme importance. Unexpectedly, as shown by Pantin, it is uptake of water rather than its loss which must be minimized and the animals control this first by choosing damp niches in rather dry habitats and then by the shape of their body. A similar shape and restriction of extensibility for the same reason is exhibited by the land nemertine *Geonemertes*. In both, the circular sectional shape is achieved by expansion of the dorsal surface, the ventral remaining small. Water loss is not a factor of significance in the life of aquatic flatworms so possession of a cylindrical body confers no special advantage on them; a flattened shape is of positive value to the animal in facilitating locomotion and gaseous exchange in an aqueous medium where oxygen tension is low. Possible water-logging is looked after by other means (see below). This is reflected by the position of *Polycelis* and *Dendrocoelum* (as well as nemertines like *Lineus*) at a low position in the graph.

Movement

Locomotion involves movement over a substratum, often uneven, and often in the presence of currents in the surrounding water. Some turbellarians orientate themselves so as to head into currents, a habit which eases the problem of maintaining contact with the substratum; this is particularly true of freshwater forms where the problem is encountered in a simple form because of the way in which rivers flow downhill. In marine species currents are more likely to set in a variety of directions and to change rapidly; this may explain why some turbellarians in the group Polyclada, which is predominantly marine, are provided with a sucker on their ventral surface. The smaller turbellarians of the sea-shore or of fresh water streams and lakes are also not without physical means of avoiding being swept away, but this takes the form of lateral strips, or anterior or posterior groups, of special mucous gland cells (known as the erythrophile glands because they are red in sections stained with haemalum and eosin), the secretion of which causes the body to adhere firmly to the substratum. Their grip is improved by the fact that they can be partially protruded from the skin. Lubrication of movement is due to the secretion of a second type of mucous cell, the cyanophile (so called because they are blue in sections stained with haemalum and eosin). In addition to helping in locomotion, the glands may be offensive or defensive, and may, in part, be responsible for the fact that turbellarians seem to form the prey of no other kind of animal – a very unusual state of affairs. The glands which are offensive are the cyanophile and they are often agglomerated at the anterior end of the body to form a frontal gland which aids in capturing prey; it has been noted, too, that flatworms may snare prey in nets of erythrophile secretion spread over the substratum.

It has been claimed that the defensive secretion is formed by the swelling of rod-shaped bodies known as rhabdites which occur in all cells of the epidermis. They are formed in cells lying within the parenchyma which migrate to become incorporated in the epidermis. Since every epidermal cell has rhabdites this seems to argue that the epidermis is merely a collection of parenchymatous cells which react to the fact that they occupy the outermost surface of the body by polarizing to form inner and outer ends and by attaining full lateral contact with their neighbours instead of the restricted contact which occurred within the parenchyma. Whilst there seems no doubt that at least in some species, or in some circumstances, rhabdites are simultaneously discharged in large numbers, swell and give a sheet of gelatinous consistency over the body (swelling being prevented by 3.75 per cent salt solution) the fact that rhabdites contain purines must also implicate them in a process of nitrogenous excretion. It is not known to what extent one function may dominate the other, or even whether each might be carried out by different kinds of rhabdite, or by the same rhabdites at different times.

The locomotion of flatworms is related to their size and to whether they are aquatic or terrestrial in habit. The smallest aquatic forms (acoels, rhabdocoels) glide, apparently by a purely ciliary action; as size increases this is supplemented by anteroposterior muscular waves of small amplitude. In the larger species (polyclads, triclads) these are the major component of the locomotor effort and ciliation is reduced over the body surface, sometimes remaining only on the ventral side. Many flatworms can swim, the smaller ones by ciliary action, the larger ones by muscular waves passing down the body, especially its lateral parts, so that the motion is reminiscent of that of a skate or ray. Terrestrial platyhelminths, whatever their size, creep by muscular action in the same way as slugs, for minute specimens of which the common British form, *Microplana* (= *Rhynchodemus*) *terrestris*, may easily be mistaken. When muscular action becomes pronounced in aquatic forms it takes the form of a leech-like looping; this is particularly apt to happen if the animal is disturbed. The massed erythrophile glands then act as points of attachment.

Feeding and digestion

The feeding of turbellarians is related to the structure of the alimentary canal and, in particular, to the type of proboscis (traditionally called a pharynx) which the animal possesses. Flatworms fall into four grades of organization on this basis:

1. They may possess no pharynx and the endodermal part of the gut opens directly to the outside (Fig. 29A). This condition is limited to the animals known as acoels (such as *Convoluta*) in which the endoderm forms a solid mass without gut cavity; on Hadži's view of the origin of acoels this might represent a cellular version of the cytopharynx of a protozoan.

2. They may possess a simple pharynx, which is an inturned part of the body surface leading to the true mouth and a hollow gut (Fig. 26B1). It is lined by a ciliated epithelium through which discharge many gland cells (gl.c).

3. A compound pharynx (Fig. 26B2) originally called a *pharynx plicatus* may arise as a ring-like fold (ph) projecting from the point where endoderm and ectoderm adjoin

(j). The pharynx projects into a pouch (ph.p) which in turn opens to the exterior. The whole of the pharynx lies in the pharyngeal pouch and, depending on the species, may face forwards, ventrally or backwards. In the first and last of these orientations it may be a lengthy tube, but in the second only a shallow bell-like structure can be packed away in the slender body of a flatworm. This type is also known as a free pharynx and is that found in the common triclads and polyclads. It is mobile, very muscular, and may be extended to great distances. It is often ciliated internally but rarely at the edges of the aperture at its free end, where there open glands.

4. The last type is a modification of the compound pharynx known as an adherent pharynx (Fig. 26B3), found in the Eulecithophora and once known as a *pharynx bulbosus*. The essential difference from the third type is that whereas in that the whole length of the structure lies in the pharyngeal pouch here only its tip projects. Its mobility is therefore less, and it is useful only in sucking.

An acoel such as *Convoluta paradoxa* feeds on very small crustaceans, on protozoans and on diatoms. These are immobilized by the worm gliding over them and pressing the endoderm (Fig. 29A, end), through the mouth (m) to make contact; they are then ingested into vacuoles in the endodermal mass and digested. A simple pharynx allows the ingestion of whole prey of suitable size, but necessitates a digestive process within the gut; this may be either mechanical (the small freshwater archoophoran *Stenostomum*) or enzymatic (*Macrostomum, Mesostoma* – also archoophoran) or perhaps a combination of both methods. The free compound type of pharynx is adapted for reaching into the body of prey too large for ingestion intact, as when a triclad pushes it into the body of a worm or arthropod and mechanically breaks the organs into particles which can be sucked into the gut for phagocytosis and intracellular digestion; or for reaching the more nutritious parts of prey, as when the polyclads *Cycloporus* and *Thysanozoon* probe colonies of the tunicates *Botryllus* or *Ciona* and suck out the contents of a zooid. In this case the pharynx is not able to distintegrate the prey sufficiently to allow immediate phagocytosis, and an enzymatic digestion has to occur followed by uptake of the dissolved foodstuffs. Finally, in the polyclad *Leptoplana* the prey (small worms or crustaceans) undergoes enzymatic digestion whilst still held in the pharynx, which therefore acts as it if were an extension of the main digestive system. Digestion proceeds only to the stage where the prey is reduced to particle size, however, and it is finished by the intracellular breakdown of the particles within food vacuoles.

Where phagocytosis and intracellular digestion occur there is a tendency for the epithelial cells of the gut wall to become syncytial and to unite across the lumen so as to convert the alimentary system into a solid mass of tissue, perhaps comparable to the endodermal mass of the acoels. When the digestive process is finished the lumen of the gut is restored by separation of the cells. Defecation occurs by the animal sucking water into the gut and then expelling it, with waste, by means of the muscles in the parenchyma. Some polyclads (*Cycloporus*) have the branches of the gut terminating in anal pores; their precise function is doubtful, and they may be exits for water taken in with the food rather than for the indigestible remains of a meal. The acoel *Convoluta roscoffensis* has established a special relationship with zoochlorellae on which it ultimately becomes totally dependent. This is discussed on p. 87.

Reproduction

The reproductive system of turbellarians is nearly always complex, and this may be traced to the fact that all are hermaphrodites. The complexity has evolved within the phylum, however, and the genital system of the most primitive forms is very elementary indeed, the reproductive cells (both eggs and sperm) arising in the parenchyma, migrating to the gut and escaping by the mouth (*Xenoturbella*, Fig. 29B). Although there are no copulatory organs appropriate to either sex in this animal, a sexual congress occurs and sperm escape from the mouth when it is applied to the surface of another animal. They make their way through the skin to the parenchyma and fertilize the eggs which, therefore, are liberated after fertilization. The next evolutionary stage involves the formation of ducts leading the gametes to the exterior. In the more primitive animals these are developed only as outlets for sperm, short tubes leading from spaces in the parenchyma. Spermatozoa migrate to these when mature, sometimes being held in a space (seminal vesicle) at the inner end of the tube before escaping. At this stage the ova are still discharged either by way of the mouth or by rupture of the body wall. There has, however, often also evolved at this stage a female duct which receives sperm and leads them to spaces in the parenchyma where they may meet eggs. When this has occurred sexual processes involve a true copulation and the animals require a copulatory organ on the male duct. This condition is characteristic of many acoels.

Further elaboration of the reproductive system includes the formation of ducts connecting the testes to the copulatory organ so that the spermatozoa no longer swim through parenchymatous spaces; the same change takes place in the female system and the eggs are then led to a special pore and no longer escape by mouth or other route. These oviducts open separately from the original copulatory duct so that at this stage there are three apertures on the body relating to reproductive activity: (1) the male opening; (2) the oviducal opening; and (3) the opening of the space (bursa copulatrix) into which sperm are received at copulation. This last may be directly connected to the oviducts so as to lead sperm towards the eggs and facilitate fertilization. Though originally independently located at a variety of positions on the body these apertures ultimately come to lie in relation to a pouch known as the genital atrium, the opening of which is the only visible reproductive aperture. It possibly arises as an invagination of the area on which the original genital openings lay.

The copulatory organ may be either a cirrus (the terminal part of the male duct capable of turning inside out and projecting from the body surface) or a penis (the terminal part of the male duct capable of projecting from the body surface by simple elongation). Whichever form it may have it is usually provided with glands, is muscular and may be armed with spines. It seems to have evolved from a type of structure known as a muscular or musculo-glandular organ (also called prostatoid, adenodactyl and pyriform organ) which often occurs scattered over the general body surface (Fig. 29C). One of these is supposed to have enlarged to form the penis, whilst the remainder vanish, but some flatworms show a condition in which the penis exists along with accessory musculo-glandular organs on various parts of the body and in the common freshwater planarians one enlarged one lies alongside the penis (Fig. 28C, aden). The function appears to be predominantly the production of prostatic secretion to supplement what is produced in the penis or cirrus.

Although the higher flatworms have separate organs (sometimes called germaria) for the elaboration of ova and others (vitellaria) for making yolk and other material, this is a complexity of structure which seems to have appeared in the evolutionary progress of the group. In simpler platyhelminths (the group Archoophora, equivalent approximately to the worms known as the Acoela) the eggs are yolky and there are no separate yolk cells; in the group Polyclada the same holds, though in other respects these animals hardly deserve to be regarded as simple. Not enough work has been done to permit it to be said whether the "yolk" granules in archoophoran and polyclad eggs are chemically and functionally comparable to the different types of granule which have been demonstrated in cells from the yolk gland of triclads.

The polyclad egg develops by a process of spiral cleavage to produce a free-swimming ciliated larva known as Müller's larva (Fig. 30A). The cilia are borne on lobes (l) which increase their number and so the efficiency of the swimming. At the end of larval life the weight of the body is too much for them to support, the animal sinks to the bottom, the lobes degenerate and growth leads to the adult form. In triclads and other freshwater forms such a free-swimming larval stage is impossible and development is direct, leading to the hatching of juvenile forms from the cocoons in which the eggs are laid.

Osmoregulation

The fact that soft-bodied animals such as flatworms are able to live in brackish and fresh waters indicates that they must exercise some control over osmotic uptake of water. This is undoubtedly in part achieved by the elimination of water through the flame-cell system but a controlled permeability of the body surface is probably also operative in preventing osmotic swelling. The most illuminating investigation of the osmotic relations of a flatworm refers to the small triclad *Procerodes* (= *Gunda*) *ulvae*.

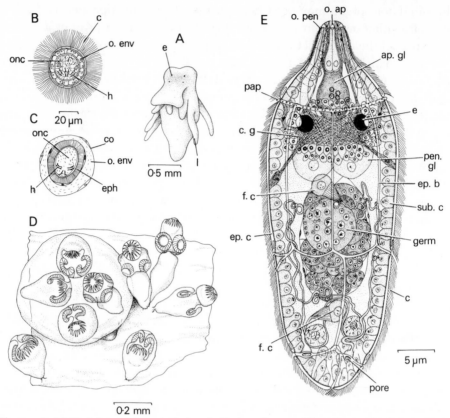

Fig. 30. A, Müller's larva of a polyclad turbellarian; B, coracidium larva of *Dibothriocephalus*; C, hexacanth of *Taenia*; D, part of the wall of the hydatid cyst of the cestode *Echinococcus granulosus*, showing the production of many scolices; E, miracidium larva of *Heronimus*. ap.gl, apical gland; c, cilia; c.g, cerebral ganglion; co, coat, reduced capsule, rapidly lost; e, eye; ep.b, space between epidermal cells occupied by ridge from subepidermal cell; ep.c, epidermal cell in optical section; eph, embryophore, the resistant layer of the egg shell; f.c, flame cell; germ, germ balls from which sporocysts or rediae will arise; h, hook; l, ciliated lobe; o.ap, opening of apical gland; o.env, outer envelope; onc, oncosphere; o.pen, opening of penetration gland; pap, sensory papillae; pen.gl, penetration gland; pore, pore of flame cell system opening to exterior between epidermal cells; sub.c, sub-epidermal cells which will form outer wall of sporocyst when epidermis is shed on penetrating second host.

This animal lives in a restricted habitat, under stones on beaches where fresh water, either in streams or seepages, is entering the sea; it lives on the upper half of the beach and is therefore bathed by stream water when the tide is out, but by sea water when the tide is in. It does not occur in every such situation and Ritchie showed that the fresh water had to be hard. A very little observation shows that the animal swells on being transferred from sea water to fresh water and decreases in volume when the reverse change is made; the locomotion is also affected and in fresh water the animal becomes sluggish. *Procerodes ulvae* can be kept permanently in 5 per cent sea water, but dies when the concentration becomes less; this is primarily due to salt loss. The increase in volume when placed in fresh water was shown by Beadle to be due to the uptake of water which was accommodated at first in the parenchyma, and it was this which affected the animal's locomotor ability; later the water is transferred to vacuoles in the epithelial cells lining the gut and the parenchyma returns to normal, allowing locomotion to take place again. Some salt is lost, reducing the osmotic gradient between worm and external fluid, but it seems likely that the epidermis is actively controlling the entry of water and it is for this that the calcium ions in hard water are required. This active control has been increased in the freshwater platyhelminths and allows them to survive even in soft waters. Flatworms in the group Acoela are confined to marine habitats and this may, perhaps, be correlated with the fact that the flame-cell system is wholly lacking in their body.

The remaining classes of the phylum Platyhelminthes, trematodes and cestodes, are wholly parasitic.

Classification of Trematoda

The trematodes are classified in three orders (sometimes elevated to the level of independent classes); Monogenea, Aspidogastraea and Digenea. The animals in the first of these live mainly on the skin and gills of fishes. Some occur in more internal, but still accessible situations such as the buccal cavity, the cloaca and the ducts which lead from it. Perhaps the best known, *Polystoma*, occurs in the urinary bladder of the frog, *Rana temporaria*. Monogenetic trematodes live on only one host during their life cycle and each generation includes a motile ciliated stage which hatches from the egg as a free-swimming larva known as an oncomiracidium. There are good reasons for believing that monogeneans may be more closely related to cestodes than they are to the other trematodes and it is for this reason that they are sometimes placed in a class of their own. They are like trematodes, however, and unlike cestodes, in possessing a gut; they are unlike trematodes in having no relationship with a mollusc at any stage in their lifetime and are presumably evolved from a rhabdocoel-like ancestor which acquired the habit of browsing on the coverings, secretory and cellular, of early vertebrates.

The order Aspidogastraea is a small one containing flukes with a large, compound sucker over much of their ventral surface. As in monogeneans the life history involves only one host and this is usually a mollusc, though experimentally, and perhaps also in nature, fish and reptiles can be parasitized. A free-swimming larva known as a cotylocidium is the infective phase.

Most trematodes fall into the order Digenea and show a life history involving two or three hosts, of which one is a vertebrate and another a mollusc. The vertebrate and molluscan hosts are not obviously related in any way, nor, in moving from one to the other, is the parasite exploiting a food chain as in trypanosomes or haemosporidians and as will be seen later, in cestodes: the only factor linking the two hosts is that of accessibility to the parasite, both living in the same habitat. This is not true of the third host, which is almost always linked trophically to mollusc or vertebrate. At each stage between hosts there is a free-living form which may be accidentally ingested by the host (particularly when this is vertebrate and the fluke is a gut parasite) or which may penetrate the tissues of the host directly. This is the way in which the mollusc is entered and it is also how vertebrates are infected by blood flukes. The eggs are produced by the mature flukes, which inhabit the vertebrate host, and hatch to give a ciliated miracidium larva (Fig. 30E) which penetrates the molluscan host. Multiplication occurs here leading to the production of a second free larval stage, the cercaria, and it is this which finds its way to the third host, or, in some cases, directly to the vertebrate.

Inspection of the life cycles of the trematode groups – putting monogeneans on one side – suggests that the animals were originally parasites of molluscs. This would explain the nigh universal occurrence of a mollusc in their life history. Since the mollusc has a soft, slimy epidermis and offers a sheltered situation in the mantle cavity, one could imagine this being an attractive place for some small rhabdocoel to inhabit, and that from some such relatively innocent association a parasitic relationship developed. This step had, presumably, already been taken before the evolution of the vertebrates. With their appearance, it must be assumed that the free-living stage acquired the habit of association with a fish-like vertebrate, either attaching to its skin, or entering the buccal cavity and pharynx, presumably attracted by mucous secretions, and finally acquiring the ability to survive within other parts of the alimentary canal.

Classification of Cestoda

The cestodes distinguish themselves from most groups of parasites in that the adults are nearly all inhabitants of the alimentary tract of vertebrates and have totally lost their own gut. Like trematodes, they have a ciliated larval stage (Fig. 30B), but this is unlike the trematode larva in that it does not make an invasive attack on the next host, but is passively ingested by it. Where circumstances permit or, as in parasites of terrestrial hosts, compel the suppression of a free-living phase and replace it with an encysted one (Fig. 30C) this can happen without loss of the parasite's ability to get itself transmitted since the process is invariably a passive one. Since passage from host to host, though dependent upon trophic relationships existing between these organisms, is ultimately due to chance uptake of free or encysted larvae, its success is partly ensured by having these produced in vast numbers. This may be why the body of most cestodes (sometimes called Eucestoda) consists of a strobila, a series of units, each called a proglottid, budded from an area just behind the point of attachment, the head or scolex, each containing a complete set of reproductive organs and liberated at intervals from the hinder end of the body. The process of budding at the head end and the shedding of proglottids at the other continue throughout the lifetime of the worm, contributing to the production of many eggs. A single specimen of the tapeworm *Moniezia expansa* has been calculated to shed a total of 40–80×10^6 fertilized eggs, suggesting odds of about sixty million to one against successful completion of the life cycle.

The cestodes can be broken into groups partly based on the structure of the scolex and partly on the arrangement of the reproductive organs. Ten or more of these groups have been recognized, some very small and others of debatable validity as distinct taxa. Four must be mentioned: (1) Tetraphyllidea; (2) Trypanorhynchida or Tetrarhynchoidea; (3) Pseudophyllidea; (4) the largest group, containing most of the common tapeworms of higher vertebrates, Cyclophyllidea or Taenioidea.

Maintenance of position by parasitic platyhelminths

All animals which have become parasitic are faced by a number of problems special to their way of life and their success depends on finding solutions which are satisfactory to themselves and not too damaging to their hosts. The problems which arise in relation to trematodes and cestodes are: (1) maintenance of position on or within the body of the host; (2) protection against chemical attacks made by the host, either by digestive enzymes or by an immune reaction; and (3) the difficulties of transmission from host to host during reproduction. Some examination of the ways in which these problems are dealt with will now be made.

Most trematodes and cestodes occur in positions where they may be dislodged by currents in the water in which the host lives, by the streaming of its ventilating current or its blood, or by peristalsis of its gut. To counter this they have evolved sucking or grasping apparatus, and much of the classification of both classes depends upon variation in its structure. As might be expected the adaptation of a parasite to the stresses imposed upon it by a particular habitat is often so close as to act as a major determinant of the hosts and situations in which it can successfully survive.

In monogenean trematodes attachment is usually achieved by an arrangement of suckers and hooks placed at the hinder end of the animal and known as the opisthaptor, plus a variety of glands and small sucking disks anteriorly, forming a prohaptor. Occasionally, as in *Polystoma*, the prohaptor forms a sucker round the mouth. In the more primitive monogeneans, a group called Monopisthocotylea, the opisthaptor is a single large sucker with some large anchoring hooks in its centre and smaller ones round its edge. The animals wander over their host, ingesting mucus and other secretions, loose cells and blood escaping where the hooks have damaged delicate membranes. In some the posterior sucker becomes divided into compartments and this is the start of an evolutionary trend leading to the development of a number of small suckers – often better termed clamps in the light of what they do – instead of one large one. Monogeneans with this type of opisthaptor are placed in the Polyopisthocotylea and include the most familiar kinds such as *Polystoma*, and the diclidophorans which live on fish gills, each with their own characteristic position and stance.

To understand the mutual geometry of fluke and fish it is necessary to know that in teleostean fishes each gill arch (Fig. 31, g.a) carries two rows of flattened gill filaments projecting into the opercular chamber and diverging slightly from one another, the anterior (ant.h) and posterior (post.h) hemibranchs. Each filament carries on its flat (dorsal and ventral) surfaces a series of small transverse ridges containing capillaries; these are the secondary gill lamellae (sec.lam) and make up the true respiratory surface. Water passes from the pharynx into the spaces between the hemibranchs borne by neighbouring gill arches; some is then deflected over the secondary lamellae into the space between the hemibranchs carried by one gill arch. The speed of movement of the water (which might wash a parasite away) is greater in the first of these spaces, less in the last, and least over the secondary lamellae. This last situation is, on this account, assumed to be the original shelter of diclidophoran flukes, and a number of genera (Fig. 31A, M) such as *Microcotyle*, *Octostoma* (sometimes called *Kuhnia*) and *Mazocraes* (originally *Octobothrium*) are to be found lying along the broad surface of a filament, across the secondary lamellae, maintaining their station by grasping the latter by both the clamps and hooks of the opisthaptor.

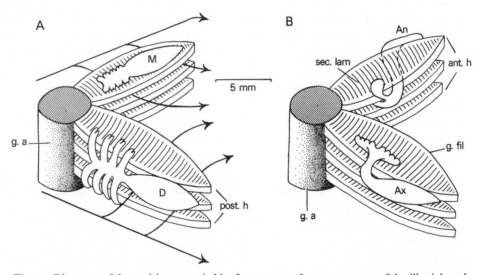

Fig. 31. Diagrams of the position occupied by four genera of monogenean on fish gills. A length of gill arch is shown with three gill filaments of anterior and posterior hemibranchs; in A arrows show the direction of the water from pharynx to opercular chamber. An = *Anthocotyle*; Ax = *Axine*; D = *Diclidophora*; M = *Monocotyle*; ant.h. anterior hemibranch; g.a., gill arch; g.fil, gill filament; post.h., posterior hemibranch; sec.lam, secondary lamellae on filament.

The flukes always lie with the suckers near the gill arch and the head downstream, so that they can move and feed without being swept away, an event which would lead to practically certain death. Since the flukes are lying directly over the lamellae which they grasp the clamps and hooks are short and sessile.

Flukes placed in the genera *Diclidophora* lie with the head downstream but the body (Fig. 31A, D) is placed along the narrow edge of the gill filaments and therefore in the more dangerous, because more vigorously ventilated, spaces between and outside the hemibranchs. They are able to maintain themselves here, however, because, in contrast to the genera just mentioned, the eight clamps of the opisthaptor are carried on long stalks and this, combined with an increase in the breadth of the fluke, allows the clamps of one side to grasp the secondary lamellae on the dorsal side of a filament while those on the other seize the lamellae on the ventral side. The grasp, indeed, is often sufficiently great to let the fluke straddle more than one filament. In this position hooks on the opisthaptor could not catch hold of secondary lamellae and they have been lost. In *Anthocotyle* (Fig. 31B, An) only an enlarged anterior pair of clamps behaves like a thumb and index finger, the others remaining small. In this genus the posterior end of the fluke is anchored to the inner edge of a gill filament; the body then lies across the same filament where its small clamps hold secondary lamellae, and its anterior part then emerges into the main space opposite

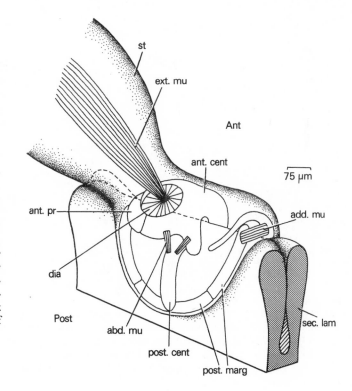

Fig. 32. The clamp of a diclidophoran grasping two secondary gill lamellae, viewed as a transparent object. (Based on Llewellyn.) abd.mu, abductor muscle; add.mu, adductor muscle; Ant, anterior; ant. cent, central sclerite of anterior half; ant.pr, proximal marginal sclerite of anterior half; dia, diaphragm; ext.mu, extrinsic muscle; Post, posterior; post.cent, central sclerite of posterior half; post.marg, two of four marginal sclerites of posterior half; sec.lam, secondary lamella of gill; st, stalk connecting clamp to body of fluke.

st
ext. mu
Ant
ant. cent
75 μm
ant. pr
add. mu
dia
Post
abd. mu
post. cent
post. marg
sec. lam

a gill slit; here it twists so as to lie along the edge of the filament, presumably to avoid the main stream of water escaping from the pharynx, which might be strong enough to sweep it away. The body twists in different directions according to which side of the fish the fluke lives on. Similar twisting occurs in other genera such as *Axine* (Fig. 31B, Ax) where the opisthaptor forms a linear series of clamps attached to the secondary lamellae on one side of a filament whilst the main body is lodged along the edge, between the hemibranchs.

Details of the mechanics of the clamping apparatus have been worked out for several monogenetic trematodes; a description of *Diclidophora* must suffice, although it is atypical in that sucking is an important component of the action. Each clamp, when open, is more or less circular in outline and can be folded transversely over a few secondary lamellae like a grasping hand (Fig. 32); during folding the posterior half is pulled against a stationary anterior half. Each half is strengthened by sclerites made of protein, those of the posterior half hinging on those of the anterior half. The central radius and the inner and outer margins of the anterior half are stiffened by three rods, the central one being extended into a flange which meets the inner along its whole length. In the middle of the clamp the central (ant.cent) and proximal (ant.pr) sclerites share in forming the lips of a circular opening plugged with a

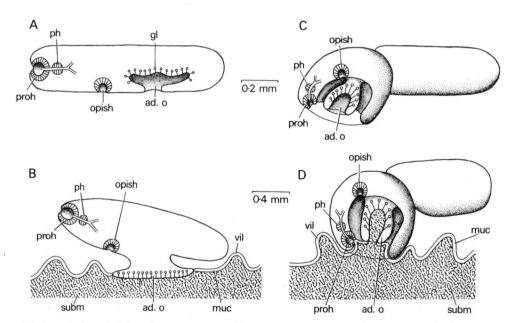

Fig. 33. Strigeoid flukes. A and B, *Cyathocotyle*; C and D, *Apatemon*. A and C show the flukes unattached, B and D attached to the intestinal wall of the host. (Based on Erasmus.) ad.o, adhesive organ; gl, glands of adhesive organ; muc, mucosa of intestinal wall, absent where gripped by adhesive organ; opish, opisthaptor; ph, pharynx; proh, prohaptor; subm, intestinal submucosa; vil, intestinal villus.

diaphragm (dia) of fibrous tissue to which there attaches a muscle (ext.mu) running into the body of the fluke. The posterior half of the clamp has a marginal semicircle of four supporting pieces (post.marg) and a single central radial one (post.cent); centrally and peripherally these articulate with the corresponding sections of the anterior skeleton. At the inner and outer edges of the disk adductor muscles (add.mu) run from the anterior to the posterior marginal sclerites whilst in the centre a pair of abductor muscles (abd.mu) links the radial ones. Grasping is achieved in two steps: (1) the marginal adductor muscles pull the posterior half down on to the anterior one, enclosing two or three secondary lamellae; (2) the main extrinsic muscle contracts, raising the diaphragm and so reducing the pressure in the space between the outer skin of the clamp and the surface of the gill. It is this action which provides the main grip. The extrinsic muscle runs partly across the body to the contralateral clamp, partly longitudinally to the anterior end.

In *Polystoma* and its relatives, which live in places where there are no projecting folds to grasp, clamps are absent; in their place the fluke has developed suckers, known as acetabula, muscular cup-shaped structures the central part of which can be raised so as to lower pressure within the cup. Frequently part of the wall to which the sucker is applied is caught within the cup; circular muscle fibres at the lip of the cup can contract and hold the tissue to increase the grip.

In digenetic trematodes the grasping apparatus has been standardized and simplified to a pair of suckers acting like those of *Polystoma*, a prohaptor situated round the mouth at the anterior end of the body, and an opisthaptor placed somewhere on the anterior half of the ventral surface (Fig. 38A, proh, opish). In Aspidogastrea the single sucker, which occupies the ventral surface and is often compartmentalized, acts similarly. This kind of attaching apparatus does not seem to limit the position of the animal in its host as does that of the monogeneans.

In one group of digenetic trematodes, the superfamily Strigeoidea, which contains flukes living in the intestine of birds and mammals, there is a further structure concerned with attachment to the host in addition to the two suckers. This is known as the holdfast or adhesive organ (Fig. 33, ad.o) and is located on the ventral side behind the posterior sucker (opish). In some genera – for example, *Cyathocotyle* (Fig. 33A, B) – it is an eversible structure which can retract with some tissue of the host held within it; in others it is lobed (Fig. 33C, D) and held within the hollow cup shape which characterizes the anterior half of the body, as in *Apatemon* and *Cotylurus*. In these animals the cup is placed over the intestinal surface and a bunch of villi are grasped by the adhesive organ. The holdfast is glandular (gl) and, although obviously important in securing the fluke to the intestinal wall, is perhaps of still greater significance as a feeding organ (see below).

In animals like *Gyrocotyle*, once regarded as the lowest cestodes but now more frequently regarded as monogeneans, the grasping apparatus takes the form of a sucker with elaborately fringed lips, the rosette (Fig. 34B, r), and some hooks (h) at

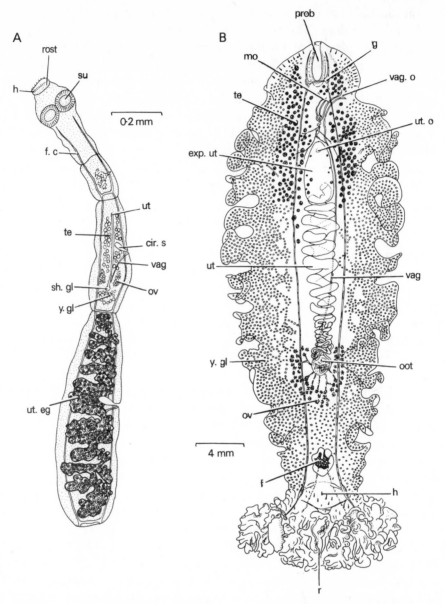

Fig. 34. A, the cestode *Echinococcus granulosus*. (After Southwell.) B, dorsal view of the mono-genean *Gyrocotyle fimbriata*. (After Lynch.) cir.s, cirrus sac; exp.ut, expansion of uterus; f, anterior dorsal opening of funnel, the posterior lips of which form the rosette, r; f.c, longi-tudinal canal of flame-cell system; g, ganglion; h, hook; mo, male opening, on ventral side; oot, ootype surrounded by receptaculum seminis; ov, ovary; prob, proboscis; r, rosette; rost, rostellum, armed with hooks; sh.gl, Mehlis's (shell) gland, round ootype; su, sucker on scolex; te, testis; ut, uterus; ut.eg, uterus full of developing eggs; ut.o, opening of uterus to exterior, on dorsal side; vag, vagina; vag. o, vaginal opening, on dorsal side; y.gl, yolk glands.

its posterior end. Though like the opisthaptor of monogeneans it is a new structure which has grown back as a funnel over the original opisthaptor of the larva which is the true homologue of the monogenean grasping apparatus.

In cestodes the attachment is by way of a head, or scolex, provided with sucking devices and often with hooks. Since this combination of structures suggests the opisthaptor of monogenean trematodes it is not surprising that there has been considerable argument as to whether the scolex is not really the posterior end of the cestode body; it is now known, on embryological evidence, to mark the true anterior end.

There are three different types of adhesive organ in cestodes. Bothridia, characteristic of Tetraphyllidea, are usually described as leaflike structures, sometimes stalked, with thin, often wavy edges; they are muscular and are able to curl the edges over folds of intestinal epithelium to obtain a grip. Bothria (Fig. 35A, C, both) are sucking grooves, characteristic of the groups Pseudophyllidea and Trypanorhyncha, and act in much the same way as bothridia. Neither bothridia nor bothria probably get as good a grip on the epithelium of the host as do the suckers or acetabula (Fig. 35B, acet) found on the scolex of the Cyclophyllidea, which are powerful muscular organs able to suck a knob of intestinal wall into their cavity and grip it tightly.

In addition to the sucking structures just described many cestodes have hooks made of keratin or some similar protein, by means of which further anchorage to the host is secured. The hooks often lie in relation to an apical part of the scolex known as the rostellum (Fig. 35B, rost) which can be protruded or withdrawn by protractor and retractor muscles lodged within the scolex. The hooks are also under muscular control and can be erected to penetrate host tissue or pulled close to the surface of the scolex. A special arrangement of hooks is met with in the order Trypanorhyncha (Fig. 35C) where they lie in rows on four long tentacle-like structures, known as proboscides (prob), which can be everted from, or retracted into, tubes known as proboscis sheaths (prob.sh). Each tube ends blindly (prob.b) and the blind end is surrounded by muscles. When the muscles contract fluid in the tube everts the proboscis; there are powerful retractor muscles to withdraw it. Unlike the other hooked structures of cestodes, which grasp the surface of the host to anchor the worm, the hooks of trypanorhynchoids are used as grapples to pull the worm over the surface of the spiral valve of elasmobranch fishes, which is their usual habitat.

Where the scolex is provided with a rostellum an intimate relationship is often established between the cestode and the topography of the intestinal wall of the host. Worms of the species *Echinobothrium brachysoma*, for example, are found attached to the spiral valve of the ray *Raia clavata* (Fig. 36A). The scolex has a central rostellum (rost) and dorsally and ventrally carries a bothridium (both) and a row of hooks (h). The scolex lies at the summit of a neck (neck) armed with eight rows of recurved teeth, each row with about fifteen teeth. From the wall of the neck two muscles

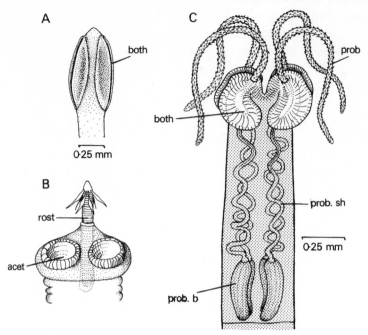

Fig. 35. Attachment devices of cestodes. A, *Dibothriocephalus*: lateral sucking grooves or bothria. B, *Hymenolepis*: hooked, evaginable rostellum and cup-shaped suckers or acetabula. C, *Gilquinia* (*Trypanorhyncha*): four leaf-shaped suckers or bothria and four retractile proboscides (with thorny spines) which can be withdrawn into sheaths by retractor muscles at the base of the scolex. acet, acetabulum; both, bothrium; prob, proboscis, carrying hooks; prob.b, proboscis bulb containing circular muscles to evert proboscis; prob.sh, proboscis sheath, with longitudinal retractor muscles; rost, rostellum.

(pro.mu) run, one dorsal, one ventral, to insert on the periphery of the rostellum; on contraction they elevate and partly evert it and, in so doing, broaden the scolex. Muscles run from the base of the teeth to the same area of the rostellar periphery and also, in the dorsoventral plane, from one set of teeth to the other, under the rostellum (lev.mu); their contraction elevates the teeth, a movement which is coordinated with protrusion of the rostellum. The rostellum is itself rich in muscle fibres, some of the more powerful of which run transversely; their contraction narrows the rostellum laterally, but lengthens its dorsoventral axis. In the gut of the ray the tapeworms are always found with the head embedded in a crypt of the intestinal wall and the strobila free in the lumen. When the rostellum is not protruded the scolex is narrow and elongated, a shape which can be increased and maintained by the contraction of transverse muscles in the head and circular muscles in the neck. In this attitude the teeth lie back along the apex of the scolex close to its surface, stiffening it so that it can be easily pushed into a crypt. Once in, protrusion of the rostellum, contraction of its muscles, erection of the teeth so that they penetrate the

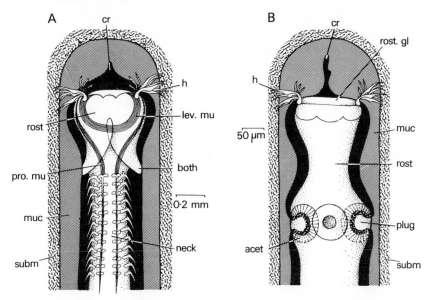

Fig. 36. A, *Echinobothrium brachysoma*, lateral view of scolex and neck embedded in crypt of wall of spiral valve of ray; B, *Echinococcus granulosus*, scolex embedded in crypt of Lieberkuhn of duodenal wall of dog; where hooks and rostellar glands occur the tapeworm is in contact with the submucous layer. (A, based on Rees, B, on Smyth.) acet, sucker; both, bothridium; cr, cavity of crypt; h, hooks; lev.mu, levator muscles of hooks; muc, mucosa of crypt wall; neck, neck of worm; plug, tissue of crypt wall gripped by sucker; pro.mu, protractor muscles of rostellum; rost, rostellum; rost.gl, rostellar glands; subm, submucous layer of crypt wall.

gut wall, all serve to anchor the worm firmly, so firmly, indeed, that it cannot be removed without tearing the wall at the point of attachment.

Among cyclophyllid (taenioid) tapeworms a protrusible rostellum armed with hooks is of regular occurrence. It is, like that of *Echinobothrium*, an important part of the means by which the tapeworm attaches to its host, being thrust deep into the crypts of Lieberkuhn, which run into the intestinal wall between the bases of the villi (Fig. 36B). Here erection of the hooks (h) causes them to penetrate the wall. Nearer the surface the acetabula (acet) engulf and contract over villi and between them these two attachments secure the worm firmly. Since there must be a certain relationship between the size of the rostellum and the crypt to allow insertion and then attachment, some tapeworms are incapable of a stable attachment in some hosts, a factor which may be responsible for defining the hosts which they can parasitize.

Feeding of parasitic platyhelminths

When sections of trematodes and cestodes were first examined with the light microscope the body seemed to be covered with a cuticle, and as this was much what was expected in creatures exposed to host fluids and needing protection against them this interpretation of the surface layer was readily accepted. The cuticle seemed to be secreted by what were described as insunk epidermal cells and the suggestion was made that the cells had retreated from the surface to avoid digestive or other fluids. More recent exploration with the electron microscope (Fig. 37) has shown that, far from being covered with a dead, protective cuticle, the surface of both trematodes and cestodes is a naked syncytial layer of living protoplasm, rich in mitochondria (mit) and pinocytotic vesicles (pin) and giving every indication of being involved in metabolic activity. Instead of the static contact between parasite and host provided by the dead cuticle there exists a living interface of interacting protoplasmic layers. The word "cuticle" is obviously a misnomer and has been replaced by "tegument". The layer just described, however, known as the matrix, is only part of the syncytium. It rests on a basement layer (bas.l) interrupted at frequent intervals and through the openings it extends to connect with a network of cells (peri.cyt), the insunk epidermis of early writers. This deep part contains nuclei (nu), Golgi apparatus (g.app), endoplasmic reticulum and secretory products (sec). In trematodes the matrix frequently encloses spines (sp) and in cestodes the surface is raised into a series of structures like microvilli (mic); each is capped with a hardened tip and is

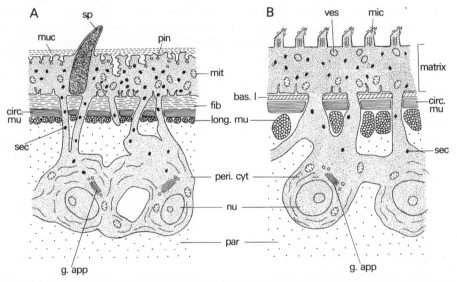

Fig. 37. A, Tegument of a trematode in diagrammatic vertical section, based on Threadgold and Erasmus. B, tegument of a cestode in diagrammatic vertical section, based on Béguin. bas.l, basement layer; circ.mu, circular muscles; fib, fibrillar layer; g.app, golgi apparatus; long. mu, longitudinal muscle; mic, microthrix; mit, mitochondrion; muc, mucopolysaccharide layer; nu, nucleus; par, matrix of parenchyma; peri.cyt, perinuclear cytoplasm; pin, pinocytotic vesicle; sec, secretory body; sp, spine; ves, vesicle.

known as a microthrix (plural microtriches or microtrichs). The structure of a microthrix is complex, as it seems to have a core of microtubules linked by septa and to
have a series of scale-like structures derived from the plasma membrane on its surface.
Because of their stiffness microtrichs may help the cestode to grip its host; because of
the cracks in the plasma membrane between the scales and their contribution to
increased surface area (calculated as a factor of x 3–6) they may contribute to the
absorption of food. The vesicles found in the tegumentary matrix may also be
involved in the uptake of food from the environment, or the excretion of material to
it, or in both. They may contribute to a layer of mucopolysaccharide (muc) lying
external to the plasma membrane which protects the worm against enzymes present
in the host. Between the matrix and the perinuclear cytoplasm are the basement
layer and the muscles (circ.mu, long.mu) of the body wall. The former is strengthened
by a layer of fibrils (fib) as in turbellarians, to which the muscle fibres attach. It is
this layer which imparts shape to the body and acts as skeleton.

The direct contact which the syncytial matrix establishes with the tissues of the
host makes an immune response by the latter to the presence of foreign protein
within its body more marked than had the matrix been cuticular. The problem of
immune reaction by the hosts of parasitic flatworms is complex, partly because each
species of worm, in the course of its life, parasitizes two or more hosts belonging to
different groups of the animal kingdom, partly because, however well the immune
responses of·vertebrates may be known, those of molluscs are still little understood.

A parasite may fail to infect a host for many reasons: it may be mechanically or
chemically incapable of effecting an entry; the metabolism of the host may fail to
supply some essential chemical requirement; it may be so surrounded by leucocytes
or other host cells as to be prevented from further development; or the hosts production of antibodies in reaction to its foreign protein may kill it. Since innumerable
parasites successfully enter innumerable hosts it must be assumed that in these
instances either the host's immune defences are inadequate or that the parasite can
overcome them or, better still, not arouse them. The last situation appears to apply
to the blood flukes, *Schistosoma*, adults of which have been shown to have a covering
of antigen identical with that of the vertebrates in which they live, so that the host
fails to recognize them as foreign bodies and treats them as part of itself. Whether the
antigen is derived from the host or made by the parasite is not known, but this is an
easy way to circumvent the defences of the host if it is within the parasite's biochemical power. Ability to match the production of this coat to the antigens of
many animals would permit invasion of all as potential hosts; inability to do so
would correspondingly reduce the range available. The maintenance of the anti-
genic cover seems essential to the continuance of the association: when it stops on
death of the parasite the host promptly surrounds the corpse with a capsule of
connective tissue. If it is disturbed – as by escape of cercariae – this may again lead
to destruction of the developmental stages within which the cercariae have grown

and which have survived in some sort of equilibrium with the host up till that moment.

The occurrence of a syncytial tegument with the perinuclear cytoplasm internal to the muscle layers of the body in both trematodes and cestodes would seem to imply true relationship between the two classes rather than convergence. It further supports a link between both and the turbellarians, in some at least of which (*Polycelis*) the epidermis seems to be formed by outward migration of cells belonging to the parenchyma. If this is so, the whole phylum appears as a genetically related grouping of animals characterized by skin of what might be called "mesodermal" origin and so unique in the animal kingdom. One word of caution is, however, necessary here. The tegument of the adult fluke is similar to that of all its developmental stages except the miracidium, which is covered by a series of large, ciliated, plate-shaped cells, relatively few in number, and with broad spaces separating each cell from its neighbours. These spaces, however, are plugged with ridges projecting from a deeper syncytial layer of cells, the perinuclear cytoplasm of which lies in the parenchyma within the muscle cells of the larva. When a miracidium achieves successful entry into a host the outer layer of ciliated cells is shed and the deeper syncytial layer forms the tegument of the sporocyst and all successive stages. The tegument is, therefore, a secondary skin and should not, perhaps, be compared directly with the ciliated epidermis of the turbellarians.

Trematodes have a gut, opening by a single aperture at the anterior end of the body, sometimes terminally, sometimes slightly ventrally. It leads through a sucking "pharynx" (Fig. 38A, ph) into the intestine (gut) which, in some monogeneans and aspidogastreans, is a simple tube, is most commonly two simple tubes, but, in most monogeneans and such digeneans as the liver flukes, branches and anastomoses so that food is taken up over almost all the body and so compensates for the absence of any circulatory system. In some of the simpler monopisthocotylean trematodes such as *Entobdella soleae*, which attaches to and creeps over the under surface of soles, the initial part of the gut consists of a protrusible proboscis with a ring of glands opening within the aperture. This is applied to the skin of the fish, the secretion of the glands discharged, a digestion of the epidermal cells occurs and the products sucked into the gut and pumped into all the intestinal branches. This appears to be the general method by which monopisthocotyleans feed. In polyopisthocotyleans, however, the attack on the host is deeper, and, although some epithelial tissue is taken, most of the food is blood. The digenetic trematodes are similar, though ingestion of cells is often replaced or supplemented by uptake of gut contents or blood, depending on where the parasite lives. Digestion appears to be predominantly extracellular.

Cestodes are peculiar in having neither mouth nor alimentary tract: all absorption must therefore be by way of the tegument and would, on this account, seem to be limited to what can enter by diffusion or by way of pinocytotic vesicles, though the presence of such structures in cestodes has not yet been put wholly beyond doubt.

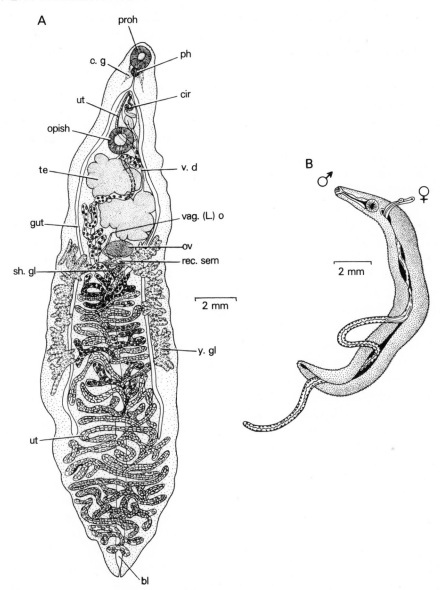

Fig. 38. A, ventral view of the digenetic trematode *Dicrocoelium dendriticum*. (After Neuhaus.) B, *Schistosoma haematobium*: male and female, the latter lying in the ventral, gynecophoric groove of the male. (After Loos.) bl, bladder of flame cell system; c.g, cerebral ganglion; cir, cirrus in cirrus sac; gut, one of the two main gut branches; opish, opisthaptor; ov, ovary; ph, sucking "pharynx"; proh, prohaptor; rec.sem, receptaculum seminis; sh.gl, Mehlis's (shell) gland, lying round ootype; te, testis; ut, uterus; vag.(L.)o, opening of Laurer's canal, used as vagina; v.d, male duct; y.g, yolk gland.

Since the body of trematodes is covered by a similar tegument, though without microtrichs, it is likely that uptake of material by this method can occur in flukes too, and pinocytotic vesicles occur abundantly. There is good evidence that the adhesive organ of strigeoid flukes is specially involved in this. Over the organ the tegument bears many microvilli, absent elsewhere; these seem to lead the secretion of numerous glands to the exterior, and at the point to which the organ is applied the host tissue is eroded so that the tegument is practically in contact with the blood vessels of the intestinal wall. In this way, it is suggested, there has come into being a structure which is functionally the equivalent of a placenta, and the fluke is able to obtain nourishment directly, almost as if it were an organ belonging to its host.

It has long been taken for granted that in a biotope such as the vertebrate intestine digested food would enter a tapeworm as readily as the intestinal villi. Though this represents an over-simple approach even to the problem of how the food enters the host, there is evidence from the use of radioactive foods that absorption from the lumen of the intestine by the parasite does indeed occur. It is not perhaps the most important route by which material can enter the body of at least some tapeworms, and the scolex as well as the strobila seems to be involved in feeding. It has been shown, for example, that the infective stage of *Echinococcus granulosus*, which normally lives in the intestine of the dog when adult, can be grown *in vitro*. The head of a worm so cultivated develops towards the production of proglottids only if two demands are satisfied: (1) that the head is in contact with a solid substratum; (2) that this is of protein. If either of these conditions is not met then the head develops a cyst-like structure rather than a strobila. The conclusion drawn is that the head can obtain nourishment from the substratum, perhaps by pinocytotic uptake, but conceivably by the secretion of digestive enzymes and absorption of the resulting products through its microtrichs. Even when fully established in a dog the scolex still seems to take up food; it is thrust into a crypt of Lieberkuhn so that the apex and sides of the rostellum make contact with its walls and it is anchored by the insertion of the rostellar hooks into the epithelium and by the suckers holding villi in their cavities. The tip of the rostellum carries gland cells which digest the host epithelium so that the tissue of the scolex is in contact with the vascularized submucosal tissues of the intestinal wall with which it can establish a placental relationship like that of the strigeoid flukes. This also happens in the tapeworm *Hymenolepis*, species of which live in man and rodents. How widespread the ability may be among cestodes is unknown; its apparent limitation may reflect nothing more than restricted investigation and it may well prove to be general.

A second factor complicating the original simple view that cestodes robbed their host of digested food by taking it from the cavity of the intestine has been the unexpected discovery that tapeworms (at least *Echinococcus granulosus* and *Hymenolepis diminuta*, but probably others as well) make use of the carbon dioxide dissolved in considerable quantities in the intestinal fluids and arising from the high metabolic activity of its epithelial cells. With this ability the utilization of glucose can be increased sevenfold and a variety of substances which act as intermediates in the Krebs tricarboxylic acid cycle can be synthetized, as well as polysaccharides and proteins.

In platyhelminths there is no vascular system. Transport of food and oxygen must therefore be by diffusion and the branching nature of the gut and the leaf-like shape of the body are both probable adaptations to minimize the difficulty of getting these substances to all parts of the body. Amongst parasitic forms the supply of oxygen may become difficult, especially in those inhabiting the gut where oxygen tension may fall to low levels because of the rate at which it is used by the bacterial flora. In such circumstances the parasite must become anaerobic and it is probable that all cestodes are forced to obtain energy in this way.

Reproduction and life history of parasitic platyhelminths

The reproductive system of the parasitic flatworms functions on a basis similar to that found in the advanced turbellarians. Like them the animals are hermaphrodite and most seem to be able to fertilize their own eggs with their own sperm though it is usually said, and there is evidence from the use of labelled sperm to support the statement that cross fertilization is preferred. The reproductive system (Figs 38A and 39) makes provision for the manufacture of eggs in an ovary (ov) usually a rather solid body; for the production of sperm in scattered testes (te); for the production of yolk in scattered vitellaria (y.gl) which also produce the material for making and tanning the egg shell; for the storage of the shelled and fertilized eggs in a uterus (ut). The male duct connects with a penis or cirrus (cir) and there is a vagina (vag) leading to a receptaculum (rec.sem) for storage of allo-sperm, though in some species the uterus may be used for copulation even when a vagina is present. In trematodes the male ducts and uterus open together to a genital atrium placed on the anterior part of the ventral surface, the vagina opening separately; in digenetic trematodes, where it is called Laurer's canal, its opening is remote (vag.(L)o). All the structures on the female side meet near a central point known as the ootype, the functioning of which is the key to the successful operation of the whole system. In cestodes only the Pseudophyllidea have a separate uterine opening; in all the others the uterus is blind and eggs escape only on disintegration of the tissues of the proglottid.

The general plan of the ducts in relation to the ootype of a trematode is shown in Fig. 40 but the same diagram could apply to a cestode (Fig. 39A(b)). The space known as the ootype (oot) receives (1) eggs (ov) from the oviduct (ovd); (2) yolk and material for making and tanning egg shells (y.cell) by way of the duct from the vitellaria (y.d) and yolk reservoir (y.res); (3) the secretion of Mehlis's glands (sh.gl, so called "shell" glands) which seems to be involved in some unspecified way with

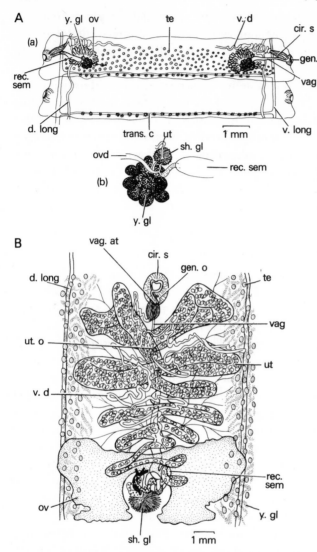

Fig. 39. A, *Moniezia*: (a), two proglottids; (b), details of central part of female system. B, *Dibothriocephalus latus*: reproductive system. (After Sommer and Landois.) cir.s, cirrus sac; d.long, dorsal longitudinal canal of flame cell system; gen.o, opening of genital atrium to which male and vaginal ducts connect; ov, ovary; ovd, oviduct; rec.sem, receptaculum seminis, expanded part of vagina; sh.gl, Mehlis's (shell) gland lying round ootype; te, testis; trans.c, transverse canal of flame-cell system; ut, uterus; ut.o, uterine pore on surface of proglottid; vag, vagina; vag.at, opening of vagina to genital atrium; v.d, male duct; v.long, ventral longitudinal canal of flame-cell system; y.gl, yolk gland.

the making of the egg shell. After these materials have been isolated in the ootype by contraction of sphincter muscles they are packeted to give rise to an egg supplied with yolk and enclosed in a protective, though not yet tanned shell. This is then passed to the uterus (ut) for fertilization and storage and the process repeated. It has been shown that in monogeneans the time taken to deal with one egg is between 15 and 60 min, whereas in the digenean *Echinostoma nudicaudatum* it needs only about 1 min, probably the minimum time in which such an elaborate process could be carried out. Final tanning of the shell occurs as the egg lies in the uterus, and the early development of the zygote takes place there too. In pseudophyllid, tetra-

Fig. 40. Diagram of ootype and surrounding structures of a trematode (based on *Fasciola*), to show how they function. Yolk cells (y.cell, distinguished by black nucleus) with granules are passed from the yolk reservoir (y.res) to meet an ovum (ov, distinguished by white nucleus) coming along the oviduct (ovd) from the ovary. In the ootype (oot) they are grouped and surrounded by secretion from Mehlis's gland (sh.gl) which also lines the wall of the ootype and neighbouring uterus (ut). From the ootype the group of cells enters the uterus, the ovum is fertilized there and the granules escape from the yolk cells to form the egg shell (sh) which will later be tanned. y.d, yolk duct; L.can, Laurer's canal from neighbouring skin used as a vagina in some flukes, in which case the ovum is fertilized before reaching the ootype.

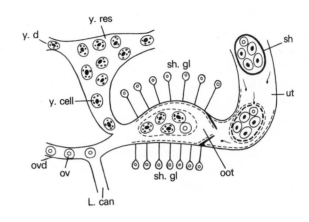

phyllid and trypanorhynchid tapeworms a thick shell of sclerotin is produced round the fertilized egg as in trematodes and the material from which it comes is enclosed in several yolk cells which are associated with each egg in the ootype. It is known as a capsule. In cyclophyllid tapeworms the sclerotized capsule may be very thin, as in *Dipylidium*, or absent, as in *Taenia*, and the number of vitelline cells associated with each ovum is correspondingly reduced to one only. The worm is not in a position to dispense with the protection given by the capsule: the hazards of exposure are as great as ever; protection is now, however, provided by a coat of keratin, known as the embryophore (Fig. 30C, eph), which is produced by the embryo itself.

If the hazards of the life history demand a greater rate of egg production than can be achieved by one ootype in each proglottid of a tapeworm, then the only way in which it can be done is to increase the number of ootypes and their accompanying apparatus. This can lead to the double set of reproductive structures seen in each proglottid of *Moniezia* (Fig. 39A) and *Dipylidium*.

The monogenean life history involves a free-swimming ciliated larva with eyes and posterior grasping apparatus, the oncomiracidium, which seeks entry to a new host within which it metamorphoses to the adult. There is no larval multiplicative stage. The free-swimming larva confines monogeneans to aquatic hosts like amphibians and fishes. The best known genus, *Polystoma*, is found in the bladder of frogs. The larva, after a brief free existence, enters the gill chamber of a tadpole through the spiracle and settles on the gills; by the time the tadpole metamorphoses and loses its gills the monogenean has also metamorphosed and migrated to the bladder. This migration was at first assumed to take place along the alimentary canal, but has been shown to involve movement over the ventral surface of the amphibian and the bladder is entered from the cloaca. Some larvae, which hatch early in the year, may become sexually mature whilst still attached to the gills of the tadpole – one of the earliest discovered examples of neoteny, the development of sexual maturity whilst the rest of the body still has a larval or juvenile form. These worms may then reproduce on the gills and die before the tadpole has metamor-phosed; most adult flukes seem to develop from such neotenously produced eggs and larvae which settle on tadpoles later in the year. Monogenean trematodes often occur in numbers, and, as they are mobile, cross-fertilization is the normal pattern of reproductive behaviour, the penis of each animal being thrust into the vagina of its partner so that mutual insemination takes place. In the genus *Diplozoon* two copulating animals become fused in this position. In having cross-fertilization, in their mobility, and in their retention of sense organs such as eyes, monogeneans still exhibit some characteristics of their free-living ancestors.

In digeneans the life history has become more complex in that a second host has been introduced; this increases the hazards of transmission, and, to compensate for the losses so caused, a series of larval reproductive phases has been introduced, so that for each larva which successfully reaches a second host a large number are produced to attempt the return journey to the first. There is still a free-swimming ciliated larval stage involved in the transfer of the fluke from the vertebrate to the molluscan host (Fig. 30E) and there is often a total inability to withstand desiccation at each stage – even when shelled or encysted – so that the choice of hosts and habitats is not unrestricted.

Three life histories may be outlined to indicate the degree of variation en-countered. The first is the well-known life history of the common liver fluke of sheep and cattle, *Fasciola hepatica*. The adult lives in the liver and its channels, feeding on liver tissue and blood. Eggs escape with the faeces and from them there ultimately escape ciliated larvae, each called a miracidium. These swim through the water, and are said to detect their next host, the pulmonate mollusc *Lymnaea truncatula*, chemically, though there seems to be a great element of chance in the event. The larva enters the snail partly mechanically, partly by digesting the epithelium by the secretion of glands at the anterior tip. On entry the miracidium sheds its coat of ciliated cells and is transformed into a form known as a sporocyst, which lives in the digestive gland of the snail. Cells within the sporocyst develop into a third stage known as a redia, from internal cells of which further rediae may form; in this way a heavy infestation is built up in the gastropod host. Finally, the rediae give rise to a fourth stage, known as a cercaria, with a heart-shaped body and a tail. Cercariae escape from the snail, swim to near-by vegetation, where they lose their tail and encyst; they are then known as metacercariae. The mammalian host is infected by eating this contaminated food. The trematode has a wide tolerance of hosts and can survive ingestion by many different mammals. The cysts are opened by the digestive processes proceeding in the duodenum, through the walls of which the young fluke bores to reach the liver.

The second type of life history is exemplified by *Clonorchis sinensis* found in the liver of man, especially amongst fish-eating populations of eastern Asia. The eggs are passed with the faeces and contain, as in *Fasciola*, a miracidium larva. This does not hatch, however, unless and until it is taken into the gut of an aquatic snail, most commonly a species of the genus *Parafossarulus*, when it penetrates the gut wall and transforms into a sporocyst in the digestive gland. As in *Fasciola*, sporocyst, rediae and cercariae follow one another, the last escaping into the water in which the snail lives. They behave differently from those of *Fasciola*, however, in that they do not encyst on vegetation, but seek and enter a freshwater fish. Many different species can be used, mostly members of the carp family. The cercariae encyst in the tissues of the fish and if this is ingested by a fish-eating mammal the cysts open, the young flukes escape and make their way to the liver.

The flukes which are perhaps the most serious trematode parasites of man are the blood flukes of the genus *Schistosoma*, and these may be taken as an example of a third type of life-cycle. They are found throughout the tropics and in some localities it seems that as much as 90 per cent of the population may be affected. Flukes of this genus are unique in that the sexes are separate (Fig. 38B), males being larger and having bodies which are C-shaped in section, the narrower, longer and cylindrical female being partly held in the groove on the male's body. They live in veins, either in those draining the intestine (*S. mansoni* and *S. japonicum*) or the bladder (*S. haematobium*). When ready to lay eggs the paired animals move towards gut or bladder, passing into smaller and smaller vessels until stopped by their diminishing diameter. Here the eggs are laid and are held in the walls by spines on their shells. Miracidia develop within the eggs and enzymes seem to diffuse from them through the shell so as partially to digest the neighbouring tissue and allow pressure (apparently exerted by the local blood vessels) to move them towards the lumen of the gut or bladder. Most eggs never reach this, but a minority do and escape in faeces or urine. The eggs hatch in water and the miracidium swims in search of the appropriate gastropod host, which is the planorbid pulmonate *Biomphalaria* in the case of *Schistosoma mansoni*, the pulmonate *Bulinus* for *S. haematobium* and the prosobranch *Oncomelania* for *S. japonicum*. If successful in its search the miracidium bores into the snail, shedding its ciliated skin to transform into a sporocyst. After a while this starts to produce cercariae and other sporocysts which, in turn, give rise to further cercariae. These ultimately escape and are capable of entering the body of the next host directly through the skin; there is no encysted metacercarial stage. The cercariae appear to be able to digest a hole in skin through which, after shedding the tail, they make their way to a blood vessel whence they are transported to gut or bladder.

The cestodes, like the digenetic trematodes have also a life history involving two hosts, though it is occasionally possible, as in the well-known dwarf tapeworm, *Hymenolepis nana*, for the cycle to be repeated in only one. The two animals involved are always ecologically related, so that one host is liable to acquire eggs or embryos from the second, and the second will normally also prey on the first. Tapeworms are more apt to be solitary parasites than trematodes and their eggs seem to be normally self-fertilized. As proglottids are formed they first develop male organs, then female, both sets finally degenerating leaving the proglottid packed with masses of fertilized eggs. It then separates from the rest of the worm and escapes in the faeces. A proglottid in the male phase can thus fertilize one in the female. The reproductive system (Fig. 39A, B) is built on the same general plan as that of the trematode and seems to operate in the same way. After fertilization the eggs are passed in a shell into the uterus. There they develop into embryos, known as oncospheres, surrounded by a variety of protective coats, which often allow them to survive extreme environmental conditions (Fig. 30C). The embryos are usually ingested by the second host whilst it is feeding and then hatch within its gut. The oncospheres contain three pairs of movable hooks (h) and, after hatching, are able to scratch their way through the gut wall and make their way to the body cavity, or the muscles, or various other parts of the host's body, where they transform into one of a variety of second larval forms which will normally rest passively until ingested by the first host in the course of its predation of the second.

The second larva may assume one of several forms, depending on the species. The most commonly studied tapeworm, *Taenia solium*, the pork tapeworm, so called because the second host is the pig, has a bladder-like second larval stage, the bladder being filled with fluid and its wall carrying a single scolex, turned inside out and projecting into the fluid. This is known as a bladder worm or cysticercus. Some worms have a similar stage with the bladder filled with tissue, not fluid; this is called a cysticercoid. Sometimes both of these larval types may produce more than one scolex and may bud off parts which can be carried round the body to settle in other sites so that a larval reproductive phase is incorporated into the life history. This type of larva (Fig. 30D) is known as a coenurus or hydatid; its effect on the host may be very serious especially if it settles in a part such as the brain.

The larva may also be wormlike, when it is called a plerocercoid. In all cases after ingestion by the first host the larval tissues are digested away leaving only one or more scolices which evaginate and attach to the gut wall and proceed to bud off proglottids. Evagination of the scolex often requires exposure first to the acid of the vertebrate stomach and then to an alkaline duodenal fluid if not specifically bile. This brings the scolex into the right form for attachment to the correct part of the wall of the intestine.

Classification of platyhelminths

Platyhelminthes
  Turbellaria
    Polycladida
    Triciadida
    [Protricladida]
    Eulecithophora (= in part, Rhabdocoelida)
    [Perilecithophora]
    Archoophora
      Acoela (plus others)
  [Temnocephala]
  Trematoda
    Monogenea
      Monopisthocotylea
      Polyopisthocotylea
    Aspidogastraea
    Digenea
      Strigeatoidea
      Echinostomida
      Renicolida
      Plagiorchida
      Opisthorchida
  Cestodaria (? part of Monogenea)
  Cestoda
    [Haplobothrioidea]
    Tetraphyllidea
    Trypanorhynchida (= Tetrarhynchoidea)
    Pseudophyllidea
    [Diphyllidea]
    Cyclophyllidea (= Taenioidea)
    [Ichthyotaeniidea]
    [Nippotaeniidea]
    [Tetrabothridea]
    [Aporidea]

7
Nemertines

THE NEMERTINE OR PROBOSCIS worms are comparable in their general organization to platyhelminths, but have advanced in a number of ways which allow a higher level of metabolic activity. The most obvious of these is the presence of an anus at the posterior end of the body, which permits a passage of food in one direction along the alimentary canal. A vascular system serves for the movement of blood about the body, presumably carrying dissolved food and oxygen. Though there is no heart to drive the blood along the vessels some of these are contractile; its movement follows the usual invertebrate pattern of forward in a median dorsal vessel (Fig. 41A, d.v) and backwards in two ventrolateral ones (vlat.v). Reversal of flow occurs as in tunicates. These two factors together give the animal the potential to lead a more active life. This is translated into actuality by a more highly organized and relatively more voluminous muscular system, with thick sheets of circular and longitudinal muscles lying one within the other and an additional external layer of longitudinal muscles in some species, giving a muscular body wall comparable to that of an earthworm (see Fig. 80). Nemertines have no body cavity, however, and between muscles and gut lies a spongy parenchyma (par) as in flatworms.

The body of the nemertine has the shape described as wormlike, that is, it is long by comparison with its breadth and more or less rounded in cross section, a shape adapted for creeping over surfaces, into crevices and for burrowing. It moves by a combination of muscular and ciliary movement over a mucous trail as does a turbellarian, but the part played by the ciliary component is predominant, especially in the small sorts. The muscle is most obviously used in an avoiding reaction when the anterior end recoils forcibly and the body shortens or rolls up in a spiral. The skin is soft and glandular, many of the glands being sunk under the epidermis as they are in flatworms. The epidermis still acts as the sole respiratory surface, and there is usually a layer of mixed connective tissue and glands known as a dermis or cutis (cutis). Rhabdites occur in the skin of some nemertines as they do in platyhelminths.

The head of the animal has increased in importance in relation to the more pronounced locomotor activity. In addition to housing the brain and the major sense organs – eyes (Fig. 41B, e) and ciliated grooves containing chemoreceptors (sl) – it is now the part of the body on which the mouth (m) lies. This may allow food to be dealt with under the direct control of major sense organs, but is probably due to the fact that nemertines, which are aggressive carnivores, capture their food by means of a special device known as the proboscis which is accommodated in the head and anterior part of the body. Because of its possession the group is often called the Rhynchocoela.

The nemertine proboscis possibly evolved as an exaggeration of a structure such as the frontal gland of some turbellarian worms, which is a papilla containing numerous adhesive glands lodged in a pocket of the head, and which can be extruded to help immobilize prey during ingestion. In nemertines the proboscis is a hollow, tubular structure with glandular (prob.ep) and muscular (prob.long.mu, prob.c.mu) walls, formed as an ingrowth of the body wall. It lies, when not in use, in a closed pouch placed dorsal to the alimentary canal and encircled anteriorly by the nervous system. This sac is the "proboscis coelom" or rhynchocoel (rcoel), filled with fluid which bathes the proboscis and provided with a wall rich in circular muscles (c.mu.prob.sh), the proboscis sheath. It seems to arise as a split in the muscular part of the invagination from which the proboscis is formed and is certainly not coelomic. The proboscis is blind at its inner end but opens anteriorly to the outside, either by way of an independent proboscis pore at the front end of the body, or by a pore confluent with the mouth; here its wall is continuous with the body wall. The proboscis is attached to the proboscis sheath only by a strand of muscle across the rhynchocoel (Fig. 41B2, ret.prob.mu) at its inner end; it is usually divided by a constriction into an inner and outer segment and in some nemertines (hoplonemertines) a stylet is developed in the wall at this point. The whole proboscis is very muscular and forms an exploratory organ and a device for snaring prey and bringing it to the mouth for ingestion. Its use has been carefully described in several species, of which the common *Lineus ruber*, an animal which can be collected from under stones between tidemarks, may be selected. When covered by water it emerges to hunt for food, mainly the small marine oligochaete *Clitellio arenarius*, though other small annelids and crustaceans may be captured, and moribund, recently dead animals or even detritus may also be eaten. Intact, moving prey is detected by sight, but moribund or dead animals chemically, perhaps by sense organs in the ciliated cephalic grooves. The proboscis is then everted in an astonishingly rapid movement and wraps itself in a tight spiral round the prey which is then dragged back towards the mouth by shortening of the proboscis and ingested. If the prey is small the whole process is over in 15-20 s, but it may take up to 30 min if it is large. To obtain purchase on the substratum both for the discharge of the proboscis and for pulling the prey towards the mouth the nemertine secretes a sticky mucus which anchors it securely to the ground.

Eversion of the proboscis is brought about by contraction of the circular muscles of the proboscis sheath, aided, probably, by contraction of the circular muscles of the body wall. This raises the pressure in the rhynchocoel until it is relieved by a sudden escape of the proboscis through the proboscis pore, turning inside out as it does so. Not the entire proboscis is everted, only the section lying anterior to the constriction, which therefore comes to lie at the apex of the extruded part. If a stylet is present this has the effect of bringing it to the tip of the proboscis when it is discharged and it may, so forcible is the attack, pierce the body wall of the animal at which it has been hurled.

At first sight it might seem that a continuous contraction of body wall muscles would be necessary to keep the proboscis everted since their relaxation would lower the pressure within the rhynchocoel and perhaps suck the proboscis back. Observation of nemertines with the proboscis out shows that these muscles are not contracted and that the body of the worm is soft, not stiff as it would be if they were. Some nemertines, too, use the proboscis as an aid in burrowing which involves alternating contraction and relaxation of the body wall muscles without the state of the proboscis being affected. This state of affairs appears to be due to the presence of a structure called the rhynchocoel villus which is a blood vessel running in the mid-ventral line of the rhynchocoel. When the proboscis is not everted it is collapsed, contains little

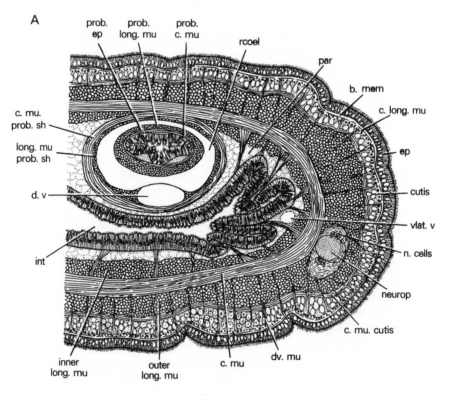

Fig. 41

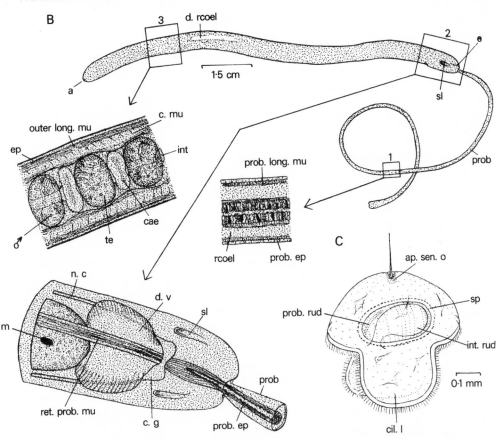

Fig. 41. A, Diagrammatic transverse section through the body of a nemertine worm, based on *Lineus longissimus*. B, *Lineus gesserensis*, with proboscis everted; areas 1, 2, 3 are drawn at higher magnification. C, pilidium larva. a, anus; ap.sen.o, apical sense organ; b.mem, basement membrance in which lies the system of fibrils controlling the worm's shape; cae, caecum of intestine; c.g., cerebral ganglion; cil.l, ciliated lobe; c.long.mu, layer of circular and longitudinal muscle under basement membrane; c.mu, circular muscle of body wall; c.mu.cutis, layer of circular muscle at base of cutis; c.mu.prob.sh, circular muscle of proboscis sheath; d.rcoel, distal end of rhynchocoel; d.v., dorsal vessel (rhynchocoel villus); dv.mu, dorsoventral muscle strands; e, eye; ep, ciliated epidermis with gland cells; inner long.mu, inner layer of longitudinal muscle; int,intestine; int.rud,rudiment of intestine; long. mu.prob.sh, longitudinal muscle of proboscis sheath; m, mouth; n.c, lateral nerve cord; n.cells, nerve cells; neurop, neuropile; outer long.mu, outer layer of longitudinal muscle (peculiar to the group of nemertines to which *Lineus* belongs); par, parenchyma; prob, proboscis; prob.c.mu, circular muscles of proboscis; prob.ep, epithelium lining proboscis; prob.long.mu, longitudinal muscles of proboscis; prob.rud, rudiment of proboscis; rcoel, rhynchocoel; ret.prob.mu, retractor muscle of proboscis; sl, head slit; sp, space (amniotic cavity) around rudiment of worm; te, testis; vlat.v, ventrolateral vessel; ♂, male pore.

blood and its walls look rather solid; on eversion of the proboscis, however, it be-comes distended with blood so that it balloons into the rhynchocoel keeping the volume of that, and so its pressure, more or less equivalent to what it was prior to eversion. In this condition its walls are stretched and become extremely thin (Fig. 41B2, d.v). Retraction of the proboscis is effected by a retractor muscle which runs from the tip of the everted proboscis to an origin on the innermost part of the rhynchocoel: its change in length between relaxation and contraction is extra-ordinarily great.

Digestion of small prey is very rapid. The animal dies almost at once and break up begins soon as the result of the secretion of enzymes from one of the two types of cell found in the intestine. A second type of intestinal cell appears to ingest granules of partially digested food and finish the process intracellularly, passing the products of digestion into the blood. The intestine possesses numerous short lateral outgrowths (Fig. 41B3, cae) which give a kind of pseudometameric appearance to the animal, especially as the gonads (te) alternate with them. These caeca are no different in their structure or function from the main part of the intestine. Some nemertines in-gest prey almost as large as themselves; when this happens the prey takes much longer to succumb to the digestive juices of the worm and may, indeed, pass through the entire gut and emerge undamaged from the anus, or break through the body wall in the course of its struggles.

Most nemertines are marine and are found, when the tide is out, sheltering under stones, or in cracks and crevices of rocks or burrowing in sand. Some, like *Tubulanus*, protect themselves by secreting a mucous tube. Their need for osmoregulation is therefore slight and is satisfactorily met by the system of flame cells which lies in the parenchyma, as in platyhelminths. As in them, it may also serve as a route for the excretion of nitrogenous waste. Because nemertines have a vascular system, however, the flame cells need not be so scattered as they are in a platyhelminth and they usually lie in a moderately compact grouping related to the lateral vessels in the neigh-bourhood of the oesophagus, forming the first approach to an organized kidney that we have encountered. A very small number of nemertines have become adapted to fresh water and the species of the genus *Geonemertes* live on land in damp areas in the warmer parts of the globe: the only British species, *G. dendyi*, is restricted to S.W. England, W. Wales and S.W. Ireland. For a terrestrial animal it shows a remark-ably persistence of habits appropriate for an aquatic mode of life. Thus it moves by means of cilia and as these function only in liquid the worm has to provide this in the form of a sheath of mucus secreted from the general surface, especially the head, through which it may be said to "swim". Whilst getting the animal over the difficulty of locomotion this generous secretion involves it in another, since much water is poured out of the body in the formation of the mucous tube. This is exaggerated by the unexpected use of the proboscis as part of an avoiding reaction: when the animal is strongly stimulated the proboscis is suddenly everted by a violent contraction of the circular muscles and shot forwards to a distance 2–3 times the length of the worm. Its tip, which is glandular, adheres to the substratum and acts as a kind of origin for the longitudinal muscle of both body and proboscis, which contract vigorously so that the nemertine is rapidly pulled forwards a considerable distance from the point of stimulation. If a worm be made to do this several times in rapid succession it can be seen to get smaller from loss of water. This lavish use of water has perhaps had several consequences, firstly in confining the animals to very humid habitats and in producing the almost cylindrical body form which minimizes surface for water loss. The flame cell system may also be involved, though its most outstanding characteristics appear to increase the elimination of excess water which (at least at times) must be picked up from the environment. The predominant water-saving device when desiccation really is imminent is the ability of the mucous sheath within which the worm moves to dry and harden to a cocoon.

Nemertines are gonochoristic, that is the sexes are separate. The gonads (te), as in flatworms, are embedded in the parenchyma and discharge directly to the outside; they are often numerous and there are no complexities of structure or accessory glands. Gametes are normally passed direct to the exterior and fertilization is external, giving rise to a free-swimming larva. Ocasionally, as in some species of *Lineus*, egg strings are laid in mucus. The development of the egg is often direct but the better-known examples of nemertine development lead to the production of a planktonic larva which is totally unlike the adult. The most famous nemertine larva is known as the pilidium (Fig. 41C), shaped like a helmet with ear-flaps. The edges of the helmet and the flaps are ciliated (cil.l) and there is an apical tuft of sensory cilia (ap.sen.o). The bulk of this is a device for keeping the developing worm afloat whilst in its central parts, around the larval gut, there is built up the rudiment of the adult animal. A drastic metamorphosis occurs, of the type known as cataclysmic. This is preceded by the development of a number of invaginations which grow inwards from the surface of the pilidium and spread round the developing worm; when these ultimately all meet and fuse (sp) the adult body is isolated from the rest of the pilidium, drops out and takes to a benthic life, leaving the discarded remains of the larva to degenerate. In this way the young animal is able to carry out simultaneously the two functions of a larval distributive phase, which are often contradictory in their needs – that of spreading the species and that of growing up.

8

Nematodes

NEMATODES are the animals popularly known as round or eel-worms, both names insisting on a shape which turns out, on closer examination, to be linked with a fundamental feature of their organization, which marks them off from most other groups in the animal kingdom. They are, on the whole, unknown to most people since nearly all are minute. Nematodes are dependent on water as an environment, even when living in soil, and they frequent places where decaying organic matter abounds. This habit has led many to adopt a parasitic mode of life and they differ from most groups of parasitic animals, and are rivalled only by the insects, in having become parasitic, apparently with equal ease, on both animals and plants. Frequently, their presence is suspected only because of their effects on the well-being of the host. Despite this unobtrusiveness, nematodes are in fact amongst the most abundant of all animals and in respect of both species and individuals are found in vast numbers in almost every kind of habitat; it has been calculated that they may number 1.25×10^6 in the top 7–8 cm of each square metre of beach sand. As a consequence of their parasitic habit the group has often, like protozoans, been studied with an emphasis on medical or agricultural importance rather than as one which offers an opportunity to see how function is correlated with a distinctly unusual form. Most of what follows is based on the genus *Ascaris*, a parasitic nematode of unusually large size, but the general organization of the body is similar in all – as expressed by Harris and Crofton, nematodes appear to the student to come in one model, in different sizes, but with endless variations in life history.

The shape of the nematode body is fusiform, i.e. more or less cylindrical and tapering to each end, and it is almost precisely circular in cross-section at all levels (see Fig. 42). Since the body is contained within a thick cuticle (cut) the circular shape could be due either to the rigidity of that structure or to a high internal pressure expanding it evenly in all directions. It is in fact the latter which is responsible for the shape, and in *Ascaris* the pressure has been shown to average about 70 mm of mercury, rising to 125 mm in some individuals. This internal pressure is exerted by the fluid which fills the space (pseudocoelom) between the body wall and the viscera

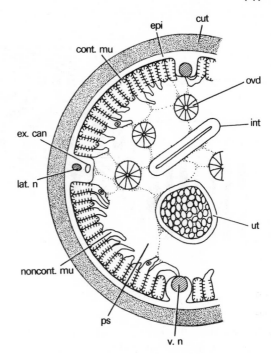

Fig. 42. Diagrammatic transverse section of half the body of a nematode. cont.mu, contractile part of longitudinal muscle cell; cut, cuticle; epi, epidermis; ex.can, excretory canal in lateral chord; int, intestine; lat.n, lateral nerve in lateral chord; noncont.mu, non-contractile part of longitudinal muscle cell; ovd, oviduct; ps, pseudocoelomic cavity; ut, uterus; v.n, ventral nerve in ventral chord.

(ps) comparable to that occupied by the parenchyma of platyhelminths. The development of some kind of space in which fluid accumulates and which acts as a hydrostatic skeleton has already been seen in coelenterates and platyhelminths and will be encountered again in coelomates. The nematode pseudocoelom is another such cavity, functionally comparable to, but morphologically distinct from, a parenchyma in being practically devoid of cells, which are limited – certainly in some species and possibly at some stages in the life history in all – to a few giant cells (four in *Ascaris*) drawn out into strands and membranes applied to the surface of the body wall and viscera, and distinct from a coelom in having no epithelial lining.

The fluid in the pseudocoelom is under pressure, partly because it is hypotonic to the tissues of the body and partly because it is compressed by a basketwork of fibres in the cuticle and by the tonic contraction of the musculature of the body wall, which is unique in the animal kingdom in being composed solely of longitudinal fibres. Cuticular structure is of great importance in the functional anatomy of nematodes (see Fig. 43). The cuticle comprises many layers and has been shown to be predominantly protein in nature and made of a mixture of collagens, albumins and a keratin-like substance, with a triple series of fibrils built into it. On the surface is a lipid layer (1) only about 100 nm thick, contributing some impermeability. The next outermost layer (2) seems to have undergone some tanning. The fibrils, which form the innermost cuticular layers (3,4,5), lie in sheets wrapped diagonally round

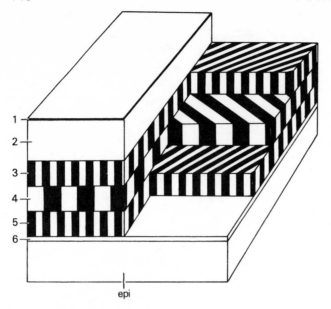

Fig. 43. Diagrammatic block of cuticle and epidermis of a nematode. epi, epidermis; 1–6, layers of cuticle; 1, lipid layer; 2, structureless layer; 3–5 triple fibrous layers; 6, basal layer resting on epidermis.

the worm and crossing (in *Ascaris*) at an angle (θ) of about 140–150°, and are functionally comparable to those already described in flatworms and nemertines (p. 103). Contraction of the longitudinal muscles, the only ones the worm possesses, which produces an increase in the angle θ, tends to reduce the volume of the animal; as this is not normally possible the result must be an increase in the internal pressure and it is this which acts antagonistically to the longitudinal muscles, replacing functionally the circular musculature of worms such as annelids, and restores the body to its original disposition when the longitudinal muscles relax and return to their normal tonic state. If the longitudinal muscles relax the pressure of the internal fluid causes the animal to increase its length, the angle θ decreases and this is accompanied by an increase in volume, such as might occur on the uptake of food. Were the angle to become less than 110° elongation would cause not an increase, but a decrease in volume; this, however, is not significant since before θ could reach this critical figure the worm would have to increase its length some 50 per cent – well beyond its power. The volume of the animal is also variable in this way if (as is very likely with parasites) it experiences fluctuations in the osmotic pressure of the external medium, especially as the cuticle, despite its lipid layer, seems to be freely permeable to water in both directions.

The high internal pressure has other consequences for the organization of the nematode. Thus tubular viscera tend to collapse and can be filled only by muscular injection of fluid, as is the gut; propulsion of material along tubes has to be done by muscle, since cilia, the favoured means of moving fluid along tubes in many invertebrate groups, are incapable of overcoming pressures of the order with which we are dealing; most hollow viscera being collapsed on to their contents by pressure, their apertures are also closed and when ejection of contents occurs it is by the contraction of dilator muscles that the opening becomes patent. "Sphincters" in nematodes are thus active dilators. The high internal pressure then ejects the contents very forcibly: Crofton recorded that a defecating *Ascaris* (in air) may eject liquid to a distance of two feet. The absence of cilia has often been quoted as indicating a possible relationship between nematodes and arthropods; it is clearly no such thing but a direct consequence of the high internal pressure.

The high pressure exerted by the internal fluid against the muscles and cuticle of the body wall has also had the effect of nearly confining increase in size to increase in length; increase in girth is much less. The body of the worm shows a number of anatomical adaptations to this elongation comparable to those found in snakes, where the same shape of body is encountered, though for different reasons. Thus organs like the gonads and their ducts tend to grow into long, narrow structures and to be serially arranged rather than lie parallel, and in extreme cases one member of a pair may abort, leaving an asymmetrical arrangement. There is another unusual feature about increase in size in nematodes: whereas in most animals growth involves a continual multiplication of the cells of which the body is composed, this activity is confined to the early stages of development in nematodes and when these are concluded and each system in the body contains a certain number of cells characteristic of the species, cell division ceases, and further growth is brought about by increase in the volume of the component cells, not in their number. Although young nematodes undergo four moults in the early stages of their life history (p. 158) at each of which the cuticle is replaced by one of larger size, this does not mean that the cuticle cannot grow. After its last moult a young *Ascaris* may be 20 mm long; a few weeks later it may measure 400 mm. A twenty-fold increase in the length cannot have been achieved by stretching: new cuticle must have been added, though details of the mechanism remain undiscovered.

The almost perfectly circular cross-sectional shape brought about by high pressure leads to an appearance of radial symmetry. This is exaggerated by the arrangement of muscles in the body wall (see below) and by the hexamerous arrangement of lips (Fig. 45), of microscopic sense organs and, in some species, of short bristles on the surface of the anterior end of the body. Some primitive nematodes have caudal glands, opening at the extreme posterior tip of the body, by means of which they attach themselves to the substratum and gently wave the body in the water in which they live in search of food. This mode of life, which is akin to that followed by many related animals (for example, some rotifers) suggests an original radial symmetry which has become secondarily bilateral by the placing of openings along one side of the body, the ventral. The openings are the excretory pore, the anus and, in females, the genital pore. The mouth is terminal (Fig. 45, m) and the gut follows a straight course to the anus which lies near the posterior end a short way in front of the tip, so that nematodes have a short post-anal tail. In males the aperture is cloacal (Fig. 46, cl.ap) since the male ducts (ej.d) open along with the gut (int), and the tail is frequently expanded into sexual claspers (ala) (copulatory bursa). In females the genital pore lies midventrally near the middle of the body and has no connexion with the alimentary tract.

The cuticle is secreted by the epidermis (Fig. 42, epi), often called hypodermis by nematologists, which is thickened along a number (usually four but varying with the species) of strips situated in mid-dorsal, mid-ventral (v.n) and lateral (lat.n) situations, to form what are known as chords, projecting a little way into the body cavity. In the more advanced nematodes the cells which secrete the cuticle fuse during development to give a syncytial layer and their nuclei are restricted to the chords. The chords also contain the longitudinal nerve bundles, those of the dorsal and ventral chords being more important than those of the lateral chords where, in many animals (see below), the longitudinal excretory ducts (ex.can) lie. Internal to the epidermis is the muscular coat of the body wall, formed, as mentioned above, solely of longitudinal fibres. These are divided into groups by the chords and are therefore arranged in four quadrants, though in a few genera each quadrant is again split, giving eight groups of muscle in all. Each muscle cell is cigar-shaped and consists of an inner (noncont.mu) and an outer (cont.mu) part: the latter contains a number of peripherally placed bundles of contractile fibrils (Fig. 44, cont.f) separated and supported by lamellae of connective tissue (con.tiss) whereas the inner has no contractile structures, but contains the nucleus (nuc), mitochondria (mit) and food reserves (lip, glyc). It is drawn out into an elongated innervation process (in.pro) which stretches out to apply its tip to the dorsal chord (chord) if the muscle cell lies in either the right or left dorsal quadrant, and to the ventral chord if it is placed in either ventral quadrant. The cell seems more correctly described, therefore, as a neuromotor unit and the arrangement is unlike that found in any other group in the animal kingdom, where it is the general rule that the nerve fibre runs to the muscle. As it approaches the chord each muscle cell divides into a number of small processes which rest on the membrane of a nerve fibre (n.f) to form a neuro-muscular junction. Synaptic vesicles occur and the nerve contains a number of very large mitochondria (mit.n) at this point. Near the chord the innervation processes of several muscle cells may rest on one another forming places where stimulation may spread from cell to cell, ensuring simultaneous contraction.

The changes in the shape and volume of a nematode described above are those which involve the simultaneous symmetrical contraction of all the muscle fibres. This is not all that is possible in the way of contraction, and ordinary locomotion, indeed, is brought about by waves of contraction passing down the body in much the same way as in a fish or snake. The usual thing is for the muscles of the two dorsal quadrants to contract anteriorly, bending the body into a C shape, concave dorsally, and for this to move down the body as a wave of contraction. The muscles of the

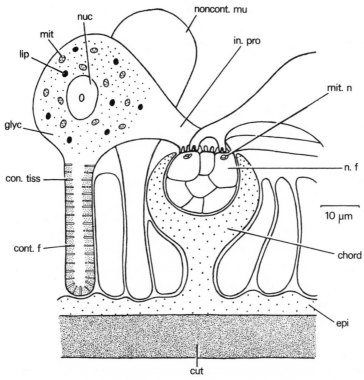

Fig. 44. Transverse section of dorsal or ventral chord of *Ascaris* showing the structure of a muscle cell and its relationship with nerve fibres lying in a gutter of epidermal tissue in the chord. (Based on Rosenbluth.) cont.f, contractile fibrils, thick in centre of area between neighbouring septa, fine near septa; con.tiss, connective tissue septum in contractile part of muscle cell; cut, cuticle; epi, epidermis; glyc, glycogen; in.pro, innervation process of muscle cell reaching to nerve fibres in chord; lip, lipid droplet; mit, mitochondrion; mit.n, group of giant mitochondria near neuromuscular juction; n.f, nerve fibre; noncont.mu, non-contractile part of muscle cell; nuc, nucleus.

two ventral quadrants then contract in the same way and a succession of waves, first on one side and then the other, passes rhythmically down the body. It does not seem possible for a dorsal and a ventral to contract simultaneously. This is probably to be correlated with the way in which both dorsal quadrants are innervated by way of the dorsal chord, and the two ventral ones by way of the ventral chord. Lateral bending, therefore, seems to be a movement beyond the power of a nematode. In water the movement just described seems competent to produce swimming, whilst in soil it causes the animal to insinuate itself between the constituent particles, against which the bending body may exert leverage. The latter is much the more efficient locomotor mechanism: in water there is little external resistance to bending and to obtain the necessary purchase on the water the animal must throw its body into very sinuous curves which use up much energy and do not result in much progression.

On the other hand, movement is much more efficient, in wet films, where the worm presses against the surface, or in soil, where the particles are used in the same way. In soil the fastest forward movement is achieved when the size of the particles is such as to prevent dissipation of energy in extensive bending and keep the animal always nearly straight. The undulations pass down the worm rapidly (about 60 mm s^{-1}), are probably initiated in the anterior ganglia since they do not occur in decapitated worms, have a wavelength about equal to the body length and cause the worm to swim. In addition, worms like *Ascaris* have been seen to carry out slow longitudinal waving which travels at 0.5–5 mm s^{-1}; forward-moving waves also occur in the intact animal and are the only ones present in headless ones; both types are associated with creeping. Nematodes also exhibit searching oscillations of the head end which stop during feeding.

The mouth, situated at the extreme tip of the head is provided with lips; there are primitively six but in parasitic forms such as *Ascaris* they tend to unite in pairs to give a dorsal and two ventrolateral lips. The mouth leads into an initial ectodermal section of the gut, lined, like the external surface, by a cuticle. This section is divisible into two parts, an anterior known as the buccal cavity and a posterior called the pharynx or oesophagus.

The detailed anatomy of the buccal region is closely linked to the kind of food eaten and to the method by which it is obtained. The simplest arrangement is found in nematodes which ingest fluid and particles the size of bacteria from their surroundings, such as *Rhabditis* (Fig. 45C), a common nematode found in the water film round soil particles and feeding on the bacteria and minute detrital particles also found there, or such as the vinegar worm, *Turbatrix aceti*, which feeds on microorganisms living in vinegar. Here ingestion of food is a simple suction and the buccal cavity is a short, cylindrical space evenly lined by cuticle. Some free-living nematodes feed on small animals or plants too large to be ingested whole; they do this by sucking fluid and small particles obtained by piercing the food and sometimes by secreting enzymes to produce an external digestion. Penetration of the body of the food organism may be achieved in two ways.

1. The cuticle lining the buccal cavity (tooth, Fig. 45D) may be elevated into sharp crests which are capable of ripping open cells so that their contents can be sucked out. This may be seen in *Mononchus*, a nematode common in fresh water and damp soil which preys on other nematodes.

2. In some nematodes which prey on animals, and in all those which attack plants, there occurs a special structure in the buccal cavity, either a hollow stylet (Fig. 45A, st) or a solid spear; a spear can be used to open plant cells, a stylet can be inserted to allow sucking of cell contents. The stylet is manufactured from an exaggeration of the tubular cuticular lining of the rhabditiform buccal cavity; the spear is usually an elaboration of a cuticular tooth. The base of the stylet expands into three knobs (bas.kn) from which muscles (pro.st.mu) run to the rim of the mouth; on contrac-

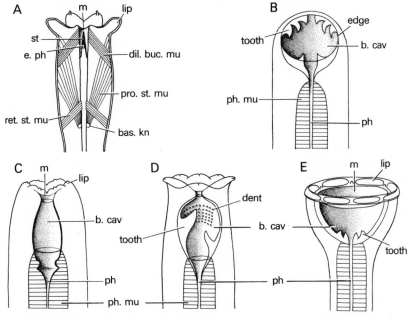

Fig. 45. Anterior ends of nematodes seen as transparent objects to show details of buccal cavity. A, *Dorylaemus*; B, *Ancylostoma*; C, *Rhabditis*; D, *Mononchus*; E, *Syngamus*. bas.kn, basal knob of stylet with insertion of protractor muscles; b.cav, buccal cavity; dent, denticle; dil.buc.mu, dilator muscle of buccal cavity; edge, rim of mouth; e.ph, entry from buccal cavity to pharyngeal cavity; m, mouth; ph, pharynx; ph.mu, radial pharyngeal muscles; pro.st.mu, protractor muscle of stylet; ret.st.mu, retractor muscle of stylet; st, stylet.

tion they project the stylet. Other muscles (ret.st.mu), which act as retractors, run back from the stylet to an origin on the body wall. The hollow in the stylet leads to the cavity of the pharynx; since its lumen may be no more than 1 μm in diameter the limitation of the nematode to the ingestion of fluid and very fine particles is obvious.

Nematodes making wounds on plants and animals must brace themselves so that the recoil from protrusion of the stylet does not push them backwards; this is usually done by pressing against soil particles or by attaching the lips firmly to the surface of the prey.

The mode of feeding of many parasitic nematodes is often identical with that of free-living relatives. An example is the whipworm *Trichuris*, found in the intestine of mammals. Its anterior end is extremely long and thin and burrows into the intestinal wall; here it repeatedly uses its stylet to mince the surrounding tissue and open up small vessels and it also secretes a digestive juice. The blood and liquid breakdown products from the damaged cells are then sucked up. Another type of feeding associated with a different buccal structure (Fig. 45B, E) is met with in such nematodes as the hookworms, *Ancylostoma* and *Necator*, found in the intestine of man, and in

Syngamus, the cause of the disease, gapes, in poultry, in which it lives in the trachea. In these worms the buccal cavity is expanded to a basin-shape and the mouth (m,edge) is wide. This permits the worm to hold a plug of host tissue in the buccal cavity and to tear it with tooth-like thickenings of the cuticle. At the same time digestive enzymes are secreted from glands lying in the thickness of the pharyngeal wall and the resulting soup is sucked into the gut.

Behind the buccal cavity lies the pharynx (oesophagus of some nematologists), one of the key structures in the functioning of the alimentary canal. Since the pressure in the nematode body cavity is high, any intake of food has to be made against the pressure and any opening of the lumen of the canal tends to eject the contents to the exterior. The pharynx is a combined pump and valve to allow entry and prevent loss of food. Anteriorly the pharynx (ph) arises from the buccal cavity (b.cav), posteriorly it communicates with the intestine and near this point there is usually an outlet valve; there is, however, no inlet valve between buccal cavity and pharynx. The pharyngeal lumen is triradiate when shut, the radii being dorsal and latero-ventral, and it is lined by cuticle. To this cuticle attach radial muscles (ph.mu) which run to the outer limit of the pharyngeal wall; this appears fixed, since contraction of the muscles dilates the lumen to a cylindrical shape if carried to completion: this rarely happens, however. On contracting, the dilator muscles expand the pharyngeal cavity against the hydrostatic pressure of the pseudocoelomic fluid plus that of the fluid making up the thickness of the pharyngeal wall and pull the pharynx half open so that its lumen is triangular in cross-section. In this position the cuticle is unstable and, on relaxation of the muscles, these forces snap the pharynx shut. It is presumed that this sudden action may act like the "click" mechanism of the insect wing, stretch the dilator muscles and cause them to contract again, thus setting up a series of contractions which do not require complete nervous control, and this underlines the ability of nematodes like *Ascaris* to open the pharynx at rates of twenty per second or more. In most nematodes the pharynx operates in two halves, anterior and posterior, and the mid-point is often marked by a swelling. In *Ascaris* the anterior half is filled with food (mainly the intestinal contents of the host) by dilatation of its walls; the posterior half then dilates whilst the anterior contracts and the food is transferred to it. Finally contraction of the posterior half expels food to the intestine, the pressure which it exerts being greater than that exerted on the intestinal contents by the pressure due to distortion of the pharyngeal wall. In *Rhabditis*, where the pharynx has an expanded middle part, this is filled from the buccal cavity, then the contents are ejected into the intestine. Since there is no inlet valve at the anterior end some is regurgitated to the buccal cavity, but it has been shown that, because of differences in the diameter of the lumen of the anterior and posterior halves of the pharynx, the loss of food in this way is negligible.

There is no vascular system in nematodes transporting material, nor does there seem to be any system of wandering amoebocytes. Food, digested in the gut and absorbed in the intestine, must therefore diffuse through the pseudocoelomic fluid. Reserve food, mainly glycogen and fat, occurs in the gut, the epidermal chords and the non-contractile parts of the muscles; as is usual in parasitic animals unable to carry oxidative mechanisms to completion, reserves are high.

Oxygen similarly diffuses through the body wall and pseudocoelomic fluid and in many small, free-living worms its supply can offer little problem. In certain habitats, however, like anaerobic muds, the gut of other animals, the supply of oxygen is inadequate to support a fully aerobic metabolism, and the problem is made worse in animals the size of *Ascaris*. Two ways of dealing with the situation occur: (1) a respiratory pigment may be used to capture what little oxygen is available; (2) anaerobic metabolic processes can occur. *Ascaris* has two different haemoglobins, one in the muscles of the body wall which acts as an oxygen carrier at very low partial pressures such as might occur in the gut of a mammal, the second in the pseudocoelomic fluid, but the latter does not seem to be involved in respiration since the oxygenated form does not dissociate in ordinary physiological conditions. In the absence of oxygen the metabolism becomes anaerobic and a variety of organic acids are the end products of the incomplete oxidation of carbohydrate which this involves.

The intestine runs straight through the body cavity and ends at a short hind-gut. It seems to be the site of the secretion of digestive juices and of uptake of digested food. In most nematodes there is no muscle in the intestinal wall, which consists solely of columnar epithelial cells with a layer of microvilli, known as the bacillary layer before electron microscopy showed its true nature, on their inner surface and an outer layer of dense connective tissue setting them off from the body cavity. Propulsion along the intestine depends on the action of the pharynx, which therefore not only sucks food in from outside but pushes it along the intestine. Movements of the body wall musculature may also help.

The hind-gut (Fig. 46, cl) is short and leads to the anus (cl.ap) and, like the pharynx, is provided with dilator muscles (dil.mu). Like the rest of the gut, this section is collapsed and the anus closed except when the muscles become active.

The central part of the nervous system is a nerve ring lying round the pharynx, with at least two pairs of ganglia in relation to it, placed laterally and ventrally, the former commonly said to correspond to the cerebral ganglia of other groups. From this ring six nerves pass forwards, primarily to sense organs placed on the lips, and a variable number pass backwards, mainly in the chords, which are gutter-shaped expansions of the epidermis into the pseudocoelom, with the nerve fibres lying in the gutter (Fig. 44). Of these the ventral one is the most important, being both sensory and motor; the dorsal nerve is regarded as exclusively motor though there is little experimental basis for either statement. Lateral and ventral nerves are ganglionated, though this is less regularly true of the dorsal nerves. At the posterior end of the body each lateral nerve expands into a lumbar ganglion (Fig. 46, lat.gang) which receives branches from the dorsal and ventral nerves. Goldschmidt has shown that in *Ascaris* –

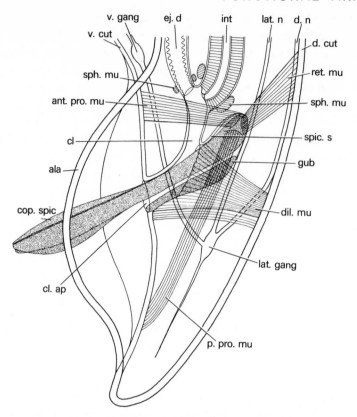

Fig. 46. Posterior end of male nematode seen as transparent object. ala, left lateral expansion
of body (expanded further to form copulatory bursa in strongyloids); ant.pro.mu, anterior
protractor muscle of spicules; cl, cloaca; cl.ap, cloacal aperture; cop.spic, copulatory spicule;
d.cut, cuticle of dorsal body wall; d.n, nerve in dorsal chord; dil.mu, dilator muscle opening
cloaca; ej.d, ejaculatory duct; gub, gubernaculum; int, intestine; lat.gang, ganglion at posterior
end of lateral nerve; lat.n, lateral nerve; p.pro.mu, posterior protractor muscle of spicules;
ret.mu, retractor muscle of spicules; sph.mu, sphincter muscle; spics.s, spicule sac; v.cut, cuticle
of ventral body wall; v.gang, ganglion on nerve in ventral chord.

and there is no reason for supposing that this animal is any different from any other
nematode – each ganglion at the anterior end has a definite number of nerve cells
within it and that the interrelationships of each nerve cell in the system seem to be
completely constant.

The economy in cells mentioned earlier as characteristic of nematode organization
seems to reach its climax in the excretory system. This consists of two longitudinal
canals (Fig. 42, ex.can) lodged in the lateral chords, which send transverse branches
to a common excretory pore placed mid-ventrally a little behind the mouth. The
canals appear to be intracellular channels within extensions of two cells, or even a
single one, the body of which lies near the pore, which is more precisely described as

the opening of the excretory cell(s). Various reduced versions of this system occur in different species of nematode, leading to the occurrence of only one longitudinal canal, or even a single cell without any canal extension. In no part of the system does anything comparable to a flame cell occur and there is therefore great difficulty in relating this system to the excretory system of any other group of invertebrates, and this is not lessened by the fact that its mode of working is not understood; its often extreme reduction, too, suggests that it is not perhaps important. When placed in solutions which are hypotonic the nematode seems to absorb water through the cuticle and when placed in hypertonic media water is lost in the same way. In species which control their water content uptake may well be by way of the cuticle but it is uncertain how water is lost, and the gut, with its extensive surface and frequent flushing, looks a better instrument for this than the excretory system. In the larvae of *Nippostrongylus muris* and *Ancylostoma caninum*, however, water is undoubtedly discharged by way of the excretory pore in hypotonic media. The loss of flame cells is, like the loss of cilia elsewhere, probably to be connected with the high turgor pressure within the body; the excretory system must therefore be of a new type to be able to function in these conditions.

In nematodes the sexes are separate and fertilization is internal. Males are often less abundant than females and in some species their number seems to be density dependent. In many groups in the animal kingdom (platyhelminths, annelids, vertebrates) the gonads often discharge to a body cavity from which the gametes escape by entering genital ducts; this arrangement is not possible in nematodes since the high internal pressure would collapse the ducts. Gonads and gonoducts are therefore necessarily continuous. In animals such as *Ascaris* or *Rhabditis* there is a single gonad in males, continuous with a duct leading to the cloaca and, in females, there are two gonads, each continuous with a duct, the two ducts joining in the immediate neighbourhood of the female opening. The gonads (in both sexes) are tubular, and germ cells are in most cases derived from a single cell or germinal area at the innermost end; the ducts are usually differentiated into lengths with varying functions. In females an oviduct and a uterus, commonly a separate receptaculum seminis, are recognizable on each duct, the two joining to form a short, median vagina before opening to the exterior. As in males, the female genital tract is provided with circular muscles, the degree of development becoming greater closer to the vulva. In males the gonad leads into a seminal vesicle and that in turn into an ejaculatory duct. Both these have layers of circular muscle in the wall. The male usually has copulatory spicules (Fig. 46, cop.spic), that is, spicular cuticular secretions arising in pockets of the dorsal wall of the cloaca (spic.s) and provided with protractor (ant.pro.mu, p.pro.mu) and retractor (ret.mu) muscles. Males are often provided with expanded cuticular plates (ala) at the posterior end of the body which help in grasping the female during copulation. In strongyloid nematodes (such as hookworms) the plates incorporate muscles and form a prominent copulatory bursa by

which the female is held so that the copulatory spicules can reach the female opening; their function is to dilate this and the vagina against the high pressure of the
internal fluids and so allow injection of sperm. Nematode spermatozoa have no
flagellum and are usually described as amoeboid, though their movement is more
like the gliding of gregarine protozoans. This seems to be related to their need to
make their way along tubes liable to collapse because of pseudocoelomic pressure, a
movement which might be impossible for a flagellated spermatozoon.

The eggs pass along the oviduct to the receptaculum, where they are fertilized.
Entering the uterus they are surrounded by a layer of protein, secreted by uterine
glands, and start developing. The zygote adds two inner coats to the egg shell, an
outer chitinous one (the only recorded use of chitin by nematodes) and an inner
layer with fat-like properties though possibly not made of lipid. The eggs are extraordinarily resistant to adverse conditions and remain viable for long periods, even
years. As might be expected, the egg production of parasitic nematodes is sometimes
staggeringly high: *Ascaris lumbricoides* (in man) is alleged to produce about 2×10^5
eggs per day and to live for several months; *Ancylostoma duodenale* (in man) does not
produce so many eggs per day, $22\text{--}24 \times 10^3$, but may live for about seven years.
The eggs hatch to give what are commonly called larvae, but a better term would
undoubtedly be juveniles since they have essentially the same form as the adult. As
each grows it sheds its cuticle four times, the fifth stage being the young adult, which
increases in length but does not again moult. The cuticle is therefore able to stretch
and elongate and moulting is not imposed on a nematode as it is on an arthropod by
its unyielding nature. The worm, however, does not appear to have the same ability
to enlarge the diameter of the cuticular tube without moulting; this may be because
such a process would destroy the fibrillar layers on which so much of the functioning
of the body depends. When ecdysis does occur a moulting fluid seems to be secreted
which dissolves the old cuticle, lifting it off the surface of the epidermis, which
thereupon secretes the new one. In these respects moulting is similar to the events in
arthropods, but whereas the elaborate endocrine control of the process is well understood in that group of animals nothing is known about it in nematodes. Amongst
juveniles a shed cuticle is sometimes retained as a protective outer covering; these
are known as encysted or sheathed larvae.

In dealing with parasitic members of other phyla it has been possible to make
suggestions about the way in which their often elaborate life history may have come
into being during the evolutionary history of the group (trypanosomes, malarial
parasites, trematodes). It is much more difficult to do this with nematodes because
the parasitic habit seems to have arisen several times, with the result that the life
history of one nematode may be quite different from that of another. Perhaps the
basic pattern of life history is like that of *Haemonchus* (found in the stomach of sheep),
where the eggs escape in the faeces, the young hatch and live there, moulting twice.
The third stage migrates on to herbage which is then cropped by a sheep. Infection

is therefore direct and by chance. In another parasite of sheep, *Strongyloides*, and in the hookworm, *Ancylostoma*, the third stage juvenile bores through the skin of the host and penetrates the vascular system. It is then automatically carried through the heart to the lungs where it is caught in the fine vessels of the first capillary bed it meets. It escapes to the lung and migrates to the pharynx, where it is swallowed and reaches its destination in the intestine. Many ascaroid nematodes have a similar larval migration but this seems to represent a different evolutionary history: the ancestral ascaroid was apparently a parasite of marine invertebrates which were infected by ingesting eggs or larvae; these passed through the gut wall into the body cavity and matured, laying eggs which might not escape until the death of the host. At a later evolutionary stage the life history became complicated by the addition of a second host, a natural predator of the first one. The original host now became an intermediate host and maturity was delayed until the parasite found itself in the gut of the definitive host. The narrow connection with a marine invertebrate and its predators was lost with time and amongst modern ascaroid worms a common pattern of life history involves a rodent as intermediate and a carnivorous predator as definitive host. There are bound to arise occasions, however, when the carnivore accidentally eats eggs from parasites in its own gut. These hatch and the larvae behave as if they were in an intermediate host, that is, they bore through the gut wall to reach the body cavity. Most, however, enter blood vessels and get swept to liver, heart and lungs where they break out and finally reach the pharynx and are swallowed. Arriving in the gut they now behave as if they had reached it from an intermediate host eaten by the definitive host, finish their development and start to reproduce. Although this life history has a superficial identity with that of hookworms it is obvious that it has been reached in a totally different way. The last type of nematode life cycle to need mention is that of the filarias which are found in the blood, lymph or connective tissue spaces of vertebrates; they are dependent on blood-sucking arthropods – fleas, mosquitoes, ticks or mites – for transmission from host to host. How this may have come into being is not known, but it is more likely to be a chance development by which organisms taken up by the arthropod in the course of feeding survived until returned to the next vertebrate than the other way round.

Classification of nematodes and their relations

Nematoda
   Secernentes (= Phasmidia)
      Rhabditoidea
      Tylenchoidea
      Strongyloidea
      Oxyuroidea

Ascaroidea
Spiruroidea
Filaroidea
Adenophorea (= Aphasmidia)
  [Plectoidea]
  [Axonolaimoidea]
  [Monohysteroidea]
  Chromodoroidea
  [Desmodoroidea]
  Enoploidea
  Tripyloidea
  Dorylaimoidea
  [Mermithoidea]
  Trichuroidea
  [Dioctophymoidea]
[Acanthocephala]
[Rotifera]
[Endoprocta (= Kamptozoa)]

9

Annelids

THE ANIMALS IN the phyla which have been dealt with in the last few chapters are popularly described as "worms"—flatworms, round worms, proboscis worms—from which it follows that the word "worm" cannot be a term implying zoological relationship, but is merely a description of a certain bodily shape. This was not always so. In the eighteenth century both Linnaeus and Cuvier established important groups in their classification of the animal kingdom which they called "Worms". Linnaeus's group Vermes included all invertebrates except arthropods; forty years later, when Cuvier established his group Vers, however, molluscs, tunicates, coelenterates and echinoderms had been removed to independent phyla and it included only the kind of long, more or less cylindrical animal still popularly known as a worm. Increasing knowledge showed that even this was an unnatural assemblage of animals similar in shape only because they all shared the habit of progressing over a surface without limbs on which to walk: much of their likeness, therefore, is due to convergence – the adoption by different animals of similar answers to the same functional problem. In particular, animals at three quite distinct levels of organization have wormlike bodies: (1) the flatworms and nemertines, totally devoid of coelom and rich in ciliated surfaces; (2) the round worms, with a body cavity of a special nature, no cilia, a thick protein cuticle and many features peculiar to themselves; (3) a group of animals with an internal coelom and bodies composed of a series of segments or metameres. This last group is regarded by many zoologists as a single phylum, the Annelida, so-called because of the ring-like (annulate) appearance which the presence of segments gives to their bodies; some, however, would regard the Annelida as a superphylum containing more than one phylum.

Classification

The annelids differentiate themselves from lower groups in being coelomate and segmented. Of these two features the second is the more important though both may well have appeared together in evolution: a coelom, it is true, separates body wall from gut wall and allows their independent activity, provides a constant environment for developing germ cells and harbours them till spawning, but most of this can be achieved by any kind of body cavity; metamerism, however, underlies a new type of locomotion based on the presence of septa which divide the coelom into separate compartments from which escape of fluid is prevented. When local muscles contract, therefore, local pressures are developed which can serve local needs, an arrangement quite unlike the diffuse spread of pressure in coelenterates or nemertines the effect of which can be circumvented only by carrying out contractions slowly so as to give antagonistic muscles time to adapt. Since it is probable that the margins of the septa primitively penetrated the muscles of body wall and gut wall to insert on the epidermis and gut lining, the primary coelomic metamerism brought a secondary muscular and epithelial segmentation in its train, reinforced by the use of the segment as the functional locomotor unit. The coelom sac within the segment, walled by waterproof septa and evenly developed layers of antagonistic circular and longitudinal muscle, acts as the hydrostatic skeleton in a unit which can vary from a short, broad shape (contraction of longitudinal, relaxation of circular muscles) to a longer, narrow shape (relaxation of longitudinal, contraction of circular muscles) without change of volume. If the animal is a burrower the broad shape provides an area of anchorage, where body wall is held firm against the substratum by friction, whilst the thinner shape allows movement within the burrow. This is much the kind of locomotion seen in earthworms and it is generally accepted that animals of similar structure and mode of life provide a reasonable starting point for the evolutionary changes which have occurred within the annelids.

The most primitive annelids alive today, which may well be most akin to the ancestral forms, are marine worms known as archiannelids. All are small, and some of their simplicity of structure may be due to this; many show a regular segmentation and an even development of muscles in the body wall, with which they can burrow through, or squirm over the surface of soft substrata. It may even be that metamerism arose in adaptation to the burrowing habit. They are primarily feeders on detritus, which they engulf by means of a buccal organ, a muscular protuberance on the floor of the first part of the gut, used like a tongue in the collection of food. One or two pairs of tentacles are commonly found on the head, but, as one would expect in an animal which often burrows, these are never long. Most archiannelids exhibit one feature in addition to their segmentation which marks them off as true annelids: this is the presence of segmentally arranged chaetae, or bundles of chaetae – slender, chitinous bristles, each growing from a single epidermal cell, and springing out of pits (chaetal sacs) placed right and left on each segment. In some genera the sacs open flush with the general body surface; in others, however, such as *Nerilla* (Fig. 47), they open on lobes projecting laterally from each segment and known as parapodia, structures which have an enormous part to play in the further evolution of one branch of the phylum.

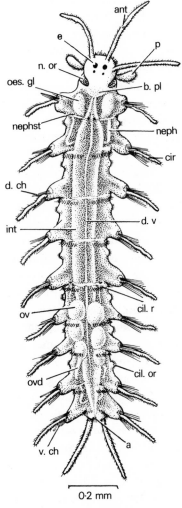

Fig. 47. *Nerilla mediterranea*, in dorsal view. a, anus; ant, antenna; b.pl, buccal plate; cil.or, ciliated organ; cil.r, ciliated ring; cir, cirrus; d.ch, dorsal group of chaetae; d.v, dorsal vessel; e, eye; int, intestine; neph, nephridium; nephst, nephrostome; n.or, nuchal organ; oes.gl, oesophageal gland; ov, ovary; ovd, oviduct; p, palp; v.ch, ventral group of chaetae.

From some such origin evolution has given rise to three large groups of worms, the polychaetes, the oligochaetes and the hirudineans. Argument still centres on whether the oligochaetes or the polychaetes are more primitive, the latter appearing much the more deserving of the title because of their marine habitat, their external method of fertilization and their free-swimming trochophore larvae. On the other hand the oligochaetes, except for modification in reproductive and osmoregulatory methods clearly correlated with their freshwater and terrestrial mode of life, have a body more primitive than that of polychaetes in its lack of differentiation into regions, its regular segmentation, its nearly complete series of septa, its unbroken layers of circular and longitudinal muscles in the body wall and in the absence of parapodia. Nearly all these features perpetuate conditions already seen in archiannelids, as do the burrowing and creeping mode of life and the habit of feeding on detritus. Indeed the oligo-chaetes could well be regarded as a group derived directly from an archiannelid ancestry which has become modified (1) in adaptation for life in fresh water or on land and (2) in more complete adaptation for burrowing to escape the rigours of terrestrial life. Such adaptations would involve suppression of head appendages and of any incipient parapodia and, since the animals now live in habitats in which broadcast gametes cannot survive, modification of the reproductive system to allow copulation, the replacement of a larval phase by direct development, and the appearance of hermaphroditism – a common adaptation in stocks whose locomotor disabilities or narrowness of habitat minimize the chances of animals meeting.

Closely related to oligochaetes are leeches (Hirudinea), animals which have taken some steps along the evolutionary road to ectoparasitism, though the more primi-tive members of the class are simple predators on small invertebrates. Like oligo-chaetes they have lost head appendages and parapodia, but have taken this trend a stage further and have lost their chaetae too. This, along with the fact that the coelomic cavity is greatly reduced and all septa lost, may be correlated with their adoption of a new locomotory method which depends on stepping with suckers located at the two ends of the body, and not at all on the development of localized areas of swelling or constriction as in oligochaetes. Like these animals, too, leeches are hermaphrodite, they copulate and their development is direct, despite the fact that most are aquatic and many marine.

The oligochaetes and the hirudineans form one evolutionary line within the annelids, the former showing a mosaic of primitive characters related to a burrowing habit and advanced ones adaptive for life in fresh water or on land. The third major class of the phylum, the polychaetes, seem to have followed a distinct evolutionary pathway because they early became committed to a different method of locomotion. From some burrowing and creeping archiannelid stock it must be supposed that polychaetes evolved by exaggerating the rudimentary parapodia already present. In this way they acquired what were in essence limbs adapted for a creeping rather than a burrowing mode of life. The use of parapodia as levers on which to walk dispelled the need for complete muscular layers in the body wall (and to a great extent the need for septa) and largely replaced them with a more complex muscula-ture related to the parapodia. The freer movement of the animals over the sub-stratum encouraged the development of sensory appendages, particularly on the head, which became a much more elaborate structure than in either of the other classes of the phylum. With this new locomotor and sensory equipment polychates were in a position to radiate into a series of new ecological niches and to give rise to a large number of species.

Some pathways through this radiation may be traced by following the work of Dales, who has used the buccal organ and its modifications as a clue (see Fig. 48). The ancestral annelid, it will be remembered, has a muscular organ on the ventral

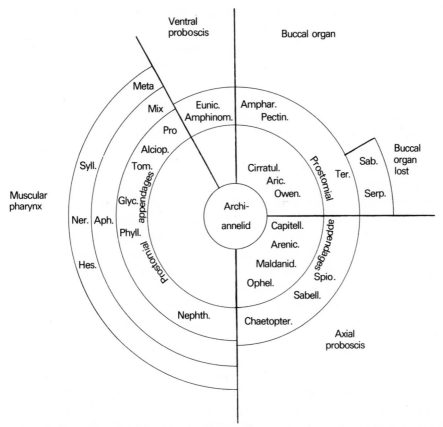

Fig. 48. Scheme of polychaete phylogeny using the stomodaeal structures and type of nephridium as guides. Pro, protonephridia; Meta, metanephridia or metamixonephridia; Mix, mixonephridia; Alciop., Alciopidae; Amphar., Ampharetidae; Amphinom., Amphinomidae; Aph., Aphroditidae (including polynoids); Arenic., Arenicolidae; Aric., Ariciidae; Capitell., Capitellidae; Chaetopter., Chaetopteridae; Cirratul., Cirratulidae; Eunic., Eunicidae; Glyc., Glyceridae; Hes., Hesionidae; Maldanid., Maldanidae; Nephth., Nephthydidae; Ner., Nereidae; Ophel., Opheliidae; Owen., Oweniidae; Pectin., Pectinariidae; Phyll., Phyllodocidae; Sab., Sabellidae; Sabell., Sabellariidae (Hermellidae); Serp., Serpulidae; Spio., Spionidae; Syll., Syllidae; Ter., Terebellidae; Tom., Tomopteridae.

wall of the first part of the gut with which it collected detritus. This organ appears to have been retained in one large subdivision of polychaetes though lost in its most advanced members, whereas in a second subdivision it has been transformed into, or replaced by, a radially symmetrical proboscis.

At the base of the first subdivision lie families which still retain a burrowing habit (cirratulids, Fig. 62A; ariciids, eunicids, terebellids, Fig. 62B) though many are also able to creep from place to place (pectinariids, Fig. 57C). These worms still retain a ventral buccal organ, sometimes, as in eunicids, enlarged to form a powerful, armoured, grasping tool or offensive weapon. Many members of this group have developed prostomial tentacles (pectinariids, ampharetids, eunicids, terebellids) by means of which they feed and respire. The acquisition of these tentacles permitted a more advanced series of worms in this subdivision (sabellids, Fig. 60A; serpulids, Fig. 61A) to lose the buccal organ because the tentacles could be modified to form a ciliary mechanism for collecting food. With such a device mobility is superfluous and the worms become tubicolous and parapodia secondarily reduced.

In the second major subdivision of polychaetes the buccal organ on the ventral wall of the buccal cavity is replaced by a symmetrically developed proboscis. There are two main evolutionary lines in this division, one (orders Capitellida and Spionida) characterized by a tendency to mud-eating, the other (order Phyllodocida) by a trend towards predation. The first line includes animals similar to earthworms in habit, eating sand or mud for the sake of its organic content and gathering it by a simple eversion of the front end of the gut; to this group belong such worms as the arenicolids (Fig. 64) and capitellids in the order Capitellida. As in the evolution of the first subdivision so too in this, a number of families in the order Spionida (spionids, sabellariids) have developed cephalic tentacles, not, however, homologous with the prostomial tentacles of terebellids, sabellids and serpulids. The possession of these structures has enabled the worms to replace unselective, bulk ingestion of mud by the selection of particles from the substratum (some spionids, Fig. 66) and ultimately to use this method exclusively (sabellariids, Fig. 61B) or replace it by a filter-feeding device (chaetopterids, Fig. 70). The buccal organ or proboscis is then lost. The second main evolutionary line in the group of worms with a symmetrical proboscis has given rise to the very large group of animals broadly known as errant polychaetes (order Phyllodocida: phyllodocids, tomopterids, Fig. 59A; glycerids, nephthydids, Fig. 58; aphroditids, Fig. 59B; syllids, Fig. 77; hesionids, nereids, Fig. 50). All are predatory, and are active creepers and even swimmers, using well-developed parapodia and all have efficient sense organs.

As might be expected in a predominantly marine class the gametes of polychaetes (whatever evolutionary line they may have followed) are broadcast, the zygote giving rise to a free-swimming, self-supporting, pelagic larva, the trochophore, which metamorphoses into the adult.

The animals mentioned so far would be accepted by all zoologists as falling within a single phylum Annelida. In addition, however, there are some other groups with clear annelid affinities, though perhaps not clear enough to place them unequivocally within the phylum. These are the echiuroids, the sipunculoids and the myzostomids.

The echiuroids comprise a small number of marine, worm-like animals which show signs of developing segments during their larval stages, but lose them later. They have, when adult, a spacious aseptate coelomic cavity, an anus at the posterior end of the body, usually one pair of stout chaetae on the ventral side anteriorly and sometimes more near the anus. They burrow, so that their lack of head appendages is

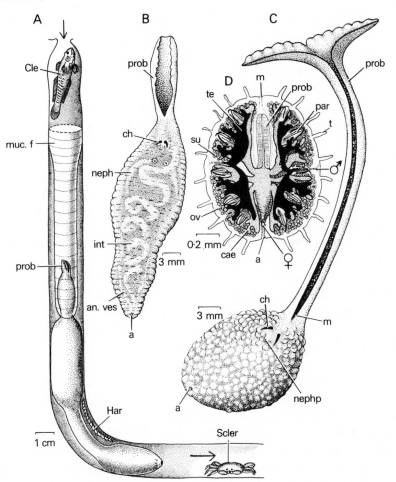

Fig. 49. A, *Urechis caupo* in the burrow which it shares with a fish, a crab and a polychaete worm; arrows show the direction of water movement through the burrow. (After Fisher and McGinitie.) B, *Thalassema neptuni*, ventral view, removed from its rock crevice and drawn so as to show some internal organs. C, *Bonellia viridis*, female, ventral view. D, *Myzostoma cirriferum*, ventral view. a, anus; an.ves, anal vesicle; cae, intestinal caecum; ch, chaeta; Cle, commensal fish, *Clevelandia ios*; Har, commensal worm, *Harmothoe adventor*; int, intestine; m, mouth; muc.f, mucous funnel; neph, nephridium; nephp, nephridiopore; ov, ovary; par, parapodium, each with chaeta and aciculum; prob, proboscis, withdrawn into proboscis sac in *Myzostoma*; Scler, commensal crab, *Scleroplax granulata*; su, sucker; t, tentacle; te, testis; ♂, male opening; ♀, female opening.

not surprising. Anterior to the mouth lies a proboscis, scoop-shaped, grooved ventrally and sometimes bilobed, by means of which detrital particles are collected for food. There is only one form found in Britain, *Thalassema neptuni* (Fig. 49B), which lives in rock crevices. The green *Bonellia viridis* (Fig. 49C) is a better-known echiuroid, nearly cosmopolitan in distribution, living in mud with only the bilobed tip of the proboscis spread on the surface. All specimens prove to be female; males do occur but are pygmies with a much simplified organization living parasitically within the genital ducts of the females. The well-known experiments of Balzer have shown that the sex of *Bonellia* larvae is indifferent. Should a larva settle in isolation it becomes female. Should it, however, find itself near an established female it is attracted to settle on her proboscis and, under the influence of her hormones, becomes male, later moving down the probsocis and through the gut to a destination in the genital duct.

Most echiuroids have a simple vascular system with part of the dorsal vessel alongside the intestine acting as heart. One of the common echiuroids of western American coasts, however, *Urechis caupo* (Fig. 49A), has no vascular system and appears to use the coelomic fluid for transport of respiratory gases. To this end the coelomic fluid contains many cells laden with haemoglobin. During periods when the worm is active the cells pick up oxygen from the hinder part of the gut which rhythmically fills and empties itself with sea water in much the same way as the cloacal chamber of a holothurian (p. 421). If the worm is inactive the oxygen stored in combination with haemoglobin permits survival for over an hour; if the haemoglobin is thrown out of action by combination with carbon monoxide the worm can live only for about a quarter of that time. *Urechis* is also of interest in that, alone amongst echiuroids, it has a filter-feeding mechanism somewhat similar to that of the polychaete *Chaetopterus* (p. 207). *Urechis* lives in a U-shaped burrow in sandy mud through which water is pumped in an anteroposterior direction by peristaltic waves passing down the body wall. The worm secretes the beginnings of a mucous funnel (muc.f) from a girdle of glands placed in the skin near the anterior pair of chaetae and attaches it to the wall of the burrow close to the mouth. It then retracts deeper into the burrow, spinning the inner parts of the funnel as it goes, and finally halts so that a cone of mucus is stretched between the base of its proboscis (prob) and the front of the burrow. Through this it filters sea water, straining off particles down to 4 nm in diameter. At intervals the posterior edge of the net is caught by the proboscis, the whole is gathered up, eaten and replaced by another. Larger particles are rejected, but are not wasted, since, not surprisingly, a number of commensals are regularly found in the tube where they obtain, with a minimum of exertion on their part, shelter, oxygen and food. Three kinds are regularly found – a pinnotherid crab *Scleroplax granulata*, a scale worm *Harmothoe adventor*, and a small teleostean fish *Clevelandia ios*.

The sipunculoids form a second group of animals allied to, though not necessarily included within the phylum Annelida. They are all burrowers in soft substrata, wormlike in shape, showing no signs of metamerism even in the larval stages and probably without chaetae, though some parts of the epidermis bear chitinous spines which might be related structures. Tentacles lie around the mouth, not always visible, since the anterior end of the body forms an introvert which can be pulled within the posterior part. The gut is U-shaped with the anus lying well forward in the mid-dorsal line, the intestine forming a long double spiral coil in the roomy coelom. Sipunculoids collect edible fragments from their surroundings by means of their tentacles. There is no blood system, but the coelomic fluid contains cells some of which contain the pigment haemerythrin and are of respiratory importance, though the oxygen demands of such lethargic animals are not likely to be high. The best-known genera are *Sipunculus, Golfingia* and *Phascolosoma,* all cosmopolitan in distribution, particularly in warmer seas.

A third group of animals similarly related to the main annelid stem is the myzosto-mids, all ecto- or endoparasitic on echinoderms (Fig. 49D). This habit has doubtless led to their flattened leaf-like shape with marginal tentacles (t), which does not at first sight suggest relationship with an annelid. That this exists, however, is shown by the presence of five pairs of parapodia (par) bearing chaetae, an alimentary tract running from an anterior mouth (m) to a posterior anus (a) and provided with pairs of lateral caeca (cae), perhaps to compensate for the marked anteroposterior compression of the rest of the gut, and a ventral nerve cord. Myzostomids are her-maphrodite (ov, te), though cross-fertilization occurs; fertilization – as in leeches – takes place internally after spermatophores have been deposited on the animal's skin. Like sipunculoids and echiuroids myzostomids have free-swimming trochophore larvae. The ectoparasitic forms are normally found on the disk of crinoids, occas-ionally on ophiuroids, both groups which feed by filtering particles from sea water on their arms and passing what has been collected along grooves to the mouth on the disk. The myzostomids help themselves to a share of this by dipping a sucking proboscis into the stream of food passing to the mouth. Endoparasitic forms also prefer crinoids as hosts and live mainly within the coelomic cavity, ingesting coelomic corpuscles or parts of the gonad. Some are gut parasites and these are also to be found in some asteroids.

What is the relation of these three groups to the undoubted annelids? Myzosto-mids are the closest, with their parapodia and numerous chaetae; echiuroids would appear to be the next nearest on account of the rudimentary metamerism which their larvae exhibit, their posterior anus and the presence of chaetae. Sipunculoids are either the least close genetically, or those most highly modified for a different mode of life and so least like anatomically, or both. There are two ideas of possible value in assessing the relationships of this collection of animals: (1) that they form a group of related organisms (sipunculoids perhaps excluded), part of the main annelid stem but distinguished from most annelids by the small number of segments in their body – "oligomerous" instead of "polymerous"; (2) that they belong to a series of relatively unsuccessful offshoots from the main evolutionary pathway leading to annelids or are parallel developments from a basic stock ancestral to all.

The first of these ideas rests mainly on the work of the Russian zoologist Ivanov, who was greatly impressed by the fact that in many polychaetes the segments of the body appear in two groups, a small number of what he called "primary" segments which appear more or less simultaneously in the trochophore, and a much larger number of "secondary" segments which are formed later. The ways in which the two series arise are different in detail, as is their structure. Similar differences in the origin and organization of segments, Ivanov believed, could be traced in other metameric groups. Ivanov read phylogenetic meaning into this developmental sequence and proposed that it recapitulated an annelid evolutionary history in which the earliest forms had been short-bodied, whereas later forms had become polymerous by adding a secondary series of segments to the posterior end of those already present in their oligomerous ancestors. The oligomerous forms are the archiannelids, the echiuroids and the myzostomids of today: the polymerous ones the annelids proper. Ivanov's ideas tend to be dismissed nowadays. The primary segments are regarded as a (possibly precocious) development to increase the trochophore's adaptations for a prolonged larval life, but not as in any respect fundamentally different from the remaining segments; the oligomerism of archiannelids is probably primitive, but partly due to small size, whilst that of myzostomids is certainly related to their epizoic mode of life.

If Ivanov's suggestion that the groups under discussion are to be unified as oligomerous annelids in contrast to the polymerous ones is rejected it is worthwhile examining the idea that they are offshoots of the main stem. At one time it was fairly generally accepted that a group of animals had originated from a single ancestral stock which had undergone some change that raised it from one grade of organization to another; further evolution at the higher grade was due to an adaptive radiation of the new prototype. This view of evolution is much less acceptable than it used to be. Instead it is held to be much more probable that several more or less closely related stocks at a lower level of organization were able to make changes – more or less simultaneously – which raised them to a higher level. If the first evolutionary mode can be represented by a tree with a single trunk emerging from the ground and repeatedly branching, the second recalls a herbaceous plant with numerous stems arising parallel with one another. Not all the steps taken to pass from one evolutionary level to the next were of equal length nor were all taken in precisely the same direction, nor were all equally successful: some led to extinction, some to resounding success. But on this basis it may be supposed that a number of different strains of animals were able to move more or less successfully towards the level of organization that we call annelidan. One group took the particular combination of steps which led to the true annelids and their abundant radiation; others took somewhat similar, but less successful changes or combinations of changes and have survived with a generally annelid-like appearance without being true annelids. Such groups would be those that we have been dealing with – myzostomids, echiuroids, sipunculoids – saved from the total extinction that must have come to other

similar groups, perhaps, by the chance that their particular evolutionary changes led
to adaptations for life in habitats avoided by more successful animals.

Organization of *Nereis*

An account of the functional anatomy of the ragworm *Nereis diversicolor* will illustrate
the basic principles of annelid and particularly polychaete organization. This worm is
common in mud flats of estuaries where it lives in a burrow with openings to the
surface. It emerges to swallow deposits of mud rich in organic remains, or to select
fragments of algae and small corpses; traces of its feeding forays can be seen radiating
from an opening (Fig. 50C). Movements within the burrow are concerned with
irrigation and backward excursions to the surface for defecation. Occasionally it
leaves its home to swim.

The body consists of a hundred or more segments and of initial and terminal
pieces, the prostomium (B, pros) and pygidium (A, pyg), which are not related to
coelom sacs and are therefore not true segments. Each segment is primitively like
every other. Its left and right coelom sacs, lined by peritoneum, lie against the
muscles of the gut wall medially, and against the muscles of the body wall laterally.
Mid-dorsally and mid-ventrally, where the sacs meet, the two sheets of peritoneum
form mesenteries suspending the gut, which is thus protected in a water-jacket.
Anteriorly and posteriorly the walls of the sacs abut against those of neighbouring
segments and form septa in which muscles develop. In most annelids the septa are
perforated and the mesenteries incomplete, and septa may be lost, so that the coelom
constitutes a more or less continuous body cavity with free circulation of fluid. It
usually remains possible, however, to isolate adjacent segments by the contraction of
septal muscles. The boundary between one segment and the next is marked externally by an annulus (B,ann) and, except for the first few, these annuli indicate the
positions of septa which subdivide the coelom. The loss of septa anteriorly is associated with the development of a proboscis for feeding. This is the modified anterior
extremity of the gut which can be protracted through the mouth, but is otherwise
hidden from view. The prostomium overhangs the mouth and forms the dorsal lip.
It houses the brain and bears the major sense organs: two pairs of eyes, a pair of
palps (palp) innervated from the anterior part of each cerebral ganglion, a pair of
tentacles innervated from their middle parts and posteriorly a pair of sensory pits
which probably have an olfactory function. Each appendage is an important receptor
organ innervated by a special nerve from the cerebral ganglion. The peristomium
(the first apparent segment lying around and behind the mouth) has four pairs of
extensile cirri (peris.cir) which are normally held so as to explore the maximal
volume of surrounding water. They are tactile organs and chemoreceptors. Their
number and innervation show that the peristomium is formed of two fused segments, the tentacular cirri representing the dorsal and ventral cirri of the parapodia

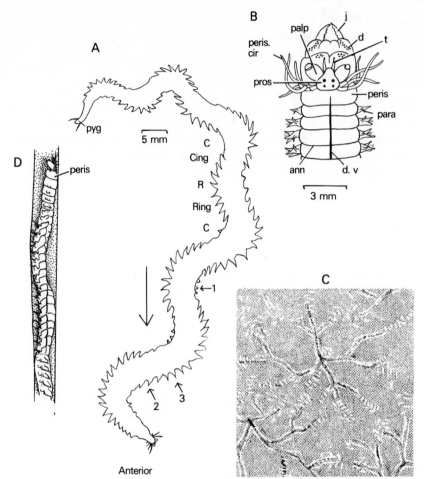

Fig. 50. *Nereis diversicolor*: A, diagram relating deflection of parapodia to lateral undulations of body. Large arrow indicates direction of movement of worm. Small arrows relate to position of parapodium; 1, approximately perpendicular to surface of segment when adjacent longitudinal muscles contracted; 2, drawing forward as muscles relax; 3, backward stroke beginning half-way up leading edge of a wave. The state of the longitudinal muscles is indicated by C, contracted; Cing, contracting; R, relaxed; Ring, relaxing. B, anterior segments, dorsal view, proboscis protruded. C, traces of feeding trails on surface of mud; radial marks made by anterior end of a worm which emerges from the burrow and explores the surface; flecks indicate jaw marks. D, *Nereis virens*, irrigating burrow. ann, annulus; d, denticle; d.v, dorsal vessel; j, jaw; palp, prostomial palp; para, parapodium; peris, peristomium; peris.cir, peristomial cirrus; pros, prostomium with 4 eyes; pyg, pygidium with sensory cirri; t, tentacle.

of these segments, the rest of the parapodia having been aborted: two pairs are innervated from small ganglia on the circumoesophageal connectives and two pairs from the suboesophageal ganglia.

Each segment posterior to the peristomium has a pair of lateral projections, the

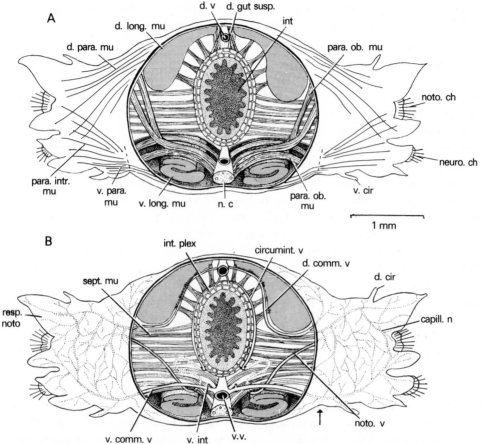

Fig. 51. *Nereis diversicolor*: diagrams of thick transverse sections of body (parapodia uncut) showing posterior surface of septum, and in A the disposition of the parapodial and inter-parapodial oblique muscles and in B the segmental blood vessels. Arrow indicates base of parapodium. capill.n, capillary network; circumint.v, circumintestinal vessel; d.cir, dorsal cirrus; d.comm.v, dorsal commissural vessel; d.gut susp, dorsal gut suspensor vessel; d.long.mu, dorsal longitudinal muscle block; d.para.mu, dorsal extrinsic parapodial muscle; d.v, dorsal vessel; int, intestine; int.plex, intestinal plexus; n.c, nerve cord; neuro.ch, neuropodial chaetae; noto.ch, notopodial chaetae; noto.v, vessel to notopodium; para.intr.mu, parapodial intrinsic muscle; para.ob.mu, parapodial oblique muscle; resp.noto, respiratory lobe of notopodium; sept.mu, septal muscle; v.cir, ventral cirrus; v.comm.v, ventral commissural vessel; v.int, cut connection with circumintestinal vessel; v.long.mu, ventral longitudinal muscle block; v.para.mu, ventral extrinsic parapodial muscle; v.v, ventral vessel.

parapodia, hollow extensions of the body wall containing muscles which bring about their movement. All are similar in structure and when the worm is active bend forwards and then back in steplike fashion. Each is biramous, with a dorsal part known as the notopodium and a ventral neuropodium, and these are further subdivided into a number of lobes and bear a bundle of chaetae (Fig. 51A, noto.ch, neuro.ch). When the parapodium is directed obliquely forwards, and in contact with the ground, the chaetae are protruded to fix the limb and one chaeta, stouter than the others and dark with pointed apex, is seen to project only a short distance from each bundle. This is the aciculum, which serves for the attachment of chaetal muscles. Each bundle of chaetae is lodged in a sac formed by an invagination of the epidermis and is protruded, retracted and turned in various directions by muscles inside the parapodium, which arise from the head of the aciculum and are inserted on the anterior and posterior parapodial walls (Fig. 52C, D acic.mu). The sensory cirri are attached to the dorsal (d.cir) and ventral surfaces of the parapodia. A pair of similar though larger cirri arises from the pygidium.

It is necessary to consider the arrangement of the somatic muscles to explain the mechanics of locomotion, crawling and swimming. The longitudinal ones can be located externally as prominent blocks, one (Fig. 51A, d.long.mu) on either side of the mid-dorsal blood vessel (d.v) and one (v.long.mu) on either side of the nerve cord (n.c) which marks the mid-ventral line of the body. The muscles of the septum (Fig. 51B, sept.mu) penetrate the dorsal blocks to insert into the more superficial circular muscles. Dorsal and ventral blocks act synergically in producing uni-lateral waves of contraction which pass down the body alternately right and left as the animal crawls. The stepping action of the parapodia is coordinated with these waves. Complete bands of circular muscles occur only between the parapodia. Elsewhere they are represented by bundles which pass laterally into the parapodium to form dorsal and ventral parapodial muscles (Figs 51A, 52C and 53A, d.para.mu, v.para.mu) which shorten the limb on contraction. Polychaetes also have segmental oblique muscles arising on either side of the nerve cord. Two pairs in *Nereis* cross the coelom posterior to each septum to insert on the dorsolateral body wall, thus acting as a brace between successive parapodia. Others pass into the parapodium to insert along anterior and posterior walls at the base of the neuropodium (Fig. 52A, ant.and p. para.ob.mu). Their area of insertion marks the hinge line of the parapo-dium, that is the bending plane between body and limb (arrow in Figure). Median to this the entrance to the parapodial coelom is crossed by diagonal and transverse muscles (Fig. 53A, diag.mu, trans.mu) which regulate its size and shape.

There are also intrinsic parapodial muscles: (1) muscles linked closely to the para-podial walls (Fig. 52C, para.intr.mu), with those lying dorsally and along the antero-dorsal surface most powerful; (2) the parapodial levator muscle (lev.mu), arising from the dorsal wall and crossing the coelom to insert above the chaetal sac at the tip of noto- and neuro-podium; (3) acicular muscles (Fig. 52D, acic.mu, neuro. prot.mu) which radiate from the head of each aciculum to lateral and ventral origins on the parapodial wall, some placed near the hinge line.

Nereid worms may move by slow creeping, by rapid creeping or by swimming. When a worm crawls forward slowly it uses its parapodia as a series of levers without waves of contraction of the longitudinal muscles of the body wall. The parapodia

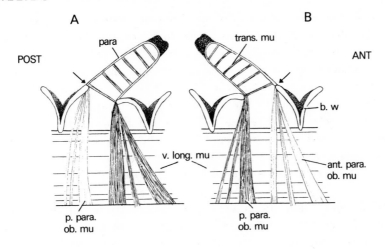

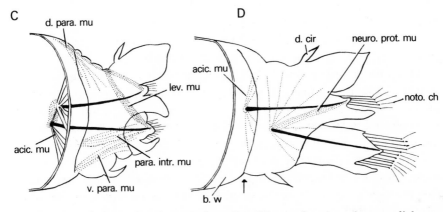

Fig. 52. *Nereis diversicolor*: diagrams to show disposition and action of parapodial muscles. (After Mettam.) A, parapodium bent forwards by contraction of anterior parapodial oblique muscles; B, bent backwards by contraction of posterior parapodial oblique muscles; C, retracted during slow creeping preparatory stroke; D, protracted during power stroke. Arrow indicates base of parapodium. acic.mu, acicular muscle; ant.para.ob.mu, anterior parapodial oblique muscle; b.w, body wall; d.cir, dorsal cirrus; d.para.mu, dorsal parapodial muscle; lev.mu, levator muscle; neuro.prot.mu, neuropodial protractor muscle; noto.ch, notopodial chaetae; para, parapodium; para.intr.mu, parapodial intrinsic muscle; p.para.ob.mu, posterior parapodial oblique muscle; trans.mu, transparapodial tensor muscle; v.long.mu, ventral longitudinal muscle; v.para.mu, ventral parapodial muscle.

do not move at random, but their stepping is coordinated to give a rhythmical pattern involving a block of 5-6 segments repeated down the body and appearing as a series of synkinetic metachronal waves, that is, passing from tail to head. At rest, each appendage is inclined slightly backwards. From this position, when crawling starts, every 5th or 6th parapodium, apparently in unison, is the first to move. It is raised and swung forwards to make contact with the ground, into which the chaetae are thrust. This movement is accomplished by contraction of the levator and wall muscles of the parapodia; because of their dorsal origin they lift the parapodia; because they are better developed on the anterior wall they pull it forwards. Protrusion of the chaetae and remotion (backward movement) of the parapodia are achieved by contraction of the muscles attached to the acicular and chaetal sacs: when all contract synchronously the chaetae are thrust out; differential contraction of the anterior muscles pulls the head of the aciculum forwards and so forces the whole parapodium backwards, bending it along the hinge line. Promotion (forward stepping) is made more efficient by a retraction of the neuropodium by means of muscles of the ventral parapodial wall so that it becomes shorter than the notopodium.

After the effective stroke the chaetae and acicula are withdrawn and the cycle repeated. Meanwhile adjacent parapodia follow the same pattern of movement. It is started by the parapodium immediately anterior when the active one has begun its effective backward stroke, and is then picked up by the next anterior limb. The right and left parapodia of any one segment are out of phase. The number of segments affected by a single wave of movement varies with the region of the body and the frequency of beat; it may be as small as 2 or as great as 8.

Cinematograph records reveal that the first parapodium to be active when crawling begins is always in an anterior group of segments and that activity spreads rapidly towards the tail, being taken up by the corresponding parapodium in each group; all this is too rapid to be seen. The metachronal waves appear to pass forwards since in a single wave the transmission of activity, at a relatively slow rate, is in an anterior direction, the direction of movement of the worm.

During rapid crawling this pattern of activity of the parapodia is coordinated with unilateral waves of contraction of the longitudinal muscles, which also pass forwards. Their timing is related to parapodial movement so that the modified circular muscles which move the parapodia are contracted at points where the longitudinal muscles are relaxed, an arrangement generally observed in annelids. Cinephotographs indicate that a wave of contraction starts at the posterior end of a group of segments at the anterior end of the worm, is picked up rapidly by similarly placed segments in succeeding groups, and then, at a slower rate spreads forwards to the anterior segments. As indicated in Fig. 50A, a parapodium is approximately perpendicular to the lateral surface of the segment (1) when the adjacent longitudinal muscles are contracted. As they relax the parapodium is drawn forwards (2) and the backward power stroke begins half way along the leading edge of the wave (3) and is maximal at its crest where the parapodium makes most effective contact with the substratum.

Swimming entails the same movements as those used in rapid creeping, though muscular waves are fewer and their amplitude greater. Since the waves are synkinetic it may be wondered how a force can be generated which will move the worm forwards. The waves are of significance only in so far as they bring the parapodia at

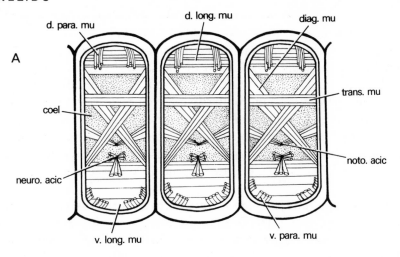

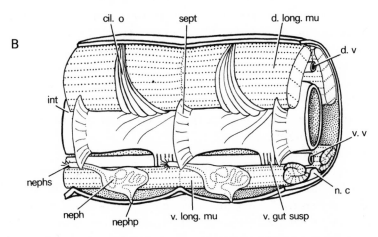

Fig. 53. *Nereis diversicolor*: A, left lateral view of 3 segments with parapodia cut away at base; B, left lateral view of 2 segments with part of body wall removed to show position of metanephridium and ciliated organ. (B after Goodrich.) cil.o, ciliated organ; coel, coelom; diag.mu, diagonal muscle at entrance to parapodial coelom; d.long.mu, dorsal longitudinal muscle; d.para.mu, dorsal extrinsic parapodial muscle; d.v, dorsal vessel; int, intestine; n.c, nerve cord; neph, metanephridium; nephp, nephridiopore; nephs, nephridiostome; neuro.acic, neuropodial aciculum with radiating muscles; noto.acic, notopodial aciculum with radiating muscles; sept, septum; trans.mu, transparapodial tensor muscle; v.gut susp, ventral gut suspensor muscle; v.long.mu, ventral longitudinal muscle; v.para.mu, ventral extrinsic parapodial muscle; v.v, ventral (subintestinal) vessel.

their crests to a position where their power stroke can create a posteriorly directed flow of water, and must be a metabolically expensive way of achieving this end. In swimming, movement of the parapodia is brought about by the action of the oblique muscles (Figs 51A and 52A) and does not involve the intrinsic and acicular muscles used in slow crawling.

Nereis irrigates its burrow; it is not then the longitudinal muscles of right and left side which work as antagonistic pairs but the dorsal and ventral ones, for the undulating waves are dorsoventral and travel from head to tail. These irrigation movements are intermittent, alternating with periods of rest, and particles in the irrigation current may be collected for food. The worm secretes a mucous net across the opening of the tube, drives water through it and swallows the net and entrapped particles; the whole process takes up to 7 minutes. The net is secreted by glands on the anterior parapodia and moulded by the chaetae into a funnel which leads to the mouth.

The gut is a straight tube leading from mouth to anus and suspended in the coelom by mesenteries (Fig. 51A, int). The intestinal region which extends through the greater part of its length is lined by a columnar epithelium which secretes a brownish digestive fluid, absorbs the products of digestion and secretes a peritrophic membrane in which faecal matter is encased. The products of digestion are passed to blood capillaries which penetrate the visceral musculature to come into contact with the base of the epithelial cells. Anteriorly the alimentary canal is modified for the ingestion of food and its preliminary trituration and digestion. The initial part, extending through four segments behind the peristomium, forms a proboscis (Fig. 54). Posterior to this is an oesophagus with a pair of anterior diverticula, resembling it histologically (oes.div), and then a thick-walled gizzard (giz). Between each section valves can close the lumen (oes.val), isolating the contents and preventing regurgitation when the proboscis is working.

The proboscis carries a pair of laterally placed jaws (j) which are brought into action when it is protruded. It also bears a number of denticles (d) characteristically arranged. They are thickenings of the cuticle and help to shift soil particles as the worm burrows or ingests food. Each jaw may be compared with a pruning hook with its cutting edge deeply serrated; it is moved by muscles embedded in the proboscis wall. When the worm is not feeding the proboscis is retracted, the mouth is closed by the oral sphincter (oral sph) and the peristomium is contracted around it. Protraction is brought about by paired muscles (prob.prot) aided by increased coelomic pressure in the anterior segments of the body; the absence of septa here allows a greater displacement of coelomic fluid than would otherwise be possible, increasing the turgidity of the head and fixing the anterior ends of the protractor muscles so that when they contract they pull the gut forwards. Food is sucked into the gut by contraction of radial muscles in the thickness of the proboscis wall (rad.mu) perhaps helped by the oesophageal diverticula. The proboscis is also used to drive a way through soft soil during burrowing as segments posterior to it anchor the worm.

The presence of coelomic sacs separating gut and body walls and the maintenance of a higher metabolic rate necessitate the development of a more efficient vascular

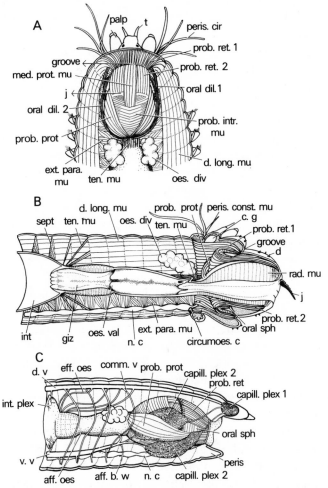

Fig. 54. *Nereis diversicolor*: anterior end. A, segments posterior to peristomium opened by mid-dorsal incision, proboscis retracted; B, sagittal half, proboscis protracted; C, plan of vascular system from right side, proboscis retracted. The anterior arrow in A represents the position of the groove, the posterior arrow that of the jaws. aff.b.w, afferent body wall vessel; aff.oes, afferent oesophageal vessel; capill.plex 1, capillary plexus over cerebral ganglion; capill.plex 2, capillary plexus associated with proboscis; c.g, cerebral ganglion; circumoes.c, cut end of right circumoesophageal connective; comm.v, right commissural vessel; d, denticle; d.long.mu, dorsal longitudinal muscle; d.v, dorsal vessel; eff.oes, efferent oesophageal vessel; ext.para.mu, extrinsic muscles of parapodia; giz, gizzard; groove, groove encircling outer wall of proboscis; int, intestine; int.plex, intestinal plexus; j, jaw; med.prot.mu, median protractor muscle; n.c, nerve cord; oes.div, oesophageal diverticulum; oes.val, oesophageal valve; oral dil. 1, oral dilator inserted on anterior margin of peristomium; oral dil. 2, oral dilator inserted on posterior margin of peristomium; oral sph, oral sphincter; palp, prostomial palp; peris, peristomium; peris.cir, peristomial cirrus; peris.const.mu, peristomial constrictor muscle; prob.intr.mu, intrinsic muscle of proboscis; prob.prot, proboscis protractor muscle; prob.ret 1, retractor of thin outer wall of proboscis; prob.ret 2, retractor of thick inner wall of proboscis; rad.mu, radial muscles; sept, septum; t, tentacle; ten.mu, tensor muscles; v.v, ventral vessel.

system than that possessed by nemertine worms. The vascular system consists of dorsal and ventral longitudinal vessels (Fig. 51, d.v, v.v) linked by segmental commissural vessels running in the body wall and a plexus of vessels in the gut wall (int.plex). There are no special hearts but all vessels are to a certain extent contractile, especially the dorsal one which maintains a postero-anterior flow. Typically a pair of ventral commissural vessels (v.comm.v) arises from the ventral vessel posterior to the septum in each segment. Each passes over the ventral longitudinal muscle block, supplies the nephridiostome and enters the parapodium to supply the ventral parapodial wall and the important respiratory area of the dorsal lobe of the notopodium. The notopodium also receives blood from a circumintestinal vessel running on the posterior surface of the septum and originating in the gut plexus (noto.v). Blood is returned from the parapodium to the dorsal vessel by a pair of dorsal commissural vessels which run over the dorsal longitudinal muscle blocks (d.comm.v). Not far from its origin each ventral commissural vessel is connected to the circumintestinal vessel (v.int) and in this way it may either supply the gut plexus or receive blood from it. The circumintestinal vessel is widely looped, which allows for the necessary freedom of movement of the intestine. The gut plexus is also directly linked to the dorsal vessel.

The segmental arrangement of the vascular system is modified anteriorly (Fig. 54C). Indications of this can be seen in the living worm where all but the four anterior pairs of parapodia are reddened by a network of capillaries. In these segments commissural vessels are absent. In the segments affected by the action of the proboscis (5–8) the dorsal and ventral commissural vessels connect directly with one another (comm.v) and the vascular supply associated with the gut is modified (eff.oes, aff.oes). The dorsal vessel continues to the prostomium where it forms a fine plexus over the cerebral ganglia (capill.plex 1). The ventral vessel extends as far as the posterior end of the proboscis where two branches on each side link with two branches from the cerebral plexus by way of extensive capillary beds (capill.plex 2), one intimately associated with the proboscis musculature. The network over the proboscis may compensate for the lack of local respiratory surfaces.

The circulation of the blood involves the integration of two patterns, a segmental one on which is superimposed a dorsal anterior and ventral posterior flow involving the whole worm. The peristaltic waves in the dorsal longitudinal vessel originate in the most posterior segments and travel forwards. In the segmental circulation (demonstrated by placing two tight ligatures around the worm to isolate a few segments) the flow is from the ventral and circumintestinal vessels, through capillaries in the parapodia, to the dorsal commissural and dorsal vessels. There is also a flow from dorsal to ventral vessel by way of the intestinal plexus. The segmental blood flow in isolated segments soon becomes irregular, indicating the interdependence of the two patterns. Contractions of the dorsal vessel are readily visible in an active worm and become more frequent with increased activity. Other vessels are contractile to some extent, having circular muscle fibres in their walls. The blood contains the pigment haemoglobin which is responsible for the transport of about 50 per cent of the oxygen consumed at high oxygen tensions (6–7 ml O_2 per litre) and an increasing percentage at lower tensions. Indeed all oxygen consumption is blocked at 3.3 ml O_2 per litre if the haemoglobin is inactivated by carbon monoxide.

Every segment of the body, except a few at each end, has a pair of excretory tubules (metanephridia) which lead from the coelom to the exterior (Fig. 53B, nephp). The main part of the organ, a much coiled, ciliated canal embedded in connective tissue, is close to the nephridiopore but its opening to the coelom, the nephridiostome (nephs), is in the preceding segment. It is a bell-shaped funnel with powerful cilia beating towards the lumen of the canal, and is suspended from the edge of an opening in the ventral part of the septum through which the nephridial tube passes. The margin of the funnel is capable of slow movement and has cytoplasmic processes which can ingest particulate waste. The cilia appear to regulate the entry of particles to the funnel as well as driving fluid into the tube. *Nereis diversicolor* is abundant in brackish water and can live where the salinity is as low as 4.0 per cent. It has been suggested that the excretory organs help to maintain its hypertonicity in dilute media not only by eliminating excess fluid but also through the resorption of salts by the epithelium of the canal. In related species restricted to sea water of full salinity (*Nereis cultrifera*) the nephridial canal is short. The segments which bear metanephridia have paired dorsal ciliated organs (cil.o). These are conspicuous areas of coelomic epithelium over the surface of the dorsal longitudinal muscles which are strongly ciliated, and their surfaces are ridged transversely. Each is triangular and from a broad base against the dorsal blood vessel tapers gradually as it passes laterally over the muscle. The organs represent the coelomostomes of modified coelomoducts and have no connexion with the exterior. The strong ciliary current which beats towards the apex of the area maintains a circulation of the coelomic fluid which would otherwise be stagnant when the worm is motionless in its burrow, and directs excretory phagocytes towards the nephridiostomes.

The sexes are separate and in all populations females predominate. In both sexes the gonads arise from a special area of ventral peritoneum in the septal region of all but a few anterior and posterior segments, and the germ cells break free at a very early stage and mature in the body cavity. Certain changes in the structure of the body are associated with their maturation. In the female the coelom becomes invaded by coelomic corpuscles which form a loose parenchyma and appear to be responsible for the deposition of yolk in the oocytes. As the oocytes grow they take up all available space including that in the parapodia; meanwhile the parenchyma disappears. Towards the time of spawning the body wall of the worm becomes thin owing to histolysis of the muscle layers. Fragments of muscle which appear in the coelom are phagocytosed and digested by wandering cells which deposit waste products in the form of bright green granules beneath the cuticle, so affecting the colour of the body. Oocytes may lie in the coelom for eight months and are released (typically in spring) by the splitting of the weakened body wall along the line of the dorsal longitudinal muscles; the ciliated funnels drive the gametes out. Unliberated ones disintegrate and their fragments are ingested by phagocytes. The green coloration of the worm is not lost until summer when the body wall becomes bronze and the muscles regenerate.

A similar series of events is associated with the sex cycle of the male, though the period during which parenchyma fills the coelom is very brief. Although histolysis of the body wall occurs, sperm are usually discharged through the metanephridia and perhaps only accidentally by rupture of the body wall. At the time of spawning a cluster of females comes to surround a single male and incite the emission of sperm. There is no epitokous swimming form (see p. 221) as in some other nereids.

Each segment has its own nerve centre in the form of a pair of ganglia situated mid-ventrally, just beneath the epidermis, which receive impulses from local receptors – peripheral sense organs related to a sub-epidermal plexus and proprioceptors within the muscles – and control the movements of the somatic and visceral muscles by way of paired nerves. The ganglia of one segment are linked by a commissure, those of different segments are united by paired longitudinal cords consisting of a mass of fibres. The two anterior pairs of ganglia, one ventral and behind the mouth, the other dorsal and above it, are differentiated from the rest on account of their size as well as their function. They are associated with the formation of a head which is concerned with the reception of stimuli from the environment and the capture of food. The dorsal pair, the supraoesophageal or cerebral, are the ganglia of the prostomium, a part of the body anterior to the mouth and not strictly segmental. They receive impulses from its sense organs and are important receptor centres. The ventral pair, the suboesophageal, innervate the segment surrounding the mouth, the peristomium, and are linked to the dorsal pair by connectives which pass around the gut.

The ganglia on the ventral nerve cord are not intrasegmental but lie with the greater part of each posterior to the intersegmental septum and a small part in the segment in front (Fig. 55, g). The part behind each septum gives off three of the four pairs of nerves in each segment, the fourth arising from the anterior portion of the succeeding ganglion (1, 2, 3, 4). All these nerves are mixed. The second is the only one to enter the parapodium, where it connects with integumentary receptors, and its efferent fibres innervate the intrinsic parapodial muscles. Near its entry into the parapodium it expands into a ganglion (para.g) which may integrate the reflex circuit within the limb. The third nerve is primarily proprioceptive, receiving fibres from the ventral longitudinal muscle band. On the lateral border of this it connects with a small longitudinal nerve which, it has been suggested, may register the pattern of body flexures associated with locomotion. The first and fourth nerves run parallel with the septal boundaries of the segment. They are associated with skin receptors on dorsal and ventral surfaces and with proprioceptors in the dorsal and

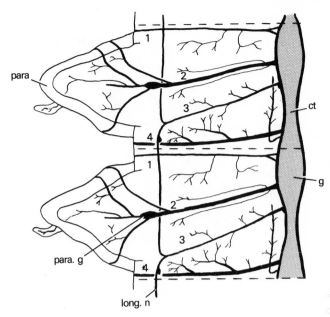

Fig. 55. *Nereis diversicolor*: diagram to show the course of segmental nerves 1–4 and their sensory distribution in two half segments, also the lateral nerve. Broken lines indicate segmental boundaries. (After Smith.) 1–4 segmental nerves; ct, connective; g, ganglion; long.n, longitudinal nerve; para, parapodium; para.g parapodial ganglion.

ventral longitudinal muscles, the extrinsic parapodial muscles and the obliques. Their motor fibres also supply these muscles. These nerves must control the coordinated movements of the muscles during locomotion.

The nerve cord is covered by a fibrous sheath from which arise the extrinsic parapodial muscles, and in which are longitudinal muscle fibres whose contraction shortens the cord. An inner fibrous sheath encloses the neuropile, a mass of fine nerve fibres and a few giant ones; at each ganglion this is surrounded ventrally and laterally by a cortex which contains the cell bodies of the motor and internuncial neurons running within the cord. The internuncial system of fibres has three pairs of longitudinal tracts (Fig. 56B, dm.f.t, dlat.f.t, v.f.t) as well as less developed transverse connexions. On entering the ganglion afferent fibres connect with one of these tracts, which presumably coordinates muscular and locomotor activities of the body from one segment to the next. Fibres entering from nerves 1 and 2, predominantly concerned with exteroceptive excitation, are mostly related to the dorso-lateral tracts; those from nerve 3, essentially proprioceptive, the dorso-median tracts, and fibres from nerve 4, with exteroceptive and proprioceptive components, contribute to both. There is thus some localization of function within the ganglion. Motor neurons (A, m.neur) lie laterally in the cortex and their axons cross in the dorsal neuropile to enter the nerves arising on the opposite side of the body.

In addition to the fibres of ordinary size there are four series of giant fibres confined to the cord: two lateral (lat.g), two paramedial (param.g) and an unpaired dorsal fibre (d.g). None of these appears to connect directly with afferent segmental fibres, and any transference of excitation must be by way of internuncial

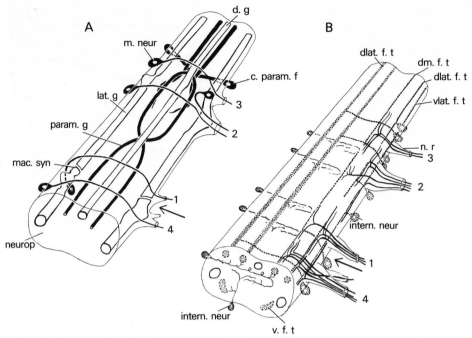

Fig. 56. A. Diagrammatic stereogram of a ganglion of *Nereis diversicolor* showing the form and arrangement of the giant fibres and the principal motor fibres of the segmental nerves. The anterior (pre-septal) end of the ganglion is at the bottom of the figure and the nerves are shown on one side only. Arrow indicates position of septum. B. Diagrammatic stereogram of a ganglion of *Platynereis dumerili* showing on the right the principal sensory fibres of the nerves (continuous lines) and some internuncial neurons with long axons supplying the longitudinal fine fibre tracts (dotted lines). On the left of the figure are some internuncial neurons with horizontal branching axons and one with a vertical axon. The position of the giant fibres is shown on the cut surface which is anterior. Arrow indicates position of septum. (After Smith.) 1–4, roots of segmental nerves; c.param.f, cell body of paramedial fibre; d.g, dorsal giant fibre; dlat.f.t, dorsolateral fine fibre tract; dm.f.t, dorsomedian fine fibre tract; intern.neur, internuncial neuron; lat.g, lateral giant nerve fibre; mac.syn, macrosynapse; m.neur, motor neuron; neurop, neuropile; n.r, nerve root; param.g, paramedian giant fibre; v.f.t, ventral fine fibre tract; vlat.f.t, ventrolateral fine fibre tract.

fibres. This is not the case with the efferent fibres: the main motor axons make a number of contacts with internuncial fibres, and with the giant fibres, as they cross the dorsal neuropile. The giant fibres, specialized for rapid responses, are associated with the same motor axons as the fine internuncial fibres, but are stimulated only by greater impulses and, if called upon, can stop the sequence of movements which they have initiated. In *Nereis virens* the lateral giant fibres have the lowest threshold and respond to an adequate stimulation of any part of the skin, the median dorsal fibre responds to stimulation of the anterior quarter of the body and the paramedial fibres to stimulation of the posterior three quarters. They conduct impulses in either direction.

It has been shown that at least some of the segmental muscles of *Nereis* are in-nervated by three motor axons, and evidence suggests that two of these produce fast and slow contractions respectively. This recalls the polyneural innervation of crustacean muscle (p. 296). No inhibitory fibre has yet been confirmed for poly-chaetes. The fast fibres in polychaetes (directly or indirectly connected to the giant fibre system) may be associated with rapid escape movements, whilst the slow ones may control the more normal movements of locomotion.

A survey of the polychaetes shows that almost all are marine. Many are errant and for the most part predacious; others are sedentary, living permanently in tubes or burrows, and predominantly microphagous. The word "errant" perhaps exag-gerates, since many species so called spend much of their time sheltering in crevices under stones, or in sand or mud and emerge only to hunt for food; a few swim freely in the sea. Sedentary species rarely come out of their homes. These contrasting modes of life have been made possible by the adaptability of the fundamental annelid body form; each has called for a different development of many of the systems of the body. Adaptations tend to develop in relation to movement, ventilation of respiratory surfaces, feeding and the passage of food through the gut, circulatory and excre-tion systems and reproduction. These will now be looked at in a more general way throughout the class.

Movement of polychaetes

Locomotor ability is best developed in errant forms and depends partly on the possession of parapodia which act as levers and so amplify the effect of the con-traction of a given volume of muscle. The other main feature of polychaete locomo-tion is the lateral undulation of the body. The worm, indeed, grips the substratum at the points where the crests of the waves lie and it is at these points that the back-ward stepping of the parapodia combines with this to move the animal forwards.

Many polychaetes which are rapid creepers also swim (phyllodocids, syllids, nereids), the swimming process being an exaggeration of the creeping process. Their body is long and with numerous segments. The most efficient swimmers have parapodia which present a large surface area to the water and this is brought about by the development of paddle-like surfaces best seen in larger worms; very small ones (e.g. syllids, Fig. 77) can manage without. The most familiar of the large, swimming polychaetes (phyllodocids) are known as paddle worms on account of the leaf-like lobes which border the uniramous parapodia. These are the expanded dorsal and ventral cirri (Fig. 57A, d.cir, v.cir); the dorsal ones are particularly large and are directed obliquely across the back when at rest but all share in the parapodial move-ment when the worm swims. When they become sexually mature a number of other-wise sluggish species undergo similar structural modifications which allow them to swim during a brief pelagic life (epitokous or heteronereid phase). The modifications

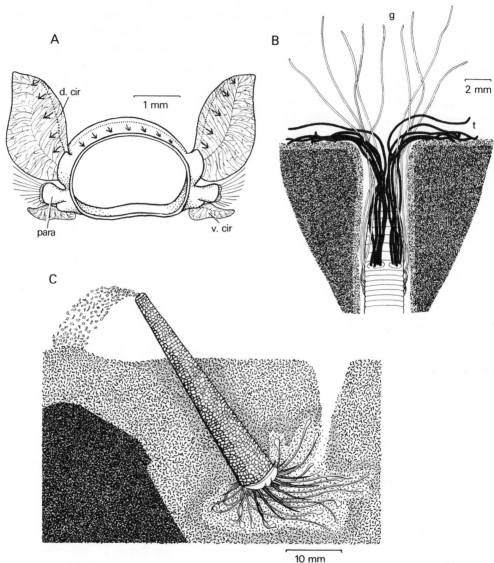

Fig. 57. A, *Phyllodoce laminosa*: posterior view of single segment. The leaf-like dorsal cirri have been raised from the dorsal body wall to show the position of the bands of cilia (dots) and the direction of the currents they produce (arrows) both on the posterior surface of the cirri and on the dorsal body wall. B, *Cirratulus cirratus*: when the head of the burrow is under water the tentacles (black) collect food from the surrounding substratum and the gills extend. C, *Pectinaria belgica*: feeding beneath the surface of the sand. Tentacles with prehensile tips extend to reach deposits rich in organic detritus and pumping movements of the body draw these deposits towards the head. Tentacles examine them for food and unwanted material is expelled from the posterior end of the tube. Digging chaetae used for burrowing are spread around the head. See also Fig. 63. d.cir, dorsal cirrus; g, gills; para, chaetigerous lobe of parapodium; t, tentacles; v.cir, ventral cirrus.

seen in *Perinereis cultrifera*, for example, include the development of lamellae on the notopodium and neuropodium, the enlargement of dorsal and ventral cirri, and the growth of special natatory chaetae which are more numerous than the normal ones and paddle-shaped. At the same time the parapodial muscles increase in bulk and the longitudinal muscles of the body wall are replaced by a special type of fibre with more voluminous sarcoplasm, perhaps related to greater swimming efficiency. *Nephthys*, the cat worm, is one of the most powerful swimmers. Its body segments are short and the closely set parapodia (Fig. 58A) sharply marked off from the prominent longitudinal muscle bands, which fling the body into many vigorous waves and so allow many parapodia to exert their power strokes. The two rami of the parapodia (Fig. 58B) are widely separated by a respiratory channel (p. 200) and each is bordered by a membranous paddle (pdl) similar to the parapodial lobes of the heteronereid. The power stroke is most pronounced in the distal half of the limb where only the acicular muscles have their insertions; they must therefore provide the chief propulsive force. No extrinsic parapodial muscles are inserted at the tip of the

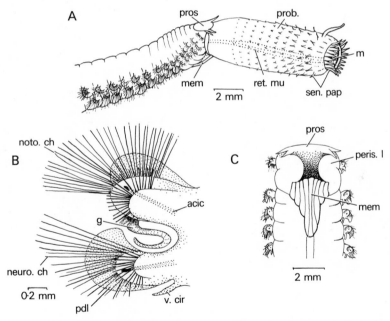

Fig. 58. *Nephthys*: A, anterior end of *Nephthys cirrosa* with proboscis protruded. Coelomic fluid between the thick inner and the thin outer wall distends the proboscis; between these two walls pass some of the proboscis retractor muscles. Two jaws embedded in the thick inner wall are not protruded. B, characteristic features of a nephthydid parapodium (anterior view). Some chaetae have been cut short. C, ventral view of anterior end with proboscis retracted. acic, aciculum; g, gill; m, mouth; mem, folded membrane joining lips; neuro.ch, neuropodial chaetae; noto.ch, notopodial chaetae; pdl, membranous paddle; peris.l, peristomial lip; prob, proboscis; pros, prostomium; ret.mu, retractor muscle; sen.pap, sensory papillae; v.cir, ventral cirrus.

limb, which may account for the inability of *Nephthys* to crawl using these appendages alone.

The creeping movements of the polynoid or scale worms, common on rocky shores (Figs 59B and 68B), are more like the stepping movements of myriapods or crustaceans than those of a nereid and wave motion along the body has been lost. Stepping is due to differential action of the acicular muscles and contraction of the muscles of the walls of the neuropodia, which are stout, conical and move independently of the notopodia which have little or no part to play in locomotion. Trunk muscles (the longitudinal muscle blocks) have no locomotory function and are

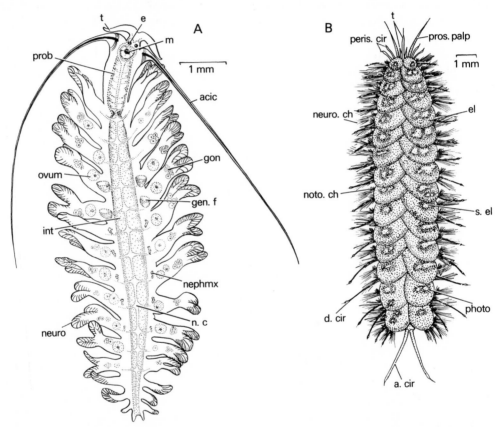

Fig. 59. A, *Tomopteris*: female, ventral view, some organs seen by transparency. The nephromixia (protonephridium and coelomostome) are similar to those of *Phyllodoce*. B, *Harmothoë lunulata*: dorsal view. The clear area near the centre of each elytron represents its surface of attachment to the body and the darker area surrounding this is the position of the photogenic tissue. acic, aciculum; a.cir, anal cirrus; d.cir, dorsal cirrus; e, dorsal eye; el, elytron; gen.f, genital funnel (coelomostome); gon, gonad; int, intestine; m, mouth; n.c, ventral nerve cord; nephmx, nephromixium; neuro, neuropodium; neuro.ch, neuropodial chaetae; noto.ch, notopodial chaetae; ovum, ovum; peris.cir, peristomial cirri; photo, photogenic organ; prob, proboscis; pros.palp, prostomial palp; s.el, stalk of elytron; t, tentacle.

reduced. A long and thin body which can be flung into waves is not advantageous for this type of locomotion and it becomes shorter and plumper. Since the parapodia function with little need of a hydrostatic skeleton mesenteries and septa are lost. These worms cannot swim – their body is too heavy, generally too short and their parapodia not adapted. The large scale worm (the sea mouse, *Aphrodite aculeata*) has a short, broad body pointed at both ends. It creeps through the surface layers of mud, the anterior neuropodia scooping away the particles which are heaped laterally as the head pushes forward; all but the tail is buried.

Errant worms which are intertidal hide at low water and sublittoral species go into crevices or burrow into soft substrata between meals. Those that enter narrow crevices do so by virtue of their flattened body (e.g. polynoids, phyllodocids) and those that burrow by virtue of adaptations mainly affecting their anterior end. *Nephthys* (Fig. 58A) and *Glycera*, both elongated worms, penetrate compacted sand with considerable agility and force, by means of the large proboscis (prob), also used to capture prey. They first swim into the sand penetrating it with a head which has become pointed in accordance with this habit and has reduced appendages and eyes. Chaetae then anchor the body and the proboscis bores a hole into which the worm moves by locomotor waves similar to those used in swimming. Greater force must be applied to penetrate sand than mud and the powerful protrusion of the proboscis of these sand dwellers is undoubtedly associated with this. To resist excessive deformation during burrowing the body wall and nerve cord of *Nephthys* are braced by a ligamentary system of special strips of connective tissue, not found in other worms, which are partly elastic. A new set of antagonists to the longitudinal muscles appears in the form of well-developed dorsoventral muscles; these control the height of the body segments and therefore their grip on the walls of the burrow by the parapodia.

The degree of mobility of sedentary worms is variable and the nature of their homes is related to the nature of their environment. Some live permanently in tubes; others live in burrows in soft substrata, their movement less restricted than that of the the strictly tubicolous forms. In a tube or a burrow the environment is the same on all sides and consequently the body tends to become symmetrical about its longitudinal axis and approximately circular in transverse section, with the muscles of the body wall more or less evenly developed all round. The majority of these polychaetes collect particles of food from near the opening of their home by spreading tentacles through the water or trailing them over the substratum. Since food can be collected with equal success in all directions the tentacles become set around the head in a radial manner (like those of a sea-anemone) so that the original bilateral symmetry may not be obvious at first glance. The mouth of a burrow or tube is the only link which the animal has with its environment and an anteroposterior differentiation of the body is a frequent consequence of this: gills, feeding organs, excretory organs are commonly concentrated in or limited to anterior segments. Since the anterior and

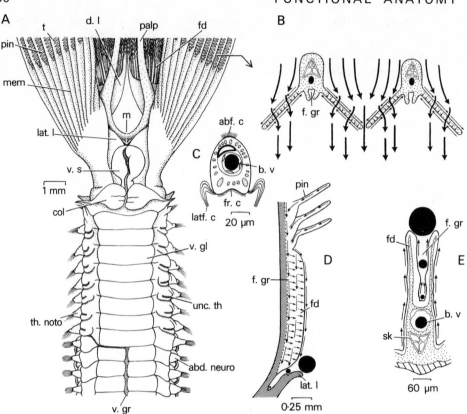

Fig. 60. *Sabella pavonina*: A, ventral view of base of tentacular (= branchial) crown and first 12 body segments. B, diagrammatic transverse section of two tentacles (as indicated by arrow on A) to show the direction of the flow of water entering the crown (large arrows) and the direction of the ciliary beat on the pinnules (small arrows). The longitudinal food groove of each tentacle is overarched by expanded bases of pinnules; particles from pinnules enter it between these bases. C, transverse section of pinnule to show ciliation. D, longitudinal half of base of tentacle, associated basal fold and lateral lip to show the ciliary currents (indicated by arrows) which direct the course of the particles and the relative sizes of the particles (black dots). (Cut surfaces hatched.) E, transverse section through a pair of basal folds to show the tracks followed by the different sized particles (black dots). Arrows indicate direction of beat of cilia. (After Nicol.) abd.neuro, abdominal neuropodium; abf.c, abfrontal cilia; b.v, blood vessel; col, collar; d.l, dorsal lip; fd, basal fold; f.gr, food groove; fr.c, frontal cilia; latf.c, laterofrontal cilia; lat.l, lateral lip; m, mouth; mem, membrane uniting tentacles; palp, palp; pin, pinnule; sk, skeletal support; t, tentacle; th.noto, thoracic notopodium; unc.th, uncini of thoracic neuropodium; v.gl, ventral gland shield; v.gr, ventral (faecal) groove; v.s, ventral sac.

posterior parts of the body subserve different functions they tend to become visibly differentiated both externally and internally, making regions known as thorax and abdomen respectively. It has not usually proved possible to bring the anus forward to the mouth of the tube, but in many tubicolous worms a ciliated groove, along which faecal matter travels (Fig. 60A, v.gr), runs from there to the head region. In sabellariids the posterior end of the body is bent forwards and the faecal pellets forcibly ejected through the mouth of the tube (Fig. 61B).

The parapodia of these worms do not form prominent lateral projections from the body and the basal part may be lost, leaving notopodium and neuropodium to arise separately and, perhaps, acquire different functions. This helps to give the characteristic circular transverse section. The attachment of the parapodium elongates dorsoventrally and notopodial and neuropodial parts differentiate in relation to the

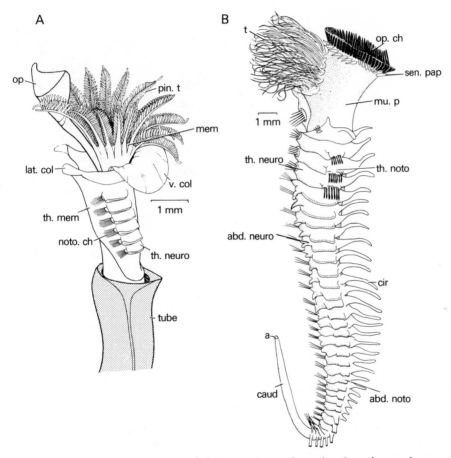

Fig. 61. A, *Pomatoceros triqueter*: extended from tube to show the thoracic membrane and parapodia. B, *Sabellaria alveolata*: removed from tube and viewed from left. a, anus; abd.neuro, abdominal neuropodium; abd.noto, abdominal notopodium; caud, caudal region; cir, dorsal cirrus or branchia; lat.col, lateral fold of collar continuous with thoracic membrane; mem, membrane uniting base of tentacles; mu.p, muscular pillar (peristomium fused with some anterior segments); noto.ch, notopodial chaetae; op, operculum; op.ch, opercular chaetae; pin.t, pinnule of tentacle; sen.pap, sensory papilla; t, feeding tentacles; th.mem, thoracic membrane; th.neuro, thoracic neuropodium; th.noto, thoracic notopodium; tube, calcareous tube; v.col, ventral fold of collar.

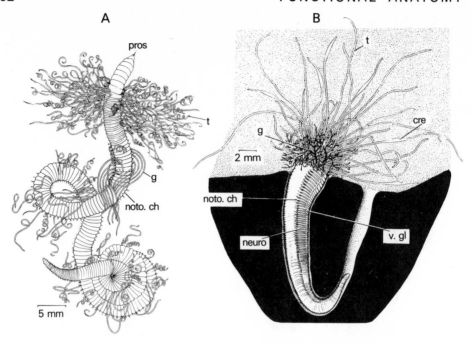

Fig. 62. A, *Cirratulus tentaculatus*. (After McIntosh.) B, *Terebella lapidaria*; extended from tube in the silt of a rock crevice and collecting food from the adjacent substratum. The ventral epidermis of the tentacle is ciliated and local areas are flattened and applied as creeping zones where the cilia drive the tentacle over the substratum. cre, creeping zone; g, gill; neuro, neuropodium; noto.ch, notopodial chaetae; pros, prostomium; t, tentacle; v.gl, ventral glands.

two major functions of gripping the walls of the tube and moving the worm along them (Figs 62B and 63). Two kinds of chaetae are associated with these functions: long bristles projecting from a conical process of the body wall which may be directed forwards and back and also invaginated and protruded to withdraw or erect them, and short, hooked, sigmoid chaetae or uncini, arising from a groove in a transverse ridge, which can be moved in various directions and rotated to a certain degree. The long chaetae lever the body along the tube or burrow, and maintain a space between it and the surrounding wall. The uncini have a gripping action. In tubicolous worms the long chaetae are located on the notopodia in the thorax but on the neuropodia in the abdomen, an arrangement which improves the mechanics of movement within the tube (Fig. 60A, th.noto, abd.neuro).

A sedentary worm emerges slowly from its shelter, but if disturbed its withdrawal is a single rapid jerk brought about by the giant nerve fibre system, particularly well developed in the fan worms (sabellids) and in the common tube worms of rocky shores (serpulids). They have one or two axons which extend the whole length of the body. In the sabellid *Myxicola* a single giant axon (up to 1 mm in diameter) links all

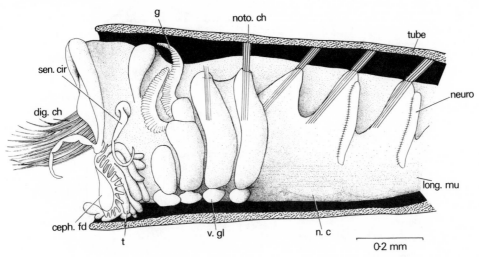

Fig. 63. *Pectinaria* (= *Lagis*) *koreni*: anterior part of worm from left, half of the tube removed. Dorsal and lateral lips of mouth surrounded by contractile feeding tentacles which, when withdrawn, are covered by the preoral cephalic fold. ceph.fd, cephalic fold; dig.ch, digging chaetae; g, gills; long.mu, longitudinal muscles; n.c, ventral nerve cord; neuro, neuropodium with uncini; noto.ch, notopodial chaetae against tube; sen.cir, sensory cirrus; t, tentacle; tube, cut surface of tube; v.gl, ventral gland shield.

the longitudinal muscles involved in the withdrawal reflex and has a conduction velocity may be as high as 20 m s^{-1}.

The more primitive sedentary polychaetes live in burrows, the more advanced in carefully constructed tubes. Burrows are to be found in the surface layers of all types of soft substrata, each species having its particular preference. Some even live in mud odoriferous with organic matter, either in a crevice or under stones. The burrow may be a more or less permanent home, its walls becoming lined by secretions from epidermal glands and compacted by the pressure of the body as it moves to and fro; or it may be a temporary shelter, for some burrowers retain freedom of movement (e.g. *Nerine, Scolelepis*) and leave the burrow from which they have been feeding to move over the surface, and then burrow again by swimming into the soil and penetrating it with the proboscis. The form of the burrow varies with the species. It may be irregular with several openings (the red rockworm *Marphysa, Nereis*), U-shaped with two (*Amphitrite, Polydora*), vertical (the red threadworm *Cirratulus*), or L-shaped (the lugworm *Arenicola*) with one. It is always considerably wider than the worm and when there is more than one opening the worm can reverse its position and feed from any one.

Arenicola, the lugworm, normally lives with its tail towards the open end of its burrow which is marked by a pile of sandy faeces since it is used by the worm at the time of defecation (Fig. 64G). The burrow runs vertically down for about a foot

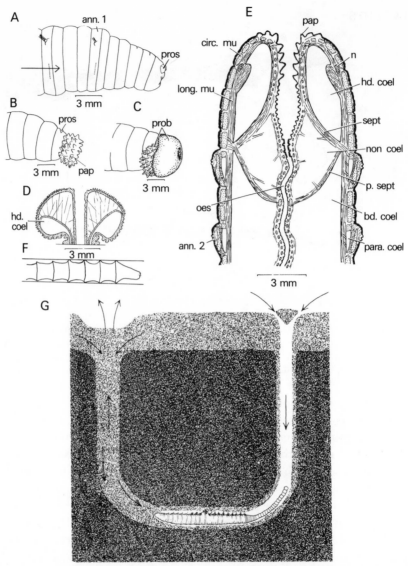

Fig. 64. *Arenicola*: A, Lateral view of anterior end; arrow indicates the level of the section E. B, proboscis partly extruded. C, proboscis extruded. D, horizontal longitudinal section through extruded proboscis. E, horizontal longitudinal section through anterior end. F, anterior segments to show chaetigerous annuli pressing against wall of burrow. G, diagram of a burrow with the head of the worm at the base of the head shaft. At the open end is a pile of faeces and two defecation channels. Yellow sand is at the surface, forms the head shaft and lines the burrow; black sand is elsewhere. Arrows show the movement of water, and, those passing down the head shaft, the movement of sand. The worm is ventilating and a dorsal bulge is pushing water from right to left. (Based on Wells.) ann.1, 1st chaetigerous annulus; ann.2, 2nd chaetigerous annulus; bd.coel, body coelom; circ.mu, circular muscles; hd.coel, head coelom; long.mu, longitudinal muscles; n, nerve; non coel, noncoelomic space in 1st septum; oes, oesophagus; pap, papillae covering surface of proboscis; para.coel, parapodial coelom; prob, proboscis; pros, prostomium; p.sept, posterior wall of 1st septum; sept, 1st septum.

then turns into a horizontal course from which the so-called head shaft runs up to the surface again to end below a depression in the sand. Typically the head shaft is not an open channel but a column of loose sand from the bottom of which the worm feeds. The permanently open part of the burrow is thus L-shaped with its wall compacted and impregnated with mucus. The sand which lines it is well oxygenated and yellow, like the surface sand and the head shaft, contrasting with the darker sand elsewhere. The worm cannot crawl, but if it is put on the surface of wet sand it begins to burrow immediately by means of peristaltic body movements and by using the proboscis. The method of burrowing has been studied in considerable detail and it will be of interest to describe this for such a common worm (Fig. 65).

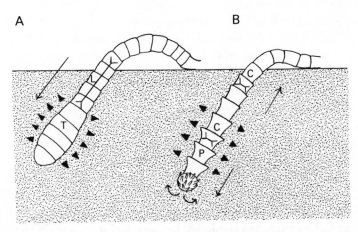

Fig. 65. Diagram to show the events of two stages in the burrowing of the worm *Arenicola*. A, the worm dilates the anterior end of the body by pressing coelomic fluid into it so that it forms a terminal anchor (T); contraction of the longitudinal muscles of some segments behind the anchor (L) pulls the worm into the sand in the direction of the arrow. B, this stage follows A; the more anterior segments form flanges constituting a penetration anchor (P) and contraction of the circular muscles between the double-headed arrows (C) narrows the body. As a result the most anterior part of the body penetrates the sand in the direction of the lower arrow whilst some more posterior segments move out of it in the direction of the upper arrow. Meanwhile the proboscis everts to give further penetration. Black arrowheads show areas of anchorage.
(After Trueman.)

When penetration begins repeated eversion of the proboscis loosens the sand and the first part of the body makes an entry into what will be the beginning of the burrow. The anterior segments grip the sand which now surrounds them, forming what has been called a terminal anchor (T). This is brought about by contraction of the longitudinal muscles in these and nearby segments (L) and the production of a high internal coelomic pressure by peristaltic waves moving anteriorly along the trunk. This phase is succeeded by a second in which certain annuli of anterior segments are dilated by coelomic fluid to form projecting flanges, the penetration anchor (P), which immobilize the front end of the worm; contraction of the circular muscles of segments anterior to this pushes the head forward and that of segments posterior causes some emergence of these segments from the burrow. During this phase the proboscis is everted to displace sand as the head advances. A terminal anchor is then reformed and contraction of the longitudinal muscles shortens the worm and pulls more segments into the burrow. Burrowing consists of the regular alternation of these two phases. The pressure of the fluid in the body coelom of a worm lying on the surface and not wriggling averages about 14 cm of water, whereas in a burrow the longitudinal muscles can produce pressures up to 100 cm of water. These high pressures enable the worm to burrow rapidly once it has entered the sand. They are dependent on the open nature of the coelom which allows the force provided by the simultaneous contraction of the longitudinal muscle of many segments to be deployed at the anterior end. The posterior coelom is septate and the pressure of the coelomic fluid there remains low. In the corresponding region of the gut faeces accumulate, being propelled posteriorly by anterior pressure changes.

In order to explain the mechanism of the proboscis its anatomy must be described. The head of *Arenicola* extends forwards from the region of the 1st chaetigerous annulus (Fig. 64A). Its coelom is separated from that of the general body coelom by the first septum which consists of two independent sheets of tissue, each with radial muscle fibres (Fig. 64E), and with valves which allow passage of fluid from the general coelom into the head coelom. The posterior sheet expands into a pair of septal pouches. The movements of the proboscis are dependent on its own circular and longitudinal muscles and on those of the adjacent body wall. When the proboscis is about to be protruded the head narrows and lengthens with the contraction of its circular muscles. The opening of the mouth and emergence of the first part of the proboscis (Fig. 64B) are brought about by the relaxation of their muscles, by contraction of the radial muscles of the posterior sheet of the septum and by an increase in pressure in the head coelom caused by these events and by an inflow of fluid from the general coelom due to contraction of the trunk musculature. Full eversion (Fig. 64C) follows, with a shortening and widening of the head as the longitudinal muscles of the body wall and extruded part of the proboscis contract.

The anterior end of the worm apparently contains some mechanism ensuring the passage of fluid from the general coelom to the proboscis at the moment of extrusion. Withdrawal is brought about by contraction of the radial muscles in the anterior wall of the septum. It may occur while the head is short and broad or as the circular muscles again contract to narrow and lengthen it. In the former case sand is sucked into the mouth; in the latter, closure of the mouth obstructs intake. Feeding also depends upon the secretion of glands on the proboscis wall which causes organic matter and sand to stick to its surface. The papillae on the wall of the proboscis may also help ingestion mechanically.

Fig. 66. *Polydora ciliata*: A, tube building: a sand grain gripped by the lateral lips will be placed on the rim of the tube (the tentacles are cut short). B, longitudinal section of burrow with worm in feeding position: between the two limbs of the burrow are accumulated layers of silt, the layering indicating the gradual deepening of the burrow. (After Dorsett.) cal, calcareous substratum; cil.gr, ciliated groove; lat.l, lateral lip; pros, prostomium; rim, rim of tube; sa.gr, sand grain; sa.muc, sand-mucus tube; silt, silt layers; t, tentacle; tube, tube projecting from substratum.

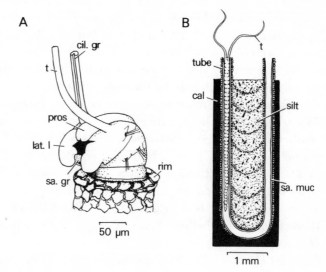

Some polychaetes live in silted crevices and this may have led to the habit of boring into rock. *Polydora ciliata* is a common polychaete, though inconspicuous on account of its size, living in calcareous rocks of varying hardness and also in mollusc shells (Fig. 66). It appears to penetrate these by mechanical and chemical methods using the powerful uncini of the enlarged 5th segment to scratch the surface whilst it is being acted on by an acid or chelating agent secreted by epidermal glands. As the worm grows it makes a U-shaped burrow, both limbs of which are lined by a tube which projects about 0.5 mm from the substratum and which is made of mucoprotein secreted by epidermal glands into which sand grains collected with the food are bound. As the worm grows the burrow penetrates deeper into the substratum.

Many sedentary polychaetes live in tubes, the organic matrix of which is secreted by glands distributed either over the whole surface of the body (*Myxicola*) or, more frequently, over a limited area. The main constituents are polysaccharide-protein complexes which give the tube the consistency of soft jelly or slime (*Myxicola*), or parchment (*Chaetopterus*), or horn (*Hyalinoecia*), but in the majority the secretion is strengthened by foreign material specially collected by the worm and fitted into place. The tubes of all serpulids are calcareous, with only 3–6 per cent organic matter (dry weight). The mineral matter is predominantly calcium carbonate in the form of aragonite (*Pomatoceros*). Tube building is a more or less continuous process, for although tubes are usually permanent homes they must be added to as the worm grows or as external conditions such as overcrowding and silting demand. It is exceptional for a worm to leave its tube and settle elsewhere and some, like *Chaetopterus*, are incapable of burrowing if removed from the tube although its American relative, *Spirochaetopterus*, can do so and form a new tube round itself. On occasions

Myxicola, the slime dweller, does this, particularly when young, swimming backwards by pulsations of its branchial crown, but its tube is little more than a mucous bag readily secreted around the body.

Building is typically carried out at the broadest region of the worm, the peristomial region, which may be the only part involved (serpulids) or may be assisted by the lips around the mouth (*Sabella* (Fig. 60A), *Bispira*, *Potamilla*, *Sabellaria* (Fig. 67), *Lanice*, *Polydora*). The lips manipulate particles of mud (*Sabella*), sand (the honeycomb worm *Sabellaria*, *Pectinaria*, *Polydora* (Fig. 66A)), and pieces of shell (the sand mason *Lanice*) which are mixed with secretions from the peristomium or adjacent areas. In addition there is often a mucous lining to the tube which comes from a series of glandular cushions, the ventral shields, occurring in every segment (the fan worm *Sabella* (Fig. 60A, v.gl), *Bispira*, *Potamilla*) or on only a limited number of thoracic ones (terebellids, Fig. 62B, v.gl) or from more scattered groups of glands (*Polydora*, serpulids).

Material for building the tube must be accurately sited around the rim and held in place until the secretion hardens in contact with the sea water. This may be done by the body itself (*Spirochaetopterus*) but often by a specially developed outgrowth carried on the peristomium, the peristomial collar (Fig. 61A, lat.col, v.col) which overhangs the rim and moulds the secretion as it flows either directly from glands on the collar (serpulids) or from sacs at the edge of the lips where it has been mixed with fine particles selected from those filtered out of the water by the branchial crown (*Sabella* (Fig. 60D, E), *Bispira*, *Potamilla*). When larger particles are used they are picked from the substratum by the tentacles (*Lanice*, *Sabellaria*, *Pectinaria*, *Polydora*) and passed to the lips which build them one by one around the edge, sometimes accurately matching the surfaces to keep the outline smooth (*Sabellaria*, Fig. 67; *Pectinaria*, Fig. 57C).

There is usually some mechanism for closing the tube when the worm withdraws. This may be an operculum formed from a modified tentacle (serpulids, Fig. 61A, op) or from the ends of the muscular pillars which bear the feeding tentacles (*Sabellaria*, Fig. 61B, mu.p). Alternatively the tube may collapse as the worm withdraws (*Sabella*, *Potamilla*) to be opened again as the branchial crown emerges.

Respiration in polychaetes

The body of the errant polychaete is bathed with water and some respiratory exchange takes place by diffusion through the integument; in some species, a current of water is maintained there by special ciliated tracts (Fig. 57A). Small errant worms get adequate oxygen in this way. Gases are also exchanged with the coelomic fluid and the blood through the wall of the hind-gut in a number of species which pump water in and out through the anus. The diffusion surface is increased by the presence of capillaries, often conspicuous on account of pigment in the blood. In large worms like *Nereis* and *Phyllodoce* the capillaries are concentrated in the parapodia

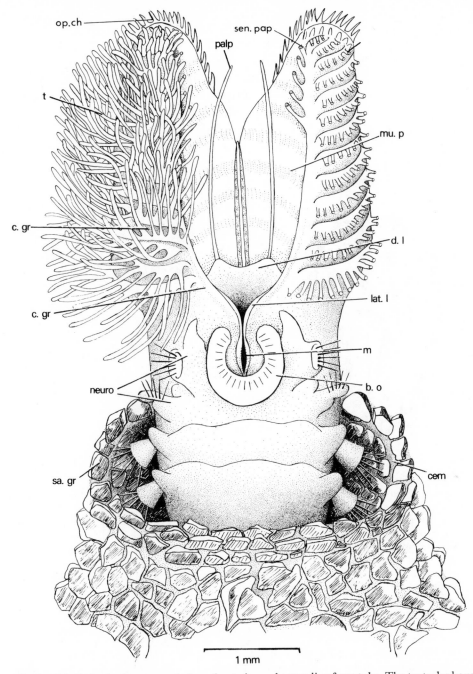

Fig. 67. *Sabellaria alveolata*: ventral view of anterior end extending from tube. The tentacles have been cut off the left half of the pillar. b.o, building organ; cem, cement; c.gr, ciliated groove leading to mouth; d.l, dorsal lip (prostomium); lat.l, lateral lip; m, mouth; mu.p, muscular pillar (peristomium fused with a few anterior segments); neuro, neuropodia (the corresponding notopodia incorporated in muscular pillar); op.ch, opercular chaetae; palp, palp; sa.gr, sand grain; sen. pap, sensory papilla; t, feeding tentacles.

(Fig. 51B, resp.noto) which, owing to their locomotor activity, are effective respiratory areas. Ciliary tracts may still be used to circulate water over less mobile
surfaces such as the dorsal integument and the adjacent expanded dorsal cirri of a
phyllodocid. When, as in worms which burrow, the mode of life restricts free access
of water to these surfaces, gills are developed, usually as parapodial derivatives, and
special mechanisms exist for renewing the water around them (see p. 203). Thus the
parapodia of *Nephthys* have gills (Fig. 58B, g) arising as ciliated outgrowths of the

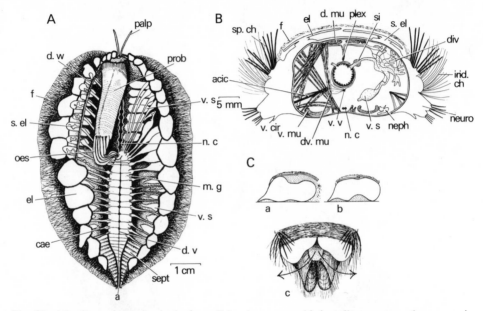

Fig. 68. *Aphrodite aculeata*: A, the body wall has been cut mid-dorsally to expose the organs in
the coelom. The elytra conceal the cut surface except anteriorly on the left where they have been
turned back to show segmental outgrowths from the body wall which on each elytron-bearing
segment form the stalk of the elytron and on other segments tubercles. B, thick, transverse
section to show the structure of the elytron-bearing parapodium, the strong development of
dorsoventral and oblique muscles associated with respiratory movements on the left, and on
the right a mid-gut caecum. C, diagrams to illustrate the circulation of water over the surface
of the body. Arrows indicate water current. a, b, transverse sections each passing through a
parapodium on the left and between two parapodia on the right. In b water (= stippled area)
has been drawn into the space between body wall and mud; in a it is being sucked up between
the parapodia into the space beneath the elytra. In c, the posterior end of body viewed dorsally,
the water is being ejected through an opening temporarily formed for this purpose beneath the
last pair of elytra. (C after van Dam.) a, anus; acic, aciculum; cae, gut caecum; div, diverticulum of caecum; d.mu, dorsal longitudinal muscle band; d.v, dorsal blood vessel; dv.mu,
dorsoventral muscles; d.w, dorsal body wall; el, elytron; f, felting; irid.ch, iridescent chaetae
of notopodium; m.g, mid-gut; n.c, ventral nerve cord; neph, nephridium; neuro, neuropodium,
oes, oesophagus; palp, palp; plex, dorsal blood plexus; prob, proboscis; s.el, stalk of elytron;
sept, septum (they are reduced anteriorly); si, sieve; sp.ch, spinous chaetae of notopodium;
v.cir, ventral cirrus; v.mu, ventral longitudinal muscle band; v.s, ventral sac; v.v, ventral
blood vessel.

notopodia: they are protected from the abrasive action of sand by being tucked into the space between notopodium and neuropodium along which a ventilating current is maintained.

An extreme modification of the parapodium is seen in the short-bodied mud dweller, the sea mouse *Aphrodite* (see Fig. 68). This involves a specialization of the notopodia characteristic of polynoid worms which prevents them from participating in locomotory activity. In this group the dorsal surface of the body is covered by two longitudinal rows of overlapping scales or elytra, each arising from a narrow stalk and spreading over two segments. Segments without elytra have a dorsal cirrus, so the two structures are probably homologous. In *Aphrodite* the elytra are hidden and protected from the sublittoral mud in which the worm creeps by a grey felting of matted chitinous threads secreted in the notopodial chaetal sacs. The up-turned tail (a posterior region of short segments not covered by elytra or felt) is the only part of the body to project into the water. A circulation of water is maintained over the body surface by a variety of muscular movements. When the ventral body wall in the anterior segments is raised water is drawn beneath the tail into a space between body and mud. The water is then passed dorsally between the parapodia by downward movement of both dorsal and ventral body walls, and through lateral intersegmental openings in the felt into a space beneath the elytra. There are no gills and the elytra are not vascular, but the dorsal body wall is thin and undoubtedly an important respiratory surface. Water is expelled as the dorsal wall and elytra move towards one another, the openings in the felt being closed. This movement spreading anteroposteriorly forces the water back to an ejection opening which appears on the tail, beneath the last pair of elytra. A jet of water passes out every 30–50 s. During expiration a fresh stream of water is drawn in ventrally and the cycle starts again. The respiratory movements are due to well-developed dorsoventral and oblique muscles. A related genus, *Hermione*, which lives on gravel or still rougher ground, has no dorsal felt, but the elytra and dorsal body wall exhibit similar respiratory movements. In other polynoids (*Harmothoe*, *Lepidonotus squamatus*) a circulation of water over the dorsal surface of the body is maintained by cilia.

Worms which have given up a totally free-ranging habit for one which involves periodic or permanent life in a burrow or tube suffer a reduction in the respiratory effectiveness of the body surface. This is compensated for by ventilation of the tube and, as the parapodial surface is almost certainly reduced, by the development of gills. An irrigation stream around the body is maintained by rhythmical pumping movements, usually intermittent, though becoming continuous at low oxygen concentrations (2 ml O_2 per litre in *Nereis virens*) and ceasing at very low concentrations (0.6 ml l^{-1}). If the tube or burrow is open at each end the normal direction of pumping can either be headward (*Amphitrite*, *Lanice*, *Pectinaria* and *Arenicola*) or tailward (*Nereis*, *Chaetopterus*, *Sabella*), but may be reversed; if it is open only at one end, this is at the head, around which water enters and leaves. The water brings

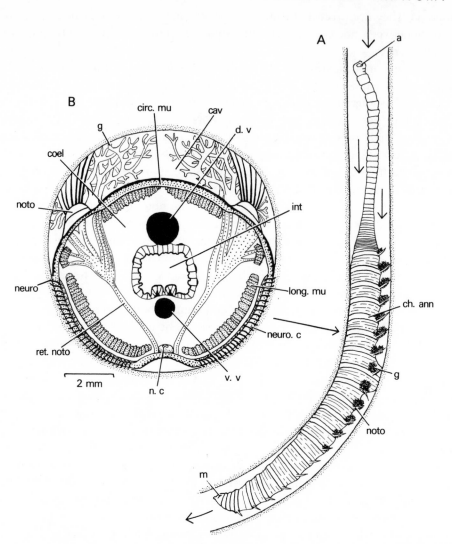

Fig. 69. *Arenicola marina*: A, lateral view of worm in observation tube: most chaetigerous segments are pressed against the wall laterally and ventrally but are separated from it dorsally by a narrow space. Waves of swelling which occlude this space travel gently headward driving water through the tube; one such wave is passing the 5th and 6th gills. Arrows indicate direction of water current. B, transverse section at level indicated by arrow of worm in its burrow, circulating water. (After Wells.) a, anus; cav, cavity of burrow; ch.ann, chaetigerous annulus; circ.mu, circular muscles; coel, coelom; d.v, dorsal blood vessel; g, gill; int, intestine; long.mu, longitudinal muscles; m, mouth; n.c, nerve cord; neuro, neuropodium; neuro.c, neuropodial coelom; noto, notopodium; ret.noto, retractor muscles of notopodium; v.v, ventral vessel.

oxygen and sweeps away carbon dioxide, urine and unwanted detritus, and in *Arenicola* is responsible for giving the sand in the head shaft a characteristic texture. In some filter feeders the tailward current is also the feeding current (*Chaetopterus*) or may strengthen the feeding current (*Sabella*). The pumping movements of the body may take the form of (1) undulating dorsoventral waves travelling from head to tail (*Nereis diversicolor, N. virens* (Fig. 50D), *Marphysa, Hyalinoecia*), (2) piston-like swellings passing forwards, but sometimes back (*Arenicola* (Fig. 69A), *Amphitrite, Lanice, Marphysa, Pectinaria*), (3) a regular lengthening and shortening of the abdomen (*Sabellaria, Pomatoceros*), or (4) the coordinated beating of a few fan-like appendages (*Chaetopterus*, Fig. 70A, d.fan). The gills, which are extended into the stream of water, may be outgrowths from a few pairs of notopodia (*Arenicola*), or from most (*Marphysa*), or they arise from the dorsal body wall (cirratulids, terebellids).

Ventilation of the gills of a burrowing worm is illustrated by *Arenicola marina* (Figs 64G and 69A). At high tide the animal drives water through its burrow in a tail-to-head direction, the water escaping through the head shaft. A good-sized animal pumps about 190 ml h^{-1}. Water is driven forward by swellings which pass along the

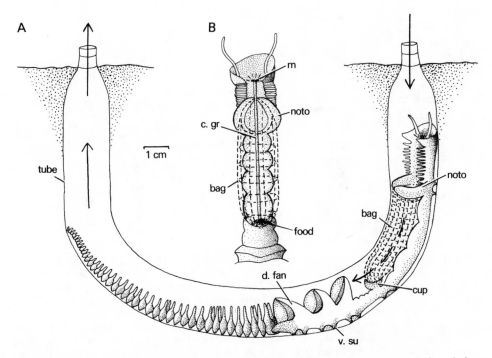

Fig. 70. *Chaetopterus*: A, feeding with mucous bag. Arrows indicate water current. B, dorsal view of anterior end. (After McGinitie.) bag, mucous bag; c.gr, ciliated groove; cup, cupule; d.fan, dorsal fan; food, food ball in cupule; m, mouth; noto, wing-like notopodia; tube, parchment tube; v.su, ventral sucker.

dorsal surface of the body and make contact with the wall of the burrow. Since at this time the ventral and ventrolateral surfaces of the body are pressed against the tube by protraction of the notopodial chaetae, the dorsal swellings force water anteriorly through the space in which the gills lie. The gills contract as a wave approaches and expand as soon as it has passed. Irrigation consists of intermittent outbursts of activity separated by periods of feeding or rest, the alternation continuing with regularity for many hours. Before each outburst the worm creeps backwards up the burrow, perhaps to test the water before irrigation begins, and sometimes to defecate. Like the outbursts of feeding activity, the cycles are regulated by a pacemaker but are not initiated by oxygen lack nor by an accumulation of carbon dioxide in the burrow. The pacemaker appears to be located in the ventral nerve cord; the presence of the brain is unnecessary. When the tide is out and the burrow partly filled with air *Arenicola* will creep backwards until most of the gills are in the air and since their surfaces are moist respiratory exchange can occur.

Polychaetes which live in burrows or tubes with one opening may obtain oxygen by extending gills or feeding tentacles out of the tube (Figs 57B, 61, 62, 63, 66, 67, 71).

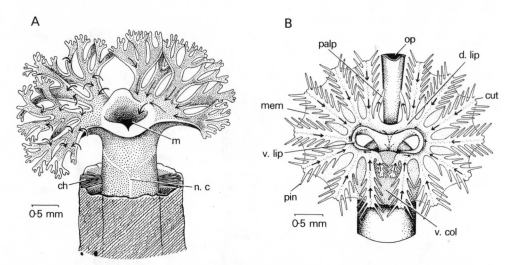

Fig. 71. A, *Owenia fusiformis*: ventral view of anterior end of worm with bilobed crown of short tentacles extended for feeding. Arrows on the left indicate direction of strong ciliary currents on the upturned edges of the tentacles which direct particles on to the oral surface; those on the right indicate ciliary currents which direct particles to the mouth. The mouth is bordered by a dorsal and two ventrolateral lips. Bundles of chaetae grip the flexible end of the tube which, when the worm retracts, almost closes the opening. B, *Pomatoceros triqueter*: diagram of oral area. The tentacles of the branchial crown have been cut off close to the membrane which unites their bases. Arrows indicate ciliary feeding currents leading to mouth on tentacles and lips and rejection currents on palps. ch, chaetae; cut, cut surface of tentacle; d.lip, dorsal lip; m, mouth; mem, membrane uniting tentacles; n.c, ventral nerve cord (by transparency); op, stalk of operculum; palp, palp arising from dorsal lip; pin, pinnule of tentacle; v.col, ventral fold of collar; v.lip, ventral lip.

The gills undergo writhing movements which accelerate the blood flow through their capillaries; the tentacles may have blood capillaries (cirratulids, sabellids, serpulids) or canals connecting with the main body coelom (sabellariids, terebellids). The most specialized filter feeders (sabellids, serpulids) have no gills, probably because even the finest divisions of their tentacles are vascularized and blood flow is maintained by rhythmical contractions of the larger vessels, though in the peacock worm, *Sabella pavonina*, it has been shown that when the tentacles are withdrawn the contractions stop. So long, however, as the tube is irrigated the tentacular crown is not an essential respiratory organ – it can be removed and the worm survive. Vigorous irrigation movements are usually maintained whether the crown is expanded or not and water is driven through the tube at the rate of about 30 ml h^{-1} (15 °C), carrying perhaps 180 mm^3 of dissolved oxygen, a large worm of 2 g requiring about 88 mm^3 O$_2$ per hour. On the other hand the tightly fitting gelatinous tube of the sabellid *Myxicola infundibulum* leaves no space for an irrigation current and respiratory exchange must be more or less confined to the tentacles and the exposed anterior segments, though strong, jerky contractions of the body periodically flush out the tube.

In addition to the tentacular crown, which provides a larger surface area relative to body size than in sabellid worms, serpulids like *Pomatoceros* have thin membranes on the thorax which increase the respiratory surface. When the crown is withdrawn circulation within it stops, but movement of blood continues in the capillaries of the thoracic membrane and body wall. Circulation of water is maintained over these surfaces by movements of the abdomen and by an inhalant ciliary current on the median wall of each thoracic membrane with a compensating exhalant one on the dorsal wall of the thorax. Ciliary currents also help to circulate water over the body surface of sabellariid worms and are especially important when the operculum more or less closes the tube.

Feeding of polychaetes

Most errant polychaetes are predacious. The method by which they obtain their food is not strikingly different from that of a nemertean except that the proboscis used for seizing is formed from the anterior part of the gut and is not an independent organ. The proboscis may be armed with teeth or jaws and is provided with muscles which assist the teeth to suck in food. Loss of teeth, such as occurs in phyllodocids, normally converts the proboscis into a purely sucking apparatus but it may still remain capable of seizing active prey as in the pelagic worm *Tomopteris* (Fig. 59A, prob). Some species have a catholic taste and take whatever opportunity offers, according to their size and strength: *Aphrodite*, for example, ingests the small worms, crustaceans and molluscs which it encounters as it moves through the mud and *Tomopteris* seizes fast-moving prey like *Sagitta* and herring larvae. Other species have a more limited diet: *Nephthys* shows a preference for other polychaete worms and

Autolytus edwardsi a liking for *Obelia*, the tentacles of which it cuts off and sucks in one by one. *Eulalia viridis* prefers barnacles, sucking out their soft tissues by inserting the proboscis between terga and scuta. These foods, sensed by the variety of receptor organs on the head, have to be accurately and speedily attacked, so that eversion of the proboscis must be rapid and its aim accurate. Eversion in almost all polychaetes is dependent on increased fluid pressure in the aseptate part of the body cavity in which the proboscis lies, and where, in errant species, a variety of protractor and tensor muscles occurs. Inversion is chiefly effected by paired retractors inserted on the muscular part of the proboscis as in nereids (p. 178) and syllids, or by longitudinal muscles associated with the oesophagus (phyllodocids) or intestine (glycerids).

In some polychaetes, mainly not predacious, a ventral thickening of the stomodaeal wall, the buccal organ, is used to draw in the food. This may appear as a muscular lower lip (cirratulids) or a more powerful muscular sac resembling the odontophore of gastropod molluscs and armed with teeth and jaws (eunicids). In the eunicid *Marphysa sanguinea*, a burrower in rock crevices which attains a length of 30–60 cm, the powerful teeth are used to scrape rocks and, with the jaws, drag weed and animal matter into the gut. *Marphysa* uses its jaws to defend itself and so does the equally large ragworm *Nereis virens*. For this reason both have been said to be carnivorous though neither is.

Sedentary polychaetes feed by one, or in some species two, of three methods: (1) engulfing the substratum in which they live for the sake of detrital particles which have settled there (*Arenicola, Ophelia, Notomastus, Capitella, Caesicirrus*); (2) selecting organic matter from the substratum (terebellids, cirratulids, *Pectinaria, Marphysa, Owenia*, many spionids); (3) collecting particles suspended in the water (*Chaetopterus, Owenia*, some spionids, sabellids, serpulids).

Those which engulf the substratum do so by means of a proboscis which has no thick-walled muscular part and is without teeth, though in *Arenicola* it bears rows of conical papillae (Fig. 64B) which help to break down the packing structure of the sand as the animal feeds from the base of the head shaft (Fig. 64G). The sand in the shaft is also loosened by regular irrigation movements of the worm and the thrusting of its head into the shaft. It is yellow surface sand, rich in microorganisms and organic matter, which slides in at the top as the worm eats from below. *Arenicola* has rhythmical outbursts of feeding alternating with other activities or with periods of apparent rest. In *A. marina* the outbursts last for about 7 minutes, with a similar interval between one and the next. This is an inherent rhythm, not dependent on any external stimulus, maintained by a pacemaker in the oesophageal wall from which activity spreads through the nervous system to those muscles of the body involved in the feeding movements.

Though many worms which eat sand or mud prefer, like *Arenicola*, nutritious surface deposits, those which use tentacles to gather surface detritus can be more particular in their selection. The food-collecting tentacles are often very extensile, stretching out to a foot or more, so that they appear as fine filaments and have a ciliated and usually grooved surface which can be applied to the substratum. Their complicated musculature enables them to move in all directions. Worms with such tentacles are sedentary, feeding from their burrows or tubes. They approach the opening and extend two (spionids) or more (terebellids, Fig. 62B; cirratulids, Fig. 62A) tentacles which creep over the surface exploring for food. Sometimes a tentacle is pulled in and put out in another direction; sometimes it explores special processes of the tube where food is trapped (*Lanice*). Food particles, entangled in mucus secreted by the tentacle, are directed orally along its ciliated groove, the transport of the larger ones being assisted by muscular movements. They are sometimes discharged at the opening of the burrow (cirratulids) or may be carried directly to the lips (spionids, terebellids). The tentacles are often wiped across the lips to be cleaned (*Amphitrite, Lanice*) and a large particle may be gripped by several tentacles and brought to the mouth, though this is generally in connexion with building the tube (*Lanice*). The collected material is sorted roughly according to size, only the smallest particles being ingested. In cirratulids the pointed prostomium pushes amongst the food deposited at the mouth of the burrow and collects small particles on its ciliated under-surface; the remainder may be ingested by the protrusible buccal organ. Terebellids either reject large particles brought to the lips or incorporate them in the tube.

A few detritus feeders do not feed from the surface but gather the less abundant organic matter which collects beneath it. *Pectinaria* (Fig. 57C) spreads long tentacles in all directions amongst the grains of sand in which it lives gathering particles from a distance and conveying them to the mouth by cilia on the tentacular grooves, or by drawing the tentacles across the lips. The food supply soon becomes exhausted in one area and the worm moves on through the sand.

Many tube-dwelling worms feed on particles suspended in the water: they either emerge to extend tentacles which trap the food (*Polydora, Owenia*, sabellids, serpulids) or use a mucous sheet to filter a current of water passed through the tube. The second method is adopted by the most highly specialized polychaete, *Chaetopterus* (Fig. 70), which lives in a U-shaped tube with the opening at either end restricted to prevent ingress of large particles. Its parapodia are modified to form (1) ventral suckers (segs 10, 13, 14, 15, 16) anchoring the worm closely to the floor of the tube (A, v.su); (2) three dorsal fans (segs 14, 15, 16) which beat at a frequency of about one stroke per second to drive water in a tailward direction (d.fan); (3) a pair of wing-like processes (notopodia of seg. 10) spreading dorsally around the wall of the tube (noto) to secrete mucus for entrapping food particles and (4) a dorsal cupule (cup) (seg. 13) which is linked with the mouth by a ciliated groove (B, c.gr). The aliform notopodia have a median ciliated groove along which strings of mucus with food are passed to the dorsal groove. Alternatively the copious mucus they secrete is blown back to form a bag (bag) the apex of which is held posteriorly by the cupule and through which the water current must pass. It has been calculated that its openings are only about 4 μm in diameter. The blind end of the bag and entrapped particles are rolled up in the cupule as more mucus is secreted by the wings. Then the secretion stops and also the pumping of the water through the tube. The net is made into a pellet in the cupule and passed along the dorsal groove to the mouth (m). The time taken for spinning the bag, collecting the food and swallowing it is about 17 mins.

Tentacles used to entrap suspended particles may be moderately extensile with a ciliated groove similar to those of detritus feeders (*Owenia*, *Polydora*, *Sabellaria*), or non-extensile with lateral pinnules (sabellids, serpulids). In the first group, in *Polydora* and spionids in general, only two extensile muscular tentacles (Fig. 66B, t) are present which are lashed vigorously through the water so that their fields of movement do not overlap at any one moment. Particles are trapped by mucus in the groove and conveyed by cilia to the prostomium and mouth. In contrast in the worm *Sabellaria* (Fig. 67) a crown of extensile tentacles is held almost motionless in the water, the combined action of the cilia drawing a current towards them: latero-frontal cilia beat at right angles to the long axis of the tentacles and direct water between them, straining it as it passes; frontal cilia in the groove direct the particles which have been collected to the base of the tentacles, where they pass along circum-scribed ciliated pathways (c.gr) to the mouth (m). This type of tentacle can also be drawn over the substratum to collect larger particles for building the tube (*Polydora*, *Sabellaria*) or for food. *Owenia fusiformis*, common in tubes on many sandy shores, also collects particles by two methods with its short tentacular crown (Fig. 71A): it is spread into the water to trap suspended food or swept over the surface of the sand; as it closes it picks up particles which are then manipulated by the lips.

The tentacles of worms in the second group, which includes the commoner kinds of filter-feeding worms, are protruded from the tube and displayed as a funnel, simple in sabellids, but rather more elaborote in *Pomatoceros* (Fig. 71B) and often beauti-fully coloured; since they have also a respiratory function they are often referred to as the branchial crown. In *Pomatoceros* and other serpulids the lips (d. lip, v. lip) are drawn out in a semicircle and form the margins of a groove. From the outer wall of the groove, that is from the dorsal lip (belonging to the prostomium), a series of tentacles arises, their bases united by a membrane (mem). The tentacles are stiff and motionless when the crown is displayed while their cilia drive water between them and carry particles towards the mouth. Each tentacle has a double row of ciliated pinnules and a ciliated groove running its length; at its base a pair of folds border the groove. These folds are weak in serpulids but very strongly developed in sabellids (*Sabella*, Fig. 60D, fd, *Bispira*, *Potamilla*); particles caught by the tentacles pass between them to the lips. The feeding current is set up in serpulids by latero-frontal cilia on the sides of the pinnules beating at right angles to their length, and drawing a current of water between them; in sabellids abfrontal cilia (Fig. 60C, abf.c) on the back of each pinnule drive a vigorous current towards the tip which is deflected between the pinnules by the laterofrontal cilia (latf.c). In either case particles are thrown by the laterofrontals on to the shallow groove on the inner face of each pinnule, carried by the frontal cilia (fr.c) there to the groove on the tentacle and so to the base of the crown.

In serpulids no sorting of food particles takes place except for a rejection of large ones; any which are small enough reach the mouth. A complex ciliary sorting mechanism, however, occurs in some sabellids (*Sabella, Bispira, Potamilla*) on the basal folds where the particles are sorted into three sizes as they pass towards the lips (Fig. 60D,E). On the lateral lips at the base of the crown (Fig. 60A, lat.l) are further ciliary tracts responsible for directing the finest particles to the mouth, medium-sized ones (to be used for tube building) to the ventral sacs (v.s) associated with the lateral lips, and the largest ones to the dorsal lip (d.l) whence they pass up two long tapering structures, the palps (palp), and are rejected.

Smaller species of filter-feeding polychaetes have a higher filter rate per unit weight than larger ones. This is not due to structural differences in the crown enabling the smaller worm to strain a larger quantity of water per unit weight and time, but to the crown being relatively larger. In the serpulid *Spirorbis* the crown (0.08 mg) represents about $\frac{1}{3}$ the total weight (0.24 mg) and in *Sabella* about $\frac{1}{9}$ (19.5 mg : 187 mg). The filtering rate of *Spirorbis* is 950 ml h^{-1} g^{-1} and that of *Sabella* 390 ml h^{-1} g^{-1}. The relatively higher filtering rate in *Spirorbis* may be correlated with the higher metabolic rate characteristic of smaller species of a group as compared with larger ones. Despite these figures annelids are, in general, less efficient filter feeders than molluscs, especially bivalves: *Crassostrea virginica*, for example, filters at the rate of about 500 ml h^{-1} g^{-1} and *Mytilus edulis* at the rate of 600 ml h^{-1} g^{-1}; sorting of the collected particles is also less efficient in polychaetes than molluscs.

Functioning of the vascular system of polychaetes

A blood vascular system made up of a segmental circulation subordinate to a longitudinal one, similar to that described for *Nereis*, is a fundamental feature of annelidan organization, though there are some polychaetes, such as the sea mouse, paddle worms and syllids, in which it is reduced and others in which its functions are taken over by a coelomic circulation. The reduction of the system in syllids may be related to small size. In the larger forms in which it is lost (glycerids, capitellids) there would appear to be some adjustment to its absence in that coelomic corpuscles carry the respiratory pigment haemoglobin. However, in at least two species which have a vascular system (*Terebella lapidaria, Travisia forbesi*) this pigment is also found in coelomic cells and in nephthydids it occurs in solution in both coelomic fluid and blood. When coelomic fluid takes over the functions of blood its free circulation through the body becomes essential. This is facilitated by reduction or loss of septa

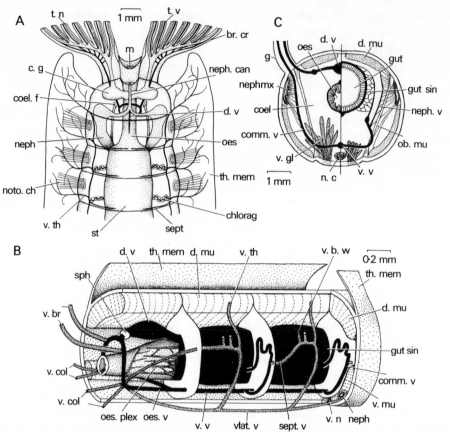

Fig. 72. A, *Protula tubularia*: semidiagrammatic representation of dorsal view of anterior segments of the serpulid polychaete to show certain structures by transparency (tentacles of branchial crown are cut short). Blood vessels are black and since the gut sinus is not indicated the dorsal vessel appears to end blindly. (After Meyer.) B, *Pomatoceros triqueter*: left lateral view of peristomial, second and third thoracic segments, with the body wall removed, to show the arrangement of blood vessels. The central blood system is shown in black. In this the blood circulates in the usual manner, forwards in the dorsal vessel (including the gut sinus) and backwards in the ventral vessel. The peripheral system of smaller, blind-ending vessels is heavily stippled; they receive blood from the central system and are alternately filled and emptied. Septa are unstippled. (After Hanson.) C, *Amphitrite*: diagrammatic half transverse sections at the level of the oesophagus (left) and of the mid-gut (right). (After Meyer.) br.cr, branchial crown; c.g, cerebral ganglion; chlorag, chloragogenous tissue; coel, coelom; coel.f, coelomic funnel; comm.v, commissural vessel; d.mu, dorsal longitudinal muscles; d.v, dorsal vessel; g, gill; gut, mid-gut; gut sin, gut sinus; m, mouth; n.c, nerve cord; neph, nephridium; neph.can, median nephridial canal; nephmx, coelomostome of nephromixium; neph.v, nephridial vessel; noto.ch, notopodial chaetae; ob.mu, oblique muscle; oes, oesophagus; oes.plex, oesophageal blood plexus; oes.v, left circumoesophageal vessel; sept, septum; sept.v, trans-septal vessel; sph, sphincter; st, stomach; th.mem, thoracic membrane; t.n, tentacular nerve; t.v, tentacular vessel; v.br, vessel to branchial crown; v.b.w, vessel to dorsal body wall; v.col, vessel to collar; v.gl, ventral glands; vlat.v, ventrolateral vessel; v.mu, ventral longitudinal muscles; v.n, ventral nerve; v.th, vessel to thoracic membrane; v.v, ventral vessel.

and by ciliated tracts of peritoneum which impose a pattern on the circulation. This pattern dominates the course of the fluid unless it is affected by such vigorous movements of body wall and gut as may be seen when *Glycera* and *Tomopteris* hunt their prey.

A considerable part of the total oxygen uptake of a soft-bodied worm occurs over the general outer surface. When this is decreased, as by the reduction of the parapodia which occurs in burrowing or tubicolous worms, gills or membranes such as the thoracic membrane of serpulids (Fig. 61A) frequently become associated with the appendages and vascularized by an elaboration of the segmental vessels (Fig. 72A,B, v.th). In terebellids (Figs 62B and 72C, g) gills are confined to a few anterior segments without parapodia, and are supplied by the dorsal vessel, blood returning to the ventral. A significant respiratory surface is also provided in these and similar worms by the hypertrophied head appendages with a circulation of blood or body fluid.

The vascular supply to the branchial crown of sabellids and serpulids is of interest in that each tentacle has only a single blind-ending vessel (Fig. 60C,E, b.v), sending blind capillaries into the pinnules, which is highly contractile and is alternately filled and emptied. The contractions which expel the blood in *Sabella pavonina* at 19°C occur about every 10 seconds. The ebb and flow in the tentacles is matched with activity in the central circulation: the branchial vessels expel their oxygenated blood and remain contracted until this is taken up by the ventral vessel; on relaxation a supply of blood is drawn from the dorsal vessel. A similar system of blind vessels associated with the central circulation serves all peripheral parts of the body: collar, palps, body wall, gonads, segmental organs and, in serpulids, the thoracic membrane. Blind capillaries project into the coelom of the larger sabellid worms which have neither membrane nor gills. These vessels, where only an ebb and flow circulation is possible, are peculiar to annelids and occur in many polychaetes: in the parapodia (*Nereis* spp., *Nephthys* spp.); associated with segmental organs (*Marphysa sanguinea, Polymnia nebula, Lanice conchilega, Arenicola* spp., *Ophelia* spp., *Travisia forbesi, Pectinaria* spp.); supplying the gonads and increasing in importance as they ripen (*Nephthys* spp., *Chaetopterus variopedatus, Polygordius lacteus*), and projecting into the coelom from the segmental and ventral vessels (*Arenicola* spp., *Magelona papillicornis, Poecilochaetus serpens*). The exploitation of this feature of the vascular system by sabellids and serpulids may be associated with their mode of life which calls for periods of activity alternating with quiescent periods when the worm is withdrawn into its tube and the peripheral circulation reduced. At such times the branchial circulation of *Pomatoceros* stops and that of the thoracic membrane is reduced. Blind vessels projecting into the coelom may also help to maintain a high oxygen concentration in the coelomic fluid and so compensate for the reduced blood supply to the somatic muscles which is found in nephthydids. Other functions appear to be associated with them in *Arenicola marina*: they are covered with dark, pigmented cells which have high concentrations of the precursors of haemoglobin and of ferritin, suggesting that this is the site of elaboration of the plasma pigment.

The blood supply to the gut of errant polychaetes is commonly provided by two or more vessels in each segment which form a network of capillaries penetrating the visceral muscles and underlying the gut epithelium (Fig. 51B, int.plex). In sedentary species the capillary system is replaced by a sinus system (Figs 72, 73, gut sin) around the stomach and intestine (the chief absorbing area) and the visceral circulation tends to lose the direct connexion with the ventral vessel which is normally present. The functional significance of this is obscure unless it forms an area into which blood can pass when the animal retracts. Young *Arenicola marina* have a plexus of well marked vessels around the gut posterior to the oesophagus and anterior to the rectum, which enlarge in older individuals to form sinuses (Fig. 73); the ventral vessel (v.v), unlike the dorsal one (d.v), has no connexion with the visceral circulation except in the region of the hearts and in the rectum. The two lateral hearts (aur, ven), placed between oesophagus and stomach, pump blood from lateral longitudinal tracts (lat.oes.v) which drain the gastric and oesophageal blood supply into the ventral vessel, and so enhance both visceral and somatic circulations. Tube worms (sabellariids, sabellids, serpulids) have gone further in emphasizing the visceral sinus system. They have a more or less continuous sinus (Fig. 72B, gut sin) incorporating the dorsal vessel, though free from the ventral vessel (v.v). A separate dorsal vessel extends briefly forward from its anterior margin. At its anterior end it is surrounded by a sphincter and has an internal valve. These structures regulate the

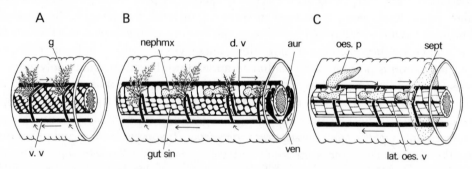

Fig. 73. *Arenicola marina*: diagram to show the plan of the blood system in three regions of the body: A, at the level of the anterior intestine; B, at the level of the stomach; C, at the level of the posterior oesophagus. The dorsal longitudinal vessel is connected to the gut sinuses throughout its length and in the intestinal and oesophageal regions with the ring or commissural vessels. The commissural vessels are always associated with the ventral vessel; in the gastric region they pass from it to the gut sinus system. Blood passes forwards in the dorsal vessel and backwards in the ventral, out of the ventral vessel into the commissural vessels in intestinal and gastric regions, and from dorsal to ventral vessel in the oesophageal region. The hearts direct blood into the ventral vessel. aur, auricle; d.v, dorsal vessel; g, gill; gut sin, gut sinus; lat.oes.v, lateral oesophageal vessel; nephmx, nephromixium; oes.p, oesophageal pouch; sept, 3rd septum; ven, ventricle; v.v, ventral vessel.

circulation in the central vessels in coordination with the ebb and flow of blood in the tentacles. A similar visceral sinus is present in other sedentary polychaetes (terebellids, ampharetids) and the anterior part of the dorsal vessel which retains its identity forms a heart which in terebellids pumps blood to the gills.

Hearts occur in association with either somatic or visceral circulations, wherever there is need to boost the blood flow and guard against back pressure. Small contractile bulbs occur at the base of the branchial capillary system of each gill of eunicids and in sabellariids on segmental vessels near their origin from the ventral vessel.

A number of polychaetes, especially sedentary and semi-sedentary species, have irregular groups of cells or spongy clumps of tissue inside blood vessels, known as intravasal tissue, possibly formed by infolding of the entire wall of the vessel including the peritoneum. In some worms the tissue is localized in the anterior part of the dorsal vessel (terebellids, ampharetids, cirratulids, amphictenids, flabelligerids) and is known as the heart body: the walls of the vessel contract against it so that at systole the lumen is obliterated. In others the tissue occurs in the gut sinus and segmental vessels (sabellids, serpulids). In all cases it slows down the rate of blood flow. For instance in *Cirratulus tentaculatus* it extends through the length of the dorsal vessel which supplies branches to the gills and perhaps controls the rate at which blood circulated through them. The reduction of flow may allow the intravasal tissue to remove substances from the blood and to secrete others into it, for certain pigments have been isolated from the heart body supporting the view that it has a haematopoietic function.

The larger polychaetes have a respiratory pigment in the blood, and sometimes in the coelomic fluid, which increases the oxygen-carrying capacity. It is lacking in some small worms in which oxygen reaches the tissues by simple diffusion. The pigment comprises a protein, which varies from species to species, and a haem or iron porphyrin compound. This may be the red protohaem of haemoglobin, which is of common occurrence, or it is chlorocruorohaem (flabelligerids, some ampharetids, sabellids, serpulids) giving the pigment chlorocruorin and making the blood green. *Serpula* has both pigments, and the genus *Magelona* has a different pigment, haemerythrin, in which the iron is not contained in a porphyrin. Chlorocruorin and haemoglobin dissolved in the blood plasma have molecules of exceptionally large size. When they are in corpuscles (capitellids, *Terebella lapidaria*, *Travisia forbesi*, *Glycera*, *Magelona*) the molecular weight is much lower.

The role of the respiratory pigments appears to be diverse, though it has been studied in only a few species. One of these is *Arenicola marina* which is regularly exposed to conditions of poor oxygen supply when its burrow is exposed for several hours and contains water left by the tide. The dissociation curve of its haemoglobin (Fig. 74) is very steep and shows that the blood reaches a high degree of saturation at low oxygen concentrations. The blood can be almost fully saturated with oxygen from the residual water even after 5 hours' exposure and the haemoglobin therefore

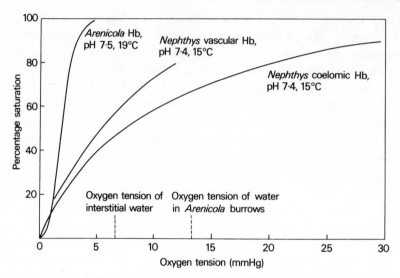

Fig. 74. Oxygen dissociation curves of the vascular and coelomic haemoglobins of *Nephthys hombergi* and the haemoglobin of *Arenicola marina*. (After Jones.)

acts as a transporter of oxygen at low partial pressures. When this is not possible the worm may resort to aerial respiration (p. 204). It was once thought that the haemoglobin had an oxygen-storing function, but this is now known to be negligible since the store is calculated to last the worm seven minutes or less. In contrast to this low-tension oxygen transport system is the high-tension one of *Nephthys*, an errant polychaete relatively unspecialized for littoral life. It moves through the sand, or over the surface, often with considerable rapidity, and has no permanent burrow. An individual at low tide has only the oxygen in the interstitial water of the sand in which it lies buried and inactive available to it. The dissociation curves of the haemoglobin in the blood and the coelomic fluid of *N. hombergi* are approximately hyperbolic and the oxygen affinities relatively low. By the time allowance is made for the gradient between external medium and blood they show that the available oxygen at low tide is at an unusable level, and since the quantity of haemoglobin in the blood is inadequate as an oxygen store, metabolism must then be predominantly anaerobic. At other times, when the worm is active, the haemoglobin will be effective. The pigments of other polychaetes – chlorocruorin in *Sabella spallanzanii*, chlorocruorin and haemoglobin in *Serpula* – function only under conditions of high oxygen concentration, that is when the worms are active and feeding.

Functioning of the digestive system of polychaetes

The gut is typically a straight tube extending from the mouth between the peristomium and the prostomium to a terminal anus. The stomodaeal part is modified for the ingestion of food by means of mobile lips or a proboscis which may have jaws and teeth. The proctodaeum is short or even absent. The alimentary canal is divisible into oesophagus, stomach, intestine and rectum, often distinguished by differences in diameter. Digestive enzymes are secreted by the stomach and sometimes by the posterior oesophagus and intestine. The intestine is the main site of absorption though this may also occur in the stomach. The rectum is concerned with the retention and elaboration of faeces.

The arrangement of the visceral muscles is similar to that of the somatic ones – circular fibres against the endoderm and longitudinal ones beneath the peritoneum. Hypertrophy of the muscles occurs where valves separate one region of the gut from another and especially where triturating regions are developed. The passage of contents towards the anus is mainly brought about by the action of muscles, either the visceral ones (though these are sometimes poorly developed) or by the squeezing action of the somatic muscles acting through the coelomic fluid. Antiperistaltic waves of activity of the visceral muscles, apparently tending to hinder the transport of the food, have been reported in some species. These are, however, valuable in that they mix the contents of the gut and counteract activities of the somatic muscles, such as the anteroposterior irrigation waves of *Sabella*, which tend to empty the gut precociously. Cilia on the gut wall help to mix secretions with the food and transport the gut contents.

The tubicolous suspension feeders, sabellids and serpulids, which eat small particles of a standard size and may feed almost continuously, have a simple gut. It is ciliated throughout, without appendages or convolutions, and the limits between the regions are ill-defined. The contents appear as a continuous string of mucus with entrapped particles which is surrounded by digestive fluid, and shows along its length the results of the progressive activity of enzymes so that posteriorly little remains but fine sand, indigestible detritus and diatom cases. They are compacted and embedded in mucus from the rectal walls to form faecal rods stout enough to pass intact along the faecal groove and be shed from the mouth of the tube. The food passes through the gut of an average sized worm in about 23 hours at 16°C. Polychaetes which select larger particles of food or ingest sand and mud show greater localization of function in the gut. In short-bodied species the secreting and absorbing areas are increased either by coiling the gut as in *Pectinaria* or by having it branched, as in *Aphrodite* and the polynoids. Specialized deposit feeders (*Amphitrite johnstoni*, *Terebella lapidaria*, *Lanice conchilega*, sabellariids) have a triturating chamber, preceded by a storage chamber (crop or stomach), in which the food is mixed with enzymes. Worms which engulf the soft substratum in which they live and have no device for sorting food particles from it must swallow large quantities in order to get adequate nourishment. Analyses of the substratum which *Arenicola* eats show that it contains over 96.5 per cent inorganic matter. In an actively feeding worm food passes as far as the rectum in about 14 minutes; it is stored and compacted there and is voided about every 45 minutes. The ingested sand accumulates in the anterior oesophagus and then the rhythmic activity of this part of the gut forces it back to the stomach, which is the main site of digestion. As food passes through the posterior oesophagus mucus and enzymes are secreted on to it and more digestive fluid is ejected from a pair of oesophageal pouches opening near the entrance to the stomach (Fig. 73, oes.p). The addition of further watery fluid from the anterior stomach combined with the low viscosity of the mucus here (due to the low pH of 5.8) make the contents more liquid than in any other region. This means that they can be mixed more easily and kept in suspension by cilia, which occur in well-defined tracts, and by movements of the body. Organic particles are thus brought in contact with the epithelial cells, engulfed and passed to amoebocytes which, while digesting the particles, wander into the blood and coelomic fluid. Owing to the absence of septa, except at the extremities of the body, both fluids circulate freely, transporting the products of digestion which they have apparently taken up from the amoebocytes. The wandering cells return waste to the lumen of the gut or deposit it in the intra-vasal tissue, epidermis or coelom, where it does not interfere with metabolic activities of the worm. Emphasis on digestion by amoebocytes is characteristic of other sand eaters (oligochaetes, spatangoids, holothurians) and is obviously a profitable means of getting a maximal amount of nourishment out of bulky and mainly indigestible food with minimal waste of enzymes.

From the stomach food is forced into the narrower intestine. Here water and the products of enzymatic digestion are absorbed; the mucus becomes more viscous and the contents firmer before being passed to the rectum. The firmness of the faecal cylinder which is coiled on the surface of the sand is due to the absorption of water by the intestine and to the mucus which binds the particles.

The most extreme modifications of the gut are found in polynoid worms where the various regions are abruptly marked from one another and there is a special arrange-ment for keeping large particles away from the delicate absorbing areas. *Aphrodite aculeata* (Fig. 68), which uses a proboscis (A, prob) to suck in animals living in the mud, may crush the prey within this very muscular organ. The oesophagus is short and the lumen at its posterior end narrow so that large particles cannot pass to the mid-gut where the food is retained while it is partially digested. Opening laterally from the mid-gut (m.g) are 18-20 pairs of caeca (cae), segmentally arranged, and with a fine sieve (B, si) guarding the entrance to each. Fluid and the finest particles are forced through the sieves by contractions of the rectum and adjacent body wall. Each caecum gives off a number of diverticula (div) as it traverses the coelom and then curves ventrally to end in a ventral sac (v.s). The epithelium has secreting and absorbing cells and some of the cells accumulate waste which leaves the body by way of the anus. The rectum is short; the faeces ejected with the respiratory stream of water need no elaboration.

Excretion and osmoregulation in polychaetes

Waste and sex cells are discharged by paired, segmentally arranged tubules, known as segmental organs, passing from the coelom and opening to the outside (Fig. 75). It is assumed that in primitive polychaetes (as in platyhelminthes and nemerteans) these were of two kinds: ducts derived from an impushing of the external epithelium, the nephridia or excretory tubules, and ducts derived from an outpushing of the coelomic epithelium, the coelomoducts or genital ducts. Both occur in polychaetes, but show considerable variation even in related families. Coelomoducts open to the coelom by a ciliated funnel, the coelomostome, and may have lost the exclusive relationship to the gonad which they had in acoelomate worms. Nephridia are of two kinds: blind protonephridia with groups of flame cells (solenocytes) which

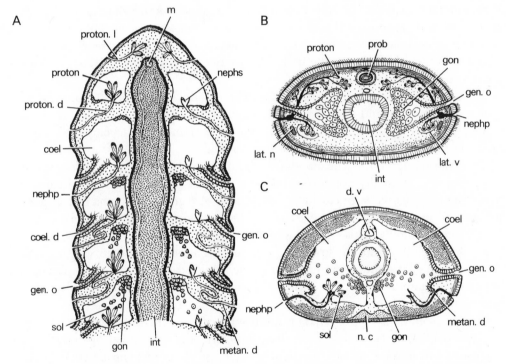

Fig. 75. A, diagram of primitive annelid in horizontal longitudinal section showing the relationships of coelomoducts and nephridia to segmental coelomic cavities; protonephridia on left, metanephridia on right. B, C, diagrammatic transverse sections to show relationships between segmental organs and other structures in nemertean (B) and annelid (C) grades of organization. (After Goodrich.) coel, coelom; coel.d, coelomoduct; d.v, dorsal blood vessel; gen.o, genital opening; gon, gonad; int, intestine; lat.n, lateral nerve; lat.v, lateral blood vessel; m, mouth; metan.d, duct of metanephridium; n.c, ventral nerve cord; nephp, nephridiopore; nephs, nephrostome; prob, proboscis; proton, protonephridium; proton.d, duct of protonephridium; proton.l, larval protonephridium; sol, solenocyte.

project into the coelom (sometimes with cilia agitating the coelomic fluid between the groups), and metanephridia with ciliated, funnel-shaped openings to the coelomic cavity, the nephrostomes. Protonephridia resemble the head kidneys of the trochophore larva which typically disappear at metamorphosis, being superseded by more posterior nephridia. Nephridia and coelomoducts relate to more than one segment: in the adult their solenocytes or funnels are associated with the coelomic cavity of one segment and the duct passes back (through the intersegmental septum if it is present) to discharge usually in the next posterior one. Their vascular supply is from segmental commissural vessels. Nephridia appear to be primarily osmoregulators: they convey waste to the outside, but there is no evidence that they are concerned with its elaboration. Little is known about nitrogenous excretion in polychaetes, but since they have access to voluminous quantities of water ammonia is likely to be the common excretory product.

Segmental organs are normally absent from a variable number of anterior segments concerned with feeding and sensation and sometimes from a variable number of posterior ones as well. Fertile segments, in which genital ducts are required, lie typically in the mid region of the body. A variety of relationships between coelomoducts and nephridia may occur here.

1. In members of only one family (Capitellidae) do independent nephridia and coelomoducts occur and open separately to the exterior; even here the permanent segmental tubules are the metanephridia and the coelomoducts develop and acquire openings to the exterior only at the breeding season.

2. In other families the coelomoducts or, more frequently, the coelomostomes (since the duct is usually not developed) fuse with the nephridia and use the nephridial duct so that each segment has a single pair of segmental organs of compound nature and of mixed genital and excretory function opening to the exterior at the base of the parapodia. This organ is a nephromixium.

    a. Occasionally, as in phyllodocids (Fig. 76), the union is between coelomostome and protonephridium, is present only in the fertile segments and the genital component remains rudimentary until the worm is mature. This structure is known as protonephromixium.

    b. More commonly nephromixia result from the fusion of coelomostomes and open nephridia (metanephridia). Each has, typically, a single funnel opening to the coelom, but in some species, the double origin of the funnel can be traced as the worm matures. Thus in syllids a genital funnel develops from the coelomic epithelium in segments where there are gonads, and fuses with the nephrostome to form a wide-funnelled genital duct in which the coelomostome and nephrostome elements may be readily distinguishable (*Odontosyllis enopla*) or ill defined. This type of segmental organ is a metanephromixium.

3. Finally, in the majority of polychaetes, the coelomostome is completely fused to the inner end of the nephridium to give an apparently simple funnel; this organ is a

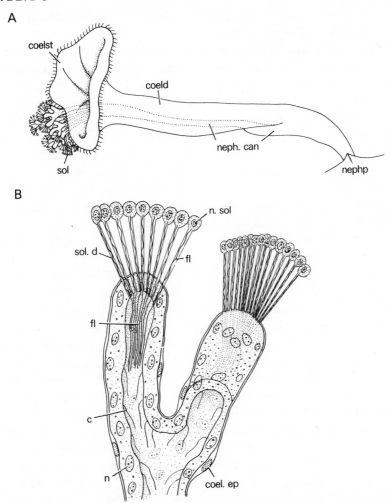

Fig. 76. *Phyllodoce paretti*: A, protonephromixium of mid-trunk region of nearly mature female. The coelomostome forms a wide genital funnel which opens to the nephridial canal by way of a coelomoduct. B, two groups of solenocytes and associated canals. The canal on the left is cut longitudinally to show the cilia and solenocytes. (After Goodrich.) c, cilium; coeld, coelomo-duct; coel.ep, coelomic epithelium; coelst, coelomostome; fl, flagellum; n, nucleus; neph.can, nephridial canal; nephp, nephridiopore; n.sol, nucleus of solenocyte; sol, solenocytes; sol.d, duct of solenocyte.

mixonephridium. Mixonephridia occur in both sterile and fertile regions of the body. They may appear as a uniform series, or the anterior ones, purely excretory, may have an elaborate duct and the posterior ones, which are also genital, an elabor-ated coelomostome. The divergence in structure has developed furthest in response to the tubicolous habit in such worms as Sabellariidae, Sabellidae, Serpulidae in which the long ducts of the single pair of excretory tubules (Fig. 72A, neph) pass dorsally and unite (neph.can) to give a median opening, well away from the current produced by the food-collecting tentacles.

In a number of errant polychaetes, some with protonephridia (Glyceridae, Nephthydidae), others with metanephridia (Nereidae, some hesionids), the coelomic funnels are no longer used for the passage of gametes. They never open to the nephridial canal or to the exterior, but the coelomostomes persist as ciliated organs (Fig. 53B, cil.o) which circulate the coelomic fluid; the sex cells escape by rupture of the body wall, and this may be aided by the ciliated organs.

Reproduction of polychaetes

Sexes are separate in most polychaetes though hermaphroditism appears to have arisen independently in a few species belonging to unrelated families (e.g. Nereidae, Syllidae, Serpulidae, Sabellidae). Gonads arise in a variable number of segments from rapidly proliferating tracts of cells on the peritoneum (Fig. 75), typically over blood vessels from which nutrients are readily available. Germ cells are freed into the coelom, where maturation is completed using nutrients available in the coelomic fluid and in cells derived from the peritoneum. The latter may initially resemble sex cells but enlarge to become nurse cells associated either singly (*Ophryotrocha*) or in groups (*Tomopteris*) with an oocyte. The coelomocytes also gather waste products of metabolism and in some species contain haemoglobin (terebellids). This pigment may help to meet the respiratory demands of the developing gametes which eventually fill the coelom. The total quantity of haemoglobin in the coelomocytes of *Amphitrite johnstoni* is about half that in the blood.

When spawning is initiated the eggs and sperm of most species are discharged through the genital ducts and the early stages of development are planktonic. The synchronous shedding of both types of gamete is probably ensured by an activator discharged into the surrounding water and in some polychaetes a substance causing spawning in males is known to be produced by the ova. It appears to be specific so that males of one species do not react to females or eggs of another. For species which are not gregarious this method of fertilization is chancy and it is not surprising that it has been modified or even abandoned for others which seem more sure. These necessitate modifications in body structure and behaviour, in some cases slight, in others considerable.

Wastage of sperm may be prevented by such modifications as spermatophores formed in the enlarged ducts of nephromixia (*Pionosyllis megalops*, some spionids); the development of special copulatory chaetae (*Capitella*) or outgrowths associated with the male duct (*Sternaspis, Pisone remota*); in the female, the development of receptacular pouches on the dorsal cirri (Alciopidae) or on the duct of the nephromixia (*Pisone remota*). Eggs may be fertilized as they leave the female and protected by her during early development. They may be retained beneath elytra (polynoids), enclosed in a membrane and attached singly to the parapodia (*Exogone, Sphaerosyllis*), or a membrane may surround a single cluster attached to the ventral body wall (*Autolytus*). In the polynoid *Harmothöe imbricata* the worms pair when mature, the male lying across the back of the female, who repels a partner when she is immature. Release of sperm over the newly formed egg mass sheltered beneath the elytra is ensured by receptors on the dorsal cirri of the male. *Nereis caudata* also forms pairs in preparation for spawning and in this species the male remains to incubate and protect the eggs. In some serpulids eggs are retained within the calcareous tube of the parent (*Spirorbis borealis*), in a special chamber formed by the operculum (*S. pagenstecheri*), or in capsules attached to the tube (*Microserpula*). In a few species large, gelatinous egg-capsules are formed and attached to stones or weeds on the shore (phyllodocids, *Scoloplos armiger*).

Their normal benthic habit is abandoned by a number of worms at the breeding season when swarms congregate at the surface of the sea for spawning; these are especially large in some nereids. The factors controlling swarming vary from species to species so that each swarm contains animals of only one species. Preparatory to their migration the adult worms typically assume a special appearance, the epitokous or hetero form, so called because the first described was taken for a new genus and named *Heteronereis*. The swarming worms swim for a short time, do not feed and may die after spawning though sometimes survive to reproduce again. The structural modifications associated with swimming, as well as maturation of the gametes, are pronounced only in the fertile segments, the first 13–30 segments and sometimes a few posterior ones of the heteronereid showing little change. Members of other families which are planktonic spawners undergo less modification and the lack of uniformity in the changes they display suggests that the habit has been evolved independently several times. Swimming is effected by enlarged or supplementary parapodial lobes (nereids, glycerids, syllids), increase in parapodial muscles (nereids, syllids), elongation of chaetae or their supplementation and even replacement by flattened, natatory ones (phyllodocids, nereids, glycerids, nephthydids, syllids, a few cirratulids) and vascularization of the parapodia to meet increased respiratory needs; the longitudinal muscles of the body may dedifferentiate and be replaced by a new type adapted for greater activity (nereids, nephthydids). Histolysis of such tissues as the muscles of the body wall, the gut and the septa is a widespread phenomenon of breeding polychaetes and not restricted to epitokous spawners though often more marked in them. The resultant breakdown products provide food for the developing gametes. Modifications in the sense organs involving the enlargement of the eyes and the development of special receptors on the parapodial cirri (nereids, syllids) are relatively uncommon.

Experiments involving the removal of parts of the brain reveal that a neurosecretory mechanism controls transformation to the epitokous form. Mature worms are prevented from maturing their gametes and assuming the special characteristics by inhibitory hormones secreted by cells in the supraoesophageal ganglia of nereids and nephthydids and only when these are withdrawn, perhaps in response to external factors, are structural changes initiated. In syllids the controlling centre appears to be located not in the nervous system but in an anterior part of the gut called the pro-ventriculus, and so to be endocrine rather than neurosecretory.

During swarming mature gametes are emitted by way of the enlarged coelomos-tomes (syllids, *Tomopteris*, Fig. 59A, gen.f), through lesions in the body wall (neph-thydids, hesionids, female nereids) or through the mouth (glycerids) which owing to histolysis of the pharynx and septa has come to communicate with the coelom. Sperm of nereids are released by a rosette of papillae on the pygidium, epitokal structures sending jets of fluid through the surrounding water on contraction of the muscles of the body wall. Males are able to detect the position of females by vibra-tion receptors on the dorsal cirri and in most nereids sperm discharge is induced by a substance emanating from the female or the discharged eggs. In *Platynereis megalops* the epitokous forms exhibit a kind of copulation, the male winding its body around the female and the posterior segments inserted into her mouth: sperm are injected into the coelom through lesions in the wall of the fore-gut and when the eggs are spawned they are already fertilized.

Palolo worms, belonging to the family Eunicidae, have spectacular reproductive activities. They are benthic, but vast swarms of the caudal halves of the worms swim tail foremost through the water on a predictable day or so of each year. For *Eunice viridis*, which lives in coral rock in the vicinity of Samoa and Fiji, this is the last day of the last quarter of the October-November moon; for the Atlantic palolo *E. schema-cephala* of the Dry Tortugas it is the third quarter of the June-July moon. The writh-ing, mature segments separated from the anterior sterile half of the worm (which regenerates a new tail) display epitokous modifications: their separation may be a fortuitous result of the extreme histolysis of the body wall. Migration of the epitokal part is associated with its sensitivity to light. In *Eunice fucata* it is photopositive to intensities between 0.005 and 54 lux, whilst the anterior part is photonegative to intensities above 1.08 lux; one epitokal change is the development of a large eye in each segment. The breeding biology of these eunicids is in some respects similar to nereids and in others to syllids. It would appear to be independently evolved.

The reproductive capacity of some polychaetes is increased by the separation of segments carrying gametes from the main part of the body, which then regenerates. The isolated reproductive part has a temporary independent life swimming in the surface waters dispersing sperm or ova. This method of reproduction is associated with a high capacity for regeneration which will be described briefly.

During the growth of a worm, segments are progressively proliferated forward from the pygidium and gradually enlarged. Proliferation appears to be dependent on a growth-promoting hormone secreted by the supraoesophageal ganglia. When the adult number of segments is reached (in some species the number is definitive, in others approximate) proliferation ceases, but the potentiality to form new segments is not lost. If posterior segments are destroyed a new pygidium is formed and growth follows the normal pattern; this is dependent on certain cells of the body remaining capable of dedifferentiation. It has been demonstrated in *Nereis diversicolor* that damage triggers off renewed endocrine activity in the brain, though it is not known whether the hormone produced is identical with the normal growth hormone. Similarly in other polychaetes nerve cells in the brain act as chemical mediators during regeneration. The capacity for regeneration varies. A variable number of anterior segments is necessary before regeneration of a lost posterior end can occur. Some species show cephalic regeneration, but this may involve the replacement of only the most essential segments: in *Sabella* it is limited to the prostomium, peristomium and one additional segment; in *Polydora flava* eight are reconstructed regardless of the number lost, while in *Chaetopterus* the segments essential to the feeding mechanism are replaced, if artificially removed, provided that the next most posterior is intact. The presence of the nervous system at the point of healing is indispensable for regeneration to occur, again indicating its general involvement in the control of growth.

In its simplest form asexual reproduction results when the body spontaneously fragments into two halves, each regenerating to the form of the original worm and capable of repeating the phenomenon (*Ctenodrilus monostylos*), or part of the body may fragment into isolated segments capable of regenerating a head anteriorly and a tail posteriorly (*Ctenodrilus pardalis*, *C. parvulus*, *Dodecaceria caulleryi*).

In some polychaetes the parent acts as a stock for the proliferation of new individuals: these may either develop the characteristics of the adult or become specialized distributors of gametes, so unlike it that they have been taken for different genera. Production of the former type of individual is exemplified by the tube-worm *Filograna implexa*, which, in spring, builds up a colony by asexual reproduction. Production of special sexual individuals appears to be under hormonal control and is associated with swarming in certain syllids, a group which shows greater variability in reproductive biology than any other. Whereas an epitokous form or heterosyllid is found in some species (e.g. *Odontosyllis*, *Eusyllis monilicornis*), in others (species of *Syllis* and *Autolytus*) the stock proliferates small sexual individuals and also participates in sexual reproduction, or it may be totally asexual. Individuals of only one sex are produced by a given stock. They develop a head after or before liberation, according to the species, and show adaptations comparable to those of other swarming polychaetes, such as enlarged eyes, no functional gut and elaboration of the parapodia for swimming. There is thus a superficial resemblance between epitoky and stolonization.

After asexual reproduction the parent proliferates new posterior segments again; these have gonads and separate to give another sexual form. In *Autolytus* this proliferation commences before the separation of the sexual zooid, and in some species the zone of proliferation produces a chain of sexual individuals attached head to tail, the oldest and the first to separate being the most posterior (Fig. 77). Chains of *Autolytus* and *Myrianida*, which has a similar method of stolonization, swim at the surface of the sea where the small sexual individuals are dispersed. The female stolons of *Autolytus* were first regarded as an independent genus, *Sacconereis*, and the male as *Polybostrichus*, terms which are still used to describe them. In some species of *Trypanosyllis* the zone of proliferation forms a cushion on the ventral surface of the last two segments. This gives rise to successive rows of sexual individuals, the youngest row being the most anterior; a single worm may produce over a hundred buds. There are other methods of proliferating new individuals in syllids and perhaps the most extraordinary is that of *Syllis ramosa*. The asexual form produces sterile as well as sexual side branches which remain attached for some time, grow to the size of the parent and branch again. Such a colony was found in the canal system of a hexactinellid sponge.

Organization of oligochaetes

The early oligochaetes, which probably had a structure and mode of life similar to those of archiannelids, evolved into two groups, one of which colonized the land (Terricola), the other fresh water (Limnicola). Recent fresh water species may either, as in the families Aeolosomatidae and Naididae, live among plants, or, like Tubificidae and Lumbriculidae, burrow in the substratum and resemble earthworms in

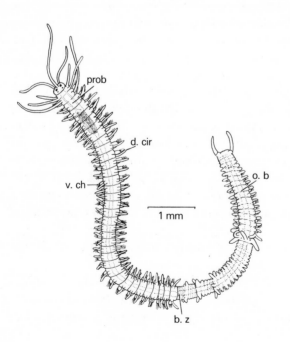

Fig. 77. *Autolytus*: budding. b.z, budding zone; d.cir, dorsal cirrus; o.b, oldest bud; prob, proboscis; v.ch, ventral chaetae.

appearance. Many of the former group are best regarded as retrogressive, displaying secondary simplification with small size, since some are not more than 1 mm long (*Aeolosoma*, *Chaetogaster*). They reproduce by fission and sexual reproduction is rare; the vascular system is reduced, with commissural vessels only in the first few segments; sometimes septa are lost (*Aeolosoma*, *Chaetogaster*); the ventral nerve cord and the cerebral ganglia lie just under the epidermis and cilia may be used for locomotion (*Aeolosoma*). One family, the Enchytraeidae (often referred to as white worms), is represented not only in fresh water but intertidally, where the worms live under stones (*Lumbricillus*), and on land, where they live in the soil amongst decaying vegetation and have a mode of life similar to that of earthworms. On account of their large size, worldwide economic importance, the ease with which they may be obtained and their organization, the terricolous oligochaetes or earthworms have been the subject of intensive anatomical and physiological study and the results of these studies will be emphasized. Some species are unusually large, for instance *Megascolides australis* and *Glossoscolex gigas* (South America) are over 3 m long and about 20 mm wide. There are about three dozen species of earthworm recorded from the British fauna, some of which are nearly cosmopolitan in distribution as they have sometimes been exported accidentally. Each species has its own preference for type of soil and food. Three very abundant species are *Eisenia foetida*, the brandling worm, particularly common in compost heaps and wherever the amount of organic matter in the soil is high, and the abundant *Allolobophora caliginosa* and *Lumbricus terrestris*, the earthworms of most garden soils and grasslands, the types usually used to introduce students to coelomate organization.

The body of the oligochaete is adapted for burrowing. It consists of an outer cylindrical tube, the body wall, separated from the gut wall by an extensive coelom (Fig. 78A, coel) subdivided by intersegmental septa (sept). It is devoid of appendages which might impede progress through the substratum, and there are no parapodia, nor head appendages, except for a median tactile prolongation of the prostomium in some naiads – *Stylaria*, *Ripistes* (Fig. 82A). Body wall and gut wall show independent peristaltic movements. The body wall muscles, using the coelomic fluid as a skeleton, generate the power needed for crawling and burrowing and their action is synchronized with movements of chaetae. The arrangement of the somatic muscles (Figs 78A and 79C) – an outer circular layer (circ.mu) and a more or less complete inner layer of longitudinal muscles (long.mu) interrupted where there are chaetal sacs – is similar to that of *Arenicola* which burrows into wet sand but cannot crawl. The more complex and variable pattern of movement in oligochaetes, particularly earthworms, is essentially due to the complete series of septa subdividing the coelom into more or less self-contained compartments. Although the septa of *Lumbricus* are incomplete ventrally where the nerve cord passes (Fig. 79C, sept), these foramina can be closed by sphincters and there is little or no movement of fluid about the body during locomotion. Moreover, outlets from the coelom (dorsal pores,

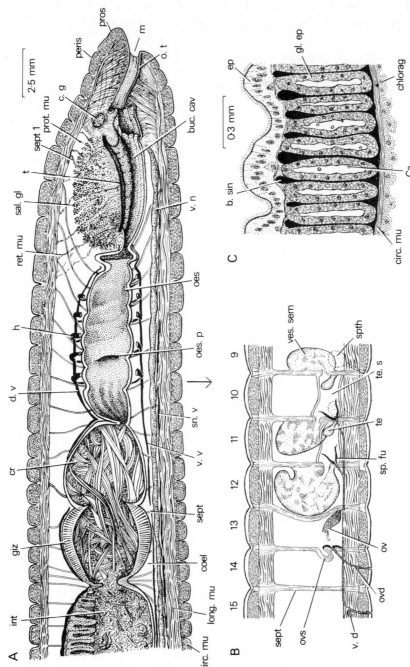

Fig. 78. *Lumbricus*: A, sagittal half of anterior segments to show the elaboration of the alimentary canal; the tongue is fully retracted. Crop, gizzard and intestine are full of plant food worked on by the triturating action of the gizzard. Arrow indicates the position of C. B, diagrammatic longitudinal section through genital segments to show relationship of gonads, ducts and sperm pouches. Segments are numbered. C, transverse section of part of calciferous glands. The glands extend through segments 11 and 12 and discharge into the oesophageal pouches. b.sin, blood sinus; buc.cav, buccal cavity; Ca, calcium spherule; c.g, cerebral ganglion; chlorag, chloragogenous tissue; circ.mu, circular muscles; coel, coelom; cr, plant tissue in crop; d.v, dorsal vessel; ep, epithelium of oesophagus; giz, muscular wall of gizzard; gl.ep, glandular epithelium of lamella; h, heart; int, intestine; long.mu, longitudinal muscles; m, mouth; oes, oesophagus; oes.p, opening of oesophageal pouch; o.t, oral tube contracted; ov, ovary; ovd, oviduct; ovs, ovisac; peris, peristomium; pros, prostomium; prot.mu, protractor

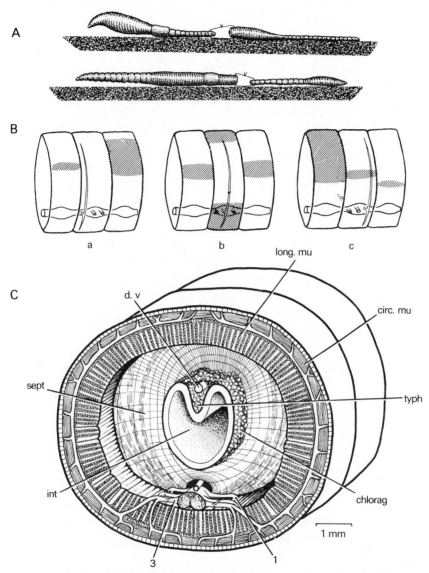

Fig. 79. *Lumbricus terrestris*: A, two earthworms cut in halves transversely and the cut surfaces connected by thread. The activity of the anterior half of each is transmitted by the pull of the thread to cause increased tactile stimulation of the posterior half and initiate a peristaltic wave. B, each diagram shows three consecutive segments, the ventral nerve cord and one segmental nerve. The epidermal sensory fields served by the first (a), second (b) and third (c) segmental nerves were determined by exploration with a needle 0.12 mm diameter: white areas are those from which responses were obtained; hatched areas represent regions from which no responses were obtained. (After Prosser.) C, thick transverse section through intestinal region viewed posteriorly. chlorag, chloragogenous tissue; circ.mu, circular muscles; d.v, dorsal vessel; int, intestine; long.mu, longitudinal muscles; sept, septum with muscles and ventral foramen through which pass nerve cord and ventral vessel; typh, typhlosole; 1, 3, first and third segmental nerves.

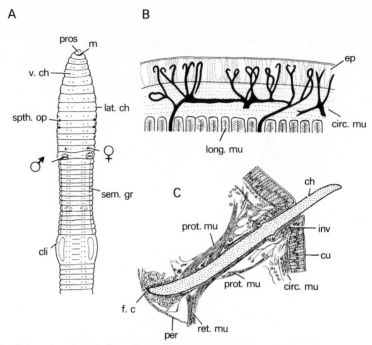

Fig. 80. *Lumbricus*: A, ventral view of anterior region; B, body wall in vertical section to show intra-epidermal capillaries (black); C, section through chaeta in chaetal follicle. (After Stephenson.) ch, chaeta; circ.mu, circular muscles; cli, clitellum; cu, cuticle; ep, epidermis; f.c, position of formative cell; inv, epidermal invagination; lat.ch, lateral chaeta; long.mu, longitudinal muscles; m, mouth; per, peritoneum; pros, prostomium; prot.mu, protractor muscle; ret.mu, retractor muscle; sem.gr, seminal groove; spth.op, spermathecal opening; v.ch, ventral chaeta; ♂, male opening; ♀, female opening.

nephridiopores and coelomoducts) have sphincters so that fluid does not escape to the exterior. This isolation of adjacent segments, and the maintenance of coelomic fluid within them, allows several peristaltic waves to pass down the body simultaneously and establishes the crawling and burrowing motion. The action of the chaetae, which have considerable versatility of movement, is also an essential component of the crawling mechanism.

The class Oligochaeta derives its name from the relatively few chaetae its members possess. Indeed, the small enchytraeid *Achaeta*, living in the surface layers of soil, and members of the family Branchiobdellidae, which resemble leeches in external appearance and habits, have none. Animals burrowing through firm substrata have short, stout chaetae used as anchors and levers. Their strength and efficiency in earthworms is illustrated by the fact that the resistance they set up in gripping the walls of the burrow is sufficient to break the worm if attempts are made to remove it. *Lumbricus* has only eight chaetae per segment, arranged in pairs (Fig. 80A, v.ch), but some earthworms secure a grip by means of a ring of chaetae per

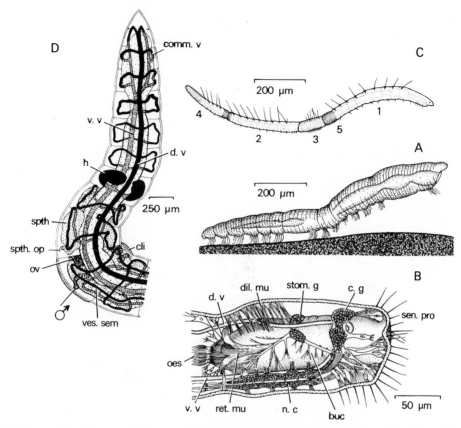

Fig. 81. A, *Chaetogaster*: anterior part of body raised on chaetae preparatory to catching prey. B, *Chaetogaster diaphanus*: internal structure of anterior segments seen by transparency. Note the primitive ladder form of the ventral nerve cord. (After Vejolovsky.) C, *Nais*: undergoing asexual reproduction by fission. The fission zones where separation of the individuals ultimately takes place are indicated by constrictions of the body. The zooids are numbered in order of appearance. (After Stolte.) D, *Tubifex rivulorum*: showing the vascular system in the anterior segments. Commissural vessels link dorsal and ventral vessels; the heart pumps blood from the supra-intestinal vessel (not shown) to the ventral vessel. buc, buccal region of gut; c.g, cerebral ganglion; cli, clitellum; comm.v, commissural vessel; dil.mu, dilator muscles of stomodaeum; d.v, dorsal blood vessel; h, heart; n.c, ventral nerve cord; oes, oesophagus; ov, ovary; ret. mu, retractor muscles of stomodaeum; sen.pro, sensory process; spth, spermatheca; spth.op, opening of spermatheca; stom.g, stomatogastric ganglion; ves.sem, vesicula seminalis; v.v, ventral ventral blood vessel; ♂, male opening.

segment, often 50–100 or even more; this perichaetal condition has been evolved in unrelated families (Glossoscolecidae, Megascolecidae). In aquatic oligochaetes, however, the chaetae may be long and hair-like, grouped in bundles which help as oars in swimming, or shorter and rod-like for walking (*Chaetogaster limnaei*) (Fig. 81A). In the genus *Ripistes* (Fig. 82A), they are used to collect particles suspended in the water. The elongated dorsal bundles of segments 6–8 are repeatedly spread fanwise and then closed, and at the same time moved backwards then forwards: the bundles are drawn through the lips and the adherent particles wiped off and swallowed.

Each chaeta is secreted by a gland at the base of an epidermal follicle (Fig. 80C, f.c) which projects into the coelom. Protractor muscles (prot.mu) inserted on the inner end of the follicle (or group of follicles if the chaetae are in bundles) radiate to origins on the body wall and move the chaeta forwards and back and from side to side, as well as protracting it. Retractors (ret.mu) are inserted along the length of the follicle, and the fibres of dorsal and ventral chaetal sacs of the same side of the

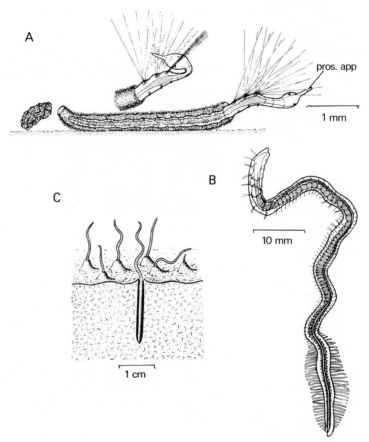

Fig. 82. A, *Ripistes parasita* (Naididae): anterior end of worm extended from tube with three pairs of chaetal bundles outspread. The upper drawing shows particles adhering to the chaetae being wiped off by the lips. The lower drawing shows a faecal pellet recently discharged. (After Cori.) B, *Branchiura sowerbyi* (Tubificidae): to show the gills restricted to the posterior segments. They are vascularized by commissural vessels. (After Stephenson.) C, *Tubifex tubifex*: the posterior ends of the worms project from the mud and their rhythmical corkscrew movements agitate the water. (After d'Udekem.) pros.app, prostomial appendage.

segment unite at their origins to form a bridge across the coelom; their action is more or less synchronized.

Movement of oligochaetes

When an earthworm crawls a wave of contraction of the circular muscles passes posteriorly along a number of anterior segments; simultaneously the chaetae of these segments are withdrawn and those immediately posterior are protracted to give a *point d'appui*. The anterior segments extend and their chaetae anchor the body as the more posterior ones are withdrawn and the longitudinal muscles draw the hinder part of the worm forward. A number of these peristaltic waves, each involving relatively few segments, may pass down the body in rapid succession, initiated at the front end; their propagation from segment to segment is due to a variety of factors which will be discussed later. The average hydrostatic pressure in the anterior third of the body of *Lumbricus terrestris* when it is actively wriggling is about 16.0 cm water and in the tail region 8.0 cm.

When the worm burrows the same movements are employed and the anterior tip of the body, which is conical and has no appendages to resist progress, is used as an awl to penetrate the soil. The prostomium (Fig. 78A, pros) and four succeeding segments are more or less solid with muscles so that there is virtually no coelom (and there are no septa) and the brain, no longer associated with head appendages, has shifted back to segment 3 (c.g). Consequently the forward thrust which is exerted will be transmitted from the septum separating segments 4 and 5. It has been calculated that this is equivalent to forces between 1.5 g and 8.5 g. If the area of the prostomium is 0.008 cm² and the thrust equivalent to 8.5 g, the pressure of the prostomium against the soil will be 1 060 g cm⁻². The prostomium is inserted through a crevice and thrust into the soil as the anterior part of the body extends. The longitudinal muscles then contract and widen the pathway. If the soil is resistant mouthfuls may be eaten until the initial part of the burrow has been formed and the worm can grip the walls. No pressure recordings have been made on worms actively burrowing. It is assumed that, as in *Arenicola*, they are higher when the worm is in a confined space since the longitudinal muscles can then exert greater force increasing the diameter of the body to overcome the resistance of the soil. It has been calculated that the longitudinal muscles of *Lumbricus* are about ten times as strong as the circular; they are bulkier than the longitudinal muscles of *Arenicola*, and strengthened by collagenous lamellae which give the characteristic feather pattern seen in transverse section.

The peristaltic movements of the body wall concerned with locomotion are, as in other annelids, transmitted by an interaction of central and peripheral nervous activity. It appears that in the earthworm either of these components is capable of maintaining peristalsis, whereas in *Nereis* the central component is essential. The ventral nerve cord of the earthworm (v.n) lies freely in the coelom, right and left halves bound together by fibrous connective tissue; between this and the outer peritoneal covering are longitudinal muscle fibres (Fig. 83D, mu). It extends back from the suboesophageal ganglia in segment 4 giving three pairs of nerves to each segment (Fig. 79C, 1, 3). The nerves pass to the body wall and describe a more or less semicircular ventrodorsal course between the circular and longitudinal muscles. They contain both sensory and motor fibres which connect with a subepidermal plexus of branching fibres and scattered nerve cells; some of the fibres end between the epidermal cells. Whereas the segmental nerves send branches to the segment in front of their origin and to the one behind, the subepidermal net is considered to be segmentally localized. Thus responses can be spread from segment to segment not only by neurones within the ventral nerve cord but also by the segmental nerves supplying the body wall and covering more than one segment (Fig. 79B).

Certain experiments elucidate the factors involved in the propagation of the peristaltic waves associated with crawling; their normal rhythm occurs only when the worm is receiving tactile stimulation from the ventral surface of the body. If the body is cut across so that only the ventral nerve cord remains intact, peristaltic waves still pass anteroposteriorly down its length. They must be propagated across the cut and

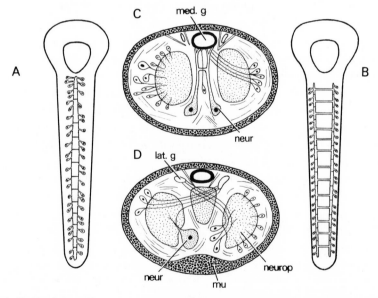

Fig. 83. A, B, *Lumbricus*: diagrams to show position of giant axons in central nervous system. A, single median; B, lateral axons. (After Nicol.) C, D, *Megascolex*: diagrams of transverse sections of ventral nerve cord to show sensory input in relation to giant fibres. In the anterior 60 segments (C) the neurons feed into the median fibre, in segments posterior to this (D) they feed into the lateral fibres. (After Adey.) lat.g, lateral giant fibre; med.g, median giant fibre; mu, muscles of sheath; neur, giant cell body; neurop, neuropile.

so from segment to segment along the cord without reinforcement by segmental reflexes. If in another worm the ventral nerve cord is severed, or even removed from several segments, and the rest of the worm is left intact, there is still no hindrance to the passage of the waves which must, in this case, be transmitted across the gap by the peripheral nerves and subepidermal plexus. A third worm may be cut into two transversely and the cut surfaces connected together by thread, and still it can crawl. The pull of the thread causes increased tactile stimulation of the posterior half as the anterior segments advance, causing the initiation there of a peristaltic wave (Fig. 79A). The muscles must react directly to changing tensions for a pull applied to the tail of an anaesthetized worm (suspended vertically so that there is neither tactile stimulation nor central nervous control) causes a peristaltic wave to be set up.

Normally tactile reflexes and stretch reflexes cooperate with the nerve cord in setting up waves and adjusting movements of the worm to external conditions, the former more important when the animal is moving over a smooth surface and the latter of increasing importance as the surface becomes irregular. Tensile stimulation may be effected through stretch receptors in the muscles. These are known to occur in arthropods and vertebrates and nerve cells described in the circular muscles of earthworms could serve this role. Stretch receptors can only propagate a movement already started by some other agency. What initiates the waves of movement is unknown. Electrical activity of the ventral nerve cord during locomotion gives a rhythm with a frequency identical with that of the body movements. It is possible that the electrical potentials coincide with the phase of longitudinal contraction but there is no proof that these rhythms initiate movement.

A rapid startle response involving an end-to-end contraction of the longitudinal muscles allows the worm to escape into its burrow if disturbed. Three giant fibres effect it: two median fibres (Fig. 83, lat.g), connected transversely, flank a median one (med.g) and run through the nerve cord from the cerebral ganglia to the pygidium. Each receives axons from large, segmentally arranged nerve cells (neur) which have fused to form a giant fibre. Impulses travel fastest along the median fibre which has anterior sensory connections as far back as the clitellar region, that is over the part of the body where exteroceptors are most abundant, and this fibre conducts posteriorly. The lateral fibres, which give a shared response, indicating that their transverse connections are physiological as well as anatomical, have posterior sensory connections and conduct anteriorly. The rate of conduction in the median fibre is 17–45 m s^{-1}, which is considerably faster than in the giant fibres of tubicolous or burrowing polychaetes; in the lateral fibres it is 7–17 m s^{-1} as compared with 0.025 m s^{-1} in the small fibres of the cord.

Although oligochaetes are without paired sensory appendages, and most are without apparent eyes, their responses to changing environmental conditions show that sensory structures must be copiously scattered over the surface of the body. The common type is a group of epithelial cells (sometimes a single cell) each with a hair-like prolongation passing through the cuticle; the process is long (Fig. 81B, sen.pro) in aquatic forms (*Tubifex*, naiads) and as short as 2 μm in earthworms (Fig. 84C, sen.pro). Depending on their situation nerve fibres pass from the receptor cells to an epidermal branch of a segmental nerve or connect with the cerebral ganglion. Whether such cells respond to more than one stimulus is not known but together with free nerve endings in the epidermis they appear to comprise the chemical and tactile receptor system of the skin and buccal region of the gut. They are most numerous in anterior and posterior segments, especially the prostomium. In *Lumbricus* the number per segment averages 1 000, far exceeding the number of epidermal sense organs in polychaetes. Earthworms are sensitive to humidity changes, burrowing

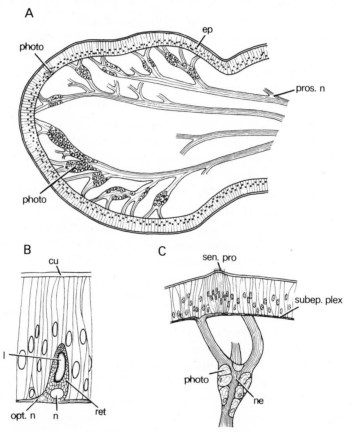

Fig. 84. *Lumbricus terrestris*: A, diagrammatic longitudinal section of prostomium to show photoreceptor cells on nerve enlargements and in the epidermis; B, epidermis of prostomium showing photoreceptor cell and optic nerve; C, nerve enlargement of prostomium with photoreceptor cells and epidermal receptor organ with hair-like processes. (After Hesse.) cu, cuticle; ep, epidermis; l, lens; n, nucleus; ne, nerve enlargement; opt.n, optic nerve; photo, photoreceptor cell; pros.n, prostomial nerve; ret, retinella; sen.pro, sensory process; subep.plex, subepidermal nerve plexus.

deeper during drought and orientating towards moist soil. The receptor organs responsible for sensing these changes are apparently restricted to the prostomium. The worms are also sensitive to light, emerging from the burrow when it is dim and avoiding high intensities. The prostomium, which emerges first, is the most sensitive area, sensitivity decreasing along the body, though increasing again at the rear end, which is exposed at defecation. Photoreceptor cells have been found in a number of species of earthworms, some scattered in the epidermis, some more deeply placed around terminal twigs of cutaneous nerves and perhaps derived from epidermal cells which, as in the leech *Clepsine*, have migrated from the surface along the course of the nerves (Fig. 84). They are absent from the ventral surface. More primitive freshwater oligochaetes (many naiads) have pigmented eyespots restricted to the epidermis of the head, but no pigment is associated with the epidermal light cells of earthworms. Each light cell contains a structure more refractive than the rest of the cytoplasm and surrounded by neurofibrillae.

The evolution of the oligochaetes has involved little structural change in body surface. As in errant polychaetes a thin cuticle covers the epidermis, allowing cutaneous respiration to occur and providing little or no barrier to the passage of water. In terrestrial species there are many epidermal gland cells. Their secretion maintains a watery layer over the surface in which oxygen can dissolve and, if the worm is subjected to desiccation they secrete more vigorously. The need for a moist skin is imperative since the skin is the only respiratory surface which earthworms have: it is, indeed, surprising that animals of such bulk are able to survive without special respiratory devices. Under extreme conditions some species also exude coelomic fluid from the dorsal pores. It is interesting to note that the pores are neither developed in oligochaetes with a marshy or aquatic habitat, nor in some terrestrial forms (Glossoscolecidae) which moisten the skin with fluid from nephridia. The epidermal secretion also lubricates the pathway through the soil and binds together soil particles, so that the surface of the worm is kept clean and the walls of the burrow are prevented from collapsing. It is a mucoprotein and about half the nitrogen lost from the body is contained in it.

Respiration in oligochaetes

The general body surface of most aquatic forms provides an adequate respiratory area and in contrast to polychaetes (many of which live in tubes) gills rarely occur. Antiperistaltic movements of the intestine draw in a current of water which may be respiratory (and also assists the forward propulsion of blood in the intestinal plexus). Gills have been independently evolved in a few members of four families. In some they are finger-like projections restricted to the anal region (*Dero*, *Aulophorus*) which lengthen under conditions of poor oxygen supply. In others they are segmentally arranged in pairs over areas of the body where oxygen is most available. Some species living in mud of river beds or swamps, where oxygen concentration is low, collect their food with the head buried in mud and the tail projecting from the surface into water, and the body agitates the water to set up currents around it. In the tubificid *Branchiura sowerbyi* (Fig. 82B) segmentally arranged gills are confined to this posterior region. The related *Tubifex tubifex* has no gills and as the oxygen tension falls a greater length of the body is extruded and ventilating movements become more vigorous (Fig. 82C); the worm survives provided $Po_2 > 1$ ml l^{-1}. Similarly *Alma nilotica* (Glossoscolecidae), living in Nile mud, has tufts of gills on the upwardly directed posterior segments. The related *A. emini* (Fig. 85A, B) of papyrus swamps in East Africa is adapted for aerial respiration: the highly vascularized posterior end of its body projects from the decomposing vegetation and (by unfolding the edges of the dorsolateral surfaces) forms a scoop which entraps bubbles of air which are carried into the mud when the worm retreats. The worm can also make use of the little oxygen available in the swamp since its haemoglobin has a very low unloading tension, and high CO_2 tensions are without noticeable effect on the dissociation curve.

The plan of the vascular system of oligochaetes is similar to that of less specialized

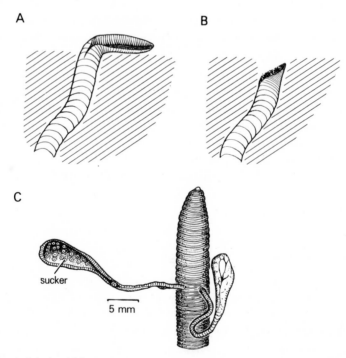

Fig. 85. *Alma emini*, the African swampworm: A, posterior end of body (spatulate by dorso-ventral flattening) pushes up through the waterlogged mud to trap air; B, worm retreats into mud with air bubbles. (After Beadle.) C, *Alma pooliana*: anterior segments, ventral view, to show copulatory appendages, and their suckers with setae. (After Michaelsen.)

polychaetes – a longitudinal vessel above and below the gut (Figs 78A and 81D, d.v, v.v) linked by vessels in the body and gut walls. Blood from the body and gut walls flows into the dorsal vessel which is contractile and maintains a postero-anterior flow. It is the main collecting vessel directing blood through a paired series of contractile hearts, which lie in the gut wall in some anterior segments (h). The hearts open into the ventral vessel in which blood flows backwards. The direction of blood flow is maintained by valves in the dorsal vessel and the hearts. Whereas in smaller oligochaetes the somatic commissural vessels are reduced in number and confined to a few anterior segments (Aeolosomatidae, Naididae, Enchytraeidae) in larger ones their importance is increased by the development of gills, which they supply, and the vascularization of the body wall so essential to soil dwellers (Fig. 80B). The blood is colourless in some oligochaetes (Aeolosomatidae, Branchiob-dellidae, some naiads, many enchytraeids), their need for oxygen being met by physical solution. The majority have the respiratory protein haemoglobin in solution in the plasma. The pigment is engaged in the uptake and transport of oxygen at all times especially when the oxygen tension outside the body is low.

Feeding and the functions of the gut in oligochaetes

Differences between the alimentary canal of the oligochaete and errant polychaete concern the elaboration of the buccal and oesophageal regions, particularly in terrestrial species. The buccal cavity, modified for collecting food, has a thin-walled anterior region and a posterior part thickened by intrinsic muscles (Fig. 78A, t), and with extrinsic muscles inserted on to the body wall (prot.mu, ret.mu). In some aquatic forms it is tubular with little distinction between anterior and posterior parts, and may be used like a pipette in sucking up food. *Chaetogaster* (Fig. 81A, B) sucks in rotifers, small crustaceans and worms, which it may collect from the canals of sponges or the mantle cavity of prosobranch gastropods. The aberrant branchiob-dellids, the only oligochaetes with buccal teeth, have elaborated this method of feeding and use the buccal cavity as a suction pump. The adult *Branchiobdella* is ectoparasitic on the gills of crayfish, attaching by an oral sucker, breaking the skin with its teeth and sucking blood. The young are not parasitic and suck up vegetable detritus and small animals.

In a number of oligochaetes, including terrestrial species, the dorsal part of the buccal wall is thickened and forms a tongue. This may be protruded through the mouth (Fig. 86) to explore the substratum and, assisted by the lips, pick up particles of decaying vegetation, withdrawing with them into the buccal cavity (*Aulophorus*, enchytraeids, earthworms). Their ingestion is assisted by the sucking action of the buccal walls, and this same mechanism underlies the ingestion of soil by worms which burrow. Oligochaetes have salivary glands discharging to the buccal cavity and especially associated with the tongue (Fig. 78A, sal.gl); in some (e.g. enchytraeids)

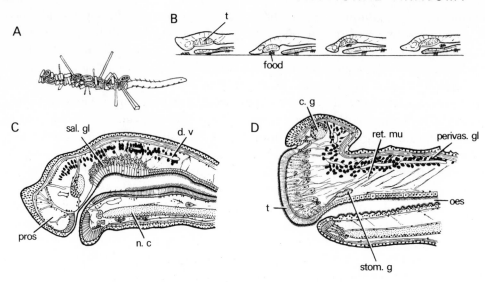

Fig. 86. *Aulophorus carteri*: A, worm emerging from tube; B, diagrams (left to right) to show the way in which food particles are collected by the tongue; C, sagittal section of anterior part of worm; D, as C with tongue protruded. (After Marcus.) c.g, cerebral ganglion in prostomium; d.v, dorsal blood vessel; n.c, nerve cord; oes, oesophagus; perivas.gl, perivascular glands; pros, prostomium; ret.mu, retractor muscles of tongue; sal.gl, salivary glands; stom.g, stomato-gastric ganglion; t, tongue on which open salivary glands.

their secretion is augmented by secretion from glands associated with anterior septa (Fig. 87, acc.sal.gl). The glands secrete mucus, which helps to secure the food particles, and in some earthworms a protease. The mucus may have other uses: the naiad *Aulophorus tonkinensis* lives in a tube, constructed of bits of leaves, wood, etc., which it carries about using its tongue covered with secretion as an adhesive sucker and the particles used to form the tube of this species, and also that of *A. carteri* (Fig. 86), are manipulated by prostomium and tongue and cemented by salivary secretion.

In aquatic oligochaetes the oesophagus is a ciliated tube leading to the intestine. Only occasionally, as in *Chaetogaster*, is there a crop dilatation to accommodate large mouthfuls of food. In contrast the walls of the fore-gut of terrestrial oligochaetes are elaborated to form calciferous glands, crop and gizzard (Fig. 78A). The gizzard (giz), with thick muscular walls, triturates the food with the help of mineral particles in the soil, and more than one gizzard is developed in some larger worms (Megascolecidae). The calciferous glands are one or more pairs of swellings of the wall within which the epithelium is thrown into deep folds or lamellae. The gut sinus penetrates the folds (C, b.sin), which thus receive blood from the absorptive region of the intestine, and a circulation between the folds is ensured by the proximity of the hearts. The glands secrete calcium carbonate concretions (Ca) into the lumen of the oesophagus – by

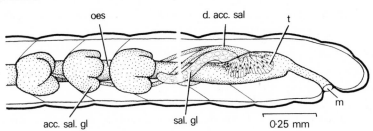

Fig. 87. *Enchytraeus pellucidus*: diagram of anterior end to show glands associated with the stomodaeal region of the gut. (After Stirrup.) acc.sal.gl, accessory salivary gland; d.acc.sal, duct of accessory salivary gland; m, mouth; oes, oesophagus; sal.gl, salivary gland; t, tongue.

way of the oesophageal pouches (A, oes.p) in earthworms – and these pass from the body with the faeces. Carbon dioxide diffuses into the worm from concentrations in the soil and is also produced by metabolic reactions within the body. An accumulation in the tissues would affect the pH of the blood and coelomic fluid. It appears that the calciferous glands control the acid-base balance of the body and maintain a stable pH by fixing a percentage of the carbon dioxide, the amount varying with that in the atmosphere. The glands are presumably also concerned in regulating the calcium content of the tissues. They are larger in specimens living in calcareous soils.

The intestine (Figs 78A and 79C, int), the longest and widest part of the gut, has a columnar epithelium which is glandular and absorptive. The visceral blood plexus (or sinus) is beneath the epithelium, between the basement membrane and visceral muscles. The peritoneal covering gives rise to chloragogenous tissue (Fig. 79C, chlorag), particularly thick in earthworms, and a deep infolding of the dorsal wall, the typhlosole (typh), increases the secreting and absorbing area and the volume of chloragogenous tissue. The food of earthworms, plant material selected from the litter layer of the ground and small fragments of organic matter from the soil, requires considerable trituration and mixing with digestive fluids. Its passage through the gut is lubricated by mucus secreted by the salivary glands and epithelial lining and the chief site of enzyme production is the anterior intestine. We have some knowledge of the digestive enzymes of *Lumbricus*. The salivary glands produce a protease as well as a lubricant, and in the anterior intestine there is an array of enzymes appropriate to the diet – amylase, lipase, chitinase, cellulase, protease. Rhythmical contractions of the visceral muscles travel in either direction along the gut to mix the contents thoroughly. These movements are controlled by a double set of nerves. One set, arising from the circumoesophageal nerve ring, forms a plexus beneath the intestinal epithelium and the second, coming from the ventral nerve cord, reaches the gut via the septa. It has been suggested that stimulation of the former is excitatory while stimulation of the latter decreases motility. Another suggestion is that the septal nerves contain both excitor and inhibitor fibres, leaving the function of the oesophageal nerves unknown. Undoubtedly there are antagonistic nerves controlling the tonus of the visceral muscles and recalling the sympathetic and parasympathetic systems of vertebrates. Double innervation of the intestine is not known in polychaetes.

The function of the chloragogenous tissue has aroused much controversy. The yellow-brown colour of the chloragocytes is due to a phospholipid allied with an un-known pigment. Within the cells there are reserves of glycogen and fat and varying amounts of ammonia and urea. Thus the tissue is involved in storage and takes part in detoxicating mechanisms. Its position in relation to the visceral blood plexus and coelomic fluid are ideally suited for these functions. Substances from the gut and body fluids can be transferred to the chloragocytes which, like the liver cells of vertebrates, are highly active in transforming and mobilizing energy sources. The chloragocytes may also have a homeostatic role in maintaining a constant level of circulating substances. Similar cells have been described in polychaetes though little is known of their functions.

Excretion and osmoregulation in oligochaetes

The segmental organs of oligochaetes are metanephridia and coelomoducts. The former, often long and coiled, occur in all segments except the first few and may be reduced or lost in the genital segments of smaller worms. The latter are confined to the genital segments and in no case do they combine with the nephridia, the isolation and complexity of which is partly associated with maintaining water balance.

It is assumed that in freshwater oligochaetes water passing into the body is ex-creted in the urine. This assumption is based on our knowledge of other freshwater invertebrates and on the water relations of the earthworm, about which we have some knowledge. Experiments have shown that the earthworm excretes a copious hypotonic urine under humid conditions (about 60 per cent of the body weight in 24 h), absorbs salt through the body surface from very dilute solutions and maintains an internal osmotic pressure above that of the environment, except in concentrated media (1.0 per cent NaCl). Water is excreted by the nephridia, and, if the uptake is excessive, by way of the anus and mouth. If the worm is subjected to long periods of drought the body becomes dehydrated and a slight degree of dehydration is the normal state of the body of an earthworm. If the soil contains less than 10 per cent water, 60 per cent or more of the body weight may be lost without irreparable harm and some species (*Lumbricus terrestris*) can withstand 80 per cent loss. Such losses interfere with the efficiency of the hydrostatic skeleton on which crawling and burrowing depend but are rapidly made good when water is available.

The metanephridia, one pair per segment, have typically a ciliated nephrostome (Fig. 88A, nephs) allowing the entrance of coelomic fluid, and a nephridial duct with a generous capillary vascular supply from commissural vessels, suggesting that there may be ultra-filtration from the blood. They are relatively small and simple in

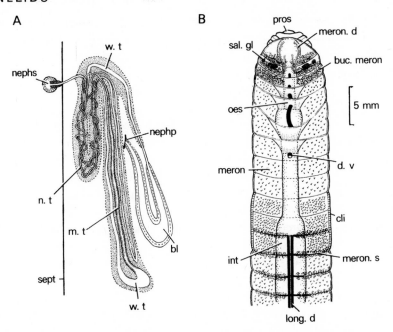

Fig. 88. A, *Lumbricus*: diagram of metanephridium. B, *Pheretima posthuma*: plan of nephridial system. (After Bahl.) bl, bladder; buc.meron, buccal meronephridia; cli, clitellum; d.v, dorsal vessel; int, intestine; long.d, longitudinal meronephridial duct; meron, meronephridium; meron.d, meronephridial duct to buccal cavity; meron.s, septum with meronephridia; m.t, middle tube; nephp, nephridiopore; nephs, nephrostome; n.t, narrow tube; oes, oesophagus; pros, prostomium; sal.gl, salivary glands; sept, septum penetrated by narrow tube; w.t, wide tube.

aquatic oligochaetes but in earthworms have a large and elaborate nephrostome and the coiled duct is divisible into a number of dissimilar regions. In some earthworms metanephridia open to the gut. Those of the posterior segments in *Allolobophora antipae* join a longitudinal duct on each side which discharges to the intestine. Such an enteronephric system is better developed in larger worms (Megascolecidae) living under hot, dry conditions and may be a means of conserving water, for water discharged into the gut can be reabsorbed. The nephridial system of the Megascolecidae shows remarkable complications affecting its efficiency. The original pair of nephridia in each segment subdivide to give a large number of small nephridia called meronephridia, which may share a common opening, or open separately. There are many variations in detail from species to species and even within one species, as illustrated by *Pheretima posthuma* which has three pairs of these multiple structures, with no coelomic openings, discharging by ducts into the buccal cavity (Fig. 88B, buc.meron, meron d). From segment 7 backwards meronephridia with no coelomic openings are scattered over the body wall and discharge separately to the outside (meron). From segment 15 backwards there are also 40–50 meronephridia with open funnels scattered over the anterior and posterior face of each septum (meron.s) and associated with a right or left duct within a mesentery (long.d); the canals open to dorsolateral longitudinal ducts with intersegmental openings to the gut.

The possible functioning of the more typical metanephridium has been studied in *Lumbricus* (Fig. 88A). The coiled duct, divisible, anatomically and physiologically, into four main regions, opens to the coelom in one segment (sept) and to the exterior by way of the nephridiopore in the segment behind (nephp). The nephrostome leads to a narrow intracellular tube (n.t) which penetrates the septum and coils in the next segment. The tube is ciliated in parts and the effective beat of the cilia is towards the nephridiopore. The middle tube (m.t.) which follows is also ciliated and has an accumulation of brownish granules in the walls. Then follows the wide tube (w.t) and muscular tube or bladder (bl) leading to the nephridiopore which is controlled by a sphincter; neither is ciliated. The nephridium voids nitrogenous waste elaborated elsewhere. The beating of the cilia of the nephrostome and ducts suggests that coelomic fluid is passed into it. This fluid contains ammonia and urea found in approximately the same concentrations in the blood of *Lumbricus* and *Allolobophora* and apparently formed in the chloragogenous tissue. They may diffuse into the coelomic fluid directly from this tissue or by way of the blood. It has been suggested that chloragogenous cells are liberated from the tissue, break down in the coelom and release their products, but this is unproven. The concentrations of ammonia and urea in the urine of these two worms are greater than in the body fluids, and that of various salts and protein is less. This suggests that a filtering process is going on somewhere in connection with urine formation. Urine may be coelomic fluid from which salts and protein have been largely removed by the nephridial duct (though it would involve

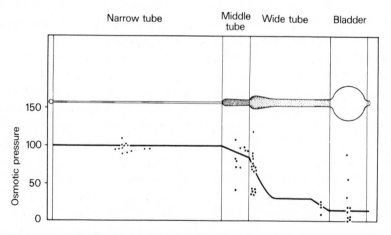

Fig. 89. To show changes in the osmotic pressure of urine in the nephridium of *Lumbricus*. The osmotic pressure of the Ringer in which the nephridium was placed is equated to 100. Points indicate the osmotic pressure of fluid collected at different levels of the nephridial tube. (After Ramsay.)

the resorption of protein across a cellular membrane). Urine could also originate by ultra-filtration from the blood and the large molecular substances such as protein be precluded from passing the nephridial wall. This is similar to the initial stages of urine formation in the vertebrate kidney. However, blood pressure in the worm is too low (4.4 mmHg–9.3 mmHg according to activity) to cause rapid ultra-filtration against the colloid osmotic pressure of the blood. It has been suggested that the hydrostatic pressure of the blood may exceed the colloid osmotic pressure so that ultra filtration can occur. The osmotic pressure of the fluid in the narrow tube of the nephridium of *Lumbricus* is probably isotonic with the surrounding coelomic fluid (Fig. 89). Water and salts may pass with equal facility through the wall in both directions. The urine becomes hypotonic in the middle tube and especially the wide tube. This may be brought about not only by resorption of substances into the blood, for which as yet there is no proof, but also by an influx of water. Perhaps the pigmented granules which accumulate in the wall of the middle tube are a breakdown product of the blood (haemochromogen).

Reproduction of oligochaetes

The reproductive processes of oligochaetes are complicated by hermaphroditism, copulation and the production of egg capsules or cocoons, all necessary adaptations for life in fresh water and on land. Paired ovaries and testes (Fig. 78B, ov,te) are associated with the septa of a few anterior segments and, in contrast to polychaetes, are strictly limited areas of proliferation. There are one or two pairs of testes, exceptionally more, and with few exceptions one pair of ovaries posterior to the testes. Sex cells are shed either into the general coelom of the fertile segment or into a special part of it, and then, in the case of the male cells, they pass into outpouchings of the septa, the seminal vesicles (ves.sem), where they mature. The testes undergo regression when the spermatogonia have been shed and while the other sex organs are fully developed. A coelomoduct is associated with each gonad, the ciliated coelomostome opening to the coelom of the fertile segment (sp.fu) and the duct typically piercing the posterior septum of the segment to reach the genital opening (v.d, ovd). The oviducts remain simple and relatively short. In some naiads, tubificids and enchytraeids the eggs have attained such a size that ducts can no longer transmit them; in fact, oviducts do not develop and they are discharged by rupture of the body wall. The sperm ducts show considerable differentiation and their arrangement forms the basis of classification of the class. Their coelomic funnels are large and convoluted in earthworms, appearing conspicuously iridescent when sperm are ripe since the sperm are stored on the walls of the funnel until copulation, orientated with their heads against the epithelium and their tails towards the lumen. Primitively the male duct (like the nephridium) coils freely in the coelom and opens to the outside in the segments posterior to the funnel (naiads, tubificids, enchytraeids); but it may open in the same segment (lumbriculids); or the two ducts on each side (from two pairs of testes) fuse and pass back a few segments before opening (lumbricid and megascolecid earthworms).

The transference of sperm from one individual to another necessitates the development of prostatic glands for the provision of fluid in which sperm are immersed and remain motile. The glands are associated with the vas deferens. Penes (90B, pe) facilitate the deposition of seminal fluid in the spermathecae (sac-like invaginations of the body wall) of the partner (Fig. 78B, spth). They are most elaborate in the genus *Alma* of Central and N.E. Africa, appearing at sexual maturity as long, slender outgrowths of the male duct which broaden distally and, in some species bear suckers, each with a spine (Fig. 85C). In *Pheretima montana* the thread-like penis has two canals, the prostatic duct and the vas deferens.

In contrast to leeches fertilization in oligochaetes is not internal but occurs when eggs from the oviduct and sperm from the spermathecae are deposited in a cocoon formed by secretion from a localized thickening of the body wall, the clitellum (Fig. 90B, cli). The clitellar glands not only provide the cocoon or egg capsule but also albumen which, together with yolk in the egg, allows development without a free larval stage. The clitellum is situated near the region of the male and female pores. In enchytraeids the vasa deferentia open to a ventral depression on the clitellar segment (Fig. 90A, cli). At copulation the worms are apposed anteriorly by their ventral surfaces, their heads pointing in opposite directions. The spermathecal region of each fits into the sucker-like depression of the partner from which two

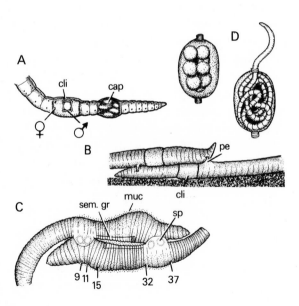

Fig. 90. A, *Enchytraeus albidus* (ventral view): to show the egg capsule being passed forward over the anterior segments. (After Michaelsen.) B, *Pheretima communissima*: anterior segments of two copulating worms. (After Oishi.) C, *Lumbricus terrestris*: anterior segments of two copulating individuals. The seminal groove leads from the male opening (seg. 15) to the clitellum (seg. 32–37). Sperm pass along it and accumulate at the openings of the spermathecae of the partner before passing into them. D, *Tubifex rivulorum*: cocoon with eggs (left) and with young emerging (right). cap, egg capsule; cli, clitellum; muc, mucus; pe, penis; sem.gr, seminal groove; sp, sperm mass; ♂, lip over male genital aperture; ♀, female genital aperture.

protrusible penes emerge and insert into the spermathecal openings. In lumbricid earthworms the clitellum is posterior to male and female openings and there is a more complicated method of copulation. This has been observed in *Lumbricus terrestris* (Fig. 90C). The clitellar region (segments 32–37) is linked with the male opening on segment 15 by a line of pigment marking the position of a seminal groove (sem.gr) apparent only at the time of copulation. The copulating worms become attached in a head-to-tail position, their ventral surfaces together so that segments 9–11, associated with two pairs of spermathecae, are apposed to the clitellum of the partner. Each worm is enclosed in a mucous tube (muc) secreted by the epidermis and extending over the genital segments (9–37). Although these tubes adhere to one another the intimate attachment of the worms is in the clitellar region, where it is assisted by secretion from special glands associated with the chaetal sacs and also by special chaetae which penetrate the body of the partner. Seminal fluid emitted from the male pores is passed along the seminal groove beneath the slime tube by peristaltic waves and becomes aggregated into masses (sp) at the openings of the spermathecae before passing into them.

Earthworms exhibit a yearly reproductive cycle. The gonads of *Lumbricus terrestris* mature in early summer and regress later in the year. The cycle appears to be regulated in the cerebral ganglia which release secretion into a capillary blood supply at the time the gonad must begin to mature. The action of these cells is mediated through a second set of neurosecretory cells in the suboesophageal ganglia. Neurosecretory cells also govern the regeneration of damaged or lost segments and possibly asexual reproduction, which in some small aquatic oligochaetes (Aeolosomatidae, Naididae) is the principal mode of reproduction. These worms produce stolons in which the new individuals are well differentiated before breaking away (Fig. 81C). In most species when an individual has attained a certain size there is an intercalation of new segments about the middle of the body. This is where the ultimate separation of the individuals will take place. Frequently, before this happens, new fission zones begin to appear in one or both components of the chain; the position of the fission zone is within limits definite for each species. By such means the population can be effectively increased when conditions are favourable. Asexual reproduction also occurs by fragmentation. For instance, *Lumbriculus* and *Aulophorus* divide into two without previously producing new segments and subsequently form a new anterior or posterior end.

Organization of hirudineans

Leeches are predators; the majority are associated with fresh water, a few are marine, and a few terrestrial, though living in damp places. Most are highly specialized in their feeding habits like the medicinal leech *Hirudo medicinalis*. It sucks blood until the gut is gorged, then abandons the host to seek another (man, domestic animal or even cold-blooded vertebrate) a few months later. The large storage reservoir provided by the crop, the high nutritive value of the food and its slow digestion necessitate only infrequent meals between which the worm is quiescent and hides. Other leeches are more ectoparasitic in habit in that they are associated with a specific host. Thus *Acanthobdella peledina* (Fig. 93) is found on salmon, and *Pontobdella muricata* (Fig. 94B) on elasmobranch fish. A few may be regarded as true parasites like *Theromyzon* which stays in the nasal passages of ducks almost indefinitely once it has got there. In contrast to these the less specialized predators ingest small

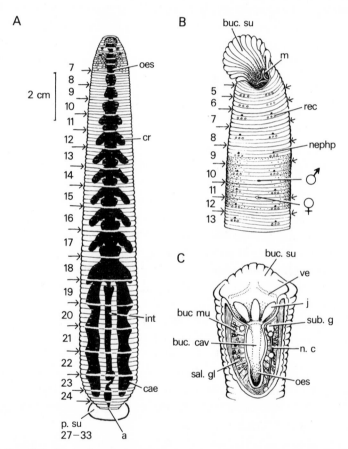

Fig. 91. *Hirudo medicinalis*: A, dorsal view, drawn as though the body wall were transparent so that the gut (black) and the radial buccal muscles are visible. The segments are numbered and their limits marked by prominent white lines crossing the gut. Paired eyes are on segments 1–5. B, anterior segments, ventral view; clitellum stippled. C, anterior segments opened midventrally. a, anus; buc.cav, buccal cavity; buc.mu, muscles of bulb; buc.su, buccal sucker; cae, last caecum of crop; cr, crop; int, intestine; j, ventrolateral jaw; m, mouth; n.c, ventral nerve cord; nephp, nephridiopore; oes, oesophagus; p.su, posterior sucker; rec, segmental receptor; sal.gl, salivary gland; sub.g, suboesophageal ganglion; ve, velum; ♂, male pore; ♀, female pore.

invertebrates such as worms, insects and molluscs; even the young of *Hirudo medicinalis* capture worms for food. Such predacious habits are common in polychaetes, some of which eat whole animals (*Nephthys*, *Glycera*) and others suck body tissue (syllids). Leeches, however, are obviously derived from oligochaetes, sharing some of their specialized characters and acquiring others which combine to make them a uniform group, easily recognizable.

The body of the leech is made up of a fixed number of segments, commonly thirty-three (Fig. 91), each with three, five or even more external annuli in the adult. The surface of the body is smooth with slime from epidermal glands and there are typically neither paired appendages nor chaetae to grip the substratum. Locomotion is by stepping, with strides often equivalent to the length of the body extended between terminal suckers used as grips (Fig. 92A). One sucker surrounds the mouth (Fig. 91B, buc.su) and a larger ventral one (Fig. 91A, p.su), formed by the last seven segments and

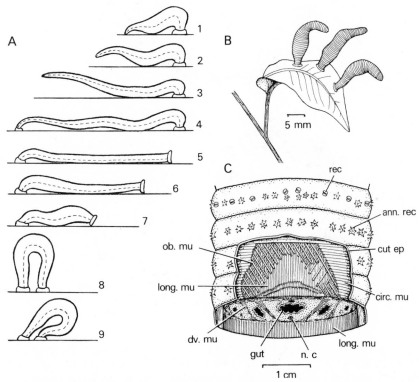

Fig. 92. *Hirudo medicinalis*. A, successive stages in forward locomotion. For explanation see p. 257. B, a haemadipsine land leech, attached to leaf by posterior sucker, awaits its prey. (After Doflein.) C, *Hirudinaria*: dorsal view of five annuli with some of the epithelium and body wall muscles removed. (After Bhatia.) ann.rec, annular receptor organ; circ.mu, circular muscles; cut ep, cut surface of epidermis; dv.mu, dorsoventral muscles; gut, gut; long.mu, longitudinal muscles; n.c, nerve cord; ob.mu, oblique muscles; rec, segmental receptor organ.

with the anus (a) immediately dorsal to it, attaches with proverbial force by means of mucus and the sucking action of concentric layers of muscle derived from the circular muscles of the constituent segments. Exceptions to these features are found in the small *Acanthobdella peledina* (Fig. 93) which, although leech-like in habit and general appearance, has chaetae on a few anterior segments and no anterior sucker. In these and other structural features it is intermediate between oligochaetes and leeches. For instance it has a perivisceral coelom (coel) and septa separating the segments, though the coelom is reduced as compared with oligochaetes by ingrowth of mesenchymatous packing tissue (mes). In other leeches this spongy tissue almost obliter-

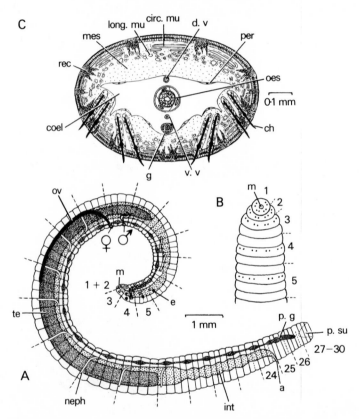

Fig. 93. *Acanthobdella peledina*: A, diagram showing general organization. Broken lines on the dorsal surface mark the limit of segments, some of which are numbered. B, ventral view of 5 anterior segments to show position of chaetae (indicated by dots). Broken lines mark the limit of segments. C, transverse section through the second annulus of segment 5. a, anus; ch, chaeta; circ.mu, circular muscles; coel, coelom; d.v, dorsal vessel; e, eye; g, ganglion; int, intestine; long.mu, longitudinal muscles; m, mouth; mes, mesenchyme; neph, nephridium; oes, oesophagus; ov, ovary; per, peritoneum; p.g, posterior ganglia; p.su, posterior sucker; rec, segmental receptor; te, testis and sac in which sperm develop; v.v, ventral vessel; ♂, male opening; ♀, female opening.

ates the coelom leaving only narrow channels, uninterrupted by septa, for the circulation of coelomic fluid. Reduction of the coelom and loss of septa are associated with the method of locomotion.

The sensory equipment of the body surface is similar to that of oligochaetes in that there are epidermal sense cells and fine nerve fibres penetrating the epidermis. Every annulus has groups of epidermal sense cells or sensillae which may protrude as papillae and are tactile or chemoreceptive (Fig. 92C, ann.rec). Light-sensitive cells occur over the cephalic region of the body, the posterior sucker (fish leeches) and among the segmental receptors (Fig. 92C, rec) on the middle annulus of each segment. Anteriorly they are grouped and backed by pigment to form four or five pairs of eyes (Fig. 91; Fig. 93A); the arrangement of these and also of the sensillae is a feature used in classification. Each photoreceptor cell (Fig. 95A, photo), like those of oligochaetes, has a large vacuole filled with a hyaline material which focuses the light rays on a fibrillar network in the cytoplasm; this makes contact with a sensory nerve fibre (sen.n). Most leeches are photonegative, but as with other predators light-avoiding reactions are modified by hunger. The sensillae (Fig. 95B) have fine non-vibratile cilia projecting from the surface and are particularly numerous anteriorly and in the buccal epithelium; muscle fibres (mu.c) enable each to be protruded as a papilla or retracted as a cup. Besides these there are other epidermal sense organs unlike those of any other annelid (Fig. 95C). Each consists of a conical cell just beneath the cuticle (con.c) with its base resting on and partly enclosed by another cell (mu.c) in which the cytoplasm has conspicuous transverse striations; the cells are related to a sensory nerve cell (sen.c). It has been suggested that they are proprioceptors and are sensitive to the contortions of the soft spongy body. Observations on living animals suggest that tactile receptors are more numerous over the general body surface, associated with a leech's strong thigmotactic behaviour, while chemo- and thermoreceptors are more numerous on the head and are used in recognizing characteristics of animals which will provide food. *Haemadipsa* (Fig. 92B), a terrestrial blood-sucking leech abundant in S.E. Asia, is attracted by the warmth of a man's hand several inches away and *Glossiphonia complanata*, a predator of freshwater snails, makes searching movements when extracts of snail tissue are added to the water in which it lives. The importance of thermoreceptors is obvious in species which suck blood from mammals or birds. Thus *Hirudo* readily attaches to a surface warmed to 33–35°C, but if the temperature is raised loosens its grip.

Feeding and digestion in hirudineans

Leeches are grouped in three orders according to the structures by which they obtain food. One, the Rhynchobdellae, stands apart in that its members have an eversible proboscis lacking teeth, which protrudes to penetrate the tissue of the prey by force whilst the anterior sucker grips the prey and stabilizes its base (Fig. 96A, C, prob). In this order are included the common *Glossiphonia complanata*, which pierces the skin of freshwater snails and has a poorly developed anterior sucker, and the piscicolid leeches such as *Branchellion torpedinis* (occuring on elasmobranchs and teleosts, Fig. 97A) and *Pontobdella muricata* (on elasmobranches, Fig 94B) with large anterior suckers. After the body of the host is penetrated the action of buccal muscles sucks in soft tissues. The other two orders are characterized by a buccal bulb (Fig. 91C) with stronger sucking action (buc.mu). In the Gnathobdellae it functions after the host has been incised by teeth carried on a dorsal and two ventrolateral jaws which leave a characteristic triradiate mark on the skin. The jaws (j) lie at the anterior end of the buccal cavity immediately behind the sucker. Each is a muscular ridge covered by cuticle thickened along its summit to form numerous minute teeth. Gnathobdellids include the medicinal, horse and terrestrial leeches; the last (*Haemadipsa*) may be seen in tropical and subtropical rain forests hanging from the foliage by the posterior sucker waiting to pounce on mammals, including man, with the anterior one (Fig. 92B). Leeches with a buccal bulb but neither jaws nor teeth are grouped in the Pharyngobdellae and members of the most important family, the Erpobdellidae, suck in whole animals. *Erpobdella*, common in ponds and streams in this country, lives on freshwater molluscs, planarians and small insect larvae, while *Trocheta*, always living near the bank, eats earthworms. These are not the only leeches to swallow small invertebrates, for some gnathobdellids with weak jaws and blunt teeth (*Haemopis*) have abandoned blood sucking and swallow such animals as earthworms and slugs. In both cases the gut is simpler than that of the typical leech.

Feeding and digestion have been studied intensively in the bloodsucking leech, especially *Hirudo medicinalis*, which will ingest up to ten times its initial weight. To obtain a meal with minimal disturbance a local anaesthetic needs to be applied, the area precisely incised, the skin opened and adjacent capillaries dilated, and an

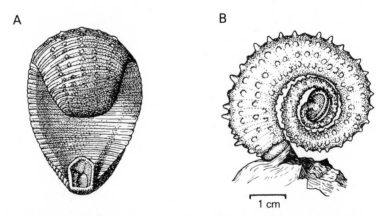

A   B

1 cm

Fig. 94. A, *Haementeria officinalis* (medicinal leech of Mexico): anterior end curved back over ventral surface displaying dorsal sensory papillae. (After Harding.) B, *Pontobdella muricata*: attached to rock by anterior sucker with body characteristically coiled.

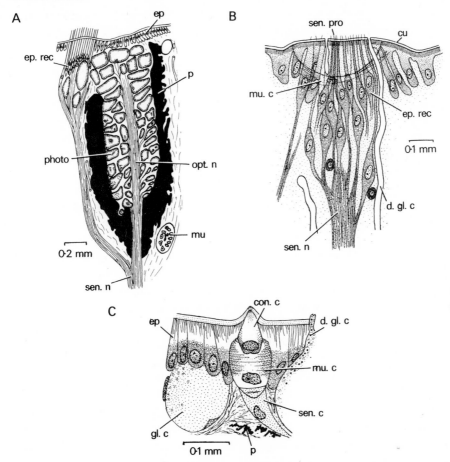

Fig. 95. A, *Hirudo medicinalis*: vertical section through eye and a group of epidermal sense cells. (After Hesse.) B, *Hirudo medicinalis*: vertical section through sensilla. (After Autrum.) C, *Glossiphonia complanata*: vertical section through epidermis to show conical cells of the organ of Bayer. (After Bayer.) con.c, conical cell; cu, cuticle; d.gl.c, duct of gland cell; ep, epidermis; ep.rec, epidermal receptor cell; gl.c, gland cell; mu, muscle; mu.c, muscle cell; opt.n, optic nerve; p, pigment; photo, photoreceptor cell; sen.c, sensory cell; sen.n, sensory nerve; sen.pro, sensory processes.

anticoagulant added to the blood to prevent it clotting. Secretions with these effects could come from the saliva (Fig. 91 C, sal.gl). So far no anaesthetic has been identified, though one must exist since the damage caused gives no pain, at least to man, even if tincture of iodine be applied to the open wound. Both a histamine-like compound, capable of dilating capillaries, and hirudin, an anticoagulant, have been identified in extracts of the head. When the leech bites, the velum (ve), or posterior wall of the sucker, is drawn back allowing the jaws to be protruded, pressed into the skin and moved by their muscles so that the teeth saw a pathway to the cutaneous capillaries. It is suggested that the histamine-like compound is meanwhile secreted by numerous unicellular salivary glands opening on the jaws and, when sucking starts, similar glands produce a secretion containing hirudin. Hirudin may not be injected into the wound though this has been alleged in the case of *Haemadipsa;* its main function is to prevent the body of the worm becoming stiff with clotted blood during and after the meal. After ingestion the food is immediately condensed by the abstraction of water and sodium chloride so that in 19 days the weight of food retained in the gut has decreased by more than 40 per cent. The fluid appears to leave the body by way of the nephridia.

The hirudinean gut has its capacity increased by paired lateral outpouchings or caeca (Fig. 91A, cae) in the region posterior to the short oesophagus. This region extends through half or even two-thirds the body length and is isolated from the oesophagus (oes) and short intestine (int) by sphincters. The caeca are comparable with the lateral segmental sacculations of the gut of oligochaetes and polychaetes. In leeches they have a segmental origin, but in the absence of septa are free to extend through more than one segment and when distended make up more than two-thirds the body volume, compressing the mesenchymatous tissue which acts as a buffer between gut and somatic muscles. This part of the gut is endodermal and in a number of species investigated has secreting and absorbing cells. Although digestion occurs here it is usually referred to as the crop because of its storage function and the slow rate of change of its contents in blood-sucking leeches. In some of the rhynchobdellids such as *Glossiphonia* and *Branchellion* the relatively short intestine also has paired caeca.

A cropful of blood lasts a medicinal leech six months and if during this time the contents are examined they are found to be free from putrefaction and, moreover, erythrocytes are recognizable. Obviously the method of digestion is abnormal: the food is not subjected to enzymes liberated from the gut epithelium; indeed peptones, milk and egg proteins are not digested if injected into the gut, nor is there intracellular digestion. The method of digestion was not known until after 1953 when a bacterium, *Pseudomonas hirudinis,* was first described by Büsing from crop contents. We now know that this bacterium, the only species occurring in the crop, is responsible for keeping the contents free from attack by other bacteria and for slow haemolysis. The haemoglobin is split into globin and haem and it is the former which is mainly utilized as food. The haem is broken down to inorganic iron and protoporphyrin which colours the intestinal contents. Thus the leech gets a steady supply of soluble nitrogenous compounds without secreting any digestive enzymes and by doing no more than maintaining the correct environment for *P. hirudinis.* The gut epithelium stores a certain amount of glycogen and fat, presumably synthetized from the products of protein digestion.

Bacteria have also been reported from the gut of bloodsucking rhynchobdellids. They are stored either in the lumen of a pair of oesophageal diverticula (e.g. *Branch-*

ellion torpedinis), or in the epithelial cells (*Placobdella costata*, occurring on tortoises), but little is known of the benefit of the bacteria to the leech. In *Haemopis*, which swallows prey whole, the food is triturated by the muscles of the gut and digestive enzymes secreted over it. No enzyme for initiating protein digestion has been found, although dipeptidase, carboxypeptidase and aminopeptidase occur; perhaps the initial stages of protein digestion are carried out by bacteria.

Functioning of the vascular system of hirudineans

Leeches have adopted a method of locomotion in which coelomic fluid is not primarily used as a local hydrostatic skeleton; the coelom has accordingly lost its segmentation and is largely replaced by mesenchyme tissue. Two fluid systems, vascular and coelomic, are no longer required for the functioning of the body and the former tends to vanish. Even in the leeches which most resemble polychaetes embryologically, the Glossiphoniidae, both are reduced (Fig. 96B). Their vascular system has a dorsal and ventral longitudinal vessel (d.v, v.v) and the dorsal one propels the blood forward, The coelom is reduced to four longitudinal vessels or sinuses – a dorsal one (d.sin) containing the dorsal vessel, a ventral one containing the ventral vessel and nerve cord (v.n), and two lateral channels (lat.sin) – and these are linked by a network of connecting sinuses (c.sin). Pulsations of the dorsal vessel circulate the fluid in the dorsal sinus. There are only a few commissural vessels and these wander through an unusual course. For instance, in *Hemiclepsis* three pairs arise from the dorsal vessel half way down the body (Fig. 97B, comm.v); two run forward to join the ventral vessel near its anterior end and the third loops backwards towards the posterior sucker and then forward to join the ventral vessel (v.v) near the outer two. The blood has no haemoglobin in this or other rhynchobdellids and the coelomic fluid is also colourless.

In the specialized gnathobdellids only the coelomic system remains. The coelom is subdivided to form a system of intercommunicating channels which is referred to as the coelomic lacunar or sinus system, and the coelomic fluid contains the pigment haemoglobin so that there is a very real resemblance to the vascular system of other annelids. A lateral longitudinal sinus on each side has muscular walls (Fig. 98, mu.sin) and circulates the coelomic fluid. Each gives off some paired, transverse, subcutaneous branches in each segment which form capillary beds in the body wall (capill), which is the respiratory surface, and others which penetrate the connective tissue between body wall and gut, and rejoin the dorsal or ventral sinuses (d.sin, n.c). The plan of the coelomic circulation thus resembles that of the blood vascular system of other annelids, two contractile sinuses of this low-pressure system replacing their single dorsal vessel. However the lateral contractile sinuses are myogenic, stimulation for contraction originating in their muscles, while the contractile vessels or hearts of those polychaetes and oligochaetes which have been studied are neurogenic.

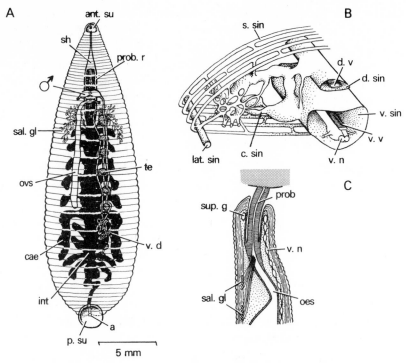

Fig. 96. *Glossiphonia complanata*: A, ventral view of alimentary and reproductive systems drawn as though the body wall is transparent; limits of segments indicated by continuous white lines across the gut. Male system, which lacks a penis, shown on right and female system on left. (After Harding.) B, diagram of coelomic sinuses in mid-region of body. (After Oka.) C, proboscis of a rhynchobdellid everted and penetrating tissues of host. (After Scriban.) a, anus; ant.su, anterior sucker; cae, last caecum of crop; c.sin, connecting sinus; d.sin, dorsal sinus; d.v, dorsal blood vessel; int, intestine; lat.sin, lateral sinus; oes, oesophagus; ovs, ovisac; prob, proboscis; prob.r, retracted proboscis; p.su, posterior sucker; sal.g, salivary gland; sh, proboscis sheath; s.sin, subcutaneous sinus; sup.g, supraoesophageal ganglion; te, testis; v.d, vas deferens; v.n, ventral nerve cord; v.sin, ventral sinus; v.v, ventral blood vessel; ♂, male atrium.

This circulation becomes the sole fluid system in gnathobdellids and erpobdellids which have no trace of blood vessels, even in development.

The lateral sinuses regulate the rate of pumping of coelomic fluid around the body which increases with increased oxygen demands. In *Hirudo* each is innervated by two nerves from each segmental ganglion: the anterior one accelerates the rate of beat and the posterior one causes a retardation. The action of the anterior nerve is inhibited by ergotoxin and increased by adrenalin, while that of the posterior is inhibited by curare and potentiated by muscarine. This suggests an adrenalin-acetylcholine antagonism as in vertebrates. In *Erpobdella* the transparency of the tissues allows direct observations to be made in relating pulsation rate with such external factors as temperature. At 17°C the average number of beats has been re-

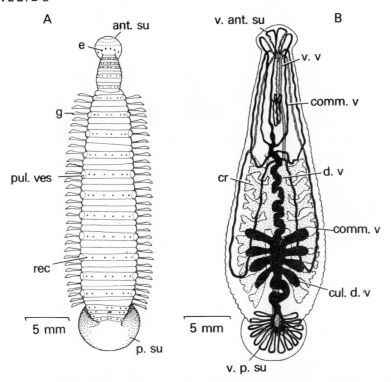

Fig. 97. A, *Branchellion torpedinis*: dorsal view. Segmental receptors and pulsating vesicles are associated with the middle of the three annuli in each body segment. (After Sukatschoff.) B, *Hemiclepsis marginata*: dorsal view to show the plan of the circulatory system. (After Oka.) ant.su, anterior sucker; comm.v, commissural vessel; cr, crop; cul.d.v, cul-de-sac of dorsal vessel; e, eye; g, gill; p.su, posterior sucker; pul.ves, pulsating vesicle; rec, segmental receptor; v.ant.su, vessels in anterior sucker; v.p.su, vessels in posterior sucker; v.v, ventral vessel.

corded as 3.7 min^{-1} and this increases to 17.1 min^{-1} at 27°C. Accessory contractile vesicles associated with the peripheral circulation and essentially concerned with cutaneous respiration occur at the junction of the lateral and transverse sinuses in *Piscicola geometra*. This is an active predator of fish with relatively high oxygen demands. The rate of pulsation of the vesicles, which project from the body surface, increases with temperature and rate of oxygen uptake. Leaf-like gills (Fig. 97A, g) are associated with similar structures (pul.ves) in the fish predator *Branchellion*.

The connective tissue (Fig. 98, c.t) which invades the coelom of the leech and is permeated with capillaries of the sinus system changes histologically as the worm ages, and accumulates haem pigments. In gnathobdellids two types of tissue are formed. The more conspicuous is the botryoidal tissue (botry) which appears as grape-like proliferations of the sinus walls. The second is the vasofibrous tissue formed by

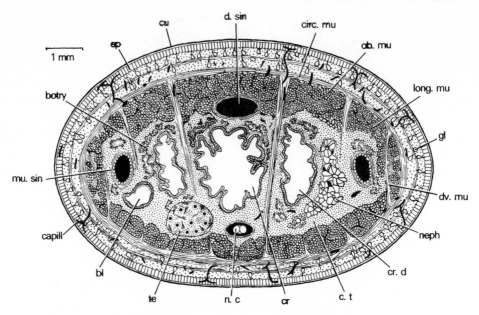

Fig. 98. *Hirudo medicinalis*: transverse section. bl, bladder of nephridium; botry, botryoidal tissue; capill, capillary in body wall; circ.mu, circular muscle; cr, crop; cr.d, last crop; diverticulum; c.t, connective tissue; cu, cuticle; d.sin, dorsal sinus; dv.mu, dorsoventral muscle; ep, epidermis; gl, gland in dermis; long.mu, longitudinal muscle; mu.sin, muscular wall of lateral sinus; n.c, nerve cord in ventral sinus; neph, nephridium; ob.mu, oblique muscle; te, testis.

enlargement of the cells lining the finer capillaries. Both take up particles from the coelomic fluid, the latter the finer particles, the former somewhat larger ones. In the rhynchobdellids changes in the connective tissue are similar, but not identical. Large pigmented cells correspond to the bunches of botryoidal cells in that they take up particles from the coelomic fluid with a diameter greater than 1 nm. Vasofibrous tissue is represented by less conspicuous cells lining capillaries which can ingest the smaller particles.

These tissues correspond to the chloragogenous tissue of polychaetes and oligochaetes which develops from peritoneum in the vicinity of blood vessels. In *Hirudo* they appear to be directly associated with the metabolism of haem derived from the fluid in the siuus system. A masked iron compound accumulates in the botryoidal tissue colouring it brown and the breakdown products of porphyrin in the vasofibrous tissue appear as a green pigment similar to bile. Porphyrin derivatives have also been identified in the large pigment cells of *Glossiphonia complanata*, which feeds on the tissues of freshwater snails. These cells, increasing in number with age, are derived from effete fat-storing cells which are a characteristic of this species.

Movement in hirudineans

Leeches move over the substratum by stepping movements like those of a looper caterpillar. They swim through water, throwing the body into dorsoventral undulations moving anteroposteriorly, or they may grip the substratum with the posterior sucker and use the undulating movements to irrigate the body surface as *Nereis* irrigates its burrow. Some stand erect on the posterior sucker, and extend the body from deoxygenated water and effect aerial respiration, and the land leech (*Haemadipsa*) supports its body in air, erect and motionless. These activities are brought about by the coordinated activity of muscular and nervous systems.

The arrangement of the muscles of the body wall is similar to that of an earthworm in that there is an outer layer of circular muscles (Figs 92C and 98, circ.mu) and an inner layer of longitudinal ones (long.mu), but it differs from that of any other type of annelid in that between these there is a double layer of spirally directed muscles fibres, one running clockwise around the longitudinal ones and the other counterclockwise (ob.mu). This trellis recalls the system of inextensible spiral fibres already described in turbellarians and nemerteans, which conditions the shape the body can assume. The spirally arranged muscles in the leech will likewise act with the circular and longitudinal ones in regulating the shape of the body, allowing the versatility of movement characteristic of the worm, and giving the rigidity essential to a soft-bodied animal supporting itself in air.

A brief description of the nervous system must precede a more detailed consideration of locomotion. The leech agrees with other annelids in having a pair of ganglia associated with each segment of the body (Fig. 99A, n.c). However the segmental arrangement is modified in the region of the suckers. Anteriorly the modification involves the ganglia of four or five short segments behind the prostomium, a number of which bear the eyes dorsally; posteriorly it involves the ganglia of the last six or seven segments. Supraoesophageal (cerebral) ganglia in the polychaete and oligochaete are joined to suboesophageal ganglia by circumoesophageal connectives. In leeches the cerebral ganglia (sup.g) are at the level of segment six (posterior to the sucker) and the ganglia homologous with the suboesophageal (sub.g) ganglia lie close to them. The nerve ring is small, the circumoesophageal connectives reduced and the ganglia of the short anterior segments are crowded together to form a large suboesophageal ganglionated mass. Nerves pass forward to eyes, sensillae and sucker (Fig. 100A, n.s.s). At the base of the posterior sucker is a second ventral ganglionated mass formed by the fusion of the ganglia of the segments which form the sucker. Along the rest of the nerve cord the segmental arrangement of ganglia is conspicuous.

The motor cell bodies of the central nervous system are concentrated in the ganglia, whereas in the earthworm they have a more random distribution along the ventral part of the cord. They are grouped together in fibrous capsules, three pairs per

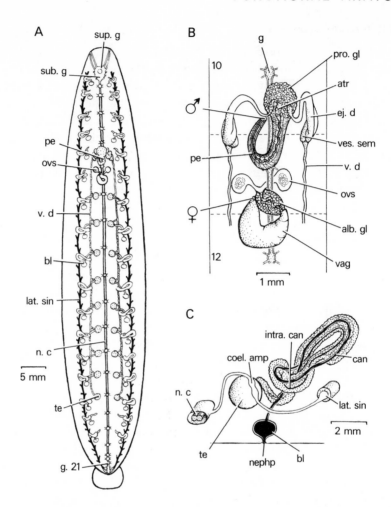

Fig. 99. *Hirudo medicinalis*: A, opened mid-dorsally and gut removed to show the central nervous system, reproductive organs and nephridia; B, details of reproductive organs in segments 10, 11 and 12; C, diagram of a nephridium and its topographical relationship with the testis and blood sinus with ciliated organs. alb.gl, albumen gland; atr, atrium; bl, bladder of metanephridium; can, intercellular canal; coel.amp, coelomic ampulla; ej.d, ejaculatory duct; g, ganglion; g. 21, 21st free ganglion; intra.can, intracellular canal; lat.sin, lateral sinus; n.c, nerve cord with segmental ganglia; nephp, nephridiopore; ovs, ovisac; pe, penis in sheath; pro.gl, prostate gland; sub.g, suboesophageal ganglia; sup.g, supraoesophageal ganglia; te, testis; vag, vagina; v.d, vas deferens; ves. sem, vesicula seminalis; ♂, male pore; ♀, position of female pore.

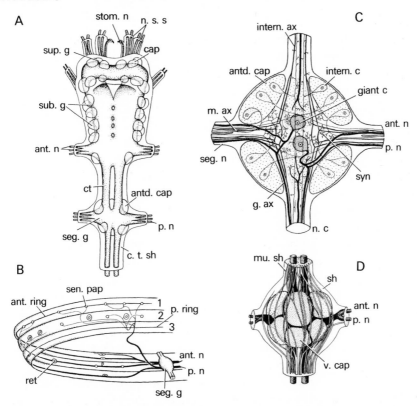

Fig. 100. A, *Branchellion torpedinis*: diagrammatic representation of the anterior ganglia. (After François.) B, *Theromyzon tessulatum*: diagram of segmental nerves (1, 2, 3 annuli of segment). (After Livanoff.) C, Diagram of a ventral ganglion of a leech to show nerve cells and their processes; the six capsules are stippled and the ventral ones demarcated by a dotted line. (After Scriban and Autrum.) D, *Pontobdella muricata*: ventral view of ganglion from mid-region of nerve cord; the capsules are stippled, the nerve trunks black. (After François.) antd.cap, anterodorsal capsule; ant.n, anterior nerve; ant.ring, anterior nerve ring; cap, capsule of neurons; ct, connective; c.t.sh, connective tissue sheath; g.ax, "giant" axon; giant c, "giant" cell; intern.ax, internuncial axon; intern.c, internuncial cell; m.ax, motor axon; mu.sh, muscle fibres in sheath; n.c, ventral nerve cord, n.s.s, nerves to sucker and sense organs; p.n, posterior nerve; p.ring, posterior nerve ring; ret, sensory papilla with retinal cells; seg.g, segmental ganglion; seg.n, segmental nerve; sen.pap, sensory papilla; sh, sheath; stom.n, stomatogastric nerve; sub.g, suboesophageal ganglion; sup.g, supraoesophageal ganglion; syn, synaptic connections; v.cap, ventral capsule.

ganglion (sup.g, cap) reflecting its double origin (C, antd.cap; D, v.cap). It is this constant form of the typical ganglion which has made it possible to analyse the composition of the nerve ring and deduce the number of ganglia forming it. Each segmental ganglion of the ventral nerve cord gives off two pairs of nerves (ant.n, p.n) which carry the main motor axones (C, m.ax) from the encapsulated cells. Other axones leave the capsules (intern.ax) to make synaptic connections with internuncial neurones running from one ganglion to the next and often extending over more than one segment. The internuncial fibres also connect with sensory fibres. In each ventral capsule of the ganglion is a giant nerve cell (giant c) with axones (g.ax) passing to the ventral nerve cord and segmental nerves. These are not giant axones, though the responses of a leech suggest that there is a fast conducting system.

The paired segmental nerves connect with a peripheral system in the body wall which receives neurones from proprioceptors in the muscles and sense organs in the skin (B, sen.pap, ret). The more anterior nerve of each pair bifurcates (ant.n) and the branches pass laterally around the body wall. One serves the ventral body wall while the other, the more anterior, connects with a nerve ring (ant.ring) which encircles the segment anteriorly between longitudinal and oblique muscles. The second nerve of the pair has one branch which serves the dorsal body wall and another associated with a posterior intermuscular ring (p.ring). Segmental and intersegmental longitudinal nerve fibres connect the intermuscular rings and from these rings fibres pass to a subepidermal system of fine nerves.

When a leech steps, starting from a position in which the posterior sucker has just attached and the anterior one is free (Fig. 92A, 1) an anteroposterior wave of contraction of the circular muscles· lengthens the body while the longitudinal muscles are relaxed (2, 3). When the anterior sucker approaches the substratum and attaches, the posterior one releases its grip (4, 5). The circular muscles are now inhibited while an anteroposterior wave of contraction of the longitudinal muscles shortens the body bringing it toward the anterior sucker (6, 7). As the posterior sucker approaches the anterior one to attach alongside it the body is flexed in a dorsoventral plane, contraction of the dorsoventral muscles in the mid-region flattening it and assisting flexure (8). The posterior sucker attaches (9) and the anterior one is simultaneously released. The two suckers are thus used alternately, the attachment of one appearing to cause the release of the other. Indeed Gray, Lissmann and Pumphrey found that if a microscope coverslip is presented to one of the suckers of a worm suspended by a thread it is immediately gripped, but dropped again as soon as a second coverslip is held by the other sucker. Attachment by the posterior sucker is invariably associated with a wave of contraction of the circular muscles, and attachment by the anterior one with a wave of contraction of the longitudinal muscles.

The sequence of events which brings about stepping is dependent on peripheral reflexes associated with the suckers and, presumably, on reflexes involving the proprioceptors in the muscles. However, the suckers may be removed or denervated without abolishing stepping, provided the ventral surface of the body is stimulated. If the suboesophageal ganglionated mass is severed from the ventral nerve cord, so that its influence and that of the brain are removed, the rhythm is effectively obliterated: the circular muscles lose their ability to contract, though the posterior sucker will still attach, and the longitudinal muscles start relaxing, indicating that fixation by this sucker inhibits their contraction. As in the earthworm the locomotory rhythm is maintained by the nerve cord which coordinates the stimuli received from the peripheral nervous system. If segmental nerves are severed in several segments in the middle two-thirds of the body conduction of the rhythm is not affected.

If a leech is dropped into water and all other peripheral tactile stimulation removed the dorsoventral muscles contract, flattening the body, and differential contractions of the longitudinal muscles maintain dorsoventral undulations in an antero-posterior direction. These swimming movements continue until the under surface of the body (it may be a sucker) is touched, when stepping movements are resumed; a dorsal stimulation has no such response but accentuates the swimming movements. Thus, as in the earthworm, tactile stimulation from the ventral surface is necessary for locomotion over the substratum. Removal of the nerve ring with the fused suboesophageal ganglia has no effect on swimming, but removal of the fused posterior ganglia inhibits it. The anterior and posterior ganglia associated with the suckers are thus essential in determining locomotory behaviour. We know little about the controlling action of the brain but presumably it modifies the patterns of reflexes according to external conditions and physiological needs.

Rhythmical electrical activity has been recorded in the ventral nerve cord of a swimming leech (*Hirudo*), but it becomes irregular if the cord is isolated from peripheral stimulation by the severance of all segmental nerves. If the body is cut in two transversely but the nerve cord left intact the two halves show coordinated swimming movements. However, if the nerve cord only is cut coordination of movement between the two halves of the worm is lost: there is increase in tonus in the circular muscles anterior to the cut, the body becoming rounded as for crawling, and a loss of tonus of these muscles posterior to the cut where the body becomes flattened and may exhibit swimming movements. It will be recalled that the body of an earthworm which is crawling may be cut in halves, rejoined by stitches and still show coordinated crawling movements. This mechanical transmission of locomotory reflexes has never been demonstrated in a leech.

Excretion in hirudineans

The segmental organs (paired metanephridia and coelomoducts), which open to the perivisceral coelom in other annelids, are present in the leech, though modified with the reduction of the body cavity. The metanephridia are built on the same general plan as those of oligochaetes (Fig. 99A, bl). They are absent in a number of anterior and posterior segments adjacent to the suckers and in some species in the clitellar region. The coelomoducts are closely associated with the gonads which comprise a pair of ovaries (ovs) followed by a segmental series of paired testes (te), usually 6–10 pairs, alternating with the gut caeca. Each gonad is enclosed in a small coelomic sac, the ovisac or testis sac, continuous with a duct which arises as an outgrowth from

its walls. The two oviducts come together mid-ventrally to open by a single pore
(Figs 91B and 99B, ♀). The ducts from the testes (vasa efferentia) fuse to form a
longitudinal duct (Fig. 99A, v.d) on each side (vas deferens) and the two pass to a
single mid-ventral opening anterior to the female pore (Figs 91B and 99B, ♂). The
primitive *Acanthobdella* is an exception and resembles oligochaetes in having two
large sacs where the sperm mature (Fig. 101C, sp.s), formed from special parts of
the coelom and extending through several segments; each communicates with the
perivisceral coelom (coel) and into each extends the funnel of a coelomoduct (sp.fu).

The ciliated nephrostome of each nephridium (Fig. 102A, cil. nephs) of the leech
projects into a small, ventral, coelomic chamber or ampulla (coel.amp) which is in
communication with other channels of the lacunar system. The nephridial canal,
long, coiled on itself and surrounded by connective tissue, is embedded in the mesenchymatous tissue which has invaded the coelom. It passes to a terminal bladder
(Fig. 99C, bl) which opens by the nephridiopore (Figs 91B and 99C, nephp) on the
ventrolateral wall. The initial part of the nephridial canal into which the nephrostome opens is expanded to form a funnel sac or capsule (Fig. 102A, cap) characteristically filled with phagocytes. The canal has no cilia along its length, is for the

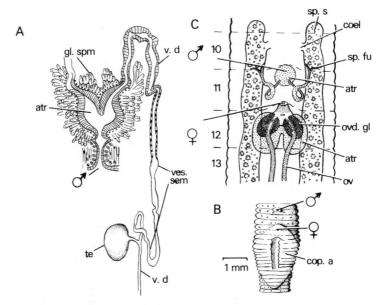

Fig. 101. A, *Piscicola geometra*: longitudinal section of male genital atrium and associated
structures; B, *Piscicola*: ventral view to show the genital openings and copulatory area where
spermatophores are deposited; C, *Acanthobdella peledina*: reproductive system; segments of the
body indicated by broken lines and numbered. atr, atrium; coel, opening to coelom; cop.a,
copulatory area; gl.spm, glands secreting spermatophore; ov, ovary; ovd.gl, oviducal gland;
sp.fu, sperm funnel; sp.s, sperm sac; te, testis; v.d, vas deferens; ves.sem, vesicula seminalis;
♂, male pore; ♀, female pore.

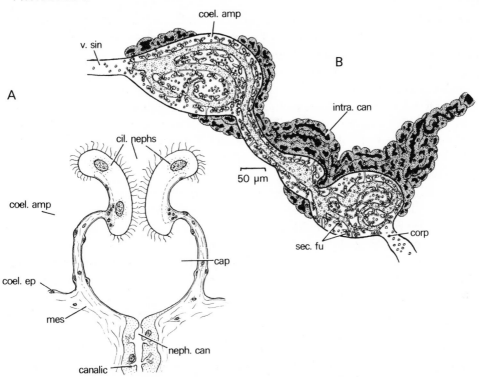

Fig. 102. A, diagram of nephrostome and capsule of a glossiphoniid projecting into the coelomic ampulla. (From Goodrich after Hotz.) B, longitudinal section of elongated coelomic ampulla of *Hirudo* and the associated capsule and nephrostome which is subdivided to form many secondary funnels opening separately to the capsule; the ampulla is surrounded by the coiled intracellular nephridial canal which does not open to the capsule. (After Bhatia.) canalic, collecting canaliculi; cap, cavity of capsule; cil.nephs, ciliated nephrostome; coel.amp, coelomic cavity of ampulla; coel.ep, coelomic epithelium; corp, corpuscles (phagocytes); intra.can, intracellular nephridial canal; mes, mesenchyme tissue; neph.can, nephridial canal (intracellular); sec.fu, multiple secondary funnels; v.sin, branch of ventral sinus.

most part intracellular (Fig. 99C, intra.can) and gives off ramifying branches ending in fine collecting canaliculi (Fig. 102A, canalic) an arrangement characteristic of leeches and known in a few oligochaetes. The unique specialization of the nephridium concerns the nephrostome or funnel and the capsule to which it opens (Fig. 102). Primitively the funnel is a ring of a few large ciliated cells (Glossiphoniidae, Piscicolidae) but in gnathobdellids it is elongated and subdivided to form numerous isolated secondary funnels (sec. fu) opening separately to a much enlarged capsule (coel. amp). The capsule leads into the nephridial canal in some primitive leeches (glossiphoniids) (neph.can), in others there is controversy over the existence of a connection (e.g. *Pontobdella*) and in the majority the inner end of the nephridial canal certainly appears to be blind (*Hirudo*).

The nephridia of *Hirudo* lie at the sides of and beneath the alimentary canal between the gut caeca. There are seventeen pairs and in segments where there are also testes the coelomic capsule, into which the multiple funnel of the nephrostome opens, is in close association with a testis (Fig. 99A, C, te, coel.amp). The ampulla communicates with both ventral (Fig.102B, coel.amp, v.sin) and lateral sinuses though closed to the nephridial duct. It is filled with phagocytes which appear healthy and may originate there; the strong outwardly beating cilia on the funnels carry the phagocytes to the coelomic circulation. Thus the initial part of the nephridium is incorporated into the coelomic circulatory system, forming an organ similar to the ciliated organ of polychaetes.

Coelomic corpuscles of leeches are known to collect excretory products from various parts of the body and in some species laden cells are seen in the nephrostomes and associated capsules. If the nephridial canal is in communication with the capsule, waste may be discharged into it directly. However, if, as in *Hirudo*, the inner end of the nephridium is blind, waste can reach the canal only by diffusion from the fluid bathing the fine canaliculi of the nephridial cells. The scavenging phagocytes may carry waste to the botryoidal and vasofibrous tissues where particulate matter accumulates and is retained. The importance of these tissues may be associated with the closure of the metanephridium.

About three-quarters of the nitrogen excreted in the urine of *Hirudo* is in the form of ammonia. This high proportion of ammonia is due to two species of bacterium in the bladder of the nephridium. One species (*Corynebacterium vesiculare*) forms a layer over the epithelium which might be taken for cilia and the other (*C. hirudinis*) floats in the vesicular fluid. Killing them with antibiotics results in a fall in the ammonia content of the urine.

The nephridia are also concerned with water balance though no detailed study of the process has been carried out. A number of species living in fresh water can go for months without food. During this time water must enter the body osmotically and be removed by the nephridia. At the same time ions will be lost, perhaps replaced by active uptake by the integument and lining of the gut. It has been demonstrated that the horse leech (*Haemopis sanguisuga*) has two separate mechanisms, one concerned with the uptake of Na^+, and perhaps other cations, and the other with Cl^-, and perhaps other anions. These normally act together. When *Haemopis* is exposed to air, loss of water is considerable. At 80 per cent R.H. the water content of the body is reduced to 20 per cent in 4–5 days at 22°C: this is lethal. If the leech is returned to water after 4 days full hydration is attained in a few hours and it regains activity. Though the surface of the body may dry, the suckers of a leech must be kept moist for successful functioning. In the land leeches (*Haemadipsa*) there is a special modification for this. The first pair of nephridia open on to the anterior sucker and the last pair to a membranous fold on either side of the anus at the base of the posterior sucker.

Reproduction of hirudineans

Leeches agree with oligochaetes in being hermaphrodite, practising copulation and producing egg capsules, a product of the clitellum. However, fertilization in the leech is internal and differences between the genital systems in the two classes are associated with this advance as well as with the relationship between the gonad and its duct. In some leeches (rhynchobdellids, erpobdellids) internal fertilization is associated with the production of spermatophores. These are manufactured in the atrium (Fig. 101A, atr, gl.spm), a muscular vestibule which receives the ducts of the two vasa deferentia (v.d) and opens to the exterior by the genital aperture (♂). When spermatophores are exchanged between two individuals (Fig. 103A,) they are deposited on the surface of the body, and the sperm escape, passing into the tissues of the recipient and making their way to the ova. A similar method of internal fertilization is practised by the primitive arthropod *Peripatus*. In other leeches (gnathobdellids) sperm are transferred to the vagina of the partner by an eversible penis (Fig. 99A, B, pe), an outgrowth of the atrial wall; one individual acts as male and the other as female.

The atrium is bicornuate and produces one spermatophore at a time; each horn marks the position of entry of a vas deferens. An extruded spermatophore reflects the shape of the atrium. It consists of a basal plate (Fig. 103B, base), filled with secretion, from which arise two sperm sacs (sp.s) which may diverge from one another or appear adpressed. Glands associated with the atrium produce the wall of the spermatophore and the secretion in the basal plate. This plate is forced against the skin of the partner and adheres. Sperm filling each sperm sac come from the seminal

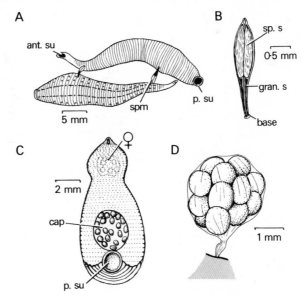

Fig. 103. *Glossiphonia complanata*: A, two individuals separating after copulation – each has left a spermatophore attached to the body of the partner; B, spermatophore; C, ventral view one hour after deposition of the first capsule; the eggs for the next are seen by transparency in the genital atrium; D, capsule fixed to substratum. (A, B, C, after Brumpt and D after Bychowsky.) ant.su, anterior sucker; base, basal attachment plate; cap, first capsule; gran.s, granular secretion; p.su, posterior sucker; spm, spermatophore; sp.s, sperm sac; ♀, female pore.

vesicle (Fig. 101A, ves.sem), a coiled region of the vas deferens anterior to the testes, where they are stored until the partners come together and intertwine their bodies. The spermatophore is then formed and extruded from the atrium on to the partner, its walls hardening on exposure.

The site of deposition of the spermatophore varies. Most frequently it is in the clitellar area, that is, near the genital openings. However, deposition may be elsewhere so that the sperm travel further, traversing the epidermis and dermal connective tissue to the circulatory system and so reaching the ovisacs. This occurs in some glossiphoniid and erpobdellid leeches. In contrast there are piscicolid leeches in which spermatophores are deposited on a special area of the clitellum ventral to the ovisacs (Fig. 101B, cop.a). This area is connected to the ovisacs by a pad of fibrous connective tissue formed by hypertrophy of their walls and the sperm pass through it to reach the ova (*Piscicola*). Alternatively the spermatophore may be deposited in the atrium of the partner which is connected to the ovisac by a special strand of connective tissue through which the sperm pass (*Callobdella lophii*), or by a duct (*Americobdella valdiviana*).

The clitellum becomes conspicuous at the breeding season with hypertrophy of glands which secrete albumen and the wall of the egg capsule. It encircles the body in the vicinity of the genital apertures and produces a girdle of secretion as the fertilized eggs are liberated from the female opening. This occurs 2–3 weeks after fertilization, though in *Hirudo medicinalis* it may be up to a few months. The eggs, liberated in secretion from the female duct, are passed into albumen from the clitellum. One egg may be liberated at a time and enclosed in a capsule (piscicolids), but

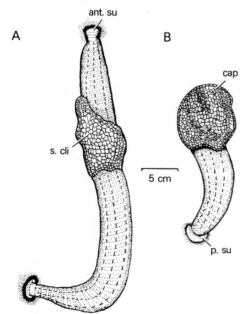

Fig. 104. *Hirudinaria granulosa*: A, capsule being formed while the body is held in position by the suckers; B, with the posterior sucker attached the anterior part of the body is withdrawn from the capsule. (From Grassé after Matthai.) ant.su, anterior sucker; cap, capsule; p.su, posterior sucker; s.cli, secretion from clitellum.

usually more, even up to fifty (*Glossiphonia complanata*). The girdle of secretion is pressed against the substratum and the outer layer, which will form the capsule wall, attaches. The leech slides backward out of it and may then mould the wall to a specific shape, implanting surface markings before the secretion hardens (Fig. 104A, B). In a few leeches (*Glossiphonia, Hemiclepsis*) clitellar glands are restricted to a small area around the genital pores and their secretions form a bag into which the eggs are discharged. In some species of *Glossiphonia* capsules are attached to the ventral surface of the body and brooded by the parent.

Classification of annelids

Annelida
  Archiannelida
  Polychaeta
    Phyllodocida
    Capitellida
    [Sternaspida]
    Spionida
    Eunicida
    Amphinomida
    Magelonida
    Ariciida
    Cirratulida
    Oweniida
    Terebellida
    Flabelligerida
    [Psammodrilida]
    Sabellida
  Oligochaeta
    Terricola
    Limnicola
  Hirudinea
    Rhynchobdellida
    Gnathobdellida
    Pharyngobdellida
Appendix to Annelida
  Myzostomida
  Echiuroidea
  Sipunculoidea

10

Introduction to arthropods

THE ANIMALS to which we now turn – crustaceans, centipedes, millipedes, insects and arachnids – are those collectively termed arthropods. The group Arthropoda is immense, not only in the number of species included in it but in the myriads of individual animals which it contains; the biomass is perhaps less than these statements might lead one to expect, since most arthropods are small animals only a few millimetres long. According to which authority one follows arthropods make up between 72 per cent and 84 per cent of the total number of animal species described.

Originally regarded as a single phylum it is now clear that arthropods are not the closely interrelated group such a classification would imply. The word arthropod is best kept to denote a grade, that is, a pattern of functional organization which has been reached by several stocks of animals following more or less parallel evolutionary pathways. It would seem – to use a human simile – as if animals at a certain level of evolutionary progress all shared similar ambitions and devised ways of fulfilling them with whatever means lay at their disposal. Since the animals are at least loosely related their final solutions to the problems often look alike, but since the relationship is not close, they differ in significant ways. Convergences in structure and function of this kind are common in arthropods and provided one of the reasons for regarding the animals as more uniform than they really are. It will make for easier understanding of the nature of the group if the main reasons for regarding it as polyphyletic and some of the convergences which have arisen are dealt with before the taxonomic divisions are discussed. This is the more reasonable in that much of the evidence rests on a functional anatomical basis and is coincident with the criteria on which the group is split into systematic units.

Classification of arthropods

The features shared by all arthropods are relatively few – all are metamerically segmented; all have modified the anterior segments to produce a head; all have a reduced coelom, with the main body cavity a haemocoel; all have a cuticular exoskeleton of chitin usually thickened on a segmental pattern to form harder pieces, known as sclerites, and linked by areas of flexible material (arthrodial membranes) at which movement can occur; all grow in a way which demands that, at intervals, the exoskeleton be moulted and a new one, of larger size, produced; all have segmentally arranged appendages usually made of a series of articulated joints (the feature from which the name arthropod is derived); all use some of the appendages to manipulate food and also for sensation, respiration and sexual functions, though their primary activity is locomotor; all have an ostiate heart (p. 291).

The starting point for the evolution of animals with this set of features must have been at a sub-annelid grade and in a marine environment, since it is only at this level that one finds the metamerically segmented body with paired appendages from which both annelids and arthropods could have come. There are, moreover, important differences, some only of degree, some fundamental, between all annelids and all arthropods, suggesting that the two groups have followed differing pathways from some common ancestor. One of these is the exoskeleton, another the nature of the body cavity, two features often supposedly interlinked, with the evolution of the exoskeleton the more basic and the reduction of the coelom consequent upon its development. In annelids, it will be recalled, coelom sacs provide local hydrostatic skeletons and the exoskeleton is a delicate cuticle. In the evolution of arthropods from annelids, it used to be argued, the exoskeleton thickened segmentally and provided a secure origin for muscles pulling directly on it and moving segment on segment and limb joint on limb joint; in such circumstances the value of the hydrostatic skeleton disappeared and the coelom was replaced by haemocoelic spaces. Since change in the disposition of parts of the body was now effected by a system of levers rather than by hydraulic means the continuous muscle layers necessary for its operation split to give the discrete muscles of the arthropod. Sclerites also encouraged the transformation of the muscle fibres from the slow, unstriated type characteristic of annelids to the faster-acting striated ones almost universal in arthropods.

There is doubt, however, about this sequence. Protoarthropods seem to have been soft-bodied animals grubbing over and in soft submarine substrata. For producing the repeated series of high pressures necessary for probing into these, either by part of the body searching for food, or by the whole animal seeking a hiding-place, a haemocoelic body cavity is even more effective than a continuous coelomic cavity; it allows a local high pressure to be repeatedly built up from the contraction of sheets of muscle locally and remotely placed without damage to blood capillaries. If machinery of this kind had already evolved it might then come about that sclerites arose, partly to protect the body during probing, partly better to transmit the thrust to the surroundings. (Adoption of the same mode of life in the same habitat by the protomollusc may have given rise to the same combination of haemocoel and exoskeleton which marks out that group.) With the (perhaps later) extension of thickened exoskeleton to the protoarthropod's appendages, converting them into powerful levers, and their migration to a ventrolateral position, the animal became able to move over the surface and to do so by means of the musculature of its limbs rather than that of the trunk, which is the force primarily involved in annelidan locomotion. As a result the metabolically expensive bending of the body was minimized or lost, a process accelerating as the growing sclerotization of the exoskeleton made the body less easily flung into sinuosities. This view would imply that annelids aud arthropods (and possibly molluscs) have arisen from some common ancestor, not one from the other, and their differences reflect their dependence on different locomotor devices.

The annelid parapodium is a biramous appendage springing from the side of the body. If it and arthropodan appendages arose from some common ancestor then these structures might also be biramous, though migrating ventrally to act as a walking limb as skeletal and muscular power increased. This is, indeed, what is found in crustaceans and it is not altogether fanciful to see in the appendages of the most primitive living crustacean, *Hutchinsoniella* (Fig. 105C), an arthropodized parapodium. From some such limb all other types of crustacean appendage, whatever their form or function, can be derived. It consists of a basal piece, the protopodite or protopod (protop) and two distal branches, a lateral exopod(ite) (exop) and a median endopod(ite) (endop); in addition there are, at the base of the limb, laterally directed plates, exites (ex), of respiratory importance, and medially directed lobes, endites (end), concerned with handling food and therefore called gnathobases. The limb is moved backwards and forwards in a pattern of rhythm and movement conforming closely to that of the annelid parapodium. Two sets of muscles attached to the edge of its articulation with the body bring this about: remotor muscles run from the posterior margin to a more posterior origin on the body wall and pull the limb backwards, and promotor muscles, from the anterior margin to an anterior origin, pull it forwards. The axis on which the limb base moves lies in the transverse plane of the body ventrally.

Functional diversification of appendage structure had presumably been present from the attainment of the arthropod grade and probably antedated it. Although trunk appendages may all be identical anterior ones have been affected by their situation on the animal's head, an appropriate setting for sensory organs, or alongside the mouth, where food is handled. The sensory appendages of the head, in all arthropods except arachnids, form structures known as antennae – long, many-jointed processes set with numerous sensory setae (ant', ant''). Alongside the mouth, appendages are modified as mouth parts, only one pair in the small group of terrestrial arthropods known as onychophorans (Peripatus), but two pairs in the millipedes and three (occasionally more) in centipedes, arachnids and crustaceans. These appendages exhibit an exaggerated ventral migration to reach the sides of the mouth and also tend to be more close-set than appendages on other parts of the body. Since the position of the mouth in relation to the series of segments has been

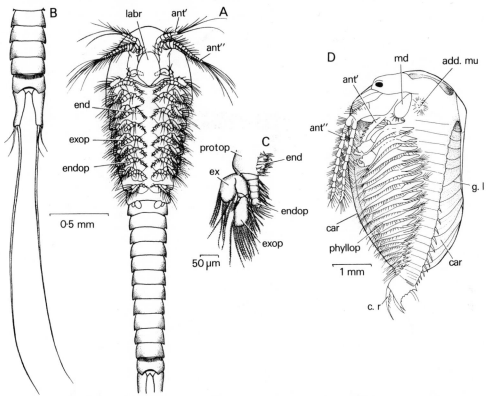

Fig. 105. A, *Hutchinsoniella macracantha*, ventral view of anterior part of body; B, of posterior end of body; C, one of its trunk limbs; D, *Estheria*, left valve of shell removed. (A, B, C, after Sanders; D, after Calman.) add.mu, muscle; ant′, antennule; ant″, antenna; car, carapace; c.r, caudal ramus; end, endite or gnathobase; endop, endopodite; ex, exite; exop, exopodite; g.l, growth line; labr, labrum; md, mandible; phyllop, phyllopodium; protop, protopodite.

shown to vary from group to group it follows that the appendages converted to mouth parts – however comparable in structure and function – need not be derived from the same segments, one example of the convergences mentioned earlier.

Along with crustaceans two other arthropodan groups, the arachnids and the extinct trilobites, show a basically biramous pattern of appendages (Fig. 106). The distal branches, however, are not the same as those of the crustacean limb, since the lateral ramus (exop) springs from the base at a level much closer to the body. In no other group of arthropods is the limb pattern biramous, nor is there any indication in development or the palaeontological history of the groups that the limb ever has been double. In these animals (onychophorans, myriapods, insects collectively known as Uniramia) the base of the limb gives rise to a single axis only, known as a telopod(ite). These facts compel the conclusion that the arthropods, originally conceived as a single phylum, must contain at least three groups which have been

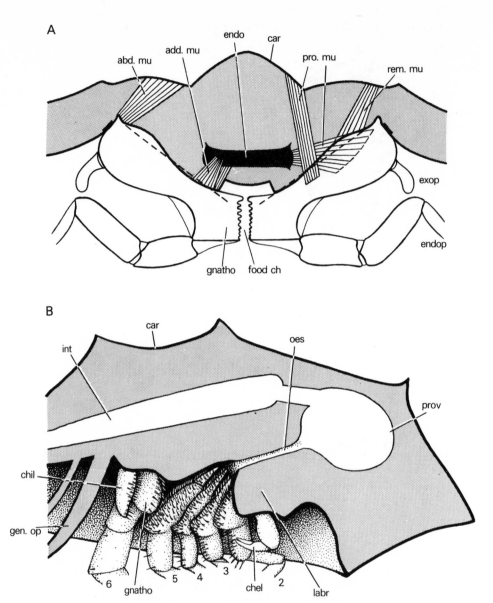

Fig. 106. A, transverse section of body of a king-crab at level of food chamber. Muscles responsible for moving the walking leg in locomotion are shown on the right, those for chewing on the left. The thick pecked line shows the axis on which the limb swings in walking, the short solid line the axis on which it moves when chewing. The edges of the carapace and the distal ends of the limbs are not shown. B, sagittal half of the anterior end of the body of a king-crab showing the food chamber bounded anteriorly by the labrum, posteriorly by the chilaria and genital operculum and laterally by the gnathobases of the walking legs. (Based on Manton.) abd.mu, abductor muscle; add.mu, adductor muscle; car, carapace; chel, chelicera; chil, chilarium; endo, endophragmal skeleton; endop, endopodite; exop, exopodite; food ch, food chamber; gen.op, genital operculum; gnatho, gnathobase; int, intestine; labr, labrum; oes, oesophagus; pro.mu, promotor muscle; prov, proventriculus; rem.mu, remotor muscle; 2–6, walking legs.

evolving as separate lines from before the time that they acquired arthropodan characteristics.

Further evidence in support of this proposition has been provided by the elaborate work of Manton on the way of working of the mandibles of the various arthropodan groups. Mandibles, appendages working to break food into pieces able to enter the gut, occur in most groups. Examination shows that they are immediately divisible into two types, the first found in crustaceans and arachnids, the second in the onychophoran-myriapod-insect assemblage. In the first of these the triturating structure is formed by gnathobases, medially projecting processes on the basal joint of the limb (endop, gnatho), the rest of the appendage not being involved, and sometimes even retaining its primitive locomotor function; in the second, the whole appendage has become the mandible. The two types must have evolved independently. Although basically similar to locomotor appendages in action, mandibles can rarely move out of phase with one another: their correct functioning as mills or cutters depends upon their doing the same thing at the same time.

Closer study of mandibular functioning allows recognition of further differences. Possibly the simplest type of action in the gnathobasic mandible is seen in the arachnidan king-crabs (*Limulus* and similar genera). In these animals the mouth lies approximately centrally on the underside of the front part of the body, opening from the anterior end of a food chamber with well-defined walls (food ch). The anterior wall is made by the upper lip or labrum (labr), although, because the mouth points backwards, this morphologically dorsal structure is topographically ventral to the opening; the side walls are formed by the basal joints of five pairs of walking legs, whilst the chamber is bounded posteriorly by two pairs of appendages, the first, two rounded lobes (chil) known as chilaria (a name meaning a boundary), the second a broad bilobed plate known as the genital operculum (gen.op) because it carries the genital openings. The chamber provides a place where the body of the prey (worms, molluscs) is broken and chewed by the approximation of the inwardly directed gnathobases of the walking legs, armed with spines and heavily sclerotized lobes. The movement which the legs execute to do this is not that which they perform in walking: then they carry out a backward-forward swing, rotating around a transverse axis (see pecked line in Fig. 106A), whereas in feeding the swing is lateromedian, rotating around an anteroposterior axis (solid line). The limbs of a pair move together, but successive pairs alternate when food is being broken and advantage is taken of the flexible nature of the ventral cuticle to press their bases close together so as to make a formidable mill, whereas in locomotion they are stretched apart to give room for the striding of the legs.

Although in structure the mandibles of all living crustaceans have departed a long way from the ancestral pattern in that no trace of the locomotor part of the limb persists, the simplest mandibular action in this group is shown by the fairy

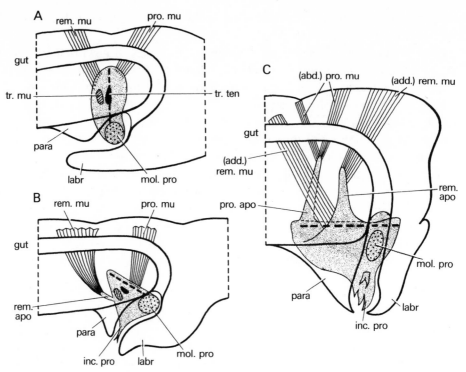

Fig. 107. Diagrammatic views of left sagittal half heads of A, *Chirocephalus*, B, *Anaspides* and C, *Ligia* to show position and action of the mandible (stippled). The thick pecked line shows the axis on which the mandible swings; note its dorsoventral direction in *Chirocephalus*, its oblique direction in *Anaspides* and horizontal position in *Ligia* resulting in a transversely directed bite. The black and hatched areas on the mandible in A and B mark the transverse mandibular tendon and muscle respectively, lost in C to allow a wider gape. C also shows how the mandible becomes entognathous, that is internally placed. Maxillae which would help to produce this condition and which would overlie the paragnath have been omitted to keep the diagram simple. (Based on Manton.) (abd.)pro.mu, abductor muscle of mandible derived from the original promotor; (add.)rem.mu, adductor muscle of mandible derived from the original remotor; inc.pro, incisor process of mandible; labr, labrum; mol.pro, molar process of mandible; para, paragnath; pro.apo, apodeme to which promotor muscle is inserted; pro.mu, promotor muscle; rem.apo, apodeme to which remotor muscle is inserted; rem.mu, remotor muscle; tr.mu, transverse muscle; tr.ten, transverse tendon.

shrimp *Chirocephalus* (Figs 107 and 108), a filter feeder collecting fine particles by means of the movements of its trunk appendages and passing them forwards to the mouth along a mid-ventral channel. The mandibles lie at the sides of the mouth, each a stout L-shaped appendage, the side arms directed medially into the mouth chamber and ending in a roughened, flattened area, the molar process (Fig. 107A, mol.pro). The mandible articulates with the side of the head along a dorsoventral line on which it is movable by a number of muscles – promotors (pro.mu) running from the anterior edge forwards to the roof of the head, remotors (rem.mu) from the

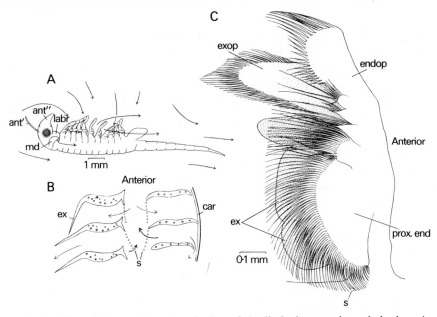

Fig. 108. A, *Chirocephalus*, to show the phasing of the limbs in metachronal rhythm. Arrows indicate the swimming-feeding current set up by the action of the thoracic limbs. One arrow emerges from beneath the large labrum and just posterior to the mandible; this is the adoral current escaping from the region of the mouth. B, frontal section of three successive limbs of two branchiopods to show how either exites (left, *Chirocephalus*) or carapace (right, *Daphnia*) control lateral outflow of water from interlimb spaces. Water is entering the upper interlimb space between the filter setae and leaving the lower one (thick arrows) to form the medial adoral outflow, slight in *Chirocephalus*, emphasized in *Daphnia* by the pivoting action of the filtratory endite. C, *Chirocephalus*: median view of trunk limb. (A, C, after Cannon.) ant′, antennule; ant″, antenna; car, carapace; endop, endopodite; ex, exite; exop, exopodite; labr, labrum; md, mandible; prox.end, proximal endite; s, setae of filter plate.

posterior edge backwards to a similar origin – both as in an ordinary trunk appendage. In addition muscles run across the body from mandible to mandible (tr.mu) and to cuticular ingrowths known as tendons (tr.ten). The effect of all these muscles is to pull the mandibles backwards and forwards on their articulations like any locomotor appendage but in phase; each time this occurs the molar processes rub on one another and food which has arrived there is crushed. An upper lip (labrum) and a pair of lateral lips (paragnaths) define the space in which the mandibles work.

Evolutionary change in this arrangement can transform the milling process into a biting one, with the two mandibles moving towards one another in a transverse plane. This is brought about by rotating the axis on which the mandible moves from the dorsoventral position which it occupies in *Chirocephalus* first to an oblique one (posterodorsal to anteroventral), as in the freshwater shrimp *Anaspides* (Fig. 107B), and finally to a more or less horizontal one (the sea slater *Ligia* (Fig. 107C), crayfishes, crabs). At the same time as this change in axis occurs the posterior end of the mandible develops a toothed incisor process (inc.pro); when the two mandibles swing in a direction which is morphologically posterior to the axis these processes are brought together in what has become a transverse bite. The original remotor muscles now function as adductors (rem.mu; (add.)rem.mu), bringing the jaws together, and the promotors as abductors (pro.mu; (abd.)pro.mu). To allow this movement the border of the mandible anterior (now dorsal) to the axis must be reduced for otherwise it would swing into the side of the head. One further improvement has occurred in the mechanism. In the more primitive crustaceans like *Chirocephalus* and *Anaspides* the presence of transverse muscles and the mandibular tendon prevents extensive separation of the biting edges and only small food can be taken. In more advanced crustaceans (*Ligia*, crayfish, crab), however, both these structures vanish and a much bigger bite becomes possible because of the increased abduction of right and left jaws.

The mandibles of arthropods in the onychophoran-myriapod-insect group are entire, miniature appendages; when they develop biting surfaces these are formed by the tips of the limbs, not by gnathobases. These two points, however, are almost the only ones in which the limbs of the three groups are alike.

The onychophorans are carnivores, feeding mainly on small arthropods. The mouth (Fig. 109) lies at the bottom of a basin-shaped depression bordered by thickened lips; since the lips arise lateral to the developing mandibles these appendages come to lie within the basin, a condition described as entognathous and evolved independently in other arthropodan groups, particularly terrestrial ones, since, as here, it provides a space where saliva can lubricate mouth parts and ease the mastication of hard food.

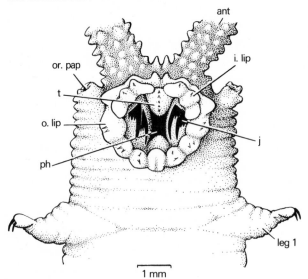

Fig. 109. *Peripatopsis sedgwicki*, ventral view of head with lips spread as for feeding. The mouth is open and shows the buccal cavity in which lie the tongue and two pairs of jaws and into which open the salivary glands and pharynx. (After Manton.) ant, base of antenna; i.lip, inner lip; j, jaw; leg, walking leg; o.lip, outer lip; or.pap, oral papilla carrying opening of slime gland; ph, pharynx; t, tongue.

Each mandible (j) carries two powerful and sharp chitinous blades which are enlarged versions of the claws carried on the walking legs. Their movements, too, are like those of the legs and consist of a backward-forward swing, the right and left jaws moving alternately just as do the two members of a pair of legs, a point in which onychophorans are more primitive than other arthropods. Any animal held within reach of the jaws is torn open and the soft parts sucked out, or is cut into pieces for swallowing. The power stroke by which this is brought about, like that of the legs, is due to remotor muscles.

It is not proposed to go into any study of the insects here, since that would be too gigantic an undertaking. It is, nevertheless, interesting to see in them the same kind of evolution in mandibular structure and function as has already been described for crustaceans, although it must be emphasized that this is a parallel evolution since the mandibles in the two groups are different structures. In primitive insects the mandible is an unjointed appendage moving on a dorsoventral axis with a backward-forward swing so as to produce a grinding of the food by molar processes. From such a starting point, and using the same devices as in crustaceans – moving the dorsal end of the articulation posteroventrally and the ventral end anterodorsally, reducing the part morphologically anterior to the axis – the articulation becomes horizontal and the movement transvere. Incisor processes replace molar processes and the result is a pair of jaws biting transversely, the gape between them becoming enlarged in the most advanced biters by the disappearance of transverse tendons. In all these steps there is a remarkable parallelism between crustacean and insect, though in detail the processes are quite different, the whole offering a most convincing example of how function dictates structure.

The last group to consider are the myriapods, and details will again be omitted. In general the myriapodan mandible bears the same relation to that of insects as does the chelicerate one (p. 273) to that of crustaceans. Whereas a transverse bite in crustaceans and insects is derived from an original fore-and-aft movement that of arachnids and myriapods is a primitive flexing of a jointed limb towards the mid-line of the body. One problem arises in myriapods, however, which did not appear in *Limulus*. In a king-crab the limb is set on the body near the mid-line and both adductor (add.mu, Fig. 106A) and abductor (abd.mu) muscles can run to origins on the exoskeleton of the broad body without difficulty. This is not true of myriapods where the mandibles primitively lie at the very lateral margin of the body: in such a position the proper siting of muscles to pull the jaws sideways is not possible. Myriapods have special devices to abduct their mandibles and each of the two major divisions (Diplopoda or millipedes, Chilopoda or centipedes) has a different way of doing this. In myriapods transverse muscles link mandible to mandible and to internal skeletal tendons and apodemes (p. 283), as happens in all primitive arthropods. Whereas these are stationary structures in other groups, they become mobile in myriapods and connect by way of an exoskeletal linkage to the mandibles in such a way that their forward movement swings the mandibles apart and their retraction allows adduction. In diplopods this is the most important factor in separating the mandibles, but in chilopods another comes into play. This is the entognathous position of the mandibles which moves them to a site where they no longer lie on the outer boundary of the body and so allows the development of muscles lateral to them to become responsible for abduction.

This review of mandibular structure and function added to the evidence from general limb anatomy leads without doubt to the conclusion that the arthropods are no single, uniform group. Living arthropods fall into three distinct taxa: (1) Chelicerata (*Limulus*, scorpions, spiders, ticks), with limbs basically biramous, with gnathobasic mandibles operated by direct adduction, without antennae on the head; (2) Crustacea (shrimps, crayfishes, crabs), with biramous limbs, gnathobasic mandibles working with an action based on a forward-backward swing, with two pairs of antennae; (3) Onychophora + Myriapoda + Insecta, with uniramous limbs, whole-limb mandibles and one pair of antennae. It is further obvious that the animals in the third division, though agreeing in the fundamental respects listed, do not form, by any means, a homogeneous group.

In arthropods the body is basically metamerically segmented as in annelids. It shows, in addition, an advance only hinted at in worms – the grouping of segments into regions specialized for one particular function or set of related functions. This is a process known as tagmatization (sometimes tagmosis), each group of segments being a tagma (plu. tagmata). Cephalization is one example of the general process and the most obvious in annelids, though in tubicolous species the anterior segments of the trunk often differ from the more posterior, allowing recognition of a thorax and an abdomen.

In arthropods head formation has proceeded further than in annelids and six segments (in addition to an unsegmented anterior tip known as the akron) are regularly found. Their fusion is so complete that they are rarely recognizable as separate except during development and through their appendages. Behind the head comes the trunk, clearly set off from the head in all sections of the arthropods except one, the Arachnida (= Chelicerata) in which an anterior set of trunk segments joins with the head to form a unit called the prosoma, whilst the other trunk segments unite to form a tagma known as the opisthosoma. In a few groups of arachnids, like the scorpions (Scorpionidea), the opisthosoma is divisible into an anterior mesosoma and a posterior metasoma, but this is not usual. In Onychophora and chilopod Myriapoda (centipedes) the trunk is undivided; in diplopod Myriapoda (millipedes), insects and crustaceans an anterior thorax can be distinguished from a posterior abdomen. Whilst these divisions in each group mark off regions modified for the more efficient performance of certain functions and so sometimes bear a superficial resemblance to one another, they are not comparable except on this functional basis. The segments which form a particular tagma, or bear a particular kind of limb, are often different from group to group not only in their position in the body but also in the number involved, as Table 3 (referring only to the crustaceans) demonstrates.

TABLE 3

| Ceph | Branchiopoda | | | Ostr | Cop | Cir | Myst | Bran | Malac | |
	Not	Dipl	An							
+	+	+	+	+	+	+	+	+	+	Akron
6	6	6	6	6	6	6	6	6	6	Head
8	11	4–16	11–19	4	6	6	6	4	8	Thorax
11	14–33	0–16	3–8	0	4	4	4	2?	7	Abdomen
+	+	+	+	+	+	+	+	+	+	Telson

Maxillopoda

Entomostraca

Abbreviations: An, Anostraca; Bran, Branchiura; Ceph, Cephalocarida; Cir, Cirripedia; Cop, Copepoda; Dipl, Diplostraca; Malac, Malacostraca; Myst, Mystacocarida; Not, Notostraca; Ostr, Ostracoda.

11

Crustaceans

TO ILLUSTRATE THE PLAN on which higher crustaceans are built as it is displayed in a benthic, freshwater form, the crayfish *Astacus* will be described.

Organization of a crayfish

The body of the crayfish (Fig. 110A) is segmented and covered externally by an exoskeleton secreted by a single layer of cells, the epidermis, and hardened, except at the arthrodial membranes, by the deposition of calcium salts; save for a thin surface layer it is a chitin–protein complex (Fig. 111D). The surface layer, the epicuticle (epic), differs from the endocuticle (endoc) in being without chitin and its amber colour is probably due to a tanned lipoprotein as in insects, conferring impermeability to water. The six segments of the head and eight segments of the thorax are fused to form the cephalothorax over which the calcified cuticle spreads as a large shield, the carapace (Fig. 110, car); ventrally some signs of separate segments are still visible and the last has a limited movement on the others and acts as a buffer between the fused segments of the thorax and the flexing abdomen.

The carapace arises as a fold of skin from the hinder edge of the head and spreads posteriorly, fused to the thorax dorsally, the lateral limit of fusion being marked by a longitudinal groove, below which it hangs freely, enclosing a cavity within which lie the gills. The six segments of the abdomen are free though movement of one upon another is limited to the vertical plane by well-developed points of articulation in the exoskeleton. The exoskeleton of each segment consists of a ring forming an arched upper part, the tergum (terg), with a short radius of curvature, and an under part, the sternum, of longer radius. The most rapid movement of the crayfish is brought about by a vigorous ventral flapping of the abdomen, when it darts backwards through the water. This is correlated with the facts that the terga overlap dorsally, the front part of each being pushed under the tergum in front when the abdomen is straight and exposed when it is bent, and that the sterna are narrow and connected

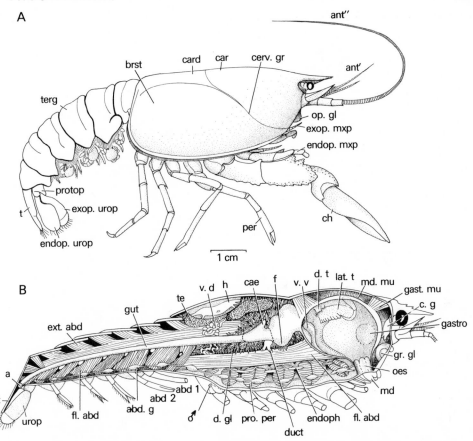

Fig. 110. *Astacus*: A, female with recently hatched young clinging to pleopods. B, lateral dissection of male. Only appendages of the left side are shown and those of the thorax are cut short. The right wall of the gastric mill has been removed; the limits of the mid-gut are indicated by broken lines. a, anus; abd 1, 1st abdominal appendage; abd 2, 2nd abdominal appendage; abd.g, 3rd abdominal ganglion; ant′, antennule; ant″, antenna; brst, branchiostegite (gill cover); cae, dorsal caecum; car, carapace; card, cardiac region of carapace; cerv.gr, cervical groove; c.g, cerebral ganglion; ch, cheliped; d.gl, digestive gland; d.t, dorsal tooth; duct, duct of digestive gland; endop.mxp, endopod 3rd maxilliped; endoph, endophragmal skeleton; endop.urop, endopodite of uropod; exop.mxp, exopod 2nd maxilliped; exop.urop; exopodite of uropod; ext.abd, extensor muscles of abdomen; f, filter chamber; fl.abd, flexor muscles of abdomen; gast.mu, gastric muscle; gastro, gastrolith; gr.gl, green gland; gut, hind-gut; h, heart; lat.t, lateral tooth; md, mandible; md.mu, mandibular muscle; oes, oesophagus; op.gl, opening of green gland; per, 1st pereiopod; pro.per, promotor muscles of pereiopod; protop, protopodite of uropod; t, telson; te, testis; terg, tergum; urop, uropod; v.d, vas deferens; v.v, ventral cardiopyloric valve; ♂, male opening.

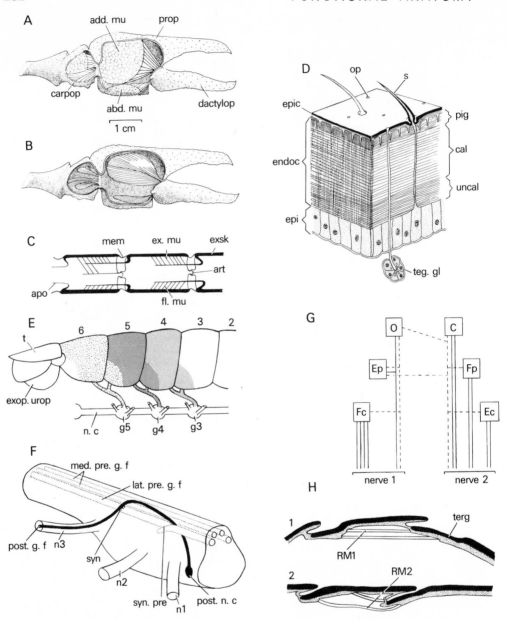

Fig. 111. *Astacus*: A, B, left chela. The penultimate segment (propodite) has a claw-like projection against which the last segment (dactylopodite) works in a pincer-like action. A, the
exoskeleton has been removed to display the muscles (closer and opener) controlling this
action and also those controlling the movement between the preceding segment of the limb
(carpopodite) and the propodite. B, the muscles have been removed to display their tendons
or apodemes. C, diagram of the two segments of a limb to show the origins and insertions of
extensor and flexor muscles. D, diagram of a block of epidermis and cuticle of a decapod.

by a wide flexible membrane which is stretched at the straightening and folded at the bending of the segments.

A pair of ventrally directed appendages is associated with each segment. They may be regarded as hollow outgrowths, jointed at intervals where arthrodial membranes allow movement. This may be limited to certain planes by articulating surfaces, though joints at the tips of the limbs have considerable freedom. Movement is effected by muscles attached to the exoskeleton. The muscles are striped and fibrils, tonofibrillae, extend their pull through the epithelial cells and even penetrate the cuticle. The tonofibrillae are formed during the deposition of the cuticle and renewed at each moult. The cuticle is often invaginated at the point of attachment of a muscle forming an inwardly directed projection, an apodeme or a tendon (Fig. 111C, apo). Similar invaginations of the ventral region of the cephalothorax give rise to a lattice of girder-like structures (Fig. 112A, endoph), which not only serve for the attachment of the flexor muscles of the abdomen (Fig. 110, fl.abd) and extrinsic muscles of the thoracic limbs (pro.per), but also provide a rigid abutment for legs which support the weight of the body. These legs, the pereiopods, spread fanwise from the last four thoracic segments. The nearest approach to the ancestral biramous limb is displayed by the abdominal appendages. In pereiopods (per) the exopodite is lost and the endopodite long and stout so that it can exert the necessary leverage. The anterior and posterior spreading of the four walking legs on each side reduces the overlap in their fields of movement. As the crayfish walks half the legs usually support the body at any one time – 1st and 3rd on one side and 2nd and 4th on the other – whilst the others are stepping. The progress may be slow and show considerable plasticity, or rapid and with a more fixed rhythm. Forward movement is helped by a rhythmical anteroposterior beat of the biramous abdo-

E, diagram of abdomen of crayfish showing the areas innervated by fibres running in the second segmental nerves. 2–6 numbers of the abdominal segments. F, diagram of an abdominal ganglion of a crayfish showing the pathways and relationships of the giant fibres. G, diagrammatic scheme of the innervation of the muscles of the last three joints of the cheliped of a crayfish; solid lines indicate stimulating axons, pecked lines, inhibiting axons. H, diagrams to show the relationships of the proprioceptor organs RM1 and RM2 of the crayfish abdomen, 1, in the flexed state; 2, extended. (D, after Dennell; E, based on Hughes and Wiersma; F, on Johnson; G, on Wiersma; H, on Alexandrowicz). abd.mu, opener (abductor) muscle; add.mu, closer (adductor) muscle; apo, apodeme; art, articulating surfaces limiting movement; C, closer of claw; cal, calcified layer; carpop, carpopodite; dactylop, dactylopodite; Ec, extensor of carpopodite; endoc, endocuticle; Ep, extensor of propodite; epi, epidermis; epic, epicuticle; ex.mu, extensor muscle; exop.urop, exopodite of uropod; exsk, exoskeleton; Fc, flexor of carpopodite; fl.mu, flexor muscle; Fp, flexor of propodite; g, abdominal ganglion; lat.pre.g.f, lateral giant pre-fibre; med.pre.g.f, median giant pre-fibre; mem, arthrodial membrane; n, nerve; n.c, ventral nerve cord; O, opener of claw; op, opening of gland; pig, pigment layer; post.g.f, giant post-fibre; post.n.c, nerve cell of giant post-fibre; prop, propodite; s, seta; syn, synapse between pre- and post-fibre; syn.pre, synapse between components of lateral giant fibre; t, telson; teg.gl, tegumental gland; terg,tergum; uncal, uncalcified layer.

minal appendages, the pleopods or swimmerets, though they are too small to move the body unaided. In the female there are five pairs of pleopods, but in the male only three, as the first two are modified as copulatory organs. The appendages of the last abdominal segment, the uropods, are concerned only with the backward swimming movement – the crayfish is incapable of swimming forward. They are larger than the others, with the rami broad and flat, and are spread out at the sides of the telson (t) to form a tail fan. When the animal is disturbed a vigorous ventral movement of the abdomen with the tail fan outspread shoots it back; if repeated, the animal may leave the bottom and swim.

The rest of the appendages are not concerned with locomotion. By far the largest pair, the chelipeds (ch), anterior to the pereiopods, are prehensile. They, and the two succeeding limb pairs, end in claws formed by the last and penultimate segments arranged to form a pincer (chela). The anterior appendages, with the exception of the antennules and antennae, surround the mouth and are concerned with feeding: those of the thorax, the maxillipeds (three pairs), assist the more anterior ones on the head, the maxillae and mandibles, in tearing, sieving and transferring food to the mouth. The maxillipeds are biramous: the endopods (endop.mxp) deal with the food; the whip-like exopods (exop.mxp) flick to and fro, reinforcing the exhalant current from the branchial chamber and directing it away from the mouth. The mandibles or jaws (md) grip the food and help to push it into the gut; the inner edge of each is sharp and toothed where it works against its partner.

The rigid exoskeleton which encloses the body isolates the living tissues from the external environment, and, as a compensation, delicate outgrowths from certain surfaces become necessary for sensation and respiration. The respiratory outgrowths are restricted to the lateral regions of the thorax where delicate plume-like gills provide a large surface area with only an extremely thin cuticle separating the blood from the surrounding water (Fig. 112A). Their protection is provided by the lateral folds of the carapace, the branchiostegites (brst). These so isolate the gills from the outside water that a ventilating mechanism (described below) becomes necessary. The sensory outgrowths of the exoskeleton form setae (Fig. 111D, s) associated with special neurons. They are fine, hollow extensions of the exoskeleton containing processes of an epidermal cell. Some are mechanoreceptors, of which there is a concentration in a statocyst on the antennule; others are chemoreceptors, which occur on the antennule and thoracic appendages; there also appear to be thermo-receptors. Other setal outgrowths are accessory to the function of the part of the exoskeleton on which they occur: for example, those which fringe the pleopods and uropods increase the effective surface; those which border the openings to the gill chamber provide a sieve for the ingoing water.

Eighteen gills are crowded into each branchial chamber which extends forwards from the posterior end of the thorax to the segment of the second maxilliped. Here it

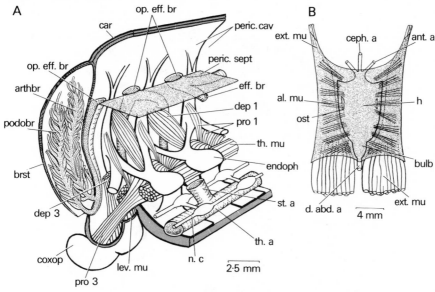

Fig. 112. *Astacus astacus*: A, stereogram of left side of thorax in the region of the pericardial cavity showing the relationships of the endophragmal skeleton to depressor, levator and promotor muscles of the appendages and to blood vessels and related structures. The heart is removed. B, pericardial septum seen from below to show the alary muscles within it, the heart seen by transparency and extensor muscles of the abdomen passing through the pericardial cavity (dotted lines). al.mu, alary muscles; ant.a, antennary artery; arthbr, posterior arthobranch; brst, branchiostegite; bulb, bulbus arteriosus; car, carapace; ceph.a, cephalic (ophthalmic) artery; coxop, coxopodite of 3rd pereiopod; d.abd.a, dorsal abdominal artery; dep 1, depressor muscle of 1st pereiopod; dep 3, depressor muscle of 3rd pereiopod; eff.br, efferent branchial vessel; endoph, endophragmal skeleton; ext.mu, extensor muscles of abdomen; h, heart; lev.mu, levator muscle; n.c, nerve cord; op.eff.br, openings of branchiopericardial vessels; ost, ostium; peric.cav, pericardial cavity; peric.sept, pericardial septum; podobr, podobranch; pro 1, promotor muscle of 1st pereiopod; pro 3, promotor muscle of 3rd pereiopod; st.a, sternal artery; th.a, ventral thoracic artery; th.mu, ventral thoracic muscle.

shallows and is limited dorsally by a constriction of the body wall (Fig. 113, const). A flow of water is set up and maintained by a pumping apparatus in a shallow prebranchial chamber anterior to this constriction, bounded laterally by the carapace, ventrally by the large epipodite of the first maxilliped (epip 1) and dorsally by a horizontal shelf of body wall; it opens anteriorly beneath the antenna (ant″). A thin muscular plate, the exopodite of the second maxilla or scaphognathite (scaph), projects into it immediately anterior to the constriction and extends to the opening at either end. This is the pump: by a rapid, complex, sculling movement it creates a forwardly directed current.

Each gill-bearing limb except the last carries an epipodial plate folded into a trough with its concavity facing backwards, successive plates fitting together to form a series of channels leading from the base of the gill chamber to its dorsal part (epip 3).

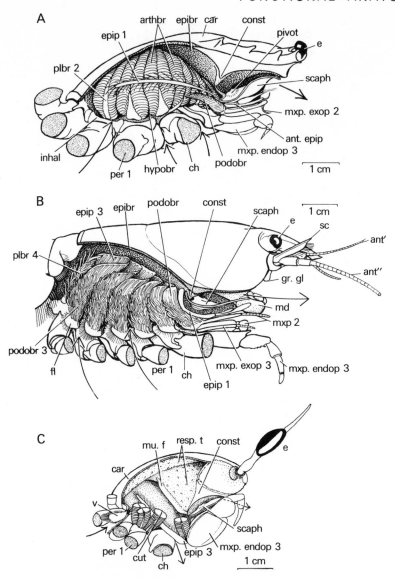

Fig. 113. A, *Carcinus maenas*, B, *Astacus astacus*, C, *Ocypoda macrocera*: lateral dissections to show the branchial chamber. Note how the chamber is divided into a larger posterior part containing the gills and, in the crabs, epipodites for cleaning, and a smaller prebranchial part containing only the scaphognathite; note also how the constriction between them is so placed that it allows the posterior end of the scaphognathite to act as a valve by sealing off the posterior chamber when expelling water. In *Carcinus* the scaphognathite is drawn in the position it occupies when pumping water from branchial to prebranchial chambers; in the crayfish it is in the position assumed when the exhalant channel is opening. The stippled areas represent the cut ends of appendages and branchiostegite. Arrows indicate inhalant and exhalant currents. Gills shown in B attach to limbs anterior to those they overlie. In C the distal parts of the four gills and the first epipodite

These channels guide water sucked under the ventral and posterior edges of the branchiostegite to the dorsal region of the branchial chamber, whence it is sucked forward by the scaphognathite, passes through the prebranchial chamber and leaves anteriorly. Entry to the channels is limited to the spaces between the legs by a flange at the base of each epipodite (fl) on which the ventral edge of the carapace rests. This edge and the flanges are beset with stiff setae to prevent ingress of particles likely to clog the gills. Respiratory exchange can occur on the surface of the vascular epipodial plates, extended into filaments on their anterior face, and on the gills which lie in the channels formed by successive epipodites. This arrangement ensures that the water is channelled over the sites where oxygen can be absorbed, into an exhalant dorsal passage leading directly to the prebranchial chamber.

The skeleton also provides the mechanical means for dealing with the food internally after it has received preliminary trituration by the mouth parts. Indeed the ectodermal ingrowths making up stomodaeum and proctodaeum form all but a very short length of the alimentary canal (which is a straight tube) and their chitinous lining, which is complicated in the fore-gut by the development of masticatory ossicles and various setal filters, is shed at each ecdysis. The endodermal mid-gut (indicated by pecked lines in Fig. 110B), although extremely short, is elaborated to form a bilobed, racemose gland, the digestive gland or hepatopancreas (d.gl), which lies in the haemocoel. This provides all the digestive enzymes, absorbs the products of their activity and accumulates reserves used to meet demands for energy and special materials during the period of moulting. A complex straining system ensures that only fluid and the finest particles enter the ducts of the gland and that the passage to the ducts is not blocked by particles from the fore- or hind-gut. A valve (Fig. 114A, d.pyl), allowing the passage of indigestible food directly from fore- to hind-gut, protects the mid-gut against abrasive particles, and so relieves the animal of the need for much selection of food which would otherwise have to take place externally.

The size of a meal varies. It may be considerable, for the first part of the fore-gut

have been cut away. ant′, antennule; ant″, antenna; ant.epip, anterior part of epipodite of 1st maxilliped; arthbr, arthrobranchs of cheliped; car, carapace; ch, cheliped; const, constriction; cut, cut axis of gill; e, eye; epibr, epibranchial space; epip 1, epipodite of 1st maxilliped; epip 3, epipodite of (B) 3rd pereiopod, (C) 3rd maxilliped; fl, flange; gr.gl, opening of green gland; hypobr, passage to hypobranchial space; inhal, ventral lip of inhalant opening; md, mandible; mu.f, muscular flap over entrance to lower part of lung; mxp 2, 2nd maxilliped; mxp.endop, endopodite of maxilliped; mxp.exop, exopodite of maxilliped; per 1, base of first pereiopod; pivot, pivot of scaphognathite; plbr 2, pleurobranch of 2nd pereiopod; plbr 4, pleurobranch of 4th pereiopod; podobr, podobranch of 2nd maxilliped; podobr 3, podobranch of 3rd pereiopod; resp.t, respiratory tufts of vascularized wall; sc, antennary scale; scaph, scaphognathite; v, setose valve guarding inhalant aperture.

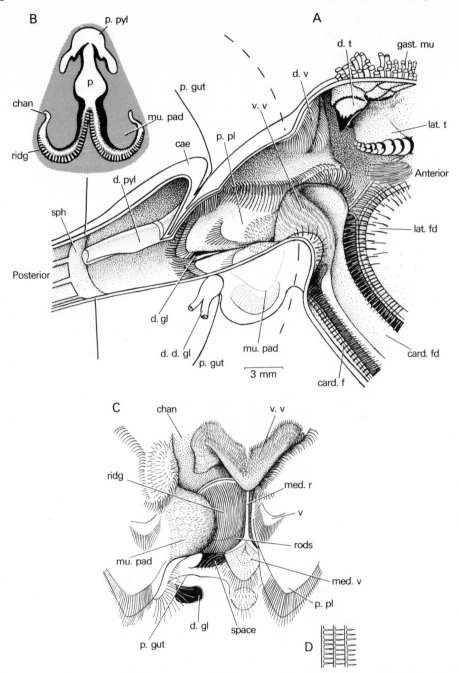

Fig. 114. *Astacus*: A, left sagittal half of the posterior part of the proventriculus with the mid-gut and beginning of the hind-gut. Continuous lines mark the limits of the mid-gut and the broken line the anterior limit of the pyloric region of the proventriculus. B, transverse section of pyloric

(the gastric mill with masticatory ossicles) is a capacious sac extending forwards into the head and back to the mid-region of the cephalothorax. It lies in a part of the body where fusion of segments gives stability, and the consequent lack of intersegmental muscles allows room for the movements of the gut and for the muscles causing them. The mill is followed by a smaller section of stomodaeum where the food is pressed, the fluid extracted from it filtered to the ducts of the digestive gland and the rest passed to the intestine. It is usually referred to as the pyloric chamber and the preceding one as the cardiac chamber of the proventriculus.

The mouth is large. Appendages round it knead the food into shapes appropriate for ramming into the oesophagus (Fig. 110B, oes), a short tube, wide and distensible, which opens on the ventral wall of the cardiac chamber. The food is transported by peristalsis though the muscular arrangement is unusual in that the circular layer is external to the longitudinal. The mouth parts do not triturate the food: this is the function of three large teeth, heavily calcified cuticular structures in the posterior part of the cardiac chamber, which together constitute the mill. One is sharp-edged and incisor-like and is placed dorsally, the others are elongated and molariform and lie one on each side (Figs 110B and 114A, d.t, lat.t). Muscles run from their bases to the body wall dorsally and laterally, one pair forwards and one pair backwards. When these muscles contract the stomach becomes longer and narrower with the effect of swinging the dorsal tooth forwards and moving the lateral teeth towards the mid-line, all three coming together at the same point and tearing and crushing any food lying there. Thick setose pads lying ventral to the lateral teeth also help to break food up. On relaxation of the muscles the elasticity of the cuticle restores the stomach to the resting position.

In addition to a mechanical attack on the food there is a chemical one, for a dark brown watery fluid containing protease, lipase and carbohydrase is passed forward from the digestive gland. It is sucked into the cardiac stomach by the action of dilator muscles from a channel running ventrally on each side and overarched by rows of closely set setae (Figs 114 and 131A, card.f). When digestion is sufficiently advanced the products, plus any particles fine enough to pass the filtering setae, are forced back to the digestive gland along the same channels by contractions of the

chamber showing press and filter. C, view of pyloric filter chamber from above. On the left the muscular walls of the press have been pulled apart to expose the filter and the channel leading from the ventral cardiac filter. D, rods of pyloric filter with setae. cae, dorsal caecum; card.f, ventral cardiac filter; card.fd, ventral cardiac fold; chan, channel from duct of digestive gland to cardiac filter; d.d.gl, duct of digestive gland; d.gl, opening of duct of digestive gland; d.pyl, dorsal pyloric valve (funnel); d.t, dorsal tooth; d.v, dorsal cardiopyloric valve; gast.mu, gastric muscles; lat.fd, lateral cardiac fold with setae; lat.t, lateral tooth; med.r, median ridge; med.v, median valve; mu.pad, muscular pad of press; p, press; p.gut, posterior limit of fore-gut; p.pl, press plate; p.pyl, posterior pyloric chamber; ridg, ridges of pyloric filter; rods, posterior end of filter plate marking transition of ridges to rods; space, space beneath filter; sph, sphincter; v, valve preventing regurgitation; v.v, ventral cardiopyloric valve.

gastric wall. The two filter channels are separated from one another by a median longitudinal fold (card.fd) which, at the entrance to the pyloric chamber, expands to form a massive valve (v.v) almost obliterating the cardiopyloric opening. The opening is further obstructed by lateral setose bulgings of the wall and by a dorsal valve (d.v), also bordered by rows of long setae.

The pyloric chamber is a narrow tubular space. The most anterior section, a relatively simple cavity entered from the cardiac chamber, is a vestibule, which acts as a kind of crop; the next is the press, distinguished by the presence of two thick bulges on the lateral walls, the press plates (p.pl), their edges and lobes beset with backwardly pointing setae. The press forms a narrow passage leading to the mid-gut and is the pathway followed by those fragments of food which prove totally indigestible. One fraction of the food which enters the press, however, is susceptible of further digestion and becomes soluble or is broken into extremely fine particles: this can be squeezed out of the press by a complex system of muscles in the wall and passed to the digestive gland. The apparatus which carries out the final extraction and filtering is located ventrally at the anterior end of the pyloric region, behind and below the ventral cardiopyloric valve (v.v). Here a pair of approximately hemispherical but somewhat helicoid depressions lie on the floor; into each fits a similarly shaped boss derived from the anteroventral end of the corresponding press plate (mu.pad). Each pad is covered with setae whilst the depression bears a cuticle raised into many fine longitudinal ridges (ridg). Posteriorly the ridges project freely to form a series of parallel rods (Fig. 114C, rods), each of which has a row of fine setae along its median edge which touch the lateral margin of the next more median rod (Fig. 114D), the whole forming a filter passing only liquid and the very finest particulate material. The two filters are separated by a median ridge (Fig. 114C, med.r) projecting backwards to form a flap-shaped valve which excludes large material from the press. The fluid and such minute particles as can pass this filter enter the most anterior and ventral part of the mid-gut in the neighbourhood of the openings of the ducts of the digestive gland, into which they can pass. From this area, too, a groove (chan) runs forward on each side between the boss and the edge of the filter plate to connect with the ventral channels in the cardiac part of the stomach: this is the route by which digestive juices reach that region and by which digested matter is led back for absorption in the gland.

The digestive gland (Fig. 110B, d.gl), which occupies the greater part of the cephalothorax posterior to the cardiac fore-gut, is made up of a mass of finely branched tubules which give a large surface area concerned with secretion and absorption. Each tubule is surrounded by a muscular net of branched fibres which contract rhythmically, expelling secretion and sucking fluid into the lumen. The process of secretion is described as holocrine since the entire cell is used up in one cycle of secretion and has to be replaced from a stock of immature cells in the tubule.

A chitinous projection from the dorsal wall of the pyloric chamber, the dorsal pyloric valve, projects into the mid-gut as a tube incomplete ventrally (Fig. 114A, d.pyl). It is long enough to span the short mid-gut and can deposit indigestible matter directly into the hind-gut where peristaltic movements force the contents towards the anus. The control of these movements and of those of the mid-gut has been ascribed to an extensive nerve plexus in the wall. The valve isolates a glandular mid-gut caecum (cae), which lies dorsal to its origin, from the faecal matter. Secretion from the caecum may protect the wall of the mid-gut and help to consolidate the faeces.

The cavity enclosed by the exoskeleton surrounds the viscera and is traversed by somatic and visceral muscles. It is haemocoelic in origin, the only coelomic spaces being within the excretory organs and gonads. The haemocoel is much less capacious than one might be led to believe on opening up a crayfish and mostly comprises chink-like spaces between organs. One of its largest parts lies in the posterior half of the cephalothorax, where a horizontal sheet of connective tissue with muscles isolates a dorsal region. This is the pericardial cavity (Fig. 112A, peric.cav) and the connective tissue sheet, known as the pericardial septum (peric. sept) bounds it anteriorly, posteriorly and ventrally. Dorsally and laterally it is adjacent to the carapace. Within the cavity, medially placed, is the heart (Figs 110B and 112B, h), whilst laterally run the extensor muscles of the abdomen (Fig. 112B, ext.mu). This position of the heart is perhaps determined by the need to be close to the gills and away from disturbance created by the gastric mill anteriorly and the flexing abdomen posteriorly. The heart has an irregularly hexagonal outline when viewed from above. It is single-chambered, with a thick muscular wall pierced by three pairs of afferent openings, the ostia (Figs 112B and 115B, ost), each surrounded by a sphincter muscle, which admit blood from the pericardial cavity or sinus. The only afferent openings to the sinus are those of efferent branchial veins from the gills (op.eff.br), so that the heart receives only oxygenated blood. Arteries (Fig. 115B) help to suspend the heart in the sinus, though this is mainly effected by bands of connective tissue, the suspensory ligaments, which are inserted at its lateral angles and pass to the rigid boundaries of the pericardial wall provided by the exoskeleton.

The arteries convey blood anteriorly, posteriorly and ventrally and branch to smaller vessels which discharge into lacunae. All blood from these irregular spaces in the body sooner or later makes its way to ventral sinuses in the thorax and then to the gills. This open circulatory system has a low arterial pressure as compared with one in which the blood is confined to vessels, but this does not imply sluggishness and an ineffective circulatory mechanism. Indeed the method by which the heart is filled at diastole is unaffected by blood pressure and volume. Systole is brought about by contraction of the cardiac muscles, whilst the diastolic mechanism is provided by the suspensory ligaments. At the end of diastole the relaxed heart

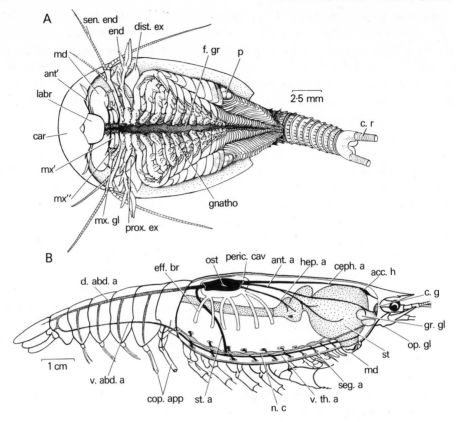

Fig. 115. A. *Triops cancriformis* (Apus): female in ventral view, with posterior trunk limbs folded back in resting position. B. *Astacus*: lateral body wall of cephalothorax removed to show certain internal organs. In the abdomen some vessels are seen by transparency. No venous sinus is shown. acc.h, accessory heart; ant′, antennule; ant.a, antennary artery; car, carapace; ceph.a, cephalic (ophthalmic) artery; c.g, cerebral ganglion; cop.app, copulatory appendages; c.r, caudal ramus, cut short; d.abd.a, dorsal abdominal artery; dist.ex, distal exite of second trunk limb; eff.br, efferent branchial vessel; end, toothed endite of second trunk limb; f.gr, food groove; gnatho, gnathobase; gr.gl, green gland; hep.a, hepatic artery; labr, labrum; md, mandible; mx′, maxillule; mx″, maxilla; mx.gl, opening of maxillary gland; n.c, ventral nerve cord; op.gl, opening of green gland; ost, ostium of heart; p, egg pouch; peric.cav, pericardial cavity; prox.ex, proximal exite of second trunk limb; seg.a, segmental artery; sen.end, sensory endite of first trunk limb; st, cardiac stomach; st.a, sternal artery; v.abd.a, ventral abdominal artery; v.th.a, ventral thoracic artery.

muscles are stretched by the tension of the ligaments, the ostia are open and valves preventing regurgitation from the arteries are closed. As the muscles contract against the pull of the ligaments, the ostia close, intracardial pressure increases, the valves open and blood is forced into the arteries. The major event in this cycle is the contraction of the heart muscles, controlled by a pacemaker in the cardiac ganglion in its wall.

The contraction not only forces blood into the arteries but, by reducing pressure in the pericardial sinus, draws blood from the efferent branchial veins. Changes in pressure in the sinus are also brought about by transverse muscles in the pericardial septum which originate at the lateral body wall and extend medially to end a short distance from the mid-line. These are the alary muscles (Fig. 112B, al.mu). When they relax at each systole the pericardial floor is bowed dorsally, and when contracted at diastole the floor flattens and so increases the pericardial volume. This increase in volume has the effect either of distending the heart or of continuing to suck blood from the gills to give a more continuous flow, or both. The movements of the alary muscles are synchronous with the heart beat, though they have an independent innervation.

From the anterior apex of the heart two pairs of arteries arise (antennary and hepatic), and a median artery (cephalic or ophthalmic) which gives a direct supply of blood to the brain, eyes and antennules, and which, before it branches to supply these organs, is enlarged to form an accessory heart worked by special somatic muscles. The pressure developed by the primary heart and supplemented by the pump forces blood anteriorly through a network of capillaries developed in and on the cerebral ganglia. Less extensive capillary beds are associated with other parts of the vascular system where exchange between blood and tissue is most critical – over the ventral nerve cord, the excretory organs, the wall of the gastric mill where the gastrolith develops and in the gills.

At the posterior end of the heart a small chamber, the bulbus arteriosus (bulb), which may provide an accessory propulsive force, lies at the origin of two arteries, the dorsal abdominal and sternal artery. The sternal artery, which is the larger, passes ventrally through the nerve cord and branches to form a subneural vessel. This gives a direct supply of arterial blood to all the appendages of the cephalo-thorax except the antennules. Both subneural and dorsal abdominal arteries supply the muscles of the abdomen, with paired branches to each segment; the former vessel supplies the first abdominal appendages and the latter the rest.

The segmental excretory organs, the green glands (Figs 110B and 115B, gr.gl), which are derived from paired coelomic vesicles, and the gonads, can, therefore, have no direct connection with the body cavity. Each is continuous with its duct which opens on the coxopodite of an appendage. The green gland, in the anterior part of the haemocoel, consists of a glandular region overlaid by a large bladder opening on the antenna (op.gl). The urine which collects in the bladder is hypotonic to the blood (see Fig. 116). The glandular parts of the organ in which this fluid is produced consist of (1) an inner end sac (sac) which has its cavity subdivided by ingrowing partitions, leading to (2) a green labyrinth of anastomosing canals (lab) and (3) a nephridial canal (neph.c) coiled upon itself and opening to the bladder (bl). The end sac has a capillary system arising from an arterial supply directly from the heart. It has been suggested that a filtrate from the blood passes into the sac (possibly substances are added to the fluid as it passes through the labyrinth) and then salts are resorbed in the nephridial canal to give the dilute urine. Thus the gland may act as a filtration-resorption mechanism. In support of this Picken has shown that the hydrostatic pressure of the blood is generally greater than its colloidal osmotic pressure and is therefore sufficient to account for a filtration into the organ: the hydrostatic pressure within the sternal sinus is 20 cm of water, whilst the colloidal osmotic pressure of the blood is 15 cm. Analyses of fluid from different regions of the gland show that the concentration of Cl$^-$ falls from about 196 mM l^{-1} in the coelomic sac (a value equal to that in the blood) to about 90 mM in the nephridial canal and 10 mM in the bladder; there is a slight increase in the labyrinth to 209 mM, suggesting some secretion there.

The average production of urine in *Astacus astacus* has been calculated as 3.8 per cent of the live weight in 24 h, but if the animal is kept in a solution of 250 mM salt the production ceases. These and other results show that the excretory organs

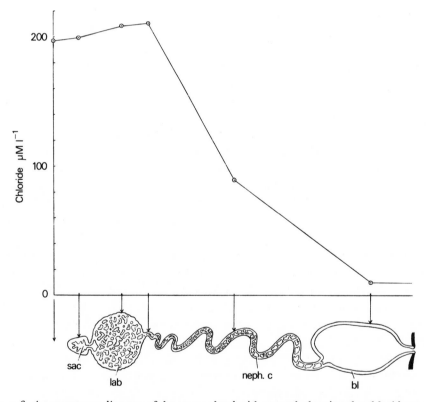

Fig. 116. *Astacus astacus*: diagram of the green gland with a graph showing the chloride content of the excretory fluid in micromoles per litre. (After Parry.) bl, bladder; lab, labyrinth; neph.c, nephridial canal; sac, end-sac.

regulate the salt and water balance and act as a defence against the continuous osmotic entry of water from the environment.

Some observations made by Maluf militate against the view that urine is secreted by extraction of salt from a filtrate isotonic with the blood. He observed that when the crayfish *Procambarus clarki* was kept in a hypertonic medium (fresh water to which salts had been added) and the formation of urine had almost ceased, apical vacuoles in the columnar cells lining the distal part of the nephridial tubule disappeared. They reappeared when the animal was returned to fresh water and the formation of urine resumed. He therefore regarded the formation of urine as consisting of the tubular secretion of a hypotonic solution. However, these apical vacuoles could equally be involed in the active resorption of salts.

The nervous system of *Astacus* is built on a plan which, in gross anatomy, is similar to that of annelid worms. Two nervous masses, the cerebral ganglia or brain, lie in the head dorsal to the gut; they are linked by connectives running round the oesophagus to the anterior end of a double ventral nerve cord which stretches back to the posterior end of the abdomen. The crustacean nerve cord is primitively, as in many branchiopods, ladder-like, with the right and left halves separated but linked in each segment by transverse commissures running between segmental ganglia. This arrangement has been considerably modified in crayfish by the same process of fusion and cephalization as affects the nervous system of other animals and which leads to functional improvement by close association of the nerve cells which control linked activities. In general, too, the plan of the nervous system reflects the degree of development of the different parts of the body, being long in those forms like crayfish which have a long body but much abbreviated in those, like crabs (p. 305), in which the abdomen is vestigial. In *Astacus* the two halves of the nerve cord have undergone a partial fusion and the commissures are internal, and there has also been a telescoping of some ganglia. The brain sends nerves to eyes, antennules and antennae and must therefore include two pairs of segmental ganglia. The first ganglion on the ventral cord, a large suboesophageal, innervates all the appendages from maxillules to third maxillipeds and so is clearly a compound structure. From there backwards, however, the original arrangement of one (double) ganglion per segment persists, each giving off three pairs of nerves.

The innervation of a segment by these nerves has been worked out only for abdominal segments of the crayfish *Procambarus* and a few other genera, but it is probably generally applicable to all macruran decapods. The third nerves contain only motor fibres and innervate the powerful flexor muscles involved in the escape reaction. The first and second nerves are mixed: the first send motor fibres to the segmental appendages and receive sensory ones from them and from setae located on the ventral surface and lateral edges of the segment. The second pair is peculiar in that it mainly innervates structures in the segment behind that of the ganglion from which it starts (Fig. 111E). These structures are the intersegmental muscles,

sensory setae over the lateral and dorsal surfaces and the stretch receptors (Fig.
111H). The last are modified muscle fibres running from tergum to tergum close to
the mid-dorsal line, two (known as RM1 and RM2) on each side in each segment.
They respond to the stretching imposed on them by the changing angle between
successive terga as the abdomen flexes or straightens. RM1 responds as a tonic organ
by continuous firing at different speeds corresponding to different degrees of
curvature whereas RM2 adapts more rapidly but requires a greater initial stretch
for stimulation and is possibly active only when the crayfish swims. The facts that
each abdominal ganglion has relationships with organs located in the two segments
and that all abdominal activities have to be coordinated indicate that interneurons
in the ventral cord must play an important role; indeed, some, known as command
neurons, are each able to set up almost the entire pattern of nervous activity leading
to the performance of a single action by the crayfish.

Crayfishes react to fright by a sudden backward movement similar to the panic
withdrawals of many tubicolous animals; as in them, the movement is brought
about by giant fibres (Fig. 111F). Four fibres, known as pre-fibres, run the length of
the nervous system in a dorsal position, two median and two lateral, the former
making synaptic contact in the brain with fibres from anterior sense organs, the
latter making most of their sensory contact posteriorly but all conducting in both
directions. The median fibres seem to be derived each from a single cell, but the
presence of segmental membrane synapses (syn.pre) in the lateral fibres like those of
the giant fibres of earthworms suggests that they are the products of the fusion of
many cells. The giant fibres make synaptic contacts with further giant motor fibres
in the abdominal ganglia, known as post-fibres. The nerve cells of the post-fibres
(post.n.c) lie ventrally in the ganglia and their axons (post.g.f) curve dorsally into
the third segmental nerve, making contacts with all four pre-fibres as they traverse
the ganglion. There are seven sets of synapses in all, one in each of the first five
abdominal ganglia and a double set in the sixth. The post-fibres supply the deep
parts of the abdominal flexors which are alone responsible for the escape reaction
and are physiologically incapable of giving more than a twitch-like contraction. The
superficial fibres of the flexor muscles are not innervated by these fibres and exhibit
graded tonic contractions required for maintaining the normal posture of the
abdomen. This double function of what appears to be anatomically a single muscle
is reminiscent of the quick and catch reactions of the adductor muscles of brachiopods (p. 450), and bivalve molluscs (p. 517).

The crustacean nervous system ensures that crustaceans behave in a way that
seems to the observer completely comparable with that of other animals, yet
physiologically it is unique on the motor side. Its main characteristics may be
summed up by saying that the innervation of muscle is polyneural and multiter-
minal, that is, that each fibre receives stimuli from more than one neuron and that
each neuron has several endings on each muscle fibre. Further, all the fibres in a

muscle may be supplied by branches of a single axon. A muscle fibre in a crustacean may be innervated by three different types of axon, each with its special effect, one causing a fast, twitch-like contraction, a second one which is slower and longer lasting, whilst a third inhibits contraction, or by various combinations of these. This contrasts markedly with the more familiar arrangement of vertebrates where each nerve fibre innervates only a group of fibres in a muscle which constitute a motor unit, where the whole muscle normally contains many units and where inhibition is wholly central.

The arrangement of the muscles and their innervation in the distal joints of the cheliped of a crayfish has been well studied and may be correlated with the kind of activity which this limb carries out. The cheliped handles food and other objects in a delicate fashion, but is also part of the animal's offensive and defensive apparatus; at that time it is thrust outwards with the chela open, ready to snap shut rapidly and powerfully on prey or predators. The muscles responsible for the movements of the distal joints of the limb include an opener and a closer of the chela (dactylopodite), a flexor and an extensor of the hand (propodite) and a flexor and extensor of the next more proximal joint (carpopodite). The innervation of these muscles is shown diagrammatically in Fig. 111G, where it can be seen that the opener of the claw and the extensor of the propodite are innervated by a single excitatory axon; stimulation of this one fibre alone can therefore bring about the thrust forward and opening of the chela. If necessary, the chela can be shut by the activity of one further axon to the closer, a muscle so powerful that it hardly matters whether its antagonist is inhibited or not. The essentials of this action, whether for seizing prey or balking a predator, can thus be brought about with outstanding economy of nervous activity, the stimulation of two neurons, or if the opener is inhibited, of three. This extreme reduction in the number of nerve cells involved in carrying out activities is one of the characteristic features of the crustacean nervous system. Whilst it eases the work of command interneurons it raises the question of how the finely graded contractions used in more gentle manipulation may be produced and controlled.

In vertebrates gradation of the response of a muscle to suit the action it is required to carry out is achieved by varying the number of motor units stimulated. In crustaceans, where all the fibres in a muscle may be innervated by branches of a single neuron, this way of modifying its power is obviously impossible and other ways must be used to reach the same end. The flexor of the carpopodite of the crayfish cheliped is innervated by four stimulatory axons; 38 per cent of its fibres are activated by stimulation of one axon, 26 per cent of two, 29 per cent of three and 7 per cent of four. This may allow a method of grading responses not unlike that met with in vertebrates. The interplay of stimulation and inhibition and the possibility of localized contraction made possible by their multiple nerve endings, however, are the main ways in which the crustacean obtains this type of muscular control. Inhibition is of great importance in obtaining the right response out of a neuromuscular mechanism that might, at first sight, seem rather inflexible. Thus the fact that the opener of the claw and the extensor of the propodite share a single motor axon and so can be flung into simultaneous activity can be overcome by the fact that each has its own inhibitor permitting each muscle to contract by itself when that is necessary. Stimulation of an inhibitory fibre is normally initiated within the central nervous system but can be totally peripheral; in the crab *Carcinus* the claw closes when its inner surface is stroked and the action of the closer muscle leads to relaxation of the opener muscle by direct stimulation of the inhibitory axon which the two muscles share (Fig. 111G). A general inhibition of movement occurs after moulting when unwanted contractions of muscle at the wrong time might deform the new exoskeleton. This appears to be the function of the inhibitor axon in the cheliped which innervates five distinct muscles since no activity has been demonstrated in it at other times.

The two gonads, ovaries or testes (Fig. 110B, te), are fused to give a pair of anterior lobes beneath and to the sides of the pericardium and a median posterior lobe which extends behind it, though not into the abdomen. Their ducts are glandular, long and convoluted in the male (v.d) where they are responsible for the manufacture and storage of spermatophores, short in the female in which they provide a thin chitinous membrane around each egg as it passes to the exterior. The position of the genital openings is associated with the method of copulation and spawning. The vas deferens opens ($\male$) on the last thoracic appendage and when spermatophores are transferred to the female its terminal region is everted and projects into the base of the groove of the first abdominal appendage (abd 1). The spoon-shaped extremity of the second abdominal appendage (abd 2) is also inserted here, moving backwards and forwards through the groove to keep it clear. The two crayfish then lie sternum to sternum, the female on her back, held in position by the chelipeds and pereiopods of the male. The copulating female must have recently ecdysed, but the exoskeleton of the male is hard. The copulatory appendages deposit the spermatophores on the pleopods and posterior sterna of the female, to which they adhere as their outer surfaces harden in contact with the water.

When, at a later date, the female spawns she lies with the ventral surface uppermost and the abdomen curved forward to form the roof of a space of which the lateral extensions of the abdominal terga edged with long setae form the sides and her body the floor. The eggs are discharged into this chamber from the oviducts which open on the coxopodites of the second pair of walking legs. The fertilized eggs become attached to special setae on the pleopods and sperm (which are then liberated from the spermatophores) adhere to the eggs by spines radiating from the sperm capsule; an explosive eruption fires their nucleus into the egg.

The protection provided for the eggs involves not only the secretion of a chitinous shell by the oviduct but also a cementing fluid from special glands in the pleopods. This first appears as a viscous fluid which fills the spawning chamber and entangles the eggs as they flow from the genital apertures in a slow, continuous stream, and it also entraps the spawn. Ultimately the fluid forms an outer membrane around each fertilized egg and a stalk for attachment to the ovigerous setae of the pleopods. It flows readily when secreted, solidifies slowly in water and its low permeability and final hardness give the necessary protection to the embryo over the several weeks of development. The fluid is allied chemically and physically to the secretion forming the superficial layer of the exoskeleton. In fact the processes involved in the formation of the egg membranes may be regarded as an exploitation of secretions otherwise involved in exoskeleton formation. In the female copulation follows ecdysis and the period of spawning may be a few days or even a month after. The secretions concerned in the formation of spermatophores may have a similar association with those concerned in the formation of the exoskeleton.

Classification of crustaceans

In common talk crustaceans are subdivided into two sections, a higher division of malacostracans and a lower division of entomostracans, but whereas the former is a natural group of related animals, entomostracans share only the negative character of not being malacostracan, and the term is convenient only when no more than that is implied. Modern taxonomists recognize seven groups of living crustaceans, plus one wholly extinct one, each of taxonomic standing equivalent to that of the Malacostraca. The seven Recent classes are Cephalocarida, Branchiopoda, Mystacocarida, Ostracoda, Copepoda, Branchiura and Cirripedia. Three of these (Cephalocarida, Mystacocarida and Branchiura) are so small that little reference will be made to them, and although Ostracoda (with about 1 000 genera) cannot be so described, they will also be disregarded. Sometimes Copepoda, Cirripedia and Mystacocarida are united into a single superclass Maxillopoda and there seems a sufficient degree of similarity in structure to justify this.

Apart from the cephalocaridan *Hutchinsoniella* (Fig. 105) the group Branchiopoda includes what are often regarded as the most primitive crustaceans; they have a long body composed of many similar segments. The limbs which they bear, however, are broad, multilobed structures of the type known as phyllopodia (Figs 108C and 115A) and this is certainly an adaptive modification for a microphagous habit, rather than primitive. Some (order Anostraca, Fig. 108A) have no carapace; others (order Notostraca, Fig. 115A) have a leaf-shaped one over their back whilst in the order Diplostraca it is bivalved. This last order contains animals of two patterns, the longer bodied Conchostraca in which the head lies within the shell (Fig. 105D), and the short bodied Cladocera, the common water fleas (Fig. 117) in which it lies outside it. Nearly all branchiopods are freshwater, can resist prolonged

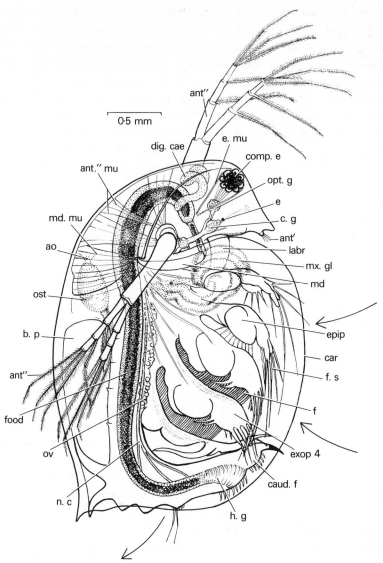

Fig. 117. *Daphnia magna*: female from the side showing the appendages and some internal organs by transparency. Arrows indicate water currents. ant′, antennule; ant″, antenna; ant″.mu, antennary muscles; ao, aorta; b.p, brood pouch; car, ventral edge of carapace; caud.f, caudal furca; c.g, cerebral ganglion; comp.e, compound eye; dig.cae, digestive caecum; e, median, simple eye; e.mu, eye muscles; epip, epipodite (branchia); exop 4, exopodite of 4th trunk limb; f, filter plate of 3rd trunk limb; food, food in mid-gut; f.s, feathered setae; h.g, hind-gut; labr, labrum; md, mandible; md.mu, mandibular muscles; mx.gl, maxillary gland in carapace; n.c, nerve cord; opt.g, optic ganglion; ost, ostium of heart; ov, ovary.

drought in the egg stage and often reproduce parthenogenetically. They are all slow swimmers and would fall an easy prey to active predators had they not adapted to life in temporary pools barred to such creatures.

Copepods (a name meaning oar-footed) (Fig. 118A) are small planktonic crustaceans; there are numerous species and abundant individuals. A few occur in fresh water but they are characteristic of the marine plankton where they constitute the basic food of many predators. Some have become parasitic and are only with difficulty recognized as copepods. Their sixteen segments are divisible into an

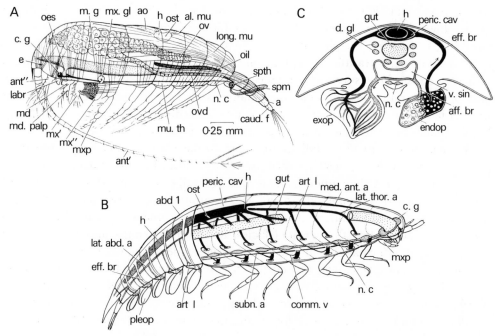

Fig. 118. A, *Calanus*: female from the side showing some organs by transparency. B, generalized isopod in lateral view with the lateral body wall removed in the head and thorax to show certain internal organs. In the abdomen some parts of the vascular system are seen by transparency. No part of the sinus system is shown. C, *Ligia oceanica*, transverse section at level of third pleopod showing exopodite on the left, endopodite on the right. a, anus; abd 1, first abdominal segment; aff.br, afferent branchial vessel; al.mu, alary muscles; ant', antennule; ant", antenna; ao, aorta; art l, artery to limb; caud.f, caudal furca; c.g, cerebral ganglion; comm.v, commissural vessel; d.gl, tubules of digestive gland; e, median eye; eff.br, efferent branchial vessel; endop, endopodite; exop, exopodite; gut, gut; h, heart; labr, labrum; lat.abd.a, lateral abdominal artery; lat.thor.a, lateral thoracic artery; long.mu, longitudinal muscles; md, mandible; md.palp, mandibular palp; med.ant.a, median anterior artery; m.g, mid-gut; mu.th, muscles of thoracic limb; mx', maxillule; mx", maxilla with filter plate; mx.gl, maxillary gland; mxp, maxilliped; n.c, ventral nerve cord; oes, oesophagus; oil, oil sac; ost, ostium of heart; ov, ovary; ovd, oviduct; peric.cav, pericardial cavity; pleop, pleopod; spm, spermatophore attached to lips of genital aperture; spth, spermotheca; subn.a, subneural artery; v.sin, ventral sinus.

anterior prosome (cephalothorax) movable on a posterior urosome (abdomen), the boundary not coinciding with the thoracic-abdominal boundary in other groups. The last few segments of the prosome are also movable on one another. The thoracic segments bear biramous appendages used in swimming.

The class Cirripedia includes the animals called barnacles, all attached to a substratum when adult, and not at first sight suggesting relationships with other crustaceans; indeed barnacles were for long regarded as somewhat odd molluscs. Their larvae, however, are immediately recognizable as crustacean. Barnacles are attached by means of the antennules and anterior part of the head which may (Fig. 119D) or may not (Fig. 119A) elongate to form a stalk. The rest of the body hangs upside down into the water within a closable bivalved carapace known as the mantle; it consists of a cephalothorax, which, except for the appendages, has lost most signs of segmentation; except in one group the abdomen is vestigial. Since barnacles are sessile a special food-collecting apparatus has had to evolve, the six pairs of setose biramous thoracic appendages creating and straining a current of water. Like other sessile organisms barnacles are hermaphrodite. Some, in a group Rhizocephala, have become parasites of decapod crustaceans.

The remaining Crustacea fall in the class Malacostraca, including those most familiar to the ordinary person – shrimps, both freshwater and marine, crabs and crayfish. The group has undergone a massive adaptive radiation and though probably exceeded by copepods so far as the number of individuals goes, excels them in the diversification of its species. The basic malacostracan features are collectively described as the caridoid facies, the more striking characters of which are: (1) the standardization of the number of segments in each tagma; (2) the presence of a carapace over the dorsal and sometimes the lateral parts of the cephalothorax; (3) compound eyes on the head, usually stalked for mobility; (4) thoracic appendages primarily for walking but often used for swimming; (5) abdominal appendages primarily for swimming. The malacostracans fall into two subclasses Phyllocarida or Leptostraca and Eumalacostraca.

Leptostraca (*Nebalia*, Fig. 120) are at a more primitive level than the eumalacostracans: they have a large carpace not attached to the thoracic segments and the abdomen has seven separate segments, with biramous limbs. They are specialized filter feeders, however, and their thoracic appendages are foliaceous.

Amongst Eumalacostraca there are two small and two large groups. The small group Hoplocarida includes the mantis shrimp, *Squilla*, an animal which burrows into soft substrata and pounces on others for prey, grasping them with long prehensile limbs (the appendages of the second thoracic segment) which act like the raptorial limbs of the praying mantis and give the animal its popular name. The carapace does not cover the sides of the body nor the hinder four thoracic segments, perhaps because of the burrowing habit. The abdomen is large and the gills occur on the abdominal appendages. Syncarida is a very small group of primitive

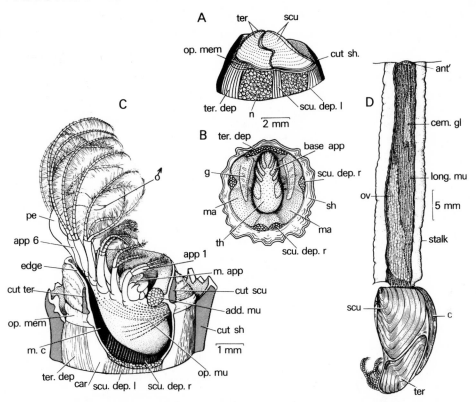

Fig. 119. *Elminius modestus*: A, from the left with part of the outer wall of plates (or shell) removed, showing the basal parts of the body and larvae in the mantle cavity through the transparent carapace or mantle; the operculum is still intact. B, seen from below as if attached to a glass plate by the membranous base which is treated as if transparent. Through the base is seen the thorax of the barnacle surrounded by the mantle cavity with a gill on each side. C, part of the mantle, tergum and scutum on the left have been removed, together with the left half of the outer shell, to show the animal lying in the mantle cavity. The uppermost part of the right half of the outer shell has also been chipped away. D, *Lepas*, the ship's barnacle, from the left. The stalk has been cut lengthways to show contents. add.mu, adductor muscle; ant', position of antennule; app 1, 1st thoracic appendage; app 6, 6th thoracic appendage; base app, base of thoracic appendage; c, carina; car, carapace or mantle; cem.gl, cement gland; cut scu, cut edge of left scutum; cut sh, cut edge of outer shell; cut ter, cut edge of left tergum; edge, mantle edge; g, gill; ma, mantle; m.app, mouth parts; m.c, mantle cavity; n, nauplii; op.mem, opercular membrane; op.mu, muscles attaching body to operculum; ov, ovary; pe, penis; scu, scutum; scu.dep.l, left scutal depressor muscle; scu.dep.r, right scutal depressor muscle; sh, outer shell; stalk, attachment stalk; ter, tergum; ter.dep, tergal depressor muscle; th, dorsal thoracic wall; ♂, penial opening.

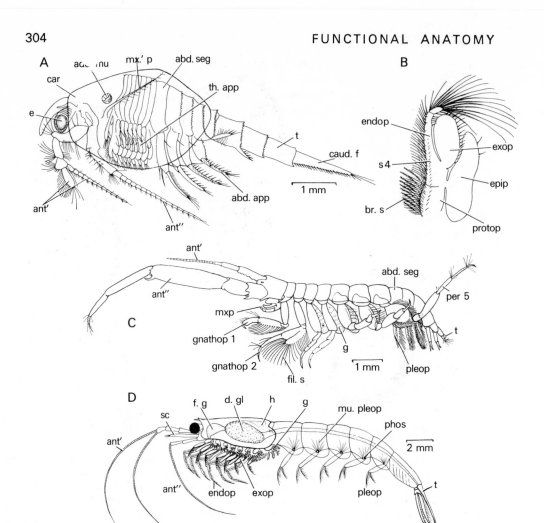

Fig. 120. *Nebalia bipes*: A, from the left; B, foliaceous thoracic limb. C, *Corophium*, from the left. D, diagram to show euphausiid organization. (B, after Cannon.) abd.app, abdominal append-age; abd.seg, 1st abdominal segment; add.mu, adductor muscle; ant′, antennule; ant″, antenna; br.s, brush setae; car, bivalved carapace; caud.f, caudal furca; d.gl, digestive gland; e, eye; endop, endopodite; epip, epipodite; exop, exopodite; f.g, fore-gut; fil.s, filtering setae; g, gill; gnathop, gnathopod; h, heart; mu.pleop, extrinsic muscles of pleopod; mx′.p, palp of maxillule; mxp, maxilliped; per 5, 5th pereiopod; phos, phosphorescent organ; pleop, pleopod; protop, protopodite; s 4, 4th row of setae; sc, antennary scale; t. telson; th.app, thoracic appendage.

crustaceans falling into one or two genera like *Anaspides*, found in clear fresh water in Australasia. They show the main features of the caridoid facies but have no carapace and appear never to have had one, whereas its absence in other eumalacostracan groups can be shown to be secondary.

Of the two large groups of eumalacostracan crustaceans the less advanced is the Peracarida (pouched shrimps), so-called because the females all possess a brood pouch, fashioned from leaflike outgrowths of thoracic legs known as oostegites, within which the fertilized eggs develop until they are miniature adults. This has allowed some isopods to emancipate themselves from an aquatic environment and become terrestrial, though weaknesses in adult physiology restrict them to humid places. There are five subdivisions of Peracarida: (1) Mysidacea (opossum shrimps); (2) Cumacea; (3) Tanaidacea; (4) Isopoda (woodlice, slaters and many marine forms without popular names); (5) Amphipoda (freshwater shrimps, marine sandhoppers and similar animals). The mysids (Fig. 121) are the most primitive and exhibit a little changed caridoid facies; cumaceans are a small group adapted for burrowing in sand and specialized for such a life by modification of carapace and gills; tanaidaceans, also few in number, burrow or live in tubes and show some reduction of the carapace and a diversification of the thoracic limbs. These trends are continued into the remaining peracaridan groups within which most species fall. Neither isopods (Fig. 122) nor amphipods (Fig. 123) have a carapace and, as their names suggest, the thoracic legs of isopods are all alike but they are split into two series in amphipods, the first five pairs being involved in feeding and the last three in the process of creeping on the side of the body which is typical of these animals. The same specialization of function also affects the abdominal appendages; those of the first three segments are respiratory pleopods whereas those of the hinder three form a hopping apparatus.

The order Eucarida includes the animals accepted as the highest in crustacean evolution. They differ from peracaridans in having no oostegites and in having the carapace fused to all thoracic segments. There are two divisions, the Euphausiacea and the Decapoda. The former (Fig. 120D) are superficially like mysids and were, indeed, long classified with them, but are clearly eucaridan, not peracaridan, the likeness merely reflecting the pelagic habit of both. They include the animals called "krill" upon which baleen whales feed. In decapods there are two main forms of body, the macruran and the brachyuran, exemplified by the prawn and the crab. The macruran types have an elongated abdomen projecting behind the cephalothorax and carrying swimming appendages. Some are pelagic and free-swimming (shrimps and prawns); others are benthic, walking over the substratum in search of food (crayfish, lobsters). In brachyurans (crabs) the abdomen is reduced and is tucked under the carapace and has lost all locomotor importance. The cephalothorax broadens sideways and becomes laterally streamlined, this being an adaptation to movement either to left or right. One of the advantages of this is that

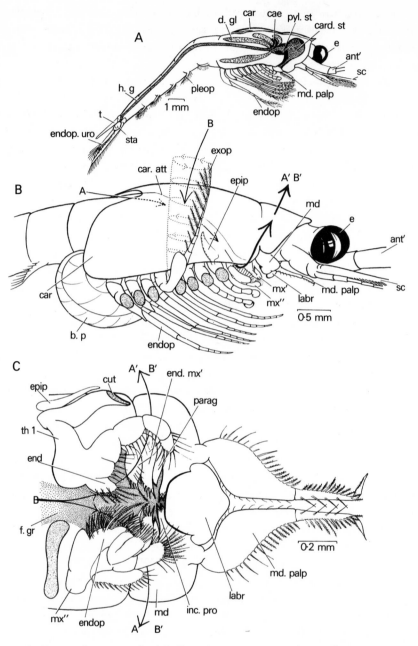

Fig. 121. A, *Praunus flexuosus*, sagittal half to show the course of the alimentary canal. The maxillae and exopods of the thoracic limbs are not shown. B and C, *Hemimysis lamornae*: B, right lateral view of cephalothorax. The exopodites of all thoracic legs have been removed except that of the fifth, the vortical movement of which is indicated by dotted lines. C, ventral view of oral region to show the structures concerned with feeding. The right 2nd maxilla has

it abolishes interference between neighbouring legs on one side during walking. This body form has been particularly successful, has been evolved in different decapod stocks and has enabled crabs to exploit many ecological niches.

Feeding of crustaceans

It is generally assumed that early crustaceans were rather small animals creeping over a soft substratum, much, perhaps, as *Hutchinsoniella* does today. Its trunk appendages were responsible for movement but also, by the action of their gnatho-bases (Fig. 105, end) passed detrital material forward to the mouth. The material came from the substratum, disturbed by the moving legs, and also from sweeping movements over the surface executed by the antennae.

From some such mechanism may be derived the locomotor and feeding mechanisms of many of the lower crustacean groups. One trend seems to have occurred often – that towards a free-swimming life and a filtratory mode of feeding. Since any filter-feeding device requires a mechanism for creating a current and another for straining particles from it, these crustaceans often seem similar but close study shows fundamental differences and compels the conclusion that most have evolved independently. In animals which live in open waters the ancestral bottom feeding is often lost, but there are frequent instances (*Chirocephalus*, mysids) where the animal has the option of using both methods. Adoption of a pelagic life leads to a general lightness of body with thin, unpigmented, often transparent exoskeleton and flesh and an exaggeration of appendages used in swimming (anostracans, cladocerans, copepods, mysids, euphausiids and some decapods). Since small size is advantageous for this life some of these animals (cladocerans, copepods) have been regarded as neotenous. Crustaceans tend to start development with few segments and add others later so neoteny is one way in which a short body can be attained by descendants of long-bodied ancestors. The basic features of filter-feeding crustaceans are summarized in Table 4. The details are dealt with in later pages, but one main point may be made now. In primitive crustaceans, whether detritivorous or filter-feeding,

been removed and the exopodite of the first right thoracic appendage; the left first thoracic appendage (maxilliped) has also been cut away. Arrows indicate (A, A′) the course of the respiratory current and (B, B′) the feeding current. (B, C modified from Cannon and Manton.) ant′, antennule; b.p, brood pouch; cae, dorsal caecum; car, carapace; car.att, line of attachment of carapace; card.st, cardiac stomach; cut, cut surface of 1st thoracic exopodite; d.gl, digestive gland; e, eye; end, endite; end.mx′, endite of maxillule; endop, endopodite; endop.uro, endopodite of uropod; epip, epipodite of 1st thoracic limb; exop, exopodite; f.gr, food groove; h.g, hind-gut; inc.pro, incisor process of mandible; labr, labrum; md, mandible; md.palp, mandibular palp; mx′, maxillule; mx′′, maxilla; parag, paragnath; pleop, pleopod; pyl.st, pyloric stomach; sc, antennal scale; sta, statocyst; t, telson; th 1, first thoracic limb, maxilliped.

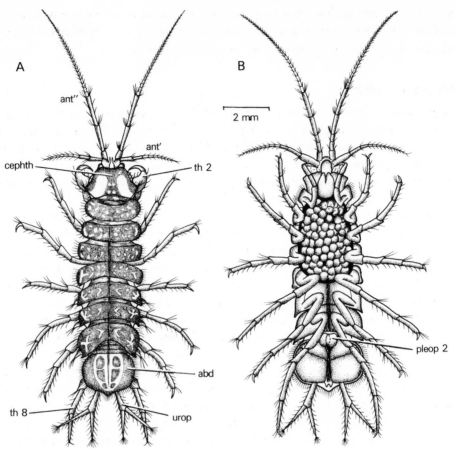

Fig. 122. *Asellus aquaticus*: A, male in dorsal view; B, female in ventral view, showing eggs developing in brood pouch. abd, fused segments of abdomen; ant', antennule; ant'', antenna; cephth, cephalothorax; pleop 2, second pair of pleopods (first pair absent in female); th 2, second thoracic appendage = first walking leg (th 1 forms maxilliped); th 8, eighth thoracic appendage = seventh walking leg; urop, uropod = sixth abdominal appendage.

the trunk limbs are responsible both for feeding and locomotion (sometimes aided by antennae), a compromise situation which limits their adaptability if not their efficiency. This occurs in branchiopods. Copepods and the eumalacostracans have escaped from this condition, however, by substituting a single maxillary filter for a series involving all the trunk limbs, and have thus set the latter free for other functions. Cirripedes have escaped in another way since by becoming sessile they do not use their limbs for any locomotory activity.

A second major evolutionary trend in crustaceans has been towards bottom dwelling, accompanied by an increase in weight achieved by thickening the cuticle and augmenting its degree of calcification, a step which also improves the efficiency

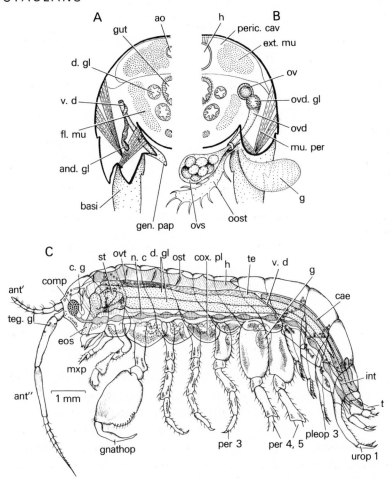

Fig. 123. A, half transverse section through 8th thoracic segment of male *Orchestia gammarella*. B, half transverse section through 6th thoracic segment of female gammarid. (After Charniaux-Cotten.) C, *Orchestia cavimana* male, from left. (After Nebeski.) and.gl, androgenic gland; ant′, antennule; ant″, antenna; ao, aorta; basi, basipodite, 7th pereiopod; cae, posterior mid-gut diverticulum; c.g, cerebral ganglion; comp, compound eye; cox.pl, coxal plate; d.gl, tubule of digestive gland; ext.mu, extensor muscles; fl.mu, flexor muscles; g, gill; gen.pap, genital papilla; gnathop, gnathopod; gut, gut; h, heart; int, intestine; mu.per, muscles of pereiopod; mxp, maxilliped; n.c, nerve cord; oes, oesophagus; oost, oostegite; ost, ostium; ov, ovum in ovary; ovd, oviduct; ovd.gl, oviducal gland; ovs, ovisac; ovt, abortive ova in testis; per, pereiopod; per 4, 5, legs elongated for hopping (cut short); peric.cav, pericardial cavity; pleop, pleopod; st, stomach; t, telson; te, testis; teg.gl, tegumentary glands; urop, uropod; v.d, vas deferens.

TABLE 4

Animal	Current maker	Direction	Sieve
Cephalocarida (*Hutchinsoniella*)	trunk limbs	P → A	trunk limbs
Anostraca (*Chirocephalus*)	trunk limbs	P → A	trunk limbs
Diplostraca	trunk limbs	P → A	trunk limbs
Copepoda	ant', ant'', md, mx'	A → P	mx''
Cirripedia	thoracic limbs	P → A	thoracic limbs
Leptostraca (*Nebalia*)	thoracic limbs	A → P	thoracic limbs
Leptostraca (other genera)	thoracic limbs	A → P	mx''
Mysidacea (*Hemimysis*)	thoracic limbs	A → P / P → A	mx''
Euphausiids	thoracic limbs mx''	P → A	mx''
Porcellana longicornis	mxped 3		mxped 2
Hapalocarcinus marsupialis			mxped 1, 2
Upogebia	abdominal pleopods	P → A	per 1, 2, mxpeds
Callianassa	abdonimal pleopods	P → A	per 1, 2, mxpeds

of the limbs as walking legs. Since this step cannot be taken whilst the legs are used in feeding it implies the previous substitution of a macrophagous for a microphagous feeding mechanism. In the small, delicately built pelagic crustacean respiratory exchange presents little difficulty: in the more heavily armoured bottom dweller this is no longer true and branchial outgrowths become of increasing importance.

The notostracan branchiopods present, after *Hutchinsoniella*, one of the more primitive feeding mechanisms. This will be described first and then those of branchiopods which have become filter feeders. The notostracan Apus (*Triops cancriformis*), sometimes called a tadpole shrimp (Fig. 115A), lives in temporary freshwater pools, spending much of its time crawling over and grubbing vigorously in mud, or it will leave this to swim through the water. The broad head shield, bearing eyes mid-dorsally, and the carapace (car) arch over the entire body except for the limbless segments of the trunk and the jointed caudal rami (c.r.). Head shield and carapace cover the appendages except for the tips of the first (sen.end) and perhaps the shorter second pair (end) which have long slender endites, jointed in the first appendage and toothed in the second. The limbs are numerous and exhibit a marked metachronal rhythm producing a continual suction of water from in front and above into the interlimb spaces and an outflow laterally between the exites. The limbs are phyllopodia, broad, flexible and weakly jointed, and the precise arrangement of lobes varies with the manner of manipulating the food. The first eleven pairs occur one per segment, but the posterior segments have up to five pairs each. When the animal swims 6–7 metachronal waves can be seen passing forward over the long series. At 19°C the rate of beat of an appendage is about 170 to the minute, which can be calculated as a quarter of a million beats per day.

The anterior phyllopodia strain the continuous stream of water entering the interlimb spaces. They have no filter plates for retaining fine particles but coarser food, sucked towards the ventral surface of the body, is prevented from escaping into the interlimb spaces by spines on the endites. These lie on the anterior and posterior surfaces of each endite, and project obliquely towards the middle line so that the posterior ones touch the anterior row on the endite of the limb behind. Large particles may be actually grasped by the endites of these and more posterior limbs and prevented from escaping. The spines on the endites of the 12th limb and those posterior to it are stronger, to triturate larger pieces of food. These limbs are set close together ventrally, which helps them to function successfully in this way.

The basal endite of every limb is a gnathobase (gnatho) armed with stout spines projecting obliquely forwards and each overlaps the gnathobase in front. As each limb beats backwards its gnathobase moves forwards pushing any material in contact with it on to the gnathobase in front. There is thus a direct transference of food from gnathobase to gnathobase forwards towards the maxillae. In contrast to the gnathobases the distal endites of the limbs push material on to the limbs behind, and at the same time it moves closer to the ventral surface of the body and is broken up. If the material breaks up readily this happens over a distance of a few segments; if it is tougher it takes a greater length, but the large number of posterior appendages provides a mill in which material can undergo a prolonged mastication. When it ultimately reaches gnathobasic level it reverses direction and moves adorally. The food consists of particles gathered from the water or broken by endites or posterior limbs.

Sometimes Apus seizes and fragments living prey such as chironomid larvae and ostracods, which are located by the long sensory first pair of trunk appendages and seized by the toothed endites of the second pair; dead animals are treated in a similar way. Other anterior trunk limbs curve towards the mouth and hold the food against the mouth parts. Their endites bite into it, as do those of the maxillules (mx′), which therefore functionally resemble those of a malacostracan rather than a branchiopod in that their distal parts can bite together as jaws. The proximal parts push food on to the mandible (md).

A method of straining fine particles from a water current occurs in the other branchiopod groups. It will first be described in the Anostraca, an order which includes two familiar genera of temporary waters, *Chirocephalus* in fresh water (Fig. 108) and *Artemia* in highly saline pools. The body is elongated, though little more than two or three centimetres in length, there is no carapace and there are eleven pairs of trunk limbs, all similar. They border a groove, absent in notostracans, which runs along the mid-ventral line to the mouth where it is overhung by a large upper lip (labr) with glands. The entrance to each interlimb space is through a sieve of setae (s) which filters the water current and retains the filtrate, the food, in the ventral groove between the two limb rows. The food is propelled forwards by an adoral current along this groove and entangled in secretion from its walls and from the labral glands. The food string so formed is manipulated by the minute maxillae and maxillules and directed into the mouth by the mandibles (md). The adoral water stream passes out on each side at the level of the maxilla and joins the lateral swimming current. If suspended particles are few the animals will swim to the bottom and stir up detrital material. *Chirocephalus* may also catch large organisms on the setae of the endites of the thoracic legs and manipulate them towards the mouth where they are dealt with by the maxillae and mandibles.

The feeding mechanism is dependent upon directing the water current created by the limbs through well-defined channels which are regulated by the flexible lobes on the phyllopodia. It has been worked out in detail for *Chirocephalus*. Each phyllo-podium has a series of backwardly projecting endites; the proximal one (prox.end) is very much longer than the rest and bears the filtratory setae (s). The endopodite (endop) forms the distal part of the main axis and, like the endites, flexes backwards against the resistance of the water on the forward beat of the limb. On the outer edge of the limb is a series of exites (ex) which similarly project backwards; the whole limb is therefore concave posteriorly. The most distal exite, the exopodite (exop), is hinged at its base and fringed distally with long setae, whilst the other lobes, which are of importance in respiration, lack setae and have a thin cuticle. The metachronal oscillation of the limbs implies that the space between any two undergoes rhythmical increase and decrease in volume. On forward movement the interlimb space is enlarged and acts as a suction chamber. The backwardly trailing exites and endites form the side walls of the enlarging space which is covered distally by the endopodite, and its floor is the body wall. All entrances to it are blocked by flaps of tissue except that along the median wall, which is spanned only by the filtratory setae of the basal endite, and water therefore enters from the mid-ventral channel through this filter plate. On the backward or recovery stroke the whole limb moves as a rigid plate, the interlimb space is reduced in volume and the water escapes between the exites and flows away posterolaterally (Fig. 108B, left side). When any one limb is near the completion of its backstroke the limb immediately behind commences to move forwards. The two, moving momentarily towards one another, force out the remainder of the water, some laterally, but also some medially into the food groove. This is said to produce the adoral current. As this current sweeps forward in a series of jerks it washes food off the filter setae. These are also cleaned by special setae on the basal endite of the next posterior limb, and on the wall of the food groove, which scrape particles off them as the limbs move to and fro.

In the animals so far dealt with movements of the trunk limbs simultaneously collect food, create a respiratory current and move the body through the water. In diplostracan branchiopods, exemplified by *Estheria* (Fig. 105D) and *Daphnia* (Fig. 117), the antennae (ant″), reduced in anostracans and vestigial in notostracans, become the primary locomotor organs and the trunk limbs are solely for feeding. This allows their enclosure in a bivalved carapace which becomes part of the feeding mechanism. The carapace (car) is open ventrally and posteriorly; in the conchostracan it also encloses the head, and its right and left halves are hinged dorsally and may be closed by contraction of an adductor muscle (add.mu). Water drawn in ventrally between the carapace folds by the beating of the trunk limbs is used solely for feeding and respiration, not for locomotion, and need be only small in amount and low in pace for this reason. As in *Chirocephalus*, food particles strained from it are directed along a ventral food channel to be entangled in labral secretion in the narrow space between the large labrum and body wall.

Estheria is a benthic form gathering its food from mud and swimming intermittently. It has 27 pairs of phyllopodia (phyllop), one pair per segment, set closer than those of an anostracan, so that each limb appears to push the limb in front forward as the wave of movement passes anteriorly. The series is divided functionally into two groups as in Apus, the anterior ones forming a filtratory system and the hind limbs grasping and cutting. All endites on the anterior limbs have filtratory setae, not, as in Anostraca, just the basal one; that forms a gnathobase sharply separated from the more distal endites. The setae on the distal ones retain particles in the food channel, whilst the actual collection for food is by the filter setae of the gnathobase. The particles are transferred to the food channel and propelled forwards by a combination of blowing and manipulation by special setae. The posterior limbs are modified for dealing with larger pieces of detritus. These are fragmented by their immense, armed gnathobases and then passed forward in the adoral current. The gnathobases have also brush setae which sweep the residue off the gnathobase behind and direct it forward along the food groove. If too great a quantity of detritus is likely to enter the carapace the valves are closed.

The Cladocera, of which *Daphnia* (Figs 117 and 124) will be taken as an example, have departed even further from the ancestral body plan. They may have arisen from an archaic bivalved branchiopod and have retained the swimming antennae, but an element of neoteny is probably involved since the number of trunk segments is considerably reduced; structural and functional differentiation of the limbs is very marked. Of the five pairs of trunk limbs which persist only two (3rd and 4th) have filter plates (f). These are carried on the enormous basal endite or gnathobase of each limb (Fig. 108B, f, right side; Fig. 124, end 3), the other endites having dwindled in size; and they span the entire space between the open edge of the carapace, at the free end of the appendage, and the food groove at its base. The setae hang posteriorly from the endite to touch the limb behind. *Daphnia* swims more or less continuously and strains off fine particles from the water through which it moves. There is no true metachronal rhythm in the limb beat. Water is sucked into the ventral opening of the carapace mainly by the beating of the 3rd and 4th pairs of limbs and thence into the interlimb spaces 3–4 and 4–5, the long feathered setae (Fig. 117, f.s) on the 1st and 2nd limbs preventing the ingress of large particles. On the backstroke of the two filtratory limbs some water from these spaces passes posterolaterally and leaves the body on either side of the posterior end of the body. The rest, however, is directed forward along the food groove by a pivoting action of the filtratory endites and also sucked forward by the forward stroke of the first limb (Fig. 124, limb 1) which is timed to coincide with the backstroke of limbs

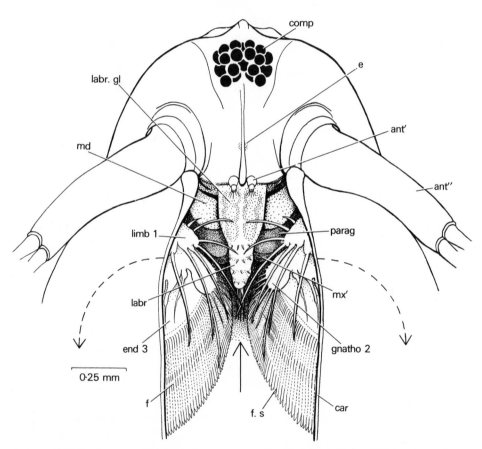

Fig. 124. *Daphnia magna*: ventral view of anterior part of body. Median arrow indicates flow of water along food groove and broken arrows the exhalant flow passing under the carapace and over the epipodites of the limbs. ant′, antennule; ant″, base of antenna; car, edge of carapace; comp, compound eye; e, median eye; end 3, endite of 3rd trunk limb; f, filter plate; f.s, filter setae; gnatho 2, gnathobase of 2nd trunk limb; labr, labrum; labr.gl, labral glands; limb 1, 1st trunk limb; md, mandible; mx′, maxillule; parag, paragnath (lower lip).

3 and 4. This stream passes laterally into interlimb space 1–2 and then posteriorly over the thin-walled epipodites (branchiae) on either side (Fig. 117, epip). The fifth limb swings forwards as the fourth is finishing its backstroke and helps to force water out of the interlimb space between them. The basal endite of the 2nd limb forms a gnathobase with modified filter setae (Fig. 124, gnatho 2) which (on the downthrust of the limb) push food particles towards the floor of the food groove, sweeping them out of the main stream of water. In this region the groove is deep. Anteriorly it is blocked by the paragnaths (parag), maxillules (mx′) and large labrum (labr), secretion from which (labr.gl) entangles the food so that it can be more easily manipulated by the maxillules and then the mandibles which direct it to the mouth.

The only other crustaceans in which the trunk limbs of the adult are used solely for collecting food from a water current, and are not concerned in locomotion, are the barnacles (Cirripedia). The mechanism by which this is done is quite different from that employed by any other group, partly because these animals are fixed to one spot and without power of locomotion. The body is enclosed in a fold of body wall, the carapace or mantle (Fig. 119A, car), and only the trunk appendages (app 1, 6) and penis (pe) can extend outside the cavity; the abdomen is reduced and has no appendages. Practically the only visible activity of the body is the sweeping of the feathery appendages through the water as a casting net. These are biramous legs known as cirri, each with two multi-articulate rami beset with setae. They form a series graded in length with the most anterior the shortest. When feeding starts the limbs lie curved with the concave surface facing the mouth; in this state they are thrust out of the opercular opening of the mantle cavity, then straighten and separate widely from one another like the fingers of a hand, swing forwards and are withdrawn again with a sort of grasping motion. Such beating movements may be continued without stopping so long as the barnacle is under water, particles being gathered by the setae which scrape one another clean as the cirri curl up with their tips, and consequently their effective beat, directed accurately towards the mouth. The legs are pulled inwards by muscular action but there are no muscles responsible for their extension, which is brought about by pumping blood into them.

The sessile barnacles, in which the body is more completely enclosed within the mantle space than in the stalked forms, show a coordinated set of rhythmical movements which pump a respiratory current through the mantle cavity while simultaneously allowing the longer cirri to capture food. The common barnacles of rocky shores (*Balanus, Elminius, Chthamalus*) display a graded series of movements. Sometimes there is a gentle rhythmical oscillation of the cirri within the mantle cavity just strong enough to maintain a flow of water adequate for respiration. This may be followed by the typical beat used to capture food particles, which, with increased vigour, may culminate in strong contractions of the whole cirral net from the completely extended state. In this last type of movement the extended part of the body – head and thorax – may pivot around its point of attachment to the shell and so sample different parts of the environment.

The barnacles around our coasts feed on a wide range of organisms. The setae on the last four pairs of trunk limbs (3–6) form a coarse net but this is incapable of entrapping particles as fine as the algae and bacteria which are found in the gut. Perhaps these are collected by the first two pairs of cirri and the mouth parts, which have the setae more closely set. Observations on *Balanus* show that if one of the four pairs of posterior cirri touches a food particle of large size it will bend immediately towards the mouth where anterior appendages scrape off the particle and pass it to the mouth parts.

The use of biramous appendages as a sweep-net is the earliest method of feeding in many species not just in barnacles, for the nauplius larva (Fig. 125A, C) feeds in this way. This type of larva occurs in all subclasses of the Crustacea, though the nauplius of some copepods, penaeid prawns and euphausiids does not feed. Swimming is effected by a simple paddle stroke of the appendages, especially the antennae (ant″) which have the most extensive movement (the antennules (ant′) are mainly for balancing). They are flexed on the forward stroke and rigidly extended on the effective backward stroke when they converge on the hinder part of the body, sweeping particles in the same direction. The particles are collected in a sweep-net formed by the ventral setae of the antennae in branchiopods or the mandibles in copepods and barnacles (s). In the ensuing forward stroke of the limbs they are carried to the mouth, which is overhung by a large labrum (labr), by two groups of short, stout setae (gnatho), one on the base of the mandibles and the other on the antennae. In nauplii which do not feed these setae are not developed.

A number of swimming, filter-feeding crustaceans possess only one pair of appendages which carry filter plates. These are situated far forwards on the maxillae, and the filters (with their elaborate setal armature) are more efficient than the others which have been described. They are present in some copepods and in some of the eumalacostracans which display most fully the caridoid facies, the most important being mysids and euphausiids. In the Eumalacostraca water is sucked through the filter by movement of the maxilla itself, and the water which is filtered is supplied by a continuous current directed anteriorly towards the chamber by the exopods of the thoracic legs. In contrast to this the maxilla (Fig. 118A, mx″) of the copepod *Calanus* is stationary during feeding and takes no part in drawing water through the filter it provides. Moreover the feeding current results from the rotary and propeller-like movements of appendages anterior to the maxilla – the antennae (ant″), mandibular palps (md. palp) and maxillules (mx′).

The abundance of marine copepods, mysids and euphausiids and their importance as food for fish and whales suggests that this method of obtaining food has been a success. They are not entirely dependent on it, however, for they can seize and deal with large food masses, though the extent to which this is done varies; it is

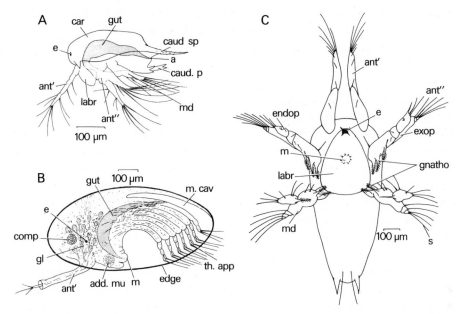

Fig. 125. *Balanus*: A, nauplius larva (4th stage); and B, cypris larva in lateral view. C, *Calanus*: 3rd nauplius in ventral view, limbs in resting position. a, anus; add.mu, adductor muscle; ant', antennule; ant'', antenna; car, carapace; caud.p, caudal process; caud.sp, caudal spine; comp, compound eye; e, nauplius eye; edge, ventral edge of carapace; endop, endopod of antenna; exop, exopod of antenna; gl, gland of antennule; gnatho, gnathobase; labr, labrum; m, mouth (under labrum in C); m.cav, mantle cavity; md, mandible; s, seta; th.app, thoracic appendage.

comparatively rare in the copepods and frequent in some mysids and euphausiids. The copepod *Calanus* is often seen to move through the water by rapid, sudden jerks resulting from the activity of the coupled trunk limbs. This kind of movement is a means of escape and not concerned with feeding. When the animal is undisturbed the antennules (ant') project laterally to act as balancers and a comparatively slow forward movement is maintained by rapid vibrations (about 600 per minute) of antennae (ant''), mandibular palps (md.palp) and maxillules (mx'). The result is a stream of water passing through the maxillary filter directed partly by the maxillipeds (mxp) and by the sucking action of the maxillulary exites. Whether this mechanism would provide sufficient food for an animal larger than *Calanus* is doubtful.

The method of filter feeding of mysids and euphausiids (for example *Euphausia superba* and *Meganyctiphanes norvegicus*) is essentially similar and is associated with their swimming movements; it involves the thoracic limbs in each group. The details have been studied in *Hemimysis lamornae* (Fig. 121), which lives on or near the bottom of shallow coastal waters and rises to higher levels at night. It swims in a horizontal position by the movements of the eight pairs of thoracic exopods (exop) which at the same time set up a feeding current. Each exopodite rotates, drawing a current of water up the axis of rotation (B) which then moves forwards along a ventral food groove (f.gr), between the bases of the thoracic limbs. The vibration of the maxilla (mx″) sucks the stream forwards, and the suction is aided by the movement of an exhalant respiratory current (AA′) produced by the epipodite of the first limb (epip) and flowing between the carapace fold and body on each side. Anteriorly this respiratory current is joined by the food stream (BB′) which has meanwhile passed through a filter formed by setae on the basal endites of the maxilla (endop). Particles retained by these, and those in the anterior part of the food groove, are scraped off by setae on the other mouths parts (end. mx′) and carried to the mandible (md). Setae on the basal endite of the first trunk limb (end) are used to scrape the filter clean.

Both *Meganyctiphanes* and *Mysis* can increase the amount of suspended material for filtration by stirring up a soft substratum. They also collect larger food masses between the thoracic endopodites and manipulate it in a basket formed between these and the mouth parts. The larger kinds produce a current whilst swimming which is strong enough to draw animals such as *Sagitta* into the basket. Remains of crustaceans may be found in the gut, though sometimes the soft parts of animals such as copepods will be eaten and the chitinous remains cast away.

The essential part of the malacostracan filter-feeding mechanism is the maxillary filter, which may have come into being to strengthen the forward flow in the ventral groove of an animal like an anostracan. When developed sufficiently it could functionally replace the trunk limbs and release them for other activities. In mysids their association with feeding could be secondary. One malacostracan, however, *Nebalia bipes*, looks, superficially, as if it had a feeding mechanism like that of a branchiopod. This, however, is not so: its limbs are not phyllopodia but modified biramous limbs and the flow of water is anteroposterior. Since its maxilla is small and not involved in filtration it must be assumed that this is a modification for a particular mode of life (probing into mud) of a feeding method originally like that of a mysid. *Nebalia bipes* (Fig. 120) is a mud-living species found under stones around our coasts. The eight pairs of thoracic limbs (th.app) are foliaceous, each with an epipodite which acts as gill (epip). They lie in a deep, bivalved carapace (car), which can be closed by an adductor muscle (add.mu), together with the first four pairs of large, biramous, abdominal appendages (abd.app) which are locomotory. When the thoracic limbs are active they move backwards and forwards out of phase, with the exopodite and epipodite (exop, epip) acting as valves in directing the water flow from the chamber between each two successive limbs, which expands and contracts in rhythmical fashion. The water which enters is filtered through setae on the protopodite and endopodite (protop, endop), the filtrate is brushed off the collecting setae and swept by still other setae towards the mid-ventral food groove along which which stout setae on the protopodite push the collection towards the mouth. Any similarity with the branchiopod stops here, for further details of the feeding mechanism reveal important differences. In *Nebalia* it is the backstroke and not the forestroke of the limb which enlarges the filter chamber. The food streams enter anteriorly and ventrally, its volume controlled by the degree of elevation of a movable rostrum, and makes its exit at the posterior end of the carapace laterally, the anterior trunk limbs being the main inhalant pumps (their exopodites and epipodites directing the water out posteriorly and preventing it passing forwards). The setae on the median part of the endopodite and protopodite form the filtering walls. They are arranged in four longitudinal rows. Of these the first and third are long and hooked and interlock with those of neighbouring limbs so that a continuous filter wall is maintained even when the limbs move apart and the interlimb space enlarges. The fourth row (s 4) comprises comb setae which clean the filter walls, and the second a row of brush setae (br.s) sweeping the food towards the food groove. The setae at the proximal end of the first row are not hooked and they direct the food anteriorly. The mouth parts both structurally and functionally resemble those of a mysid, though the mandible is reduced.

The more advanced peracaridans (isopods and amphipods) and eucarids (decapods) are specialized for macrophagy, the thoracic limbs having evolved a variety of distal grasping and biting structures. Interlimb suction chambers are absent and without the need to form valves to close these, the limbs become slender. Their ability to deal successfully with larger food may also be associated with the larger size which some crabs and lobsters attain. A few, however, are microphagous, totally or partly, though retaining the equipment of tools needed to collect and subdivide large scraps. They live in silty or muddy places where detritus is plentiful, collecting this from the substratum or filtering particles in suspension in the water. Their methods of doing so are novel and not a persistence of the ancestral filter-feeding devices. The squat lobster, *Galathea*, feeds on bottom detritus and is also raptorial, seizing food with the chelae and maxillipeds. More usually the dense tufts of setae on the endopodite of the 3rd maxillipeds sweep over the substratum loosening and collecting fine particles in front or under the body. The limb bends towards the mouth and the setae are then swept clean by tufts of setae on the 2nd maxillipeds. The food is passed to the anterior mouth parts where some sorting may take place. The related *Porcellana longicornis* (Fig. 126A) may be seen with one 3rd maxilliped outstretched (endop.mxp) so that its long bipectinate setae form a spoon-shaped net to filter the water; the filtrate is collected by the 2nd maxilliped (mxp) when the limb is flexed. The hermit crab *Eupagurus bernhardus* resembles *Galathea* in using the 3rd maxilliped to collect sediment but it also scrapes up food with the smaller claw and passes it to the 3rd maxilliped. This crab and other hermits feed more commonly on larger food collected by the small chela. *Hapalocarcinus*, a crab which lives in galls in corals – chambers connected by small openings to the outside which it has induced in the coral skeleton – projects its maxillipeds through the opening to fish for plankton.

The burrowing prawn-like decapod *Upogebia* lives in burrows in sand or mud, and in contrast to the hermit crab its abdomen is large and muscular, with well-developed pleopods. These beat continuously to maintain a current of water from which detritus is collected by the fringed setae on the first two pairs of thoracic legs. They form a collecting basket which is swept out by the maxillipeds. Fragments of food are also picked up from the bottom and passed to the mouth parts for manipulation.

A second malacostracan living in mud produces and filters a water current in a similar manner and is, in addition, a selective deposit feeder; indeed, it obtains most of its food by this second method. This is the amphipod *Corophium volutator* which shows peculiarities in its structure associated with its way of life (Fig. 120C). It is abundant in certain estuarine mud flats where it may be found in burrows near the surface or crawling over the surface, selecting organic particles with the gnathopods

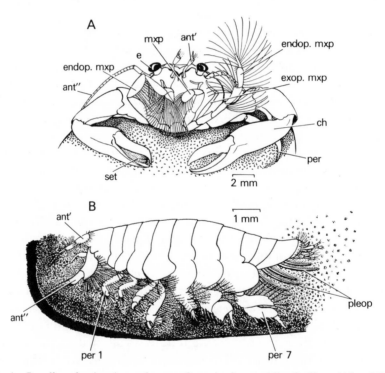

Fig. 126. A, *Porcellana longicornis*: male seen from in front whilst feeding. (After Nicol.) B, *Haustorius arenarius*: burrowing through sand. (After Dennell.) ant', antennule; ant'', antenna; ch, cheliped; e, eye; endop.mxp, endopodite, maxilliped 3; exop.mxp, exopodite, maxilliped 3; mxp, maxilliped 2; per 1, 7, 1st and 7th pereiopods; per, pereiopod; pleop, pleopods; set, setose pad.

(gnathop 1,2), especially the first pair. The food may be conveyed directly to the incisor processes of the mandibles and agglutinated by secretion from the oesophagus. Much time is spent in a U-shaped burrow with two openings, which extends an inch or two down from the surface. The amphipod excavates this with the large antennae (ant″) and first two pairs of pereiopods, purchase being obtained during digging by the long, specialized 5th pereiopod (per 5). When the animal is in its burrow a current of water is drawn in by the beating of the pleopods (pleop). This carries only the finest particles, which are filtered through a net of setae fringing the 1st and 2nd gnathopods. The collected food is passed by the maxillipeds to the mandibles.

The majority of Malacostraca feed on weeds, decaying vegetation, dead or disintegrating animals and a few are predators. Amphipods and isopods are vegetarians or omnivorous and, with few exceptions amongst the amphipods, eat only large fragments. A brief description of the feeding mechanism of an isopod, a most highly evolved peracaridan, will indicate how this is done. *Idotea emarginata*, a shallow water species, feeds on seaweed. This is held by the first thoracic legs against the mouth (Fig. 127E, m) which is delimited by the labrum (labr) in front, a mandible (mol.pro) on either side and the bilobed paragnath (parag) posteriorly; maxillules (mx′), maxillae (mx″) and maxillipeds (mxp) lie below with their tips directed forwards. The incisor process (inc.pro) and lacinia mobilis (lac.mob) of each mandible bite the weed whilst spines on the maxillules and maxillipeds (scr.sp) move over it and rasp the surface. The loosened pieces are pushed on to the molar processes of the mandible for chewing. They are directed there by the spines on the mandible (sp.md) and similarly toothed spines on the maxillules (sp.mx′). The mouth parts are combed by spines on the palp of the maxilliped (palp) and on the first leg and fragments are also brushed forward by setae (br.se) on the endites of the maxillipeds, maxillae and maxillules. The maxillae are reduced in size and complexity as compared with those of *Mysis*, and are hardly sclerotized. The maxillipeds are plate-like and hooked together in the mid-line (h). When the animal is not feeding they shut off the oral region and its appendages.

Ligia oceanica, another marine isopod, lives higher on the shore, hiding in damp crevices and emerging at night to feed as the tide retreats. It is a scavenger, but prefers brown weed and is said to invade the *Fucus* zone to obtain this. The animal feeds vigorously in the limited time available, passing a continuous supply of weed through the gut and producing faeces with much undigested material.

Most decapods are well known as predators, though much of their food is often carrion or detritus rather than living prey, and some also take vegetable matter. They are not the only crustaceans which are carnivores, for the freshwater cladoceran *Leptodora* (Fig. 128D) feeds on copepods, some marine copepods devour fish larvae, caprellid amphipods eat hydroids and bryozoans and the amphipod *Phronima sedentaria* eats salps and then lives in the empty test. The shore crab, *Carcinus maenas*

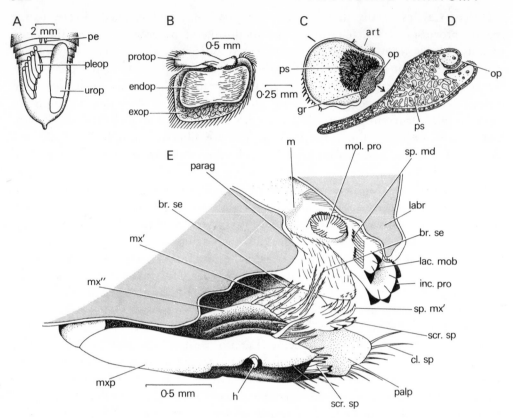

Fig. 127. A, ventral view of abdominal region of *Idotea* to show fusion of segments. On the right of the diagram the uropod (6th abdominal appendage) is shown; on the left it has been removed to expose the pleopods. B, *Ligia oceanica*, a pleopod. C, *Porcellio scaber*, exopod of first pleopod to show, by transparency, the bunch of pseudotracheae lying inside the appendage. D, a section of the same appendage along the line indicated. E, *Idotea emarginata*, part of sagittal half of head to show the position of the mouth parts. Stippled areas represent cut surfaces. (C and D, after Stoller; E after Naylor.) art, articulation between exopod and base of limb; br.se, brush setae; cl.sp, cleaning spines; endop, endopodite; exop, exopodite; gr, grooved area of surface; h, hook coupling left maxilliped to right; inc.pro, incisor process of mandible; labr, labrum; lac.mob, lacinia mobilis; m, mouth; mol.pro, molar process of mandible; mx′, maxillule; mx″, maxilla; mxp, maxilliped; op, external opening of pseudotracheae; palp, palp of maxilliped; parag, paragnath; pe, penis on last thoracic segments; pleop, pleopod; protop, protopodite; ps, pseudotracheae; scr.sp, scraping spines; sp.md, toothed spines on mandible; sp.mx′, toothed spines on maxillule, pushing food upwards; urop, uropod (6th abdominal appendage) forming cover over pleopods.

(Fig. 129A), seizes prey such as polychaete worms by the great chelae and passes it directly to the mandibles (md) between the 3rd maxillipeds (mxp 3) which move laterally to permit this. The food is gripped and held by the mandibles proximally, and distally as the 3rd maxillipeds come together again in the mid-line upon it. They then move outwards, stretching the food, which may break, and the part gripped by the mandibles is pushed into the mouth by the mandibular palps, the endites of the 1st maxillipeds and the maxillules. If the food is too tough to break in this way it is cut by movements of the two maxillipeds, which work up and down upon it, and the powerful maxillules. The maxillae apparently do nothing. Pieces which are severed are pushed into the mouth, for the food is not finely divided. Eventually the mouth parts are cleaned by one rubbing against the other and fragments swept away by currents set up by the exopodites of the maxillipeds. The mandibles, perhaps contrary to popular belief, neither chew nor crush the food unless it is extremely soft.

In the prawn *Leander* the 2nd maxillipeds are more important parts of the feeding mechanism. Their endopods grasp food collected by either the 1st and 2nd pereiopods, both chelate, or by the 3rd maxillipeds which are leg-like and will scrape it up from the bottom. Pieces are torn off by deeper mouth parts, if the food is large, or passed directly to the incisor process of the mandibles. These rotate inwards and upwards tucking it into the mouth and tearing it as it goes. The pieces then pass between the molar processes on their way to the oesophagus and are doubtless pounded by them. The mandibular palp, first maxilliped and maxilla appear to be of no importance in manipulating food; perhaps the sole function of the small, median lobes of the maxilla is in helping to regulate the movements of the scaphognathite, for they are in incessant motion. The exopods of the maxillipeds vibrate to maintain a strong forward current from the mouth which carries away water from the gill chamber, excreta from the green glands and rejected food particles.

Exoskeleton

The crustacean exoskeleton consists essentially of two layers, the epicuticle (Fig. 111D, epic) and endocuticle (endoc.) The endocuticle is a chitin-protein secreted by the epidermis and its thickness and composition vary considerably: it forms hard sclerites in some regions of the body, but retains its primitive elasticity in areas like the arthrodial membranes and the lining of the gut. The hardening is due to the deposition of calcium salts which, however, are not present in that part of the cuticle immediately overlying the epidermis, where some freedom of movement must be maintained. The chitin (B type) differs in its molecular pattern from that (A type) produced by coelenterates, annelids, molluscs and brachiopods. The extremely thin epicuticle consists of a lipid-impregnated protein and has no chitin.

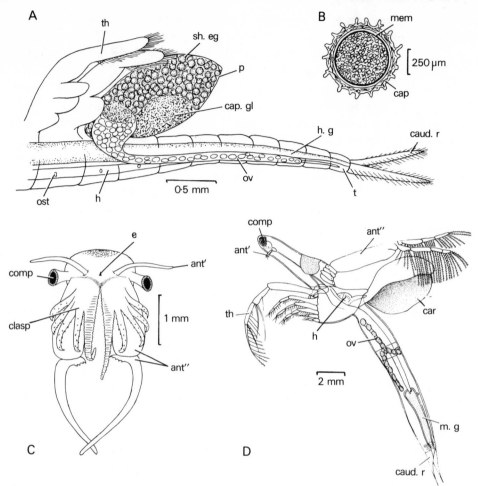

Fig. 128. A, B, C, *Chirocephalus diaphanus*: A, posterior part of body of female; B, section through shelled egg; C, frontal view of head of male to show clasping organs (modified antennae). D, *Leptodora kindti* female. ant′, antennule; ant″, antenna; cap, capsule; cap.gl, capsule gland; car, carapace (used as brood pouch); caud.r, caudal ramus; clasp, clasping organ; comp, compound eye; e, median eye; h, heart; h.g, hind-gut; mem, egg membrane; m.g, mid-gut; ost, ostium; ov, ovary; p, egg pouch; sh.eg, shelled egg; t, telson; th, thoracic limb.

It contains a polyphenol oxidase, as in insects, and undergoes hardening by quinone tanning. It is also calcified.

In most branchiopods the exoskeleton is relatively soft, for there is little or no calcification or tanning, and the shape of the body and limbs is maintained largely by the pressure of the body fluids. The transparency of the fairy shrimp, *Chirocephalus*, is due to the simple nature of the cuticle. By contrast, the weight of the exoskeleton of the benthic decapod counteracts the buoyancy of the rest of the body. Moreover the endocuticle is strongly pigmented and together with the chromatophores in the subepidermal layers, is mainly responsible for the characteristic colour of the animal.

The exoskeleton protects the living tissues and reduces permeability, and both functions are enhanced where calcification occurs. The epicuticle has no waxy layer which might prevent rapid transpiration in species coming on to land, as it has in terrestrial arthropods, but despite this the integument appears to be less permeable in more terrestrial forms. Increase in body size can occur only at ecdysis and results from a copious intake of water through the gut.

Ecdysis involves the liberation of a moulting fluid, containing a chitinase and alkaline phosphatase, which frees the old cuticle from the epidermis. This then secretes the new one which is thin and stretchable, consisting of an uncalcified endocuticle and an epicuticle. At this stage the lipid-containing epicuticle is the sole barrier against permeable substances and protects the endocuticle from the digestive attack of the moulting fluid. This softens the old cuticle which, in certain well-defined areas where there is a particularly marked local breakdown of the organic matrix, splits open and allows the body to swell. Not all the constituents of the old exoskeleton are lost, for some resorption occurs prior to secretion of the moulting fluid; the amount varies from species to species. *Carcinus* resorbs 79 per cent of the organic matter and 18 per cent of the inorganic, though only about 5 per cent of the calcium is saved. In other decapods (*Astacus*, *Homarus* and certain land crabs) appreciable amounts of calcium salts are resorbed and are concentrated in the gastric epithelium to form the familiar gastroliths.

Beneath the epidermis are multicellular glands (Fig. 111D, teg.gl) with narrow ducts which penetrate the cuticle and open at the surface. It has been suggested that these integumental glands secrete the epicuticle, though it is difficult to reconcile this function with their irregular distribution. Nevertheless their distribution may be related to its properties, for in decapods they are abundant in the regions marked by extensive tanning, such as the terminal joint of the walking leg of crabs. They are also plentiful in the gizzard. Histologically similar glands are known to have other functions both in the decapods and less advanced Crustacea. After each ecdysis they supply the cement for attaching sand grains (statoliths) to the sensory setae of the statocyst, and in female decapods, for attaching the eggs to the pleopods. In cirripedes their secretion attaches the barnacle to the substratum.

The exoskeleton produces setal outgrowths (s), some features of which have already been discussed, and it, and the epidermis, are modified to form components of luminescent organs. The photogenic cells of the superficial photophores of decapods are modified epidermal cells and the lens which overlies them is a thickening of the cuticle. Similarly the lens which caps each ommatidium of the compound eye is a transparent bit of exoskeleton.

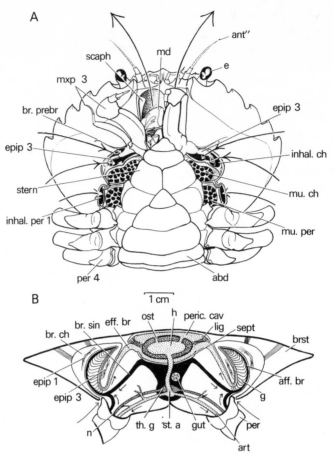

Fig. 129. *Carcinus maenas*: A, female in ventral view to show the anterior inhalant and the exhalant branchial openings and their relationships to some of the appendages. Chelipeds and first pereiopods removed. The right 3rd maxilliped is divaricated to show that this makes a double inhalant opening at the base of the cheliped. The right first and second maxillipeds have been cut off to expose the tip of the scaphognathite in the prebranchial chamber and the endopodite and endites of the maxilla lying over the mandible. Arrows indicate inhalant and exhalant water currents. B, diagrammatic transverse section through the thorax of a crab to show the blood system. Arrows indicate inhalant openings to branchial chambers and direction of blood flow. abd, abdomen; aff.br, afferent branchial vessel; ant'', antenna; art, artery to thoracic limb; br.ch, branchial chamber; br.prebr, boundary between branchial and pre-branchial chambers; br.sin, branchial vessel; brst, branchiostegite; e, eye; eff.br, efferent branchial vessel; epip 1, epipodite of 1st maxilliped; epip 3, epipodite of 3rd maxilliped; g, gill; gut, gut in sternal sinus; h, heart; inhal.ch, inhalant passage above cheliped (= Milne Edwards's opening); inhal.per 1, inhalant passage above 1st pereiopod; lig, suspensory ligament; md, mandible; mu.ch, muscles of cheliped; mu.per, muscles of 1st pereiopod; mxp 3, 3rd maxilliped; n, nerve to thoracic limb; ost, ostium; per, pereiopod; peric.cav, pericardial cavity; scaph, anterior tip of scaphognathite; sept, pericardial septum; st.a, sternal artery; stern, thoracic sternite; th.g, thoracic ganglion.

Locomotion of crustaceans

It is thought that although the ancestral crustacean was small it had many segments and possessed flattened, biramous, paddle-like trunk limbs which were more ventrally placed than those of annelids and were moved forward and backward by different means. When an annelid or arthropod creeps or walks the appendages move in a stepping action and their forward motion is most effective when they are lifted clear of the ground. In errant polychaetes the bulk of the movement of the parapodium is due to contraction of longitudinal trunk muscles and is therefore necessarily associated with the undulations which they cause. Thus energy is wasted in moving the body in a transverse plane. The position of the parapodium relative to the body may also be changed by muscles attached to an aciculum, a supporting rod resisting any shortening of the limb which would bring it off the ground at each forward step. This lifting of the appendages, however, is achieved in Peripatus, crustaceans and the higher arthropods, where propulsive force is provided by extrinsic limb muscles. It is effected by a telescopic movement of the limb in Peripatus brought about by haemocoelic fluid and by that and a flexing of the joints in other arthropods. In these animals lateral undulations of the body are no longer part of the locomotory mechanism, and the alternate phasing of the movement of the appendages of any one segment is therefore lost and the paired appendages may move together. This advanced type of movement is seen in the swimming of branchiopods, in the coupled, biramous feet of copepods, which enable an animal 1 mm long to dart through the water at a rate of 20 cm s^{-1}, and in most swimming crustaceans.

The primitive limb may have been a biramous paddle similar to the posterior limbs of *Hutchinsoniella* (Fig. 105). These limbs have on the protopodite median and lateral lobes (endites and exites) which are all approximately similar, except on the anterior limbs of the fossil branchiopod *Lepidocaris*, which resemble the phyllopodia of modern branchiopods in that one lobe, the basal endite, is larger than the rest and presumably forms a gnathobase for manipulating food The antennae of *Hutchinsoniella* (ant', ant") are well developed and perhaps aid the rhythmical beating of the trunk limbs. Indeed, antennae may have been one of the primitive locomotor organs of the Crustacea, for this method of swimming is widespread in modern members of the group which are of small size, such as nauplius larvae and adult diplostracans (Fig. 105D).

The phyllopodia of modern branchiopods (Fig. 108) have presumably been derived from the ancestral type by exaggeration of exites and endites and by having the cuticle everywhere flexible and movement not restricted to joints. They beat in a characteristic metachronal rhythm transmitted posteriorly through the central nervous system and are used for swimming, respiration and food-collecting. The exites of the anostracan limb, such as that of *Chirocephalus*, can alter the phase of beat and also move in a rotatory fashion so that the body may be propelled forward, or move slowly backward, or rise or fall through the water. It is a type of limb with apparently little possibility of evolutionary change as compared with the biramous type, the stenopodium, possessed by the majority of crustaceans. This is a jointed system of tubes moved by internal muscle bundles, some running entirely within the limb, others passing into the body. Each muscle has its origin on an area of exoskeleton and typically inserts on a more distal segment so that on contraction the joint between the two segments is affected. The degree of freedom of movement at each joint along the limb is determined by the physical properties of the exo-skeleton, its local configuration and the special properties of the associated muscles. Many joints allow movement only in one plane and then two antagonistic muscles can regulate every possible position. Others allow movement in more than one direc-tion, even a rotation, and this may involve extra muscles with special properties. To reduce the neuromuscular complexity hydrostatic devices are often used: flexing may be due to muscle but extension to distending the limb with blood. This mechanism may derive from days when arthropods had not yet acquired a continuous series of sclerites.

Primitively, as in copepods, the biramous limb is used as a swimming paddle. Its modifications in association with other types of locomotion, which permitted the evolution of a larger body form, can be traced in the Malacostraca. In the promalacostracan the transition from thorax to abdomen was probably gradual and the number of segments and appendages may have been larger than in modern forms. Adaptation to a life in which creeping over a solid substratum alternated with periods of swimming led to the evolution of a sharp differentiation between an anterior ambulatory thorax and a posterior natatory abdomen. If the gait of the promalacostracan was sinuous like that of a centipede, the legs would tend to get in one another's way. This has been avoided by reducing the number by transforming some into maxillipeds, by differential lengthening, by approximating their bases by shortening and broadening and, in extreme cases, fusing the segments which bear them, or by all these devices simultaneously. The result is a fan-like arrangement. These changes, however, affected the anterior part of the body only, the posterior part retaining more primitive features and unspecialized limbs.

In the more primitive Eumalacostraca, like the syncarid *Anaspides*, the appen-dages on the thorax are still a series of more or less identical limbs each with endopodite, exopodite and gill – the exopodites renewing the water over the gills and aiding the abdominal appendages in swimming, the endopodites acting as walking legs. Two more familiar groups which display the caridoid facies well, one eucaridan (Fig. 120D, Euphausiacea) and the other peracaridan (Fig. 121, Mysidacea), have made some progress towards the adoption of a fully pelagic habit. In the euphausiid the rotatory movements of the thoracic exopods assist the pleopods in swimming, whilst in mysids the pleopods are – rather oddly – reduced and only

the exopods are used. The main evolution within the subclass has been an exploitation of the benthic mode of life and the macrophagy which it permits. This seems
to have occurred several times and given rise to groups as diverse as the stoma-
topods, amphipods, isopods and decapods. In the most primitive decapods, the
natant macrurans, the caridoid form is still well displayed: the abdominal segments
have freedom of movement and their natatory pleopods are large; the thorax is
strengthened by the fusion of its segments and the five posterior pairs of appendages
are typically ambulatory – exopods are even retained in some primitive prawns.
The reptant macrurans have pleopods which are too feeble for swimming and use
the flexing abdomen and the fan-like uropods to make backward evasive movements. Finally in the brachyurous forms only the last three or four pairs of thoracic
appendages are concerned with locomotion.

With the exception of flight every type of locomotion is to be found in the
Eumalacostraca, associated with extensive modification of the jointed limb and of
the form of the body. Often there are several types of movement in a single species,
as in amphipods, which have a compressed, elongated body with no carapace and
all the free segments movable on one another. Gammarids swim with the first three
pairs of abdominal appendages, jump with the last three pairs, crawl with the last
three pairs of thoracic appendages and walk with the anterior ones. Some species
can burrow, and this involves the coordinated activity of several sets of appendages.
Thus in *Haustorius* (Fig. 126B), which lives in sand, the broad, short body (with the
lateral compression and flexion typical of other amphipods almost completely gone)
is pressed into the soft substratum by the long, stiff, heavily armoured and
backwardly projected pereiopods 5–7 (per 7), while the antennules (ant′) are bent
back ventrally to protect the mouth parts. The antennae (ant″), acting as two
ploughshares, are then forced into the substratum, whilst behind them the anterior
four pairs of pereiopods scratch, rake and push the sand particles backwards into a
tunnel-shaped space enclosed laterally by the broad bases of the limbs and their
setae. A powerful current of water set up by the metachronal beating of the
pleopods (pleop) passes through the tunnel and casts out the sand grains posteriorly.
Often the animal leaves the bottom and swims on its back. These activities all
depend upon the freedom of the body segments and this is also important in other
crustaceans. The sudden extension of the abdomen of talitrid amphipods, assisted
by a strong backward movement of the posterior abdominal appendages, enables a
leap through the air to be made with such force that an animal may reach a height
of 20 times its own length. Similarly, in certain decapods such as *Homarus*, *Astacus*
and *Galathea*, a strong ventral flexure of the abdomen shoots the animal backwards
through the water, and repeated flexure in *Homarus* forces the body back at a speed
of 8 m s^{-1}. Nothing approaching this speed can be achieved by walking. One of the
fastest stepping frequencies is attained by the isopod *Ligia oceanica* when running
over dry rocks. Each of the seven pairs of thoracic legs may execute over 16 steps

per second and a specimen 35 mm long will move at a speed of about 50 cm s⁻¹. The
legs are long, radiating fanwise from the central region of the broad body, and
stepping in different planes so as to avoid interference with one another. They are
folded under the body in a pattern which lets the animal cling to stones or rocks and
penetrate narrow crevices. There is no carapace and the segments of the body are
free, facilitating movement over uneven surfaces.

Another group which has specialized in walking is the crabs. Their loss of the
primary habit of swimming has led to a reduction of the abdomen, effectively
absent as a separate unit of the body, to a shortening and broadening of the
cephalothorax and to the retention of a relatively small number of walking legs,
which are formed of long, robust endopodites. The anterior thoracic appendages
(Fig. 129A) are transformed into maxillipeds, still biramous, which border the
ventral mouth and assist the mouth parts in tearing, sieving and transferring food;
the third (posterior) maxillipeds (mxp 3) form an operculum which can close over
the buccal region and its appendages. Accompanying these changes a reorganization of the mechanics of the internal skeleton of the cephalothorax has resulted in
a structure which is robust and efficient. Crabs have exploited this in a wide range
of adaptive radiation. There are four pairs of pereiopods arising ventrally from the
broad, flattened cephalothorax so disposed that the planes in which they lie are
inclined to intersect close to the animal's centre of gravity. The habit of walking
sideways, occasional in *Astacus* and other macrurans, has become a highly
specialized adaptation and most crabs go forward or diagonally only when changing
direction. The common shore crab *Carcinus* walks up and down with the tide, the
lateral surfaces facing the current and the chelipeds held against the cephalothorax.
In walking sideways the legs are swung in a transverse plane and the action of those
on the two sides is quite different; the leading legs pull by flexing and the trailing
ones push by extending. This direction of stepping is probably helped by the
narrowness and lateral elongation of the thoracic sternites, which have been pressed
backwards to make room for the mouth parts; it allows the movement of a leg to
take place most readily in the same plane as that in which it lies.

Although crabs have lost the pleopods some have a secondary method of
swimming by a transverse paddle movement of the thoracic legs. In one group, the
portunid swimming crabs, the adaptation has been taken further. Only three pairs
of pereiopods are used for walking and the fourth pair are flattened paddles fringed
with sensory setae. They are held off the ground during walking and beat so as to
help slightly in the movement. When the crab is swimming it moves sideways and
these two pereiopods beat backwards and forwards together well above the carapace
and at right angles to the direction of movement. Their blades rotate at the end of
each half beat so as to provide a propulsive force during both forward and
backward strokes. Meanwhile the chelipeds are pressed against the body and the
remaining legs execute a kind of transverse dog paddle which aids sideward

movement and presumably helps to cancel anteroposterior components inherent in the stroke of the last pair. Portunids swim backwards by the last pair of legs executing a slow figure-of-eight movement, the forward beat alone being effective and the limb being feathered on the return.

The paddles steer the portunid through the water with considerable accuracy, as may be seen when it pursues and captures a fish which is swimming towards it from the right or the left. When it attacks the chelae grasp only rapidly moving objects. It is one of the fastest swimming crustaceans, with a sideways swimming rate of 1 m s^{-1}. Some other crabs also pursue actively swimming prey, like the shore crab *Carcinus*, which is known to attack prawns (*Palaemonetes*).

The shape of the cephalothorax of a crab may be related to its mode of life (Fig. 130A). In running (1) and swimming (2) crabs the cephalothorax is streamlined so as to facilitate lateral movement through the water in either direction, and approximates to a spindle elongated from side to side; swimming crabs naturally show this better than the runners. Crabs from the bottom of deep, still waters or from algal and other growths in moving water (spider crabs) do not show streamlining (3) nor is it exhibited by burrowing crabs, the shape of whose body is related to their particular way of burrowing (4). The masked crab *Corystes* burrows in clean fine sand. It first sits upright on the surface with the rather narrow, elongated carapace tilted forwards, displaying the dorsal sculpturing which bears resemblance to a human face. The elongated talon-like claws of the last pair of legs (per) then dig into the sand, working medially across the longitudinal axis of the body and scooping it towards the ventral surface. The three anterior pairs of walking legs pass the sand upwards and outwards so that the combined activity of the legs pulls the crab down and obliquely backwards; this reduces the pressure of sand against the mouth. The long chelae also thrust sand away from the buccal region. Ultimately only the approximated tips of the antennae project from the surface. These appendages are as long as or longer than the body and are held together to form a tube (ant″.tube), for each is fringed by two longitudinal rows of setae which interdigitate with those of the opposing limb. The tube leads to the gill chambers and is the passage for the inhalant respiratory water stream. The crab burrows as a means of protection, which is otherwise scarce on a sandy shore, and usually emerges to feed at night.

Functioning of the crustacean gut

The crustacean gut extends as a straight tube through the length of the body, with mouth and anus ventral and subterminal. The fore-gut and hind-gut are ectodermal with a chitinous lining shed at each ecdysis. The mid-gut is endodermal and is concerned with secreting digestive fluid, absorbing products of digestion and storing such substances as glycogen, fat and calcium. It also produces secretions which compact the faecal matter. The gut may possess one or more pairs of diverticula which, with few exceptions, are outgrowths of the mid-gut. These include the digestive diverticula formed by hypertrophy of the mesenteric glandular area, their pattern varying with the size of the animal. In some small Cladocera, Copepoda and Ostracoda digestive diverticula are not developed; in others, like *Daphnia*, they comprise a single pair of short caeca, and in larger branchiopods and in cirripedes they are lobulated; in the Malacostraca they take the form of pairs of long tubules or of the voluminous, bilobed, racemose gland of decapods. Whatever their form, they open into the anterior part of the mid-gut by a single pair of ducts, laterally placed.

Other caecal outgrowths of uncertain function may occur: the most conspicuous are two coiled tubules arising from the anterior part of the mid-gut and a single one from the hind-gut in most crabs. The anterior ones appear to be homologous with a glandular caecum opening dorsally into the mid-gut in some other Malacostraca, which is known to produce a peritrophic membrane enclosing the faeces so that they

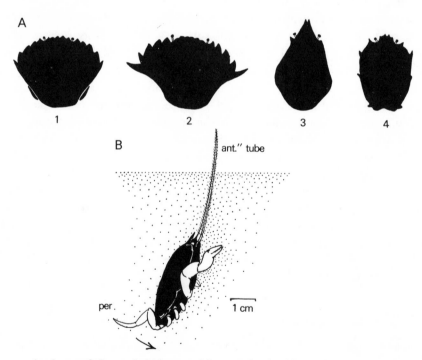

Fig. 130. A, shape of the cephalothorax of four crabs, in dorsal view. 1, the shore crab, *Carcinus maenas*; 2, the portunid, *Bathynectus longipes*; 3, the spider crab, *Hyas araneus*; 4, the burrowing crab, *Corystes cassivelaunus*. B, *Corystes cassivelaunus*: female buried in sand. The three anterior right pereiopods removed; arrow indicates direction of digging movement of 4th. ant″. tube, antennal tube; per, 4th pereiopod.

are voided in packets. A long posterior caecum, as well as a short anterior one, arises from the mid-gut of *Homarus* and may increase the absorbing area.

The arthropodan lack of cilia together with the often extensive cuticularization of the gut and consequent paucity of glands give rise to problems in (1) the movement of food, (2) the lubrication of this movement, and (3) the consolidation of faeces. The first is solved by chitinous structures worked by the underlying musculature and by extrinsic muscles, which direct the contents along the gut. Movement of the food is aided by the rhythmic swallowing of water, especially when small amounts are eaten, and defecatory reflexes are certainly initiated by raised hydrostatic pressure in the hind-gut set up by anal intake of water. Waste is consolidated by secreting a peritrophic membrane around it.

Entomostracans which are setose feeders eat only particles already finely sorted and not needing mechanical treatment, and the gut remains simple, retaining the proportions of parts characteristic of animals in other phyla. In *Calanus*, for example, the short and narrow oesophagus (Fig. 118A, oes) opens ventrally into a mid-gut (m.g) which extends to the last segment of the body. In the living animal the contents of the broad anterior half are seen to be propelled to and fro by visceral muscles and mixed with enzymes secreted by the epithelium. The narrow posterior half is probably absorptive but it also secretes the peritrophic membrane. In *Daphnia* the short digestive caeca (Fig. 117, dig.cae) are the chief absorptive areas, though some absorption also occurs in other parts of the mesenteron. Soluble products of digestion and particles enter the caeca, but the latter are expelled again by muscular action; they are never big enough to block the lumen. The faecal particles are compacted by secretion from glands in the posterior part of the mid-gut.

Some entomostracans have the chitinous lining of the oesophagus thickened to give a few longitudinal ridges and near its inner end, where it enters the mid-gut, these may be broad and setose. In malacostracans this inner region is enlarged to form the "stomach" or proventriculus where the food is broken up, mixed with secretion from the digestive gland and fluid and fine particles filtered from the larger pieces.

In the ancestral malacostracan (Fig. 131) it is likely that the tubular oesophagus had four longitudinal thickenings (long.th), one dorsal, one ventral and one on each side, and that these continued along the spacious proventriculus as even broader folds, ending in four projections or valves (d.v, v.v), together forming the base of the so-called funnel (d.l, l.l, v.l). This structure, formed of backwardly directed projections of the chitinous lining of the fore-gut into the lumen of the mid-gut (m.g), enclosed the faecal material, keeping it away from the openings of the digestive gland (l.d, r.d) and dorsal caecum (cae). Indigestible particles, which in other invertebrate phyla are kept together in mucous secretions, are thus transferred

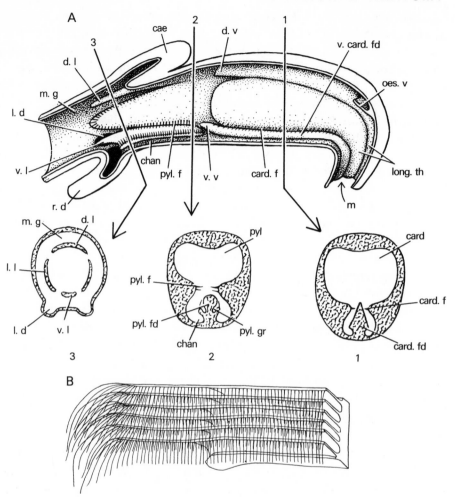

Fig. 131. A, diagram to show the basic features of the proventriculus and mid-gut of Malacostraca. Arrows indicate the levels of sections: 1, passes through the cardiac chamber, 2, the pyloric chamber and 3, the mid-gut in region of funnel. B, *Cancer pagurus*: part of the pyloric filter to show the fine filter setae on the ridges and rods. cae, mid-gut caecum; card, cardiac chamber; card.f, cardiac filter; card.fd, cardiac fold; chan, channel from digestive gland to cardiac filter; d.l, dorsal lip of funnel; d.v, dorsal cardiopyloric valve; l.d, left duct of digestive gland; l.l, lateral lip of funnel; long.th, longitudinal thickenings of gut wall; m, mouth; m.g, mid-gut; oes.v, oesophageal valve; pyl, pyloric chamber; pyl.f, pyloric filter; pyl.fd, ventral pyloric fold; pyl.gr, pyloric groove; r.d, right duct of digestive gland; v.card.fd, ventral cardiac fold; v.l, ventral lip of funnel; v.v, ventral cardiopyloric valve.

to a region of the mid-gut where they become enclosed within a peritrophic membrane.

Trituration of the food in the proventriculus is accomplished by means of intrinsic and extrinsic muscles. The latter insert on the longitudinal ridges and originate on the dorsal exoskeleton. In those malacostracans where the head constitutes a small proportion of the total body length (e.g. peracarids) so does the proventriculus, but in decapods with relative increase in cephalic size it becomes enlarged.

Adaptive modifications of the fore-gut throughout the subclass rest mainly on specialization of the longitudinal thickenings of the wall. The deepening of the lateral folds separates a roomy dorsal passage (card, pyl) from a smaller ventral one subdivided into ventrolateral channels by the ventral fold (card.fd, pyl.fd). Entrance to these channels from above is guarded by filtering setae (card.f, pyl.f) so that only fluid and small particles enter, and the whole forms the ventral filter. Posteriorly, in the pyloric region, along the surface of the ventral fold itself, one or more pairs of secondary longitudinal gutters (pyl.gr) develop, also overarched by closely set setae, and so a more complex filtering mechanism is developed; the secondary channels lead to the ducts of the digestive gland (l.d, r.d) and are known collectively as the gland filter.

In the branch of Malacostraca leading to the decapods the pylorus became a distinct chamber, separated from the anterior cardiac chamber by a marked constriction and by valvular projections derived from the dorsal and ventral longitudinal folds. This may be correlated with the evolution of macrophagy. The primitive filter-feeding malacostracan may have fed more or less continuously. The development of a cardiac chamber in which its food could be collected and mixed with enzymes from the digestive gland would ensure a better extraction of nourishment and the mixing might be combined with some trituration. When malacostracans took to a bottom-living existence it would be natural to continue to feed on suspended matter but also to pick up detritus from the bottom and eventually become active predators. Then some mechanical treatment of the food in the fore-gut would be of greater importance. The degree of treatment which the food receives varies. In *Ligia*, which feeds on weed and detrital material at low tide, mainly at night, quick extraction of nourishment is necessary. Food is cut up by the mouth parts and, in both cardiac and pyloric regions, is exposed to powerful enzymes from the digestive gland and squeezed by the longitudinal folds. Liquid from it, and small particles, pass through the ventral filter to the ducts of the gland. This comprises three tubes on each side which contract and relax rhythmically, forcing secretion into the fore-gut and intestine and withdrawing the liquid food. The food is directed from the stomodaeum to the long intestine where further digestion occurs, the peristaltic and antiperistaltic movements of the tube (especially its anterior part) mixing food and enzymes and expressing fluid. A continuous

stream of food passes through the gut and there is no delay due to its retention in
the cardiac region, which merges without marked separation into the pyloric
fore-gut and is devoid of masticatory ossicles. Indeed equal-sized pieces are found in
both fore- and mid-gut.

In contrast to this the fore-gut of another peracaridan group, the Mysidacea,
approaches that of the decapods. The food – small particles either filtered from a
water current or obtained from larger masses by the cutting and grinding activity of
the mandibles and of the maxillulary endites – may be retained in the cardiac
chamber for some time and mixed with enzymatic fluid. In some, such as *Praunus
flexuosus* (Fig. 121A), an omnivorous species living near the bottom, this chamber
(card.st) is large, with chitinous plates and ridges armed with teeth and setae all
concentrated at the junction with a smaller pyloric chamber (pyl.st). At this point
the lateral folds bear a pair of cusped teeth which project across the lumen; strong
barbed spines arise dorsally, and ventrally the median ridge projects from the floor,
separating two ventrolateral grooves which are overarched by spines arising from
the lateral folds ventral to the teeth. In addition to working on the food these
structures also regulate its passage towards the mid-gut, the lateral teeth and setae
keeping it away from the ventral passage, and the barbed dorsal spines forcing it
into the dorsal cavity of the pyloric chamber. This dorsal food channel is separated
from the pyloric filter channel by a thickening of the lateral fold on each side. Bands
of intrinsic circular muscle work antagonistically to a pair of dorsal extrinsic dilator
muscles to constrict the food mass and squeeze out fluid which is directed ventrally
to the filter. The food is passed to the mid-gut where it becomes enclosed in a
peritrophic membrane secreted by the dorsal caecum (cae), the entrance to which is
protected by a pyloric valve. The mid-gut is long and here the food may still be
churned by antiperistaltic movements of the wall. Digested and partly digested
products are drawn into the tubules of the digestive gland (d.gl) by the rhythmic
contraction of muscle bands which occur along the tubes.

In decapods (Fig. 110B) the two parts of the fore-gut are more effectively
separated than in mysids. The cardiac chamber occupies half the length of the
cephalothorax in the larger forms and, except in natant macrurans, its cuticle is
thickened and calcified along certain areas to give supporting ossicles as well as
teeth. Because of its heavy armature this part of the gut is often referred to as the
gizzard or gastric mill. In natant macrurans such as *Palaemon*, where lightness is
important, the food is broken up externally and there is no gastric mill. The
mandibles are highly efficient masticatory appendages with effective incisor and
molar processes and contrast with the simpler appendages of reptant forms, which
triturate the food internally. The large size of some of these benthic forms may be
correlated with their highly specialized alimentary system which enables them to
deal with larger meals.

The pyloric chamber is characterized by the elaborate nature of the press and ventral filter. In other malacostracans the filter comprises lateral longitudinal gutters along the ventral fold overarched by closely set setae. With increase in the height of the fold there may be increase in the number of filter channels. There are two pairs in isopods, three in some mysids. In the decapods, however, the filter is a much more elaborate structure with innumerable channels. Its increased complexity and the development of a powerful press may be associated partly with a more efficient extraction of nourishment from the food and partly with the need for a more effective protection of the channels leading to the digestive gland. The dorsal fold of the pyloric chamber of the fore-gut roofs over the food channel and bears a chitinous projection into the mid-gut, the dorsal pyloric valve, which has developed with the need to cover the entrance to the dorsal caecum. In *Carcinus*, which has two dorsal caeca opening near the anterior limit of the mid-gut, the valve is broad, covering the entrance to both, and is relatively short. The mid-gut is longer than in *Astacus*, extending to near the posterior end of the cephalothorax. Its length varies considerably within the Eumalacostraca for in lobsters and in mysids it ends at a short distance from the anus, whereas in isopods it reaches only a little past the openings of the ducts of the digestive gland – perhaps the minimal length for the production of the peritrophic membrane which, at least in *Idotea*, is known to be secreted by its epithelium.

Functioning of the crustacean vascular system

The Crustacea, like other arthropods, have an open vascular system in which the circulatory fluid is not confined to vessels and so separated by membranes from the interstitial or lymphatic fluids; for this reason it is often called haemolymph. The open system is further differentiated from a closed one by a lower arterial pressure and higher blood volume. The haemolymph, unlike that of tracheate arthropods but like that of molluscs, transports oxygen as well as nutrients and waste. In most crustaceans the circulation is maintained by a systemic heart, dorsally placed, and assisted by movement of the appendages and other organs, and sometimes by one or more accessory hearts, local dilatations of a blood vessel with muscles which decrease their volume on contraction. Barnacles have only this type of heart, situated between the adductor muscles of the shell and the oesophagus, where the acceleration of venous flow due to the beating of the cirri is least.

The heart is a single chamber suspended in a pericardial space by suspensory ligaments and by arteries which penetrate the pericardial wall. The pericardial space is itself a blood sinus separating the heart (with which it communicates through ostia) from the rest of the haemocoelic cavity. Primitively the heart is tubular, as in the Anostraca (*Chirocephalus, Artemia*) where it extends through the thorax and abdomen and has a large number of paired, segmentally arranged ostia. The heart may also be elongated in the lower malacostracans but the ostia are not segmental and may be reduced to a single pair (*Anaspides*). Since blood enters the pericardial cavity mainly from the respiratory organs, the heart tends to become restricted to the part of the body where these organs occur; thus in mysids, amphipods and eucarids (Fig. 120D, h) it is thoracic, but in isopods (Fig. 118B, h) it extends into or is confined to the abdomen. The heart of decapods (Figs 110B and 112B, h) is large, though short, with the walls thickened by several layers of crossing muscle strands: in this it contrasts with the tubular hearts of lower forms with a single muscle layer.

Systole is effected by the cardiac muscles which, on contraction, expel blood into the arteries where there are valves preventing backflow. Pressure in the pericardial cavity is lowered and, as a consequence, blood is drawn into it from visceral sinuses or branchial vessels. The heart then tends to return to a resting shape, but a diastolic mechanism is provided by the suspensory ligaments which distend the heart and open the ostia by their elastic pull. This mechanism (characteristic of all arthropods but of no other phylum) is adequate to maintain circulation in a low-pressure system and even during the increases in blood volume and body size at times of ecdysis. In the majority of crustaceans the heart is controlled by nerve cells situated in a dorsal cardiac ganglion or scattered through the cardiac nerves. In branchiopods evidence suggests that the pacemaker is muscular.

Changes in the volume of the pericardial cavity influence the rate of circulation and these are brought about by movements of the body and of the pericardial septum. In ostracods, copepods (Fig. 118A) and Eumalacostraca (Fig. 112B) alary muscles (al.mu) originate on the lateral body wall and extend medially into the septum, ending a short distance from the mid-line, which it crosses as a nonmuscular elastic membrane. By their contraction they flatten the normally convex pericardial floor and so enlarge the cavity and draw in blood from surrounding spaces. The only afferent openings to the pericardial space in higher crustaceans are the branchiopericardial sinuses or efferent branchial veins from the respiratory organs, and the action of the alary muscles speeds up the circulation through these and the channels in the gills. In Ostracoda, Copepoda and Cladocera the pericardial cavity is less completely closed and may have large openings to the visceral sinuses.

In lower crustaceans only an anterior aorta (Fig. 118A, ao) leaves the heart. It leads over the gut towards the brain and may open directly into the haemocoel without branching, as in branchiopods and copepods. The blood flow, as in other arthropods and annelids, is anterior in the dorsal and posterior in the ventral channels, and oxygenation occurs in gills, appendages or carapace just before the haemolymph returns to the heart by way of the sinus system. Malacostraca are larger and have appendages, often with specialized gills, which can support the body; associated with these changes is the development of an arterial system more like that of *Astacus*. Their arteries have elastic, nonmuscular walls and in certain places branch to give capillary beds. The arterial blood is discharged into lacunae and then collected into venous sinuses, which are spaces typically bounded by a membrane. It ultimately flows to the sternal sinus and so to the gills and heart. Anterior and posterior arteries lead from the heart (Fig. 118B, med.ant.a, lat.thor.a, lat.abd.a) together with a variable number of lateral vessels (art.1), segmentally arranged; or the lateral vessels may arise from the aortae. Branches from the segmental vessels supply the viscera and limbs and ventral ones pass medially to join a longitudinal artery (subn.a) associated with the ventral nerve cord and musculature. This ventral vessel, not present in *Nebalia* nor in amphipods (which have a much reduced arterial system), becomes more prominent in decapods. Where special respiratory organs occur, segmental veins (eff.br) return blood from them to the pericardium (peric.cav) and the series of segmental arteries tends to be interrupted.

The arrangement of vessels is modified in the decapods to give a more direct arterial supply to the ventral region of the body and, in particular, to the thoracic appendages, which are of increasing importance in these benthic forms. Thus much of the blood which passes posteriorly from the heart enters the sternal artery (Fig. 115B, st.a). This artery is, perhaps, derived from fusion of a segmental pair. It passes ventrally to the subneural vessel, now known as the ventral vessel (v.th.a), from which segmental branches pass to all the appendages of the cephalothorax except antennules and antennae.

The adaptation of isopod species to aquatic, amphibious and terrestrial habitats is associated with an increasing efficiency of the circulatory system. This culminates in land isopods, which have a relatively larger heart, stronger alary muscles and strictly defined, often vessel-like lacunae which, with the true vessel, form an almost complete circulatory system ensuring a better regulated and swifter transport of blood.

Despite the possession of an open vascular system some of the higher Malacostraca are able to cast off injured appendages without loss of blood or damage to soft tissues. In crabs chelipeds and pereiopods have each a preformed breakage plane high on the limb, near the line of separation between basipodite and ischiopodite. Here a vertical invagination of the epidermis produces a two-layered septum which leaves a small opening through which run blood vessels and nerve, but no muscle. The autotomy of a limb at this plane is reflexly initiated and is brought about by the extreme contraction of a muscle (the autotomizer muscle) which is at other times involved in the normal activity of the limb. The outer layer of the septum is cast off with the limb, the inner persists and its opening sealed by a valve which is closed by blood pressure within the stump. A sheet of new cuticle is secreted over the whole surface. Regeneration of the limb begins at the next moult.

Functioning of the excretory organs of crustaceans

The excretory organs of Crustacea are limited to certain segments of the body. Two pairs usually occur during larval and adult life, the antennary and maxillary glands, closed tubules opening at the base of the appendage from which they take their name. The adult organ is the maxillary gland in entomostracans and some malacostracans, and the antennary gland in amphipods, mysids and eucarids; in larval stages the other gland is usually functional. A few crustaceans (ostracods, *Nebalia* and the mysidacean *Lophogaster*) have both glands in the adult. Vestigial excretory organs may be present in other segments of the body, indeed in the mysid *Praunus* paired end sacs are said to occur in all eight thoracic segments. Such evidence suggests that the segmental organs of Crustacea are the remains of a once continuous series such as is found in annelids. Their reduction in number and the transference of their function to other organs and tissues is associated with the reduction of the coelom.

The excretory organs are of the type found generally throughout the arthropods. Each consists of an end sac lying in the haemocoel and opening into a narrow excretory canal which discharges to the exterior by a short ectodermal duct. The end sacs are the persistent remains of a few of the series of paired, coelomic sacs which make a transient appearance during development. The canal may be a coelomoduct (*Estheria, Nebalia*), homologous with the genital duct, though more frequently it originates from ectoderm (branchiopods, ostracods, *Homarus*) or ectomesoderm.

In branchiopods extensive coils of the excretory canal are lodged in the narrow cavity of the carapace fold, for which reason the maxillary gland is often referred to as the shell gland (Figs 117 and 118A, mx.gl). In higher malacostracans the canal loses its simple tubular form and at least part of it develops into a compact gland which is most elaborate in decapods (Fig. 116); the end sac (sac) may remain simple or become subdivided. The initial part of the canal forms a spongy mass traversed by a system of fine canals and offering a large surface area to the haemolymph; this is the labyrinth (lab) which, with the end sac, constitutes the glandular section of the organ. The distal part forms a distensible bladder (bl) which may remain a simple sac, as in *Astacus*, or develop lobes and chambers which extend back even into the abdomen, as in crabs. Freshwater crustaceans have special adaptations which enable them to excrete a hyposmotic urine. In members of the Astacidae this takes the form of an elongated part of the labyrinth, a whitish cord of spongy substance, convoluted upon itself, forming the nephridial canal (neph.c) leading to the bladder; in the freshwater peracaridans *Gammarus pulex* and *Asellus aquaticus* it is an extra segment which, either by secreting water or absorbing salts, allows their specialized excretory glands (antennary in the former and maxillary in the latter) to produce a hyposmotic urine. In other crustaceans, including freshwater species, the modification of the excretory fluid by the canal of the antennary gland is much less pronounced. Regulation of inorganic salts occurs resulting in a conservation of Ca^{2+} and K^+ and the elimination of Mg^{2+} and SO_4^{2-}. The urine is isotonic with the blood, and even in species like *Carcinus maenas*, in which the blood remains strongly hypertonic when the animal is placed in external salinities as low as 4 per cent sea water (1.2 ‰), the segmental organs are not involved in the osmoregulatory mechanism.

No mention has been made of the excretion of nitrogenous waste, since this does not appear to be a primary concern of these organs: their main function is the maintenance of the ionic balance of the body fluids by the selective treatment of ions. The nitrogenous end products are chiefly ammonia with some urea and uric acid and a considerable amount of amino nitrogen, which may be excreted in the urine. It is surprising to find that in terrestrial peracarids excretion is still predominantly ammonotelic. Ammonia and urea are lost through the permeable surface of the gills and a percentage of the former is excreted in volatile form. Uric acid may be deposited in various parts of the body which act as kidneys of accumulation, including the exoskeleton.

Loss of water and salts by way of the antennary gland is balanced by the uptake of water and ions through the gills and Na^+, K^+, Ca^{2+}, and Cl^- often must be taken up against concentration gradients. It is by this means that crustaceans, like other invertebrates, can maintain a concentration of ions in the blood plasma differing from that of a passive equilibrium with the external medium. Decapods and peracaridans living in brackish or fresh water suffer loss of salts by outward diffusion as well as excretion and make good the loss by active absorption from the dilute medium with the expenditure of energy. This uptake of salt takes place over certain areas of the gill leaflets (Fig. 132, ab.a) in decapods, but apparently over the whole gill in gammarids, where both the basal and apical surfaces of the cells are highly infolded, with large numbers of mitochondria in the basal folds to provide energy for the ion pumps by which salt is obtained, cations of sodium and potassium probably being exchanged for hydrogen and anions for hydroxyl.

Artemia, the brine shrimp, can survive in a range of salinities varying from 10 per cent sea water (3.5‰) to crystallizing brine (salinity 600‰). At and above 25 per cent sea water the blood is hypotonic to the external medium, and hypertonic below. Hypotonicity is maintained by a process similar to that found in marine teleosts – water with its dissolved salts is swallowed, and the excess salts are excreted by mitochondrial pumps located in certain cells on the epipodites of the ten anterior swimming limbs. There is evidence that salt elimination is also helped by the fact that *Artemia* produces a urine hyperosmotic to the blood.

Respiration of crustaceans

Although the crustacean body is enclosed in a skeleton strong enough to withstand

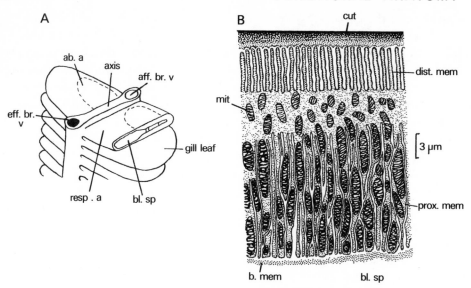

Fig. 132. A, Stereogram of small part of a gill of a freshwater crab to show areas of each leaflet concerned with respiration (resp.a) and salt uptake (ab.a). B, fine structure of the epithelium over the absorbing area. (After Copeland.) ab.a, area absorbing salt; aff.br.v, afferent branchial vessel; axis, gill axis; bl.sp, blood space; b.mem, basement membrane; cut, cuticle; dist.mem, membrane over outer surface of cell; eff.br.v, efferent branchial vessel; gill leaf, gill leaflet; mit, mitochondrion; prox.mem, membrane at inner surface of cell.

the pull of muscles, no special respiratory surface is required by animals below a certain size; thus in *Calanus* gaseous exchange occurs over the general surface of the body. When gills become necessary they are derivatives of the appendages. Respiratory pigments develop in small species which live in an environment deficient in oxygen and in larger forms which have elaborate gills. Thus *Daphnia* and Apus have haemoglobin in the blood, the amount varying inversely with the amount of oxygen in the water, and haemocyanin characterizes the decapod, but not the nondecapod malacostracans, which are almost always smaller. In addition to appendages the carapace fold figures largely in the respiratory mechanisms of such groups as possess one. Arising from the maxillary segment, it stretches over the thorax and whilst the outer wall is thick the inner one is thin; the cavity which it encloses is ventilated by the movements of appendages. The inner wall is well vascularized and forms the chief respiratory surface in mysids, land crabs and barnacles. It is usually simple, but in barnacles is produced to form a single gill on each side, washed by a flow of water, set up by beating of the cirri, which enters the mantle cavity right and left of the head and passes out posteriorly.

Gills are typically lateral outgrowths of the basal region of thoracic legs, known as epipodites. The simplest is a lobe or lamella (Fig. 117, epip) as in branchiopods and *Anaspides*, bathed by currents produced by these limbs as they perform locomotory or feeding movements. In amphipods, the habit of crawling on one side and the absence of a carapace to protect respiratory organs make lateral gills impractical, and the epipodites are median outgrowths (Fig. 123, g); in females they arise ventral to the oostegites (oost), which are modified from a second series of epipodial plates. They lie in a channel between the limbs deepened by coxal plates projecting from the margins of the thoracic segments, and ventilated by the beating of the pleopods, the first three pairs of abdominal appendages. A current of water is passed over the gills even when the shrimp is not moving, facilitated by the longer posterior thoracic legs raising the hinder end of the body. In the littoral sandhopper, *Talitrus saltator*, the gills function only in moist air kept in motion by pleopods, and the animals drown in water.

The eucarid Malacostraca have more complex gills with a larger area for respiratory exchange. Each consists of an axis which bears numerous branches. In euphausiids they undergo further division to give feathery structures (Fig. 120D, g) not covered by the carapace but hanging freely at the sides of the body (one attached to each thoracic leg) and washed by the swirl of water set up by the exopods during swimming. In decapods, a group which includes the largest crustaceans, typically benthic, the gills are enclosed in a branchial chamber ventilated by the scaphognathite propelling water forward with a rapid sculling movement. The scaphognathite of the prawn *Palaemon serratus* maintains a rate of 294 beats per minute at 17°C, though this is slow compared with the thoracic appendages of calanoid copepods which range from 60–2640 beats per minute. Perhaps because the gills are less efficient when enclosed they then increase in number. The maximal number on each side of a segment is four: two arthrobranchs attached to the articular membrane between the coxopodite of the appendage and the body, one pleurobranch dorsal to this on the body wall, and, on the leg itself, one podobranch, which may be joined to an epipodial plate carried by the coxopodite. Probably all arise as outgrowths of the limb, for a dorsal migration from this position to the more protected one beneath the branchiostegite occurs in the development of *Penaeus* and *Crangon*. The axis of the gill bears numerous lateral branches which may be filiform or, in crabs, flattened plates. In both euphausiids and decapods the axis contains the main afferent and efferent vessels linked by channels running through the leaflets. The cuticle covering the respiratory surface is extremely thin and in *Carcinus maenas* is penetrated by invaginations which extend into the underlying epidermal cells and so come close to the blood. It is probable that they increase the efficiency of the gill.

The degree of isolation of the gill chamber varies considerably within the decapods and may be correlated with the mode of life. It is least, as might be expected, in swimming macrurans in which the entrance is a narrow slit along the margin of the branchiostegite, protected against the entry of silt by setae along the edge of this fold and on the coxopodites of the limbs. It is greatest in crabs where the cephalothorax is flattened from above downwards and the lateral portions of the branchiostegite bend abruptly inwards and are closely applied to the ventrolateral region of the thorax, except near the bases of the legs. This gives freedom for their movement as well as a series of inhalant openings guarded by setose tracts. The branchiostegite now forms a lateral margin to the cephalothorax, thickened, often armed with teeth, and projecting in the direction of habitual movement. The adaptations, which primarily protect the gills from the dangers of silting associated with benthic life on a wave-washed shore, not only enable the decapod to exploit more silty or sandy situations but provide preadaptive features for colonization of the supralittoral zone and terrestrial environment. The portable shelter of hermit crabs, typically a gastropod shell, enhances the isolation of the gill chambers, and the carapace becomes membranous behind the cervical groove where it is covered by the shell.

In *Carcinus* and *Cancer* there are nine gills in each branchial chamber. Except for the podobranch of the second maxilliped (Fig. 113A, podobr), which is directed. horizontally backwards, they curve upwards and inwards from their points of attachment so that their tapering tips converge along the dorsomedian wall of the chamber. The gills are closely applied side by side and separate a shallow hypobranchial space towards the body wall from an epibranchial space under the branchiostegite. The latter leads to the shallow exhalant passage or prebranchial chamber in the cephalic region where the movement of the scaphognathite (scaph) to and from the roof directs water towards an anterior exhalant opening, bordered behind by the endopodite of the first maxilliped. Water enters (Fig. 129A) the hypobranchial space, mainly at the base of the cheliped, passes between the gill lamellae to the epibranchial space and then anteriorly to the exhalant opening. In *Carcinus maenas* the entrance channels comprise a relatively large anterior opening (inhal.ch) in front of the cheliped and a narrow slit above each leg (inhal.per 1), including the cheliped. The size of the channels is affected by the position of the appendages: when a leg is moved backwards and upwards the opening above it is naturally almost closed. The size of the anterior inhalant opening is regulated by the basal segment and epipodite of the third maxilliped (epip 3) in such a way as to allow maximal intake of water and use of all the gills at times when these appendages are in use, as in feeding, or at other times, when oxygen is short. If the third maxillipeds are at rest the broad bases of their epipodites and coxopodites block the anterior part of the opening and bar the passage of water forwards towards the more anterior gills. If they are moved apart water gains access to these gills. The respiratory flow is also influenced by changes in the beat of the scaphognathite and the activity of the exopods (flagella) of the maxillipeds (Fig. 113B, mxp.exop 3), which flick to and fro extremely rapidly and so reinforce the exhalant current from the gill chambers, turning it outwards away from the mouth. This occurs when the crab feeds. After this the mouth parts clean themselves, are folded away and the oral field almost completely enclosed by the third maxillipeds.

The two scaphognathites may change both the rate and the direction of the water current, which is typically postero-anterior. They work independently and either may cease beating for a while or change the rate and force of its beat. In a more or less regular fashion the effective beat is reversed and water is drawn through the exhalant openings into the epibranchial space for periods lasting a few seconds; the current then reverts to normal. Reversal always occurs in *Corystes* when it burrows into sand and is maintained until it emerges. In deoxygenated, shallow water *Carcinus* can raise the normally exhalant openings to the surface and draw through them better oxygenated surface water or air which bubbles out in front of the coxae of the chelipeds.

Despite the fringe of setae at the openings to the branchial chambers a certain amount of particulate matter passes through the filter and might close the passages between the gill leaflets were it not for the activity of the epipodites (sometimes called mastigobranchs) on each maxilliped. These are long, flexible and stoutly setose, and can reach almost every part of the gills, moving constantly over their surface and brushing it clean. The epipodite of the first maxilliped lies above the gills and sweeps their outer surface, those of the second and third lie in the hypobranchial space and sweep the inner surface.

In other decapods epipodites have a similar function though they are not so well adapted. In these the branchiostegite fits less tightly. As a consequence a pereiopod can be inserted into the gill chamber to clean the gills, making the development of epipodites less essential, and the removal of silt by the reversal of the respiratory current more effective. In *Galathea* (Fig. 133A) and *Porcellana*, both inhabiting silty places on the shore, the last pereiopod is specialized for this function and is not used for walking; it may be found folded closely against the body, cleaning the grooves of the carapace or brushing the gill lamellae. To this effect the two terminal segments of the limb are beset with stiff setae, some serrated, giving it the appearance of a bottle brush. In *Porcellana* there are no epipodites and in the larger species, *P. platycheles*, a very thick curtain of bipectinate setae guards the inhalant openings to the gill chamber; in *Galathea* epipodites are present and the setal filter less developed. In both, podobranchs, the most ventral gills of the malacostracan series, are missing, perhaps because they would be too easily silted.

Crustaceans which have come on to land are members of groups well represented in the upper littoral zone where there live species showing morphological and physiological features associated with a terrestrial habitat. Most of those on land avoid the rigours of a truly terrestrial existence and venture into dry places only for short periods. The groups which are represented are isopods, crabs and, in some tropical regions, amphipods. All have the respiratory organs (which are areas of rapid water loss on land) tucked away from exposed surfaces. The only apparent modification affecting the respiratory apparatus in amphipods is a tendency for the pleopods to be reduced to stumps and (in some species) an increase in gill size. In isopods the respiratory organs, the pleopods, display considerable adaptive differences (Fig. 127). They have broad rami flattened against the abdomen and in the simplest arrangement within the order all are similar (except for sexual modifications); the metachronal beating of both rami serves for respiration and swimming. Some isopods can walk and swim; when out of water the delicate rami are protected by opercular structures. The exopodites of *Ligia* are stouter than the endopodites, though sharing the respiratory function, and cover them when the animal is not immersed, whereas in the freshwater *Asellus* and its relatives (a group called Asellota) and the marine *Idotea* and its relatives, the Valvifera, a single, stouter opercular flap covers the appendages on each side; in the former it is formed by the exopodite of the third pleopod and in the latter by the endopodite of the uropod. In woodlice (Oniscoidea) the efficiency of the pleopods as respiratory organs is increased by the development of diffusion lungs or pseudotracheae in the exopodites. Each is a tuft of air-filled tubules which opens to the exterior by a slit with

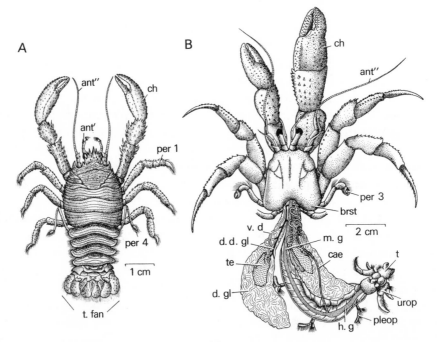

Fig. 133. A, *Galathea squamifera*, in dorsal view. B, the hermit crab, *Eupagurus bernhardus*, removed from its shell and its abdomen dissected, in dorsal view. ant′, antennule; ant″, antenna; brst, branchiostegite; cae, caecum of hind-gut; ch, cheliped; d.d.gl, duct of digestive gland; d.gl, left lobe of digestive gland; h.g, hind-gut; m.g, mid-gut; per, pereiopod; pleop, pleopod; t, telson; te, testis; t.fan, tail fan, composed of central telson flanked by uropod on each side; urop, uropod (6th abdominal appendage); v.d, vas deferens, opening on last pereiopod which is normally kept within the branchial chamber.

no closing device. The woodlice *Porcellio* and *Armadillidium* have two pairs which appear as conspicuous white patches on the first two pairs of abdominal appendages. In dry air they are more effective respiratory organs than unmodified pleopods. *Trichoniscus* and *Cylisticus* have five pairs of pseudotracheae which are less developed and probably less effective; they seem to have evolved independently. In *Oniscus* pseudotracheae are absent and their place is taken by a system of air-filled spaces immediately beneath the cuticle through which air diffuses. The spaces have no communication with the exterior. An interesting re-establishment of the original metachronal activity of the pleopods can be observed when a terrestrial isopod is immersed in water.

Surfaces other than those of the pleopods are also used for oxygen uptake in terrestrial isopods. *Ligia* and *Oniscus* kept in moist air and with the pleopods blocked take up about 50 per cent of their normal consumption and this must be through the general integument; less is taken up by *Porcellio* and *Armadillidium*. Desiccation inhibits the absorption of oxygen and this is one of the factors controlling the distribution of these animals. They can survive where there is water to drink to counteract the effects of transpiration from the integument, and where the relative humidity of the air does not fall below about 85 per cent and they are in contact with damp surfaces, for then the rate of water uptake is sufficient to offset loss. In *Oniscus* and more advanced isopods there is a system of capillary channels over the body surface; water picked up by the uropods from the substratum by capillarity is conducted along these and may reach the respiratory surface of the pleopods.

The branchial cavities of land crabs are enlarged and the gills reduced in size or number (Fig. 113C). The gills have a strengthened chitinous covering of their lamellae, making them less prone to collapse in air, but less efficient as gills. The inner surface of the branchiostegite is rich in blood vessels and may be produced into tufted papillae. This adds to the respiratory surface. The cavities may remain partly filled with water, as in land crabs of the families Gecarcinidae and Ocypodidae, or have only a moist surface as in *Birgus latro*, the robber crab. In both, movement of the scaphognathite ventilates the branchial cavity. *Birgus* is related to hermit crabs, though it does not hide its abdomen and has dorsal plates on the abdominal somites. Its carapace is rather free and its margin can be raised for the inlet of air by the contraction of muscles and by the fifth pereiopod lifting the posterior part from within the branchial cavity. The respiratory surfaces are kept moist by secretion from glands in the wall and are also damped by water when the crab drinks, which it does three or four times a day.

The frequency and amplitude of the beat of the appendages which propel the respiratory water current in Crustacea are controlled by centres within the nervous system and influenced by external factors, the most important being temperature, lack of oxygen and excess of carbon dioxide. In decapods the respiratory centre is in the suboesophageal ganglia. The range of frequency of beat is considerable. The pleopods of *Asellus aquaticus* (at 14°C) maintain 18–167 beats per minute and those of *Gammarus pulex* 12–140 beats per minute. The higher rates are associated with the low oxygen partial pressures which these freshwater crustaceans are likely to encounter; very low tensions, less than 10–25 per cent normal, result in decreased ventilation. In related marine species the ability to regulate the rate of beat of the pleopods with changes in oxygen tension is only moderately developed and it is least developed in semiterrestrial littoral species (*Ligia, Orchestia*). The effects of excess carbon dioxide are not known for certain and their study is complicated by an alleged reaction with the calcareous skeleton.

Astacus, like other aquatic Crustacea, is sensitive to high temperatures which influence the ventilation of the respiratory organs. For *Astacus astacus* it has been shown that although the frequency of the beat of the scaphognathite increases from 33–72 per minute over the temperature range 6–30°C, and results in an increased volume of water flowing through the gill chambers, the volume of water affected by a single stroke is maximal around 8–10°C, and decreases above this temperature. A considerable volume of water washes over the gills of the larger decapods. *Homarus gammarus* uses 4 l h^{-1} at 5°C with a linear increase to 11.8 l h^{-1} at 21°C.

Reproduction of crustaceans

The gonads of arthropods arise in development from coelomic pouches continuous with ducts which are coelomoducts. Typically the sexes are separate in crustaceans, though in one class, the Cirripedia, hermaphroditism is general and it also occurs sporadically among the Malacostraca and notostracan branchiopods. The gonads are usually paired and spread through several segments of the body. In malacostracans and some entomostracans male and female ducts open on different segments. This suggests that there may have been a primitive segmental arrangement and that the retention of different members of the series is associated with different functional needs in the two sexes. The development of the genital system in terrestrial isopods supports this: the primordium of each gonad of *Porcellio* is a tube with branches lying in six thoracic segments. In the male the three anterior tubules become testes, two become supporting structures and the last, together with the main duct, becomes the vas deferens, which extends to the last thoracic segment of the adult and joins its partner from the other side to give a single genital papilla. In the female the elongated tube becomes the ovary and one tubule becomes the oviduct which opens on the sixth thoracic segment; the others become suspensory.

The sexes are usually distinguished by secondary characters associated with the transfer of sperm and the protection of eggs and young. There are extreme cases in which the sexes show marked discrepancy in size: in some copepods, cirripedes and parasitic isopods the male is a dwarf and lives, sometimes parasitically, on the body of the female.

The male deposits sperm either in the vicinity of the female opening or within it, and may have appendages modified for clasping the female during copulation. The appendages used vary from group to group: in *Chirocephalus* it is the antenna (Fig. 128C, ant″), in talitrids the enlarged second gnathopod (Fig. 123C, gnathop) and in some mysids an elongated pair of pleopods. Sperm may be transferred directly from the genital papillae of the male (cladocerans, gammarids, mysids, *Crangon*) or by modified appendages which are adjacent to the male opening. In the isopod *Porcellio* sperm are deposited in each oviduct by one of a pair of copulatory styles formed from the endopods of the second pleopods. Each is grooved longitudinally and the genital papilla projects into the base of the groove to discharge seminal fluid.

In malacostracans hormones control the differentiation of the primary and secondary sexual characters of the male. The gland responsible, the androgenic gland, was first demonstrated in the amphipod *Orchestia gammarella*, where it lies in the haemocoel alongside the vas deferens. A corresponding gland has been found in other amphipods, some decapods, including *Carcinus maenas*, and isopods. It does not seem to occur in arthropods other than crustaceans. In females differentiation of the gonad is not under hormonal control, though, as in vertebrates, the ovary secretes a hormone which is responsible for the development of temporary characters associated with the incubation of the eggs. Secretion occurs during the period of deposition of yolk in the ova, and controls the enlargement of the oostegites forming the brood pouch in isopods (they are reduced between breeding periods), the lengthening of the marginal setae on the oostegites in gammarids (in which otherwise they remain unchanged) and the development of egg-carrying setae on the pleopods of decapods. The deposition of yolk is itself controlled by a hormone secreted by an endocrine organ in the eyestalk known as the X-organ which limits egg production, and hence the appearance of secondary sexual characters, to intervals characteristic of each species. This recalls the role of the hypophysis in vertebrates.

In some amphipods, such as *Orchestia cavimana* (Fig. 123C), the gonad, in early developmental stages, starts by containing both male and female cells; in males the presence of the androgenic gland suppresses further development of eggs, which remain abortive and fail to lay down yolk (ovt); in females absence of the gland prevents development of the primordial male cells. The presence of the androgenic gland in decapods appears to explain the feminization of male decapods parasitized by rhizocephalan cirripedes such as *Sacculina*. The gland is gradually destroyed by the parasite; spermatogenesis ceases and the external secondary sexual characters are lost at a subsequent moult and replaced by more feminine ones. The suppression of yolk formation in parasitized females inhibits the appearance of temporary sex characters related to incubation.

Environmental factors may also influence the determination of sex. This occurs in

Cladocera in which, during periods of overcrowding, adverse temperature, and food scarcity, or even when the nature of the food changes, the parthenogenetic females give rise to males. The last of these factors operates through an increase in the lipid content of the algae on which the cladoceran feeds.

In most Crustacea the sperm are not motile and do not have the typical filiform shape. Since they are often just deposited on the surface of the female and might be washed away, they are transferred in packets or spermatophores, manufactured in the vas deferens by special glands. The normal rule is for copulation to occur after the female has ecdysed and while her exoskeleton is still soft and easy to manipulate. The sexual behaviour of gammarids demonstrates the importance of the accurate timing of these two events. The male is attracted to a ripe female before she ecdyses and swims around holding her, sometimes for days. She moults in this position, the male helping to remove the old skin, and they copulate. In *Gammarus duebeni* the male turns the female so that she lies on her back below and across him at right angles, and with strong thrusts of the abdomen directs sperm packets from the genital papillae into the brood pouch, the transference being assisted by movements of his pleopods. One to four such matings between the two individuals may occur in a period of 10–30 minutes, after which the male abandons his partner. As the eggs pass along the oviduct they are embedded in albuminous secretion and shortly after copulation enter the brood pouch in two groups, one from each duct. The secretion protects the eggs only for the short period of maturation and the timing of its dispersal is synchronized with that of the secretion binding the sperm. Fertilization then occurs.

The complexity of the spermatophores varies from group to group and shows specific differences that may be related to mode of life or to devices for the discharge of the sperm. In the planktonic copepod *Calanus* the spermatophore (Fig. 118A, spm) is an elongated sac nearly 0.5 mm long and at copulation one end is attached near the oviducal opening on the genital segment of the female, adhering by its sticky outer wall. The sperm pass into the spermathecae (spth), two sacs which lie, one on either side, just within the median genital papilla. The spermatozoa are stored there until they meet ova approaching the female opening (ovd) to be discharged.

The spermatophores of decapods are simple sacs or rod-like structures. In some (hermit crabs) they are stalked (Fig. 134C, D) and become attached in large numbers to the pleopods of the female. The stalk often fastens them to a basal ribbon (r) which acts as a conveyor belt from their site of manufacture in the spiral coils of the vas deferens. It keeps them together as they pass through the genital aperture to the tube formed by the apposition of the first and second pleopods which deposits them on to the female. The spermatophores also remain external to the female in macrurans. In the lobster (*Homarus*) the gelatinous transparent rods up to 25 mm long are directed by the copulatory appendages into a pouch between the bases of the last walking legs of the female. They thus lie across the route taken by

the ova passing from the genital opening to the pleopods. In *Astacus* the spermatophores are shorter rods deposited on the pleopods and ventral surface of the
abdomen.

Copulation in crabs results in sperm being transferred directly into a spermatheca
on each oviduct, where they are stored. Consequently, the spermatophore has lost
much of its protective importance. In *Carcinus maenas* the copulating pair lie sternum
to sternum with the abdomen unflexed, the female under the male who grips her
with the anterior pereiopods. The right and left genital papillae (Fig. 134A, g.pap) of
the male are inserted into the basal grooves of the first abdominal appendage
(abd.app 1) and the appendages enter the oviducts. Sausage-shaped spermatophores, freed from the genital openings, are then pushed into the female by the
piston-like action of the second abdominal appendage (abd.app 2). The sperm may
remain in the spermatheca for months, sealed in by a plug of secretion provided by
the male, before fertilizing the ova.

The spermatozoa of decapods have a remarkably complex structure which varies
from species to species. Their complexity is linked with their immobility. The
simplest, found in natant decapods, have the nucleus surrounded by undifferen-
tiated protoplasm prolonged into a single spine comparable to the tail of a motile
sperm. In reptant forms the spermatocyte gives rise to a capsule which explodes to
deposit the male nucleus in the cytoplasm of the egg. Figure 135 shows the way in
which fertilization occurs in the hermit crab *Diogenes pugilator*. The spermatozoon
(similar to that of the lobster) comprises a capsule, beneath this a central undiffer-
entiated mass of protoplasm from which arise three long spines for anchorage (sp),
and then the nucleus (n). The wall of the capsule consists of a rigid outer membrane
(out.mem), an extensible inner one (inn.mem) and fluid between. It encloses a
central canal shut by an operculum at one end (op), and enclosing at the other an
acrosome (ac) comparable to a firing-pin and having frail protoplasmic connections
with the operculum (proto). The sperm attaches itself to the egg by means of the
spines, with the tip of the capsule towards the egg membrane (mem). The
operculum breaks, severing its connection with the firing-pin. With this release of
pressure the inner membrane expands, folding over the outer, by way of the
opening made by breakage of the operculum and it presses on the surface of the egg,
rupturing the membrane. The capsule becomes bell-shaped as the canal opens out
and then the firing-pin penetrates the cytoplasm of the egg, dragging the nucleus
behind it, the spines being carried passively behind the nucleus. Later all parts
except the nucleus are resorbed. In crabs the sperm nucleus is in the form of a
cupule moulded to the base of the capsule, so that the two are intimately associated;
in sperm which are transferred directly to the oviduct, spines are frequently
reduced.

In contrast to such specialized arrangements for ensuring successful fertilization is
the condition in barnacles. In these sessile forms the sperm are filiform and motile

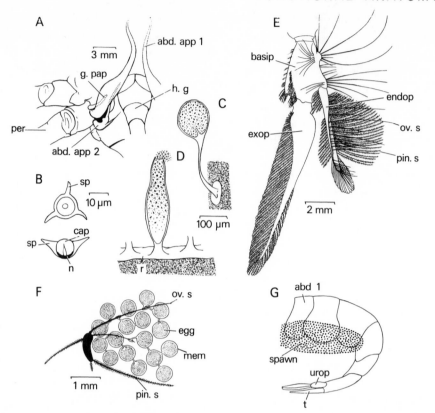

Fig. 134. *Carcinus maenas*; A, copulatory appendages of male as seen on underside of abdomen. B, spermatozoon in (upper figure) apical and (lower figure) lateral view. C, *Diogenes pugilator*, spermatophore. D, *Eupagurus bernhardus*, spermatophore dehiscing. *Crangon vulgaris*: E, first right pleopod of female in breeding dress; F, transverse section of endopodite of this pleopod to show eggs attached to ovigerous setae; G, lateral view of abdomen to show that the spawn mass does not interfere with its flexure during backward escape movements. (C, after Bloch; D, after Jackson; E, F, G, after Yonge.) abd, abdominal segment; abd.app 1, first abdominal appendage; abd.app 2, second abdominal appendage, its tip inserted in cavity of first; basip, basipodite; cap, capsule; endop, endopodite; exop, exopodite; g.pap, genital papilla in cavity of first abdominal appendage; h.g, region of hind-gut; mem, egg membrane; n, nucleus; ov.s, ovigerous setae; per, last pereiopod; pin.s, pinnate setae; r, ribbon to which spermatophores are attached; sp, spine; t, telson; urop, uropod.

and transferred from one individual to the next by a long flexible penis (Fig. 119C, pe). There are no spermatophores. The penis is placed accurately into the mantle cavity of the partner where the spermatozoa are discharged and the eggs fertilized. To do this it may more than double its length.

The eggs may pass from the oviducts into the sea but are more usually retained on the body, either entangled in a special secretion or enclosed in a pouch, during some or all of their developmental stages. Those which are freed, predominantly in

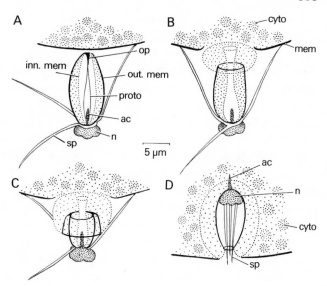

Fig. 135. *Diogenes pugilator*: fertilization. A, spermatozoon attached to egg membrane by spine. B, C, capsule open and everting. D, eversion complete bringing firing-pin (acrosome) and nucleus into cytoplasm of ovum. (After Bloch.) ac, acrosome; cyto, cytoplasm of egg; inn.mem, inner membrane of capsule wall; mem, egg membrane; n, nucleus of sperm; op, operculum; out. mem, outer membrane of capsule wall; proto, protoplasmic connection between acrosome and operculum; sp, spine.

marine planktonic species (e.g. *Calanus*, most euphausiids, penaeid prawns), are relatively smaller and more numerous. Membranes covering the eggs may be produced solely by the oviduct, which characteristically secretes a thin transparent covering of chitin around each (indicating its ectodermal origin). Additional secretions from the duct differ considerably from group to group or may be absent; they afford an amount of protection varying with the requirements of the young and the method of incubation. Gammarids protect the young in a brood pouch where each completes its development within a transparent membrane. In copepods an oviducal secretion cements the eggs together to form an ovisac – really just a mass of eggs – which hangs from the genital segment and retains the young until they have reached the nauplius stage. The most lasting covering is found in branchiopods (e.g. *Chirocephalus, Artemia, Triops*) in which the protection of the eggs during periods – perhaps years – of prolonged desiccation is essential to the survival of the species. *Chirocephalus diaphanus*, an inhabitant of temporary freshwater pools, has a median pouch (Fig. 128A, p), projecting ventrally from the twelfth segment (immediately behind the phyllopodia), where the eggs are carried for a short time before liberation. It is formed by the union of the two oviducts and has considerable concentrations of glands (cap.gl) which secrete a resistant outer wall around each egg after it has been fertilized. This wall is thick, spiny and not chitinous. It resembles the second egg covering of decapods (see below) which is allied by its chemical and physical properties to the epicuticle covering the outer surface of the body, and has the same selective permeability. In *Artemia*, the brine shrimp, the eggs are retained in the ventral brood pouch under favourable conditions and hatch as nauplii; under less favourable conditions they are laid and either develop rapidly or lie quiescent. Apus has a long series of phyllopodia extending beyond the genital segment, and a median egg sac would interfere with their functioning. Each oviduct has its own sac (Fig. 115A, p) formed from the two exites of the phyllopodium of the genital segment (11th thoracic) and in no way interfering with the functioning of the other parts of the appendage. Copulation occurs after ecdysis and within a few hours a batch of eggs is extruded into each brood sac. They are freed immediately before the next ecdysis and their sticky surfaces anchor them to plants or stones.

Similarly in higher crustaceans parts of appendages not vital to other activities of the animal are used to carry the eggs. The pleopods of female decapods support the relatively enormous volume of a single spawn mass beneath the abdomen until the young have reached an advanced stage of development and are ready to hatch as larvae, or even as miniatures of the adult (*Astacus*). This may take several months. Indeed the exploitation of some habitats by the more advanced crustaceans is largely due to the retention of the developing eggs and the protection afforded them by the integumental secretion. It comes from cement glands related to special ovigerous setae, and not only forms the outer egg membrane but also the stalk by which the eggs are attached to the seta. The placing of the spawn in decapods varies with the mode of life and the use to which the abdomen is put. In species in which it is an important locomotor organ, as in *Crangon*, *Leander* and *Palaemonetes*, its rapid flexing for escape movements would be impeded by a spawn mass extending along its length, and eggs are not carried beyond the 4th pair of pleopods, and are, moreover, concentrated on the protopodites of the pleopods and carried close to the body. The position of the spawn in the shrimp (*Crangon vulgaris*) is still further influenced by the burrowing habit of this species. The egg mass (Fig. 134G, spawn) is conspicuously long and shallow and is attached to unbranched setae arising from the protopodites of the last two pairs of thoracic appendages, as well as those of the first four pairs of pleopods and from the endopodites of the first pleopods. This leaves the exopodites of the abdominal appendages free to carry out burrowing activities.

Only the thorax is concerned with locomotion in crabs and the appendages of the reduced abdomen are concerned solely with reproduction. The abdomen shows marked sexual dimorphism. In the female it is broad, all the segments are movable and it is not so closely applied to the thorax as the abdomen of the male. The first pair of pleopods at the angle between thorax and forwardly directed abdomen is reduced, but the endopodites of pleopods 2–5 bear ovigerous setae. They retain the eggs in the space between posterior thorax and upturned abdomen, and the exopodites embrace the mass externally; in a large *Cancer* this may comprise 3×10^6 eggs. The abdomen of the hermit crab, *Eupagurus*, is protected by a coiled gastropod shell and both rami of the pleopods carry eggs. The asymmetry and configuration of the shell are reflected in the development of the pleopods, which are present only on the left side, and in the fact that no eggs are attached to the last pleopod in the narrower apical region of the shell.

In most reptant decapods female secondary sexual characters, such as the broad abdomen with deep sides and the ovigerous setae with their special glands, are acquired at maturity and appear to persist. In natant forms a number of these characters are seasonal and in shrimps and prawns (*Crangon, Leander, Palaemonetes*) the setae appear only when the female moults into the egg-carrying condition (Fig. 134E, F). Other setae associated with the breeding dress may function only for a brief period during the spawning process. Some of these appear on the posterior sterna of the thorax and guide the eggs from the genital openings to the place of attachment of the spermatophores and towards the pleopods. The abdomen is turned forward beneath the thorax to form a chamber into which the eggs are shed and other special setae temporarily close it at the sides to prevent their escape. The fertilized eggs are propelled into it by setae on the first pleopod and rotated in the cementing fluid which, owing to its low surface tension, flows evenly over each, binding all to the ovigerous setae.

Some hermit and true crabs, abundant in the tropics, have the gill chambers adapted for breathing air and in this respect have gone far towards becoming land dwellers. They live far from the sea but must return at intervals since, as in other decapods, the eggs hatch as larvae. The only crustaceans which can live and reproduce entirely on land are members of the Oniscoidea, a suborder of the Isopoda. In these, as in other peracaridans, the young are retained by the female until they have the form of the adult. The pouch in which they develop lies between the ventral surface of the thorax and thin-walled plates, the epipodites or oostegites (Fig. 123B, oost), attached to the inner surface of the coxopodites of some thoracic legs. It receives the openings of the oviducts (ovd). Movement of the plates allows interchange between the medium within the pouch and that outside, and in mysids their pulsation maintains a flow of water over the developing young which supplies oxygen and carries away waste. In terrestrial isopods the eggs develop in a watery medium and are by no means cleidoic. The system of capillary channels over the surface of the animal by which external water is conducted to the respiratory surface of the pleopods (p. 347) also serves the brood pouch. The developing embryos may receive nourishment from the parent by way of four vascular folds of tissue, the cotyledons, which, in *Oniscus* and *Porcellio*, project into the brood pouch from the ventral body wall. Cotyledons and oostegites are discarded at the moult which follows liberation of the young and the mother reverts to the nonbreeding dress. However, if another batch of eggs has begun development, new cotyledons and oostegites appear with the new integument.

In some Crustacea part of the cavity enclosed by the carapace is used as a brood chamber in direct communication with the oviducts and away from interference by the feeding and locomotor activities of the limbs. The lower part of the mantle cavity of a barnacle accommodates several hundred shelled eggs which are agglutinated by secretion from the distal part of the oviduct. They hatch as nauplii. A rather more circumscribed part of the cavity is used to accommodate the eggs of water fleas such as *Daphnia* and *Simocephalus*. Again it is dorsal, between body wall and carapace, but it is closed posteriorly by two strong processes projecting into it from the trunk. Movements of the trunk draw these away from the opening, and allow a circulation of water through the pouch. Thin-shelled, parthenogenetic eggs, laid during favourable conditions, complete development in 2–3 days and are liberated. At other times the eggs must be fertilized before they will develop. Only two are produced, a thick resistant shell is secreted around them and they are cast off when the female moults. At this moult a saddle-shaped, dorsal area of the carapace, the ephippium, becomes especially thick, and separates from the rest to form a case around the resting eggs. There are some cladocerans, such as *Moina*, which have a more intimate association between embryo and parent: the brood pouch is closed and the embryos are nourished by secretion passed into it from the parent.

The life cycle of crustaceans agrees with that of other arthropods in being made up of instars, separated by ecdyses or moults. Most species are marine and their early stages are usually planktonic larvae, spreading the species and growing up at the same time. The more primitive groups, the branchiopods and free-living copepods, have the more primitive life histories in which larva and adult have much the same habitat and lead much the same sort of life, and transition from one to the other is gradual. In more advanced and specialized groups larva and adult frequently pursue a different mode of life, for the adult may be benthic, as in crabs, sessile, as in barnacles, or parasitic, as in some cirripedes, copepods and isopods. The transformation from early larva to adult then involves changes in body form at certain stages which are sudden and radical and constitute a metamorphosis as striking as that of an insect.

The embryo either hatches as a nauplius or, if hatching is delayed, a naupliar stage is frequently recognizable during development. The nauplius larva (Fig. 125A, C) is characteristic of the class, occurring in both Entomostraca and Malacostraca and suggesting the monophyletic origin of a very varied assemblage of forms. It resembles the trochophore of annelids in consisting essentially of cephalic and anal regions in juxtaposition. There are three pairs of jointed appendages which will become the first three pairs in the adult; a large labrum (labr) overhangs the mouth (m) and dorsally there is a cuticular shield (car) through which can be seen a simple median eye (e). The antennules (ant') are mainly balancing organs, the antennae (ant") the principal locomotor organs, whilst these and the mandibles (md) may share the function of food collecting and have special spines and processes for this activity. Each moult introduces additional somites and limbs (Fig. 136) which develop the characteristics of the adult, differentiation proceeding backwards

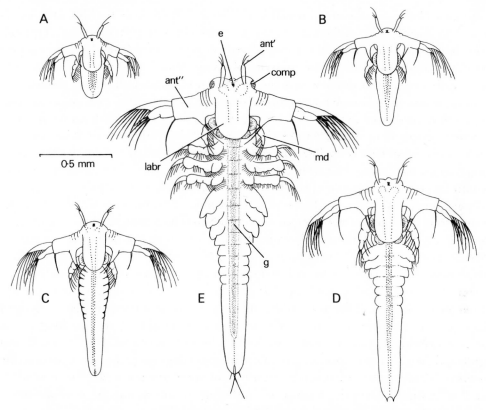

Fig. 136. *Artemia salina*: larval stages showing progressive increase in the number of segments and appendages. (After Sars.) ant´, antennule; ant″, antenna; comp, compound eye; e, median eye; g, gut; labr, labrum; md, mandible.

along the trunk. This pattern of development is best displayed by the branchiopods in which changes between one instar and the next are of more or less equal importance, and so small that the development may be called continuous. In the Copepoda, however, changes in body form, both external and internal, are more pronounced after certain ecdyses. In *Calanus* the first six stages, all naupliar, have only three functional pairs of appendages, though the masticatory bristles on the basal joints of the antennae are not well developed until the third stage, for the first two do not feed; at the fifth, maxillules are present and at the sixth, maxillae. The next moult brings greater transformations. The larva then has maxillipeds and two pairs of biramous swimming feet on the thorax, and a hind region from which all remaining somites of thorax and abdomen will later differentiate; there are also several important internal advances towards adult organization. This stage is known as the first copepodid and by a series of instars (six in *Calanus*) leads without further abrupt transition into the adult.

In barnacles also there are two planktonic forms which constitute the initial part of the life history but differences between them are more striking than between those of *Calanus*. The nauplius, as in all cirripedes, is distinguished by anterolateral spines on the dorsal shield. After a series of six naupliar stages which show steady increase in size, in the number of setae on the appendages and, at the sixth stage, the appearance of rudimentary maxillules, the next ecdysis gives a cypris larva (Fig. 125B), so named because of its superficial resemblance to the ostracod *Cypris*. The carapace (edge) is bivalved, with an adductor muscle (add.mu) drawing the two halves together so that it encloses the whole body; through it can be seen a pair of compound eyes (comp) as well as the nauplius one (e). The large antennae, mandibles and caudal spine of the nauplius are reduced and six pairs of biramous thoracic appendages (th.app) function in swimming. The antennules (ant') enlarge, project from the front of the carapace, and each develops a sucker-like disk bearing the opening of cement glands (gl) which help in attaching the larva when it settles. Development to this stage, with a body form essentially similar to that of the adult, takes about three weeks. The cypris swims for a time and then attaches itself by the antennules to a stone or rock in the vicinity of barnacles of the same species. Metamorphosis into the sessile adult involves a reorientation of the body and changes in the proportions of its parts: the thorax comes to lie across the long axis of the larval carapace, which is lost and replaced by another enveloping the body and reinforced by calcareous plates; the thoracic appendages lengthen and their setose covering is elaborated. In this life history metamorphosis takes place in two steps, the first (nauplius – cypris) leading to the adult organization, the second (cypris settling) orientating this to suit the definitive mode of life.

Essentially the same kind of life cycle is exhibited by many higher malacostracans with benthic adults. Since in them, however, the nauplius is represented by an embryonic phase the start of free life is at a higher level of organization. This is the zoea (Fig. 137A), the larval stage of the prawns, shrimps and crabs. It already shows certain malacostracan features. There is a cephalothorax somewhat laterally compressed and covered by a carapace (car) fused with the thoracic terga, and a relatively large abdomen, with five (later six) distinct segments, ending in a telson (t). There are lateral compound eyes (comp) and sometimes a nauplius eye. All the head appendages (ant', ant", md, mx', mx") and the first two (crabs) or three (prawns) pairs of thoracic appendages are developed, but none posterior to these; the rest of the thorax is still rudimentary with segments barely differentiated. The carapace is produced into spines: a rostral spine (r.sp) is of frequent occurrence, together with one dorsal (d.sp) and a lateral spine (lat.sp) on each side in crabs (though rarely in other decapods), or with only a pair of posterior lateral spines which are long in squat lobsters (*Galathea*, *Porcellana*) and short in hermit crabs. The larva swims with the thoracic appendages (mxp) which will become the maxillipeds of the adult; they are biramous, the rami being long and setose. The abdomen assists in swimming and is rhythmically flexed beneath the thorax and then stretched out behind. In some macrurous forms, like the common shrimp *Crangon*, the last pair of abdominal limbs, the uropods, develops precociously and forms an effective tail fan with the telson. Certain structures of the larva – spines on the carapace, large forked telson, elongated maxillipeds – fulfil larval needs only and are later modified or lost. The larva lives in the surface water feeding on minute planktonic organisms and growing in size and complexity with repeated castings of its skin; the differentiation of its thoracic segments and appendages occurs in a regular anteroposterior succession.

The common edible crab *Cancer pagurus* has five zoeal stages each lasting a few days and distinguished by an increasing number of setae on the swimming legs as the larva grows. Rudimentary pleopods appear at the third stage when rudiments of the thoracic appendages are obvious. At the fifth moult the zoea metamorphoses to a form known as a megalopa (Fig. 137B), transitional between zoea and adult, which lasts for about twelve days. This post-larval stage, characteristic of all brachyuran crabs, exhibits most of the external features of the adult: the cephalothorax is depressed; the maxillipeds are solely mouth parts and no longer natatory; the four posterior pairs of thoracic appendages are walking legs held close to the body whilst the animal swims, and the great chelae have developed claws; the long spines on the carapace have disappeared, though there is still a short rostral spine and backwardly projecting dorsal spine; the abdomen is relatively large and projects horizontally behind the carapace, exposing the pleopods (pleop) which are the functional swimming organs and bear long swimming setae. The megalopa, although still planktonic, tends to sink through the water, eventually reaching the benthic habitat of the adult where it secures its position by strong spines (sp) projecting from the basal joints of the chelae and three anterior pairs of pereiopods as well as by the claws of these appendages.

The life cycle of some natant eucarids (euphausiids and penaeid prawns) includes both the basic crustacean larva, the nauplius, and the malacostracan larva, the zoea, followed by a post-larval stage with abdominal propulsion which merges into the adult. Larvae and adults live in the same habitat so there is no marked metamorphosis. The egg has sufficient yolk to make it unnecessary for the nauplius to feed and this stage consequently lacks mouth and food-collecting processes on the antennae and mandibles, which are solely swimming appendages. In the prawn the developmental stages are well marked by changes in the type of swimming organs, for whereas these are at first cephalic appendages, they are thoracic in the zoea and abdominal in the post-larva. The penaeid zoea has all the thoracic segments distinct and equally developed, bearing appendages with swimming exopods. This is in contrast to the euphausiids (except *Euphausia superba*) and higher decapods, like the crab already described, in which thoracic segments and appendages are much delayed in their appearance. In respect of its thoracic structure the penaeid zoea

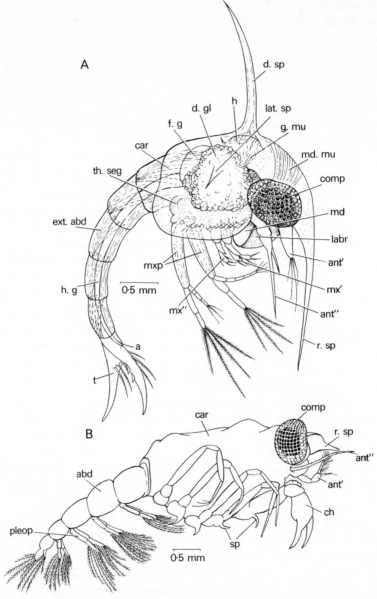

Fig. 137. A, *Carcinus maenas*: first zoea. B, *Pilumnus hirtellus*: megalopa. a, anus; abd, abdomen; ant′, antennule; ant″, antenna; car, carapace; ch, cheliped; comp, compound eye; d.gl, digestive gland; d.sp, dorsal spine; ext.abd, extensor muscles of abdomen; f.g, fore-gut; g.mu, muscle of gastric mill; h, heart; h.g, hind-gut; labr, labrum; lat.sp, lateral spine; mx′ maxillule; mx″, maxilla; md, mandible; md.mu, mandibular muscle; mxp, maxilliped; pleop, uniramous pleopod; r.sp, rostral spine; sp, spines; t, telson; th.seg, thoracic segment developing.

resembles the later larval stages of the caridean shrimps and prawns, and also the form which hatches from the egg in the reptant macrurans such as *Nephrops* and *Homarus*. It may therefore be regarded as the type of zoea characteristic of the macrurans, sometimes referred to as the "schizopod" or "mysis" stage on account of its biramous thoracic legs, both names which, unfortunately, imply false relationships. In the post-larval stage the pleopods become setose and functional and exopods disappear from the legs.

In a number of malacostracans the embryonic period of development is extended and the young hatch in a form which does not differ essentially from the adult. This tendency to shorten the life history is sporadic in the decapods, reaching a climax in the freshwater crayfish in which all larval stages are suppressed, but it is general throughout the Peracarida. Females of this group have a brood pouch or marsupium (from which the name Peracarida is derived) in which the young develop to resemble miniature adults. In the isopods, however, they leave the pouch with the last pair of thoracic legs still undeveloped, and in this same order some members (the epicaridean and gnathiid isopods) have special secondary larval stages associated with their habit. In other parasitic Crustacea the typical larval stages occur with varying degrees of modification. In *Sacculina* the nauplius and cypris stages are identical with those of barnacles except that there is no gut. In the copepod *Chondracanthus* the copepodid larva is neotenic in the male and attached, like a secondary parasite, to the body of the enormously larger female.

Classification of arthropods

Arthropoda
  Onychophora
  Crustacea
    Cephalocarida
    Branchiopoda
      Anostraca
      Notostraca
      Diplostraca – Conchostraca and Cladocera
    [Mystacocarida]
    [Ostracoda]
    Copepoda ————————————Maxillopoda
    [Branchiura]
    Cirripedia
    Malacostraca
      Phyllocarida (= Leptostraca)
      Eumalacostraca
        Syncarida

Hoplocarida
Peracarida
   Mysidacea
   [Cumacea]
   [Tanaidacea]
   Isopoda
   Amphipoda
Eucarida
   Euphausiacea
   Decapoda
Chelicerata (= Arachnida)
Myriapoda
  Diplopoda
  Chilopoda
Insecta

12

Land arthropods

IT IS CUSTOMARY to group the onychophorans, myriapods, insects and arachnids together as "land arthropods", but it must be clearly understood that these are in no sense a closely related collection of animals. Indeed the arachnids (or chelicerates) have almost certainly a totally different origin from the others, as is indicated by their possession of biramous rather than uniramous limbs, whilst within the Uniramia the onychophorans represent one primitive evolutionary line and the myriapods and insects other though closer lines. Nevertheless, the name "land arthropods" has the advantage of emphasizing that these animals, along with all other terrestrial creatures, have solved a series of problems relating to support and locomotion in a medium as little dense as air, and to water control, respiration and reproduction in a frequently desiccant environment. Many of their solutions have proved similar, probably because they alone were within the competence of animals all at the arthropodan grade of organization, and this has contributed to the collection of convergencies which misled early students into believing that arthropods were more closely allied than they really are.

To deal in this book with land arthropods on the same scale as the other taxa would double its size since they constitute an overwhelmingly large proportion of all animal species (see p. 8) and this in turn implies a correspondingly large range of adaptive radiation. This chapter, therefore, is no more than a glance at the ways in which they have adapted for movement, water conservation, respiration, excretion and reproduction, the major problems of all terrestrial animals.

As indicated above (p. 269) the earliest arthropods may well have been soft-bodied, without the chitinous exoskeleton which most present day forms possess. This is still true of the onychophorans (Peripatus) which have a slightly spiny cuticle but are without sclerites on limbs and body. Because of this they must be taken as the most primitive land arthropods and probably the relics of the original arthropodan invasion of the land. In other groups sclerites are present on the body and limbs and, on the assumption that these had appeared whilst the animals were still aquatic, they represent a pre-adaptation for terrestrial life in that they make the limbs sufficiently powerful not only to propel the animal but to lift it clear of the ground in doing so. In most arthropods the legs are long enough for their proximal parts to form a sling suspending the body at a low level; this keeps the centre of gravity low and makes the animal stable, an important advantage since many land arthropods are small enough to be affected by relatively slight movements of the air.

When a leg steps it is raised, moved forwards and extended to make contact with the substratum and then forced backwards, shortening at first and then elongating once more. (In this respect the arthropod leg differs from the parapodium of an annelid which ends the forward stroke more or less at right angles to the body and shortened; during the backstroke it elongates and becomes longer still because the aciculum is protruded.) During the backward stroke the leg exerts a force on the substratum which propels the animal forwards. As with a motor car the power which the limb exerts is greater when the speed is low (what has come to be called "bottom gear") and less when the speed is high ("top gear"). When an animal like Peripatus starts to move (Fig. 138) it employs bottom gear, when it has got under way it uses a middle gear and when it needs to move rapidly it may employ a top gear. Bottom gear is characterized by the fact that the backward, propulsive stroke lasts longer than the forward one; this may be expressed by the ratio 3.7 : 6.3, representing the fraction of the time of one whole step occupied by the forward and backward strokes respectively. It is also characterized by the fact that, since the backward stroke is lengthy, more of the animal's legs are in it and in contact with the ground at any instant, so increasing the force exerted on the substratum to get the animal into movement. As the animal picks up speed the step is altered to a middle gear (5 : 5; half the legs on ground, half off) and ultimately to top gear

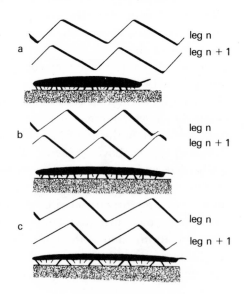

Fig. 138. To show gait of Peripatus. Each diagram shows the outline of an animal moving to the right—a, in bottom gear; b, in middle gear; c, in top gear. Above each animal is a plot against time of the movement of two successive legs, the thick part of the line representing the backward propulsive stroke, the thin part the forward recovery stroke. Note the differing number of legs in contact with the ground in each of the three gears, their different lengths and the change in the length of the whole animal. (After Manton.)

leg n
leg n + 1

leg n
leg n + 1

leg n
leg n + 1

(6.5 : 3.5) when each limb is in contact with the ground for a shorter fraction of the step time, though the duration of a full step (backward plus forward strokes) is still more or less the same (about 0.75s). Because of the short stay on the ground fewer limbs (less than half) are in contact with it at a given moment and the load which each carries is correspondingly greater. Other devices may be used to acquire speed as in our own walking: a longer limb has this effect as, too, does the longer step produced by swinging the leg through a greater angle. The second device may be used by Peripatus to accelerate in middle gear but is impractical with top gear because of the heavier load which each limb would be required to carry, particularly when obliquely extended at the end of each step. Elongation of the limbs is also part of this acceleration, but because longer swings tend to interference of the movement of neighbouring limbs it is accompanied by an elongation and narrowing of the body to separate them. These changes are possible only because of the lack of sclerites and involve the continuous muscle sheets of body and limb walls acting on the haemocoelic fluid as a hydrostatic skeleton.

Peripatus, even in accelerated middle or top gears, cannot creep more rapidly than about 8–9 mm s^{-1}: this and its lack of exoskeletal armour expose it to attack by predators. It has some protection in its nocturnal and self-effacing behaviour and in the entangling secretion of a pair of slime glands opening on the oral papillae (Fig. 109), but the softness of the body wall combines with a thick elastic subepidermal layer of connective tissue and extremely extensible muscles (unstriped and so unlike those of other arthropods) to permit an outstanding ability to squeeze the body through very narrow apertures leading to retreats inaccessible to predators without equally deformable bodies.

The myriapods contain two familiar kinds of animal, the Chilopoda or centipedes and the Diplopoda or millipedes, together with two smaller groups, the pauropods and symphylans, to which no reference will be made. The mode of life of the two major groups tends to be different in that centipedes are carnivores and, like most such animals, have to be fast-moving to overcome their prey's efforts to escape, whereas millipedes are vegetarians, burrowing into soil, fallen timber and leaves in their search for food. Some centipedes, like the common *Geophilus*, also burrow, but employ a mechanism unlike that used by millipedes. There are therefore two different adaptations to examine, that leading to the ability to burrow and that leading to the ability to run fast.

Speed may be gained in several ways: by accelerating stepping; by using top gear; by lengthening legs; by longer strides. In the fast-running centipedes some or all of these devices may be employed. *Cryptops*, a small, blind centipede of gardens, can reach a speed of about 300 mm s^{-1}. This is at first due largely to acceleration of the rate of stepping but a decreased duration of the propulsive stroke (that is, a change to a higher gear) is added to give the faster rates. During most of the increase of speed the length of the stride remains unchanged but is, in effect, reduced when the animal is most active because the strong contractions of the remotor and promotor muscles, moving the legs backwards and forwards and out of phase on the two sides of the body, tend to fling it into lateral waves. In the common centipede *Lithobius* (Fig. 139A), all the devices are used, initial acceleration being due primarily to speeding up the stepping rate together with some increase in the angle through which the legs swing; the fastest rates are due to further increase in this angle and a decreased duration of the power stroke.

The extension of a leg forwards and backwards can be increased by a process called rocking, a twisting of the segments of the limb along its long axis. In centipedes (Fig. 140C), when the leg is swung forwards its dorsoventral axis twists backwards; when the leg moves in a backstroke its dorsoventral axis rotates forwards. These movements bring the hinge joints between the segments of the leg into a position where the action of the depressor muscles is facilitated and since their contraction straightens the leg they have the effect of making it into a longer lever. The possession of longer legs not only increases the length of the lever by which the animal is propelled but, if they are of different size, allows longer striding without the movement of one limb interfering with that of its neighbours. The possibility of tripping is further reduced by an evolutionary shortening of the body and so of the number of legs. This trend reaches its climax in the centipede *Scutigera* which has a length of about 2 cm, only fifteen trunk segments, limbs ranging from about 10 to 18 mm in length and which can run at speeds up to 420 mm s^{-1}, fast enough to let it catch flies. Like *Cryptops*, both *Lithobius* and *Scutigera* tend to waste energy in lateral undulations of the body when running rapidly. Two features seem to minimize this: the tergites are alternately long and short and tergite–tergite joints do not normally lie directly over those between successive sternites (Fig. 139B).

Centipedes in a section known as Geophilomorpha, of which the long, thin, yellow *Geophilus* is a familiar example, can burrow as well as creep over a surface. Oddly, this turns out to be effected by dilatation of the body in much the same way as in earthworms – a strange ability in an arthropod, where normally the body is made rigid by the exoskeleton. In centipedes, however (Fig. 139B), there lies, on each side of the body, between the tergites and the sternites, an area of flexible cuticle (fl.cut) from which the legs (l) spring and in which the spiracles (sp) lie. It is this which allows the geophilomorph body to change diameter and so burrow, and it is this which gives to centipedes like *Lithobius* the freedom of movement of the legs essential for their pursuit of speed. As in all burrowing animals, from earthworms to moles, lateral projections are minimized to prevent damage against the walls of the burrow and the legs and antennae are short. The fact that shortness of the legs inhibits speed is not of much significance to burrowers and with loss of speed the shortness of the body found in *Lithobius* and *Scutigera* becomes unimportant; geophilomorph centipedes, indeed, have become secondarily elongated.

Most millipedes burrow but their methods of so doing are unlike that of

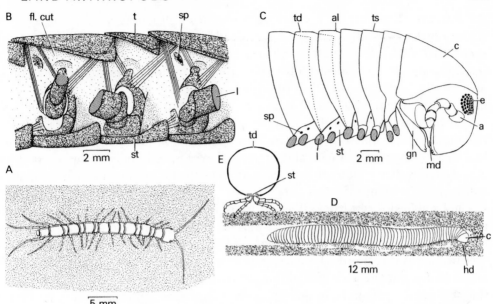

Fig. 139. A, *Lithobius* in dorsal view. Note the alternation of long and short tergites. B, *Lithobius*, three segments in right lateral view to show the weak sclerotization of the side wall of the body; the legs have been cut off. C, a juliform millipede, head and first seven trunk segments in right lateral view. The legs have been cut off. D, a juliform millipede in the process of burrowing. E, transverse section of body. a, antenna; al, anterior limit of telescoping; c, collum; e, eye; fl.cut, flexible cuticle; gn, gnathochilarium; hd, intucked head; l, leg base; sp, spiracle; st, sternite; t, tergite; td, tergite of diplosegment; ts, tergite of single segment. (Based on Manton.)

geophilomorphs and depend on the movements of legs rather than of trunk musculature. There are two ways of burrowing and the organization required is different for each, producing two types of body, one exhibited by the diplopods classified as Oniscomorpha and Juliformia, the other by the Polydesmoidea and Nematophora. The former are typified by the familiar, more or less cylindrical millipedes *Glomeris* and *Blaniulus*, while the latter includes what may be called "flat-back" millipedes like *Polydesmus*. All burrow by pushing into the substratum or such material as the leaf litter on the floor of woodlands. Oniscomorph and juliform millipedes (Fig. 139D) push by means of a specially armoured tergite, known as the collum (c), on the first segment behind the head (hd); it acts like a bulldozer and is pushed forwards by the action of the legs behind. The flat-back millipedes raise themselves on their legs and press upwards by means of the tergites of the trunk segments; since they deal with litter with a lamellar structure this action separates the layers and opens up spaces into which they may then creep.

In the burrowing of juliform millipedes considerable power is necessary but the legs, like the other appendages, must be short, the one requirement cancelling the other. Power may be increased by multiplying the number of legs, that is by raising the number of segments; by itself, however, this produces an animal which is inefficient because of its length and slender proportions and because of the probability of buckling in transmitting forces along a considerable distance of jointed body. The millipedes solve this problem by increasing the cross-sectional area of the segment and then uniting segments in pairs to escape from the narrowness and multiplicity of intersegmental joints that would result if fusion did not take place. The unit of the millipede body is therefore a diplosegment formed by the union of two embryonic segments; it reveals this origin by bearing two pairs of legs and two pairs of spiracles (Fig. 139C). Power is also obtained by use of a low gear gait, involving many appendages applied to the ground in their propulsive strokes at any instant.

The combined thrust from all legs is transmitted by way of the exoskeleton to the collum. During burrowing the head is turned backwards on the ventral side so that the collum becomes the most anterior part of the body and it is then pressed into the substratum; as it is also the broadest part of the body it acts like a tunnelling shield and creates a burrow wide enough for the rest of the body to enter. The first three rings behind the collum bear only one pair of legs apiece; this leaves a space for the intucking of the head during burrowing. Whether these segments are single or simplified diplosegments is not clear. In a mechanism of this sort there is danger of loss of power in forward transmission from the posterior legs to the collum, leading to telescoping of segments; bending would be largely prevented by contact with the walls of the burrow. Overshortening is prevented by the relationship of the exoskeletal rings of successive diplosegments. In each the tergite is C-shaped, the gap being mid-ventral and filled by the sternite (Fig. 139E). The anterior border of each tergite projects into the posterior half of the ring in front and forms a tapering, partly inturned surface which articulates with the tergite of that segment, forming a ball and socket joint or cone-in-cone arrangement. Movement of the posterior into the anterior segment is arrested by a ringlike constriction on the more anterior tergite. This produces an apparatus which telescopes only to a limited extent but which can twist dorsoventrally or laterally as required, successive tergites being able to rotate round one another along transverse, dorsoventral or sagittal axes.

The flat-backed millipedes push in similar fashion, but upwards, not forwards, straightening their legs to raise the expanded and flattened tergites of their trunk segments. Juliform millipedes burrow into tunnels: their body is rounded to fit the tunnel, smooth, with no projections, and the legs, though not especially short, appear so because they have a nearly mid-ventral attachment and hardly project beyond the outline of the body. In the flat-backed polydesmoids the tergites are drawn out ventrolaterally into keels roofing a space in which the legs lie. These are longer than in juliform animals and the whole body is flattened in section. These features, however, in no way detract from the efficient movement of animals living in the low but otherwise relatively extensive spaces of leaf litter.

The remaining land arthropods are the insects and arachnids, in which the number of pairs of walking legs never exceeds three (insects) or four (arachnids); in the latter group, indeed, one pair of legs is often not used in locomotion, making them all hexapodous. In most of these walking consists of alternate stepping with two sets of three legs, the animal being supported on one triangle whilst the other legs step. The first and third on one side work along with the opposite middle leg. The advantages of reducing the number of legs are twofold – they may be longer and therefore both more powerful levers and capable of greater striding, and, linked with this, their movements may be so arranged as never to interfere with one another. For mechanical stability and the suppression of undue lateral curving, however, it is essential that the legs should arise close together: it is easy to see, for example that if the three pairs of legs of a fly were to be placed one pair at each end and the third in the middle, stepping would certainly fling the animal into curves facing alternatively left and right, and the rigidity required to prevent this would either make the body unnecessarily heavy, or interfere with other activities, or both. In insects and arachnids, therefore, the legs spring from one part of the body only, the thorax in insects, the prosoma of arachnids, and the posterior part of the body carries none.

These resemblances, however, prove once more to illustrate the fact that mechanical requirements have often caused an ultimate similarity of originally dissimilar structures, because, when examined in detail, the movements of insect and arachnid limbs in stepping are found to show one fundamental difference. In insects, as in myriapods and crustaceans, anteroposterior limb movement takes place between their basal section (coxa) and the sclerites of the body: in arachnids (except for *Limulus* and to a certain extent spiders), and only in arachnids, the coxa is immovably fixed to the body and movement of the limb occurs at more distal joints. From this it must be concluded that arachnids represent an evolutionary line distinct from the onychophoran-myriapod-insect group, an idea already suggested by the uniramous form of the limb in the one and its biramous character in the other. The movement of the arachnid leg varies from group to group: in some a backward and forward movement comparable to that of other arthropods takes place, though the joint at which this occurs is never between limb and body but always at some point within the limb (the coxa-trochanter joint in ticks and mites; the trochanter-femur joint in harvestmen). In scorpions (Fig. 140A, B) no real promotor-remotor movement is found; instead the limb bends dorsally and then straightens with its plane of bending about an axis transverse to the long axis of the body instead of vertical to it. (If the promotor-remotor movement of an ordinary arthropod leg may be compared to the movement of an oar in rowing, the scorpion's mode of stepping is like the action of the pole in punting.) Were the stepping of the scorpion leg to be executed solely in this fashion the leg would move extensively upwards during flexing and this would interfere with burrowing. To

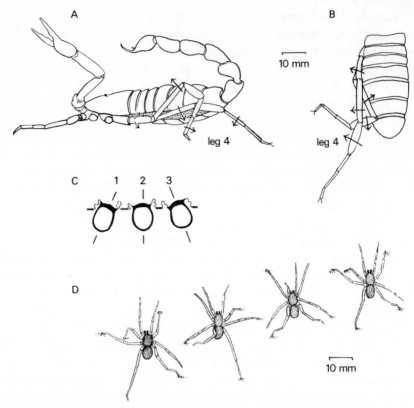

Fig. 140. A, scorpion in side view; B, hind end of prosoma in dorsal view. In each diagram the fourth leg is shown in both the flexed and extended state. Arrows indicate direction of rocking. C, diagram of myriapod leg cut across dorsally to show rocking. In position 1 the leg is at the end of a forward stroke and its basal joint has rocked backwards. In position 3 the leg is at the end of a backward stroke and its basal joint has rocked forwards. D, four successive positions of a spider running to the top of the page to show leg movements. These involve (1) some promotor-remotor rotation of the coxa; (2) a punt pole-like action of the legs. Legs in contact with the ground indicated by shading. Right leg 1 does not step. (Based on Manton.)

prevent this the limb rocks as it bends and so lies over the animal's back in a more horizontal plane rather than projecting into the air. If it affected the whole length of the limb this rotation would cause the distal tip to lose contact with the substratum and the whole propulsive effort would be wasted; this tendency must therefore be corrected by a counter-rotation keeping the tarsal joints firmly on the ground (see arrows in Fig. 140A, B). Spiders (Fig. 140D) retain a little mobility between the coxa of the leg and the body which permits some backward and forward motion and some rocking at this joint. Most of the movement, however, is achieved by the same kind of punt pole action as in the scorpion, posterior limbs pushing and anterior limbs pulling. The muscles primarily responsible for straight-

ening the limbs are their depressor muscles. Rocking and counter-rocking motions
ensure maximal extension and retention of a grip on the substratum.

Alone amongst invertebrates most insects fly. Some, however, are wingless; this
may be a secondary state as in fleas and lice, part of an adaptation to a particular
mode of life, or it may be primary as in collembolans (springtails) and thysanurans
(bristletails) which have never evolved wings. The organs of flight are two pairs of
wings carried on the second (mesothoracic) and third (metathoracic) segments of
the thorax. In some orders the wings are, actually or effectively, reduced to a single
pair – in beetles because the mesothoracic pair have become protective elytra
covering the hinder pair which alone move; in Hymenoptera because the metathoracic pair are coupled to the anterior pair and are driven by them, whilst in Diptera
the posterior pair are converted to halteres, organs involved in the control of flight
but not in its production.

Like birds and bats insects fly by flapping their wings so as to create a current of
air from which they gain lift. The direction of the air-flow may be modified by the
orientation of the wings so that the insect can hover, or fly forwards or backwards;
sometimes wing movement occurs without locomotion, as when standing bees
ventilate their hive. The versatility cf the wing is due to its relative flexibility and to
the action of a number of muscles which run from its base to the side walls of the
segment on which it is borne. Some of these lower it, others raise it and these and
others can twist it in different ways. These muscles are known as the direct flight
muscles. They are not, however, except in dragonflies, the main source of power for
wing movement; this is to be found in the indirect flight muscles, so called because
neither their origin nor their insertion is related to the wing base. There are two
main indirect muscles: (1) the dorsal longitudinal and (2) the dorsoventral muscles.
The former (Figs 141, 143A, dl) run between apodemes at the anterior and
posterior ends of the segment and so, on contraction, tend to shorten it and arch its
roof, an action which has the effect of depressing the wings; the latter (dv) run from
the tergum to the sternum or the basal joints of the leg. On contraction they lower
the roof of the segment and are thus directly antagonistic to the longitudinal
muscles. This movement raises the wings and, incidentally, pulls the legs up in flight.
Wing flapping is, therefore, primarily a response to the shape of the segment and
the stresses in its walls brought about by the action of the indirect muscles, but the
details of the movement are decided by the effect of the direct muscles. In addition
to the two main indirect sets of muscle there are minor ones which tense the main
components of the exoskeleton in relation to one another. One of these accessory
indirect muscles, the pleurosternal (Fig. 141, pst.mu), which runs on each side from
the lateral wall of the segment to its floor, will be referred to below in relation to the
production of high flapping rates.

In birds and bats the movement of the wings is controlled by a neuromuscular
mechanism comparable to that by which we move our legs in walking – each time

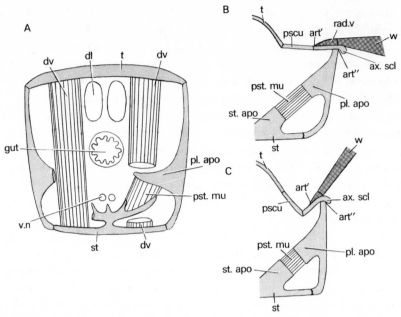

Fig. 141. A, transverse section across the thorax of a locust to show some of the muscles involved in flight. The dorsoventral indirect muscle has been cut on the right to show the pleural and sternal apodemes between which the pleurosternal muscle runs; it is shown intact on the left. B, C, diagrams to illustrate the click mechanism: B shows the base of a wing and its articulation with the thorax when the wing is horizontal, C, when it is at the end of an upstroke. Note the widening of the thorax in B due to the fact that the parascutum and second axillary sclerite lie in the same plane and its narrower section in C where they lie at an angle to one another. Widening has the effect of storing elastic energy in the thoracic wall and stretching the pleurosternal muscle. The narrowing when the wing is raised releases this energy and, along with the contraction of the muscle in response to stretching, snaps the wing rapidly to the raised position. A similar click acts on depression. art′, articulation between wing and tergite; art″, articulation between wing and pleuron; ax.scl, second axillary sclerite in base of wing; dl, dorsal longitudinal muscle, cut across; dv, dorsoventral muscle; gut, crop; pl.apo, apodeme of pleural wall; pscu, parascutum; pst.mu, pleurosternal muscle; rad.v, radial vein of wing; st, sternite; st.apo, apodeme of sternite; t, tergite; v.n, ventral nerve cord; w, wing.

the wing is depressed a volley of nerve impulses passes to the depressor muscles whilst the levators are inhibited, the reverse occurring on elevation. Every wing stroke involves a repetition of this nervous activity. A similar mechanism underlies flight in some insects, particularly those classified as exopterygotes (grasshoppers, locusts, dragonflies, earwigs, mayflies and the like) and in Lepidoptera amongst the endopterygotes. Their flight is characterized by a low flapping rate, the upper limit being set by the rate at which nerves can carry repeated impulses and muscles execute individually excited contractions. Because of the agreement in number and timing of the nerve impulses and the muscular contractions this flight rhythm is called synchronous or neurogenic.

As was first made clear by the investigations of Pringle, no similar concordance can be found in the flight mechanism of dipterans, hymenopterans and some beetles. It is true that a surge of nervous impulses coincides with the start of flight and presumably initiates it, but, thereafter, though wing movement is regular, nervous activity is not and there is no obvious relationship between the two. This is therefore an asynchronous rhythm and it turns out to be myogenic. It is capable of producing high rates of movement (up to about $1\ 000\ s^{-1}$), well beyond the capacity of any neurogenic mechanism. The rate of vibration of insect wings can be assessed by the pitch of the note produced as they beat, the rather slower movement of the larger bee producing its familiar buzz, whilst the rapid oscillation of the smaller gnat or midge wings gives these animals' high hum. The mechanism by which the asynchronous rhythm is maintained exploits the mechanical properties of the thoracic exoskeleton. When one set of indirect muscles contracts the thorax is deformed, on their relaxation there is an elastic rebound which stretches the second set of indirect muscles and stimulates their contraction. This sets up a second deformation which in turn stretches and stimulates the first group. Once in operation – and this is the function of the initial burst of nerve impulses – this mechanism is more or less self-perpetuating, though irregular firing of nerves helps to keep it in action.

In many insects the flying rhythm is accelerated by a device known as a click mechanism and this is particularly important in those with a myogenic control. In dipterans (Fig. 141B, C) this depends upon the lateral walls of the segments, the position of which is partly controlled by the pleurosternal muscle (pst.mu), contraction of which narrows the segment. The details of the articulation of the wings with the thorax are such that when they are in a horizontal position the thorax is widened and the pleurosternal muscles stretched, whereas the raised and lowered positions of the wings coincide with a narrower cross-section. Elevation and depression therefore involve first (as the wings move to the horizontal) a widening of the thorax and stretching, and so stimulation, of the pleurosternal muscles; once the wings have passed the horizontal position, however, the muscles and the elasticity of the thorax pull them, in a rapid click, to the end of their beat. The speed of the last part of their motion accelerates and accentuates the stretching of the indirect muscles and so induces a faster rhythm of wing movement than would otherwise be possible.

In all terrestrial animals there exists the danger of desiccation and this is magnified by the fact that where the surface is permeable to respiratory gases it also proves to be permeable to water vapour. Whilst the general body surface can be waterproofed the respiratory surface can not; its ventilation therefore always involves some degree of compromise between the need to control water loss and the need for oxygen. The excretory system is another site of possible water loss since water is the vehicle in which nitrogenous waste is normally voided. These

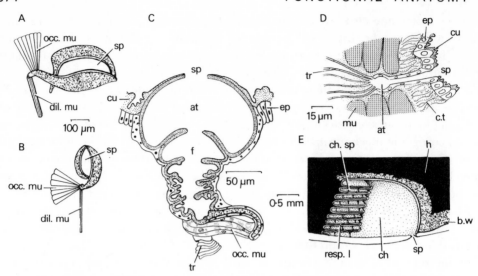

Fig. 142. Diagrams to show the closing and opening mechanism of A, the first, B, the second abdominal spiracles of a locust. C, vertical section through the abdominal spiracle of the sheep ked, *Melophagus*, showing the filter and the closing apparatus. D, section showing spiracle and tracheae of Peripatus. E, diagram of a longitudinal section through a lung-book of an arachnid. (A, B, after Albrecht; C, after Webb.) at, atrium; bw, body wall; ch, respiratory chamber; ch.sp, chitinous spines keeping leaflets apart; c.t, connective tissue; cu, cuticle; dil.mu, dilator muscle of spiracle; ep, epidermis; f, filter; h, haemocoel; mu, muscle fibre; occ.mu, occlusor muscle of spiracle; resp.l, respiratory lamella; sp, spiracle.

inter-related problems are presented in rather acute form in land arthropods, partly because their small size means not only a relatively great surface area, but also an absolutely low total content of water, and partly because flight is the most demanding of all locomotor methods in its need for oxygen.

As in crustaceans the surface of land arthropods is covered by a cuticle, partly sclerotized in its outer layers for mechanical strength, and tanned to impart some degree of waterproofing. This appears to be all that is present in myriapods and it compels them to seek humid areas for survival. In insects, spiders, mites and ticks there is added a layer of waxy material, orientated lipid molecules lying in the epicuticle, the very thin, outermost layer of the exoskeleton. This greatly increases the degree of waterproofing and confers on its owners more or less total freedom of their environment.

Respiration in land arthropods is usually subserved by a tracheal system, though organs known as lung books (Fig. 142E) may supplement it in spiders or replace it in scorpions. Tracheae are epidermal tubular invaginations opening from the skin at spiracles and running into the body. Since they are epidermal in origin they are lined by a cuticle which is shed at each moult. Though not usually waterproofed the cuticular lining is sclerotized and typically strengthened by ring or spiral thickenings (taenidia) which make the tracheae almost incompressible but do not interfere with change in length accompanying muscular action. In Peripatus (Fig. 142D) numerous minute spiracles lead into tufts of short tracheal tubes without spiral thickening and neither branching nor anastomosing, but in most land arthropods (some primitive insect orders and spiders excepted) the tracheae originating at one spiracle normally unite with those from others to form an anastomosing and branching network permeating the entire body.

Except in Peripatus and spiders the spiracles are placed segmentally, though not necessarily on all segments. They permit passage of gases, including water vapour, and are also liable to let dust enter. To prevent the latter there is often a filter (Fig. 142C) and to control the former there is usually a device allowing closing and opening (Fig. 142A, B, C). Peripatus lacks any such control and this is one factor preventing it from living anywhere but in humid situations. Spiracular control is usually moderated by the oxygen and carbon dioxide content of the atmosphere to which the animal is exposed and of the gas in the tracheal system; in fleas, for example, the spiracles normally open in response to oxygen lack but close in relation to the amount of carbon dioxide present. In resting animals most spiracles are shut.

Any internal respiratory surface must be ventilated. This may happen by simple diffusion, or it may result from pressure changes induced by the animal – any movement within the gut or muscles, for example, has some effect on the volume of the tracheal system and expels or sucks in air. It has long been known, since calculations by Krogh in 1920, that diffusion is by itself capable of supplying all the oxygen which an insect requires, provided the spiracles stay open. The difficulty confronting the animal is that, in meeting this condition, it will desiccate. In many insects, therefore, many spiracles remain closed and, especially in larger forms, this is compensated by a ventilation of the tracheal system usually achieved by alternate telescoping and elongation of the abdominal segments. This may be related to the fact that the thorax has become more or less rigid as part of the mechanical requirements for flight. Nevertheless, the movements involved in flying do produce a degree of rhythmical change in the volume of the thorax as the wings flap; in such insects as grasshoppers, dragonflies and moths this seems to be the only way in which the thorax is ventilated; in others, such as locusts, this is augmented by abdominal pumping, whilst in hymenopterans and dipterans the latter is the sole ventilating mechanism. The rate of pumping is correlated with the requirements of the moment; thus the desert locust pumps about $40 \, l \, kg^{-1} \, h^{-1}$ at rest, but raises this to about $250 \, l \, kg^{-1} \, h^{-1}$ when oxygen demand is maximal and a bee can increase the oxygen supply by a factor of about 400. These extraordinarily high figures reflect the heavy demands made by flight and the fact that insect muscles do not normally acquire an oxygen debt.

In vertebrates and most animals, including the land arthropods with lung books, the respiratory surface is richly vascularized and the blood transports the oxygen to

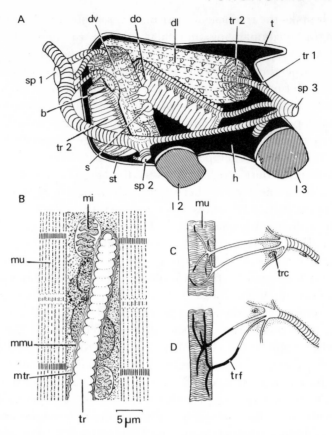

Fig. 143. A, diagrammatic representation of the tracheal supply to four thoracic muscles in the water bug, *Hydrocyrius*. B, Diagram of longitudinal section of flight muscle fibres in the wasp *Polistes* to show how tracheoles may invaginate the fibre to lie close to mitochondria. C, tracheolar supply to active insect muscle. D, tracheolar supply to resting insect muscle. b, basalar muscle; dl, dorsal longitudinal muscle with central primary trachea; do, dorsal oblique muscle with superficial primary trachea; dv, dorsoventral muscle with central primary trachea; h, haemocoel; l2, second leg; l3, third leg; mi, mitochondrion; mmu, outer membrane of muscle fibre; mtr, outer membrane of tracheal end cell; mu, muscle fibre; s, air sac at end of secondary trachea; sp1, 2, 3, spiracles; st, sternite; tr, tracheole; tr1, primary trachea; tr2, secondary trachea running radially from central trachea in dorsal longitudinal and dorsoventral muscles and on surface of muscle from superficial trachea in the other two muscles; trc, tracheal end cell; trf, part of tracheole filled with fluid. (A, after Miller; B, based on Smith; C, D, after Wigglesworth.)

the rest of the body. This may be partly true, too, in Peripatus and spiders, but in insects the blood plays no significant part in the transport of oxygen. Instead, the tracheae run to each organ and end in close relationship to the cells of which it is composed; in the case of the flight muscles, indeed, the innermost ends of the tracheal tubes invaginate themselves into the muscle fibres (Fig. 143B) so as to bring air to within a microscopically short distance of the mitochondria of the muscle cell. The last part of the supply pathway must involve diffusion through liquid and, since this is a slower process than diffusion through a gas, it is to the animal's advantage to minimize it when oxygen is in demand. The innermost parts of the tracheal system (Fig. 143C, D) are intracellular channels (tracheoles) within extensions of a single cell, the tracheal end cell. Tracheoles tend to fill with fluid, but when the cells to which they run are actively metabolizing the build-up of end products raises their internal osmotic pressure, sucks back some of the tracheolar fluid and so reduces the distance which oxygen has to diffuse through liquid.

The dimensions of tracheoles are so minute (1–0.2 μm) as to defy ventilation and the ultimate arrival of oxygen depends on diffusion: this may be one, or perhaps the, critical reason for the small size of land arthropods. The greater the length of the pathway between spiracle and tissue that can be ventilated, however, the more efficient is the tracheal system: this seems to be the main value of air sacs – dilatations of the tracheal trunks, often of considerable size, though they incidentally function as air stores and lighten the body.

In molluscs and vertebrates the action of the heart directly or indirectly produces a filtrate of the blood plasma, containing ions and small molecules, which is altered by the kidney so as to eliminate water and waste but conserve all else. In annelids cilia drive coelomic fluid into nephridial canals which behave like vertebrate kidneys. Neither of these mechanisms is available to land arthropods: cilia are invariably absent and the heart and circulation are so weak as to be incapable of producing any blood filtrate. The mechanism used by insects is a combination of a series of tubules known as Malpighian tubules, which are responsible for the production of a liquid derived from the blood, and rectal glands, which are responsible for regulating the contents of this liquid before it leaves the body. The Malpighian tubules are blind, unbranching outgrowths of the alimentary tract, placed at the point where mid-gut and hind-gut join. They vary in number from two to several hundred according to the species and lie in the body cavity amongst the other viscera. In the absence of hydrostatic pressure to drive fluid into the tubules the insect uses osmotic pressure to achieve the same result. Along the length of the tubule (Fig. 144) potassium ions in particular, are actively transported through the epithelium into the tubule, probably with sufficient anions to maintain electroneutrality. In this way an osmotic gradient is set up which sucks water and solutes out of the body cavity into the excretory organ. Without further treatment this would be an extravagant excretory fluid in that a permanent drain of water, potassium and other substances would be set up; to excrete this would be as wasteful as to allow vertebrate glomerular filtrate to escape. The corrective reabsorption of metabolically valuable material, however, does not occur in the Malpighian tubules but in the rectum to which this fluid and the contents of the intestine are passed. The wall of the rectum is thickened in many insects into glands

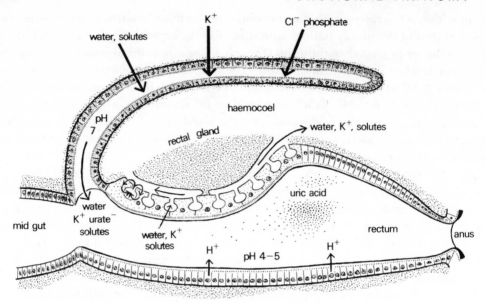

Fig. 144. Diagrammatic representation of excretion and osmoregulation in an insect. The diagram shows the hind-gut and a single Malpighian tubule opening to it at the junction with the mid-gut. Arrows indicate the flow of materials as described in the text.

and it is here, or in the ordinary rectal epithelium if glands are absent, that reabsorption of water and salts takes place. In the glands the cells lining the rectum are rich in microvilli under the cuticle. There is also present within the thickness of the epithelium a series of spaces between neighbouring cells which connect basally with other spaces linked in turn to the haemocoel. The same mechanism of an osmotic pump works here, but in such a direction as to extract water and solutes from the rectum and return to the haemocoel. The possibility of waste being recycled in the same way as the water, potassium and other useful material is prevented by the fact that the main nitrogenous end product of the insect is uric acid. This is represented in the tubule fluid (pH 7.0) by the soluble salt potassium urate; in the rectum the pH is lowered to about 4.0–5.0 with the result that solid uric acid is precipitated. This removes it from the possibility of reabsorption and leaves the potassium free to be taken back.

One further problem which has confronted all terrestrial animals is that of reproduction on land. Its main difficulties concern the fertilization of the egg and the protection of the earliest developmental stages against desiccation, whilst ensuring that they have access to food and oxygen. Internal fertilization overcomes the first of these and the answer to the second is provided by the cleidoic egg, a waterproofed box forming an almost self-sufficient system in which the zygote lies with food and water to see it through the early phases of its development. No store of oxygen is present, however, and – once again – a compromise situation is found in which a minimal loss of water is accepted as the cost of the porosity of the egg shell necessary to allow diffusion of oxygen from the surrounding air.

Alone amongst terrestrial animals many insects have a life history reminiscent of that of their marine ancestors. There, it will be recalled, the zygotes often develop into free-swimming larvae with a form and mode of living totally unlike those of their parents, and whilst in some, like the crustacean *Artemia* (Fig. 136), there is a gradual progress from the small, simple body of the young to the more complex form of the adult, in others the acquisition of adaptations fitting the larva for its planktonic habitat leads to an organization so far from that of the adult body form that the transformation can be achieved only by a cataclysmic metamorphosis. The advantages of such a life history are at least twofold: larval and adult populations are not in competition with one another for the same environmental resources, and the planktonic habitat may be kinder to immature animals than that occupied by the adults. In the more primitive types of insect young stages and adults live side by side, occupying broadly the same ecological niche, and differing from one another mainly in size and maturity. The more advanced orders, however, have larval and adult forms often so totally dissimilar that it is only familiarity that blinds us to the staggering change in form and habit that separates the caterpillar from the butterfly and the gentle from the bluebottle. Since these specialized larvae are found in advanced orders this state of affairs represents a rediscovery of the advantages of the situation and is not a simple transfer to the land of a life history evolved at sea. Whilst the two examples mentioned illustrate how this life history separates the young from competition for living space with the adults other examples emphasize more plainly, perhaps, ways in which the young came to develop in an easier habitat: in many insects such as mayflies, dragonflies, gnats and midges, the larval stages are secondarily aquatic and are so secured from such dangers as desiccation and the coldest winter weather.

Whilst the aquatic habit solves some problems, however, it creates others because insects are air-breathers. A host of devices have come into being to overcome this difficulty. Like secondarily aquatic vertebrates aquatic insects may surface to breathe, breaking through the surface film to expose their spiracles to air. For an animal as small and weak as an insect the surface film may be difficult to penetrate but the action is usually facilitated by having hairs or the cuticle around the spiracles oily or waxy so as to repel water. Other insects, such as the water boatman, may retain a bubble of air amongst a forest of hairs set on the body surface to which the spiracles open; as in pulmonate molluscs this can act as a physical gill into which oxygen diffuses from the water as that in the bubble is used in respiration. To be effective this so-called "plastron" respiration requires a high concentration of dissolved oxygen in the water. Not all aquatic insects have open spiracles: in many they are shut though the tracheal system is filled with air. It is used much as if it were a closed blood system in which diffusion and local pressure changes circulate the gas as the beating of a heart would move blood. In these animals the body bears the equivalent of gills (on the dorsal surface of the abdomen in mayfly larvae, on the lining of the rectum in dragonfly larvae) which are ventilated so that oxygen may diffuse from the water into the equivalent of a capillary bed of tracheae which they contain.

13

Echinoderms

THE ECHINODERMS, which include sea lilies, starfish, brittle-stars, sea-urchins and sea-cucumbers, differ from all other invertebrates in having a body pattern displaying a pentamerous symmetry assumed at the metamorphosis of a bilaterally symmetrical larva. Combined with the brilliant colour of the body their unusual shape, often five-rayed, makes them some of the most beautiful of sea-creatures. Unlike that of other invertebrates their skeleton is a mesodermal endoskeleton and its structure is unique in that each component consists of a single crystal of calcite formed initially within one cell. As the grain of calcite within the skeleton-forming cell enlarges the nucleus subdivides, and growth and repeated subdivision of the nuclei result in a syncytial mass containing a calcite deposit which has grown by branching in many directions to give a crystalline lattice which can be added to or, if necessary, eroded so as to reach a particular final shape. This reticulate type of skeleton is light and the holes in it paradoxically strengthen it as fibre glass or concrete is reinforced by providing lodgement for connective tissue binding the bars of the lattice together, and by interrupting the cleavage plane of the crystal and so limiting damage if a breaking force is applied. In sea-urchins, however, the skeletal elements come to form a rigid theca or test of interlocking plates covered only by the epidermis; because of the continuity of the elements thecae are well preserved in fossils. The echinoderm skeleton fossilizes easily and retains much of its crystallographic structure, consequently we have a wealth of fossils going back as far as the Lower Cambrian.

Classification

Two contrasting body forms are evident in the most familiar echinoderms: one with arms radially divergent from a central disk and the other without. The first pattern is evident in starfish and brittle-stars, which are collectively known as sea-stars (Figs 155B and 160A). The body consists of a central disk from which five arms radiate, though in some starfish such as *Solaster*, the sun-star (Fig. 154B), the arms are more numerous. The anus (absent in brittle-stars) lies on the upper surface of the disk and the mouth on the under surface. Since there is doubt as to whether the under surface corresponds to that of other animals it is preferably called the oral surface, and the upper the aboral. The arms merge into the disk in starfishes, but in brittle-stars are sharply marked off from it. From the mouth an area may be traced along each arm, the ambulacral area, from which project large numbers (up to 200) of soft, blind-ending structures known as tube feet. These appendages are characteristic of echinoderms and nothing approaching them is found in any other phylum. They are the locomotor organs of starfish and most other echinoderms, but this is not their function in brittle-stars, the liveliest of all echinoderms, which move by sinuous flexures of the arms, a snake-like movement from which the class name Ophiuroidea is derived. The movement is effected by a series of ossicles, or vertebrae, each hinged to its neighbour and worked by muscles which, with the vertebrae, make up most of the substance of the arm. The ease with which the arm fractures between one vertebra and the next has led to the common name "brittle-star".

In sea-urchins (Fig. 160B) the body pattern is different. In one large grouping, known as the regular or symmetrical forms, the body is spherical with mouth and anus at opposite poles, and the five equidistant ambulacral areas, with their rows of locomotory tube feet, pass from the mouth towards the anus – an arrangement which could be brought into being if the arms of the starfish were stitched together, their tips meeting aborally. This pattern is modified in a second grouping known as the irregular urchins (p. 410). Similarly in sea-cucumbers, which are sausage-shaped (Fig. 163D, E), the ambulacral areas are meridional, running from the mouth to the anus, though in this group the mouth is directed forward and not down as in the urchin, and the anus lies behind. In contrast to the urchin the body of the cucumber is limp and slimy, for the skeleton in the tough, muscular body wall is, at the most, a collection of separate spicules.

The first echinoderms to appear in the fossil record belong to none of the groups just described, but to the crinozoans. They were initially globular, but arms evolved during the Lower Cambrian and are prominent in the one surviving class, the Crinoidea, comprising the sea-lilies and feather-stars. Palaeozoic and Carboniferous crinoids were sessile, with a stalk anchoring the body to the sea-bed, as are the Recent sea-lilies. The central disk has a cup-shaped test or theca the base of which is formed by a single ossicle, the centrodorsal (Fig. 153B, cd.oss), and arms bearing a number of lateral branches called pinnules arise from its edge. The basic number of five arms is often increased by splitting into ten or more. The upper surface of the disk is soft and flexible with the mouth and anus directed upwards from its central area. Five ciliated grooves continuous with a groove along each arm, which in turn receives branches from each pinnule, converge on the mouth. The sea-lilies are the only Recent stalked forms. They live in deep water and are now apparently in the process of extinction. Before the Cretaceous stalked crinoids were largely superseded by species which became secondarily free during their life history, though retaining in the adult a means of temporary attachment in the form of clawed cirri arising from the centrodorsal ossicle (the only trace of the stem to be retained in living forms); like their forebears they were microphagous, using the arms as a feeding net. These are the feather-stars or comatulids such as *Antedon rosacea*, which are stalked during some larval stages (Fig. 152D) and capable of active swimming and temporary attachment in the adult.

Echinoderms have long been thought to comprise two distinct subphyla: (1) Pelmatozoa, sessile animals including all stalked forms, mostly extinct; (2) Eleutherozoa, free-living forms displaying the most recent trends in the evolution of the phylum. This grouping has been challenged by some zoologists who trace close affinity between sea-stars and sea-lilies and between sea-urchins and sea-cucumbers. They recognize four groups each of which may be regarded as a subphylum: (1) the Palaeozoic Homalozoa with no suggestion of radial symmetry; (2) Echinozoa, armless, ranging from the Carboniferous to the present day and now represented by members of the classes Holothuroidea (sea-cucumbers) and Echinoidea (sea-urchins); (3) Crinozoa, which have had arms or ambulacral feeding appendages from the Lower Cambrian onwards and are all extinct except for members of the class Crinoidea; (4) Asterozoa, including the Asteroidea and Ophiuroidea, with radial divergent arms, known from the Lower Ordovician to the present day and perhaps descended from attached Cambrian forms.

Although the name Echinodermata (meaning "spiny skinned") is appropriate to modern asteroids, ophiuroids and echinoids there is no evidence that spines were present in the earliest echinoderms – they do not occur either in the crinoids or in the holothurians, which have a very reduced skeleton retaining larval characteristics.

Echinoderms are all marine and the sea provides them with calcium carbonate for the skeleton. Their absence from other habitats may be related to the direct connection which exists between their coelomic spaces and the outside medium. As a result the fluid which fills the intricate coelomic cavities permeating the body is almost identical ionically with sea water, but has a higher level of potassium. The most conspicuous part of the coelom is the perivisceral cavity surrounding the main organs; there are also two systems of small tubes following the pentamerous body plan and each consisting of a central ring around the gut near the mouth and five radial branches. The most important of these is the water vascular system, concerned with the activity of the tube feet, which are hydraulic organs with feeding, sensory and, in eleutherozoans, locomotory functions. Their walls are not impermeable and some fluid passes through them to the exterior during movement. In the past it has been held that water was sucked into the water vascular system through a perforated ossicle on the body surface (the madreporite, so called because it superficially resembles a madreporic coral) to make good this loss; it is now believed that the fluid comes from neighbouring coelomic spaces. The second set of coelomic canals comprises the perihaemal system, which surrounds and isolates strands of a special tissue known as haemal or lacunar, which also follows the pentamerous body plan. The haemal system comprises connective tissue with intercommunicating channels (said to be of blastocoelic origin) which lack definite walls. It has communication with the coelom and its fluid, which is rich in coelomocytes, does not differ essentially from coelomic fluid. Water vascular, perihaemal and haemal systems are all interrelated at one place where the perihaemal system forms an enlarged space, the axial sinus, connected to the circumoral ring and running through the oral-aboral axis in an interradius. Within the sinus lies the canal leading from the madreporite, referred to as the stone canal since its walls are strengthened by calcite spherules, and also the axial gland or organ, an enlargement belonging to the haemal system and of enigmatic function.

Echinoderms are also unusual in having no excretory organs. Some waste from the body is collected and discharged to the outer surface by wandering phagocytes, but this may not be the sole method of getting rid of waste, and it has yet to be proved that these cells play an active part in the removal of unwanted nitrogenous compounds. The principal nitrogenous products, ammonia and urea, are soluble and pass directly to the exterior by way of respiratory surfaces.

The reproductive system is simple and, owing to the absence of secondary sexual characters, the sexes are not usually distinguishable externally. Primitively there is a single gonad opening in the same interradius as the madreporite and anus – this is the condition in holothurians. In other living echinoderms the gonads are more numerous, their arrangement conforming to the radial symmetry of the body. Fertilization is external. The spawning of one individual induces others to follow and the hordes of liberated sperm encounter the ova by chance. Some echinoderms brood their young but in the majority the eggs develop in the plankton to a larval stage which undergoes a dramatic metamorphosis.

Organization of a starfish

The major features of the rather unusual animals gathered in the phylum Echinodermata have been summarized in the above paragraphs. A more detailed look will now be taken at the common starfish, *Asterias rubens*, probably the most familiar of them all and an abundant form around our coasts. Five arms radiate from a central disk, each tapering gradually to a blunt tip. The mouth (Fig. 145, m), at the centre of the under surface, is surrounded by a muscular fold of tissue, the peristome (peris), and in the centre of the aboral surface is the anus (a) which, however, is minute and rarely visible. The five ambulacral areas take the form of grooves, one running along the oral surface of each arm, and bordered on each side

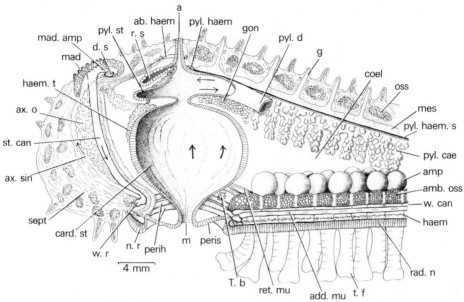

Fig. 145. *Asterias rubens*: starfish cut vertically through madreporic interradius and the proximal part of the opposite arm from which one branch of the pyloric caeca has been removed. Arrows show the direction of some important ciliary currents. a, anus; ab.haem, aboral haemal ring; add.mu, adductor muscle of ambulacral ossicles; amb.oss, ambulacral ossicle; amp, ampulla; ax.o, axial organ; ax.sin, axial sinus; card.st, cardiac stomach; coel, perivisceral coelom; d.s, dorsal sac; g, gill (papula); gon, gonad; haem, radial haemal strand in perihaemal coelom; haem.t, haemal tuft; m, mouth; mad, madreporite; mad.amp, madreporic ampulla; mes, mesentery; n.r, nerve ring; oss, ossicle in aboral body wall; perih, perihaemal rings; peris, peristome; pyl.cae, pyloric caecum; pyl.d, duct of pyloric caecum; pyl.haem, pyloric haemal ring; pyl.haem.s, pyloric haemal strand; pyl.st, pyloric stomach; rad.n, radial nerve; ret.mu, retractor muscle of cardiac stomach; r.s, rectal sac; sept, interbrachial septum; st.can, stone canal; T.b, Tiedemann's body on water vascular ring; t.f, tube foot; w.can, radial water vascular canal showing openings of transverse canals to tube feet; w.r, circumoral ring of water vascular system.

by a double row of tube feet or podia (Figs 145 and 146, t.f) each with a terminal sucker. Lateral to the podia stout adambulacral spines (Fig. 146, ad.sp) project downwards towards the substratum or bend across and protect the groove. Each consists of a calcareous ossicle attached by muscle to an underlying skeletal plate (ad.oss). The adambulacral spines bear clusters of small pincer-like bodies (ped) which seize, hold, paralyse and remove small animals coming into contact with the body. These are pedicellariae. They occur, but are less numerous, elsewhere on the starfish, which on account of its relatively slow motion and firm surface invites settlement of larvae, small predators and detritus. Each pedicellaria consists of three ossicles: two form pincer-like jaws (Fig. 147C, oss) and the third is a basal ossicle (bas.oss) from which arise two groups of antagonistic muscles (abd.mu, add.mu) passing to insertions on the jaws; the adductors bring the jaws together

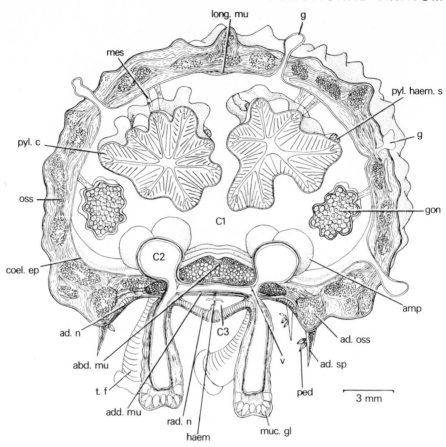

Fig. 146. *Asterias rubens*: thick transverse section of arm. abd.mu, abductor muscle of ambulacral ossicles; add.mu, adductor muscle of ambulacral ossicles; ad.n, adradial (= marginal) nerve; ad.oss, adambulacral ossicle; ad.sp, adambulacral spine; amp, ampulla; C1, perivisceral coelom; C2, coelom of water vascular system; C3, perihaemal coelom (radial perihaemal canals); coel.ep, coelomic epithelium; g, gill (= papula); gon, gonad; haem, radial haemal strand; long.mu, longitudinal muscle; mes, mesentery; muc.gl, mucous glands of tube foot; oss, ossicle of body wall; ped, pedicellaria; pyl.c, pyloric caecum; pyl.haem.s, pyloric haemal strand; rad.n, radial nerve; t.f, tube foot; v, valve at junction of podium and transverse canal.

and the abductors swing them apart. The pedicellariae on the adambulacral spines have flexible stalks which increase their radius of operation; those on the aboral and lateral surfaces of the body may be stalked or sessile. At the base of the spines projecting from the body wall are groups of even smaller pedicellariae of a different kind (Fig. 147B, ped). These are the crossed pedicellariae, so called because the bases of the two jaws overlap within the skin. They have a flexible stalk capable of movement in every direction. If the jaws of either sort are touched on their outer

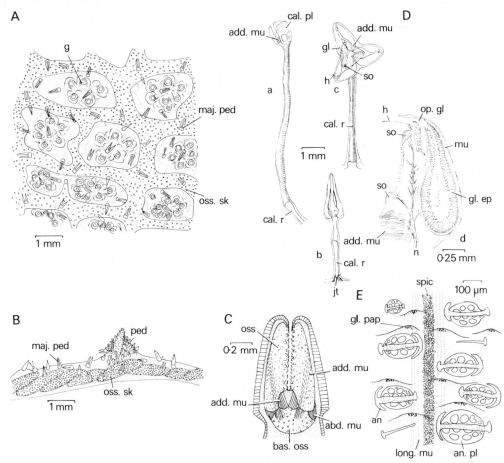

Fig. 147. A, B, C, *Asterias rubens*: A, view of skin from aboral surface to show papulae or gills in areas between ossicles. B, oblique view of thick section showing ossicles of skin and minor (= crossed) pedicellariae characteristically grouped around base of spine. C, diagram of major (= straight) pedicellaria. D, *Echinus esculentus*: pedicellariae. a, trifoliate; b, tridactyl; c, glandular or gemmiform; d, vertical section of one valve of c. (d, after Hamaan.) E, *Lepto-synapta inhaerens*: surface view of skin. abd.mu, abductor muscle; add.mu, adductor muscle; an, anchor; an.pl, anchor plate; bas.oss, basal ossicle on which adductor and abductor muscles arise; cal.pl, calcareous plate; cal.r, calcareous rod; g, gill; gl, gland; gl.ep, glandular epithe-lium; gl.pap, sensory and glandular papilla; h, inoculating hook; jt, joint between pedicellaria and ossicle; long.mu, bundle of longitudinal muscle; maj.ped, major pedicellaria; mu, muscle coat; n, nerve; op.gl, opening of poison gland; oss, one of two ossicles which form pincer-like blades of pedicellaria; oss.sk, ossicle of skin; ped, crossed pedicellaria at base of spine; s.o, sense organ; spic, spicule.

surfaces they spring open, and contact with animals or parts of animals on their inner surfaces brings them together with great speed.

The starfish moves slowly forward at the rate of a few cm min^{-1} propelled by the stepping of the tube feet, 3–10 steps per minute; typically one arm leads though occasionally an interradius does. When the animal is walking on a horizontal surface its thousand or so feet give an impression of confused activity, but each is found to be carrying out a similar stepping cycle and all are coordinated in respect of their direction and pace of movement though not in the phases of their stepping. From a retracted starting position (Fig. 148B, position 1) the foot is orientated in the line of advance, then protracted (position 2) and the sucker applied to the substratum. With the sucker as fulcrum the foot is swung back through an angle of 90° (position 3) levering the body foward; it then withdraws (position 4), while others maintain their hold, and reorientates for the next step. The rate of movement in a stepping series of feet may change abruptly and should another radius or interradius take the lead there has to be a change in alignment involving the resetting of all the feet. On vertical surfaces a starfish pulls itself up by means of attachment and subsequent contraction of the podia without any bending.

The body wall (Figs 146 and 150A) encloses the perivisceral coelom (C1) and is covered on the outside by a simple epidermis with mucous glands and sense cells and on the inside by a squamous coelomic epithelium (coel.ep). The intervening thickness contains ossicles (oss) embedded in dense connective tissue together with sheets of circular and longitudinal muscle which lie close to the coelomic epithelium (mu.bw) and are concerned with altering the shape of the body. On each arm the ambulacral groove (Figs 145, 146 and 149) is partly roofed and bordered on each side by a row of ossicles (amb.oss) with staggered pores between them (amb.p) for the passage of tube feet. These ossicles alternate regularly with marginal or adambulacral ossicles (ad.oss) bearing long spines (ad.sp), and the right and left rows of ambulacrals are joined by abductor (abd.mu) and adductor (add.mu) muscles which can open and close the groove. The repetitive arrangement of these structures along the oral surface of the arm imparts a pseudo segmentation which also affects the nerve supply of the podia. The rows of ossicles extend to the central disk, where four ambulacrals and two adambulacrals belonging to each arm spread around the mouth to form a peristomial ring. In contrast, the aboral body wall has an irregular array of ossicles.

If the coelomic surface of the body wall is examined innumerable small pits will be seen between the ossicles. Here the body wall remains microscopically thin and extends outwards to form small tufts or bladders bathed by sea water. These are gills or papulae (Figs 146, 147A and 150A, g), which compensate for the absence of respiratory surfaces elsewhere, except for the tube feet. When gills and feet are extended they are filled with coelomic fluid. The gills, unlike the feet, move little and fluid is circulated over their outer and inner surfaces by cilia.

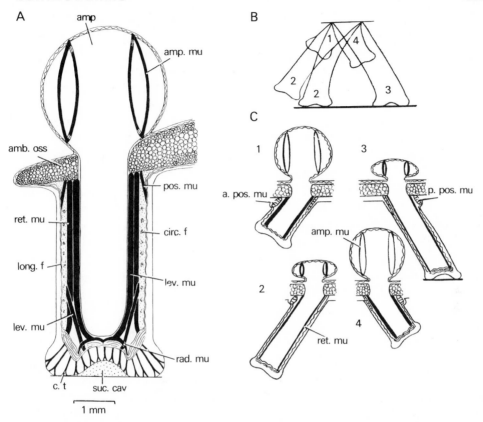

Fig. 148. *Asterias rubens*: A, diagrammatic longitudinal section to show the arrangement of muscles in a foot and ampulla. Only two rings of the ampullary muscles are illustrated. B, 1–4, the successive phases of the ambulatory step: 1, withdrawn; 2, protracted and gripping substratum; 3, swung back. C, 1–4, show the state of contraction and relaxation of the muscles of the foot during a step. Numbers correspond to those in B. The anterior postural fibres (a.pos.mu) orient the foot as it moves forwards, the posterior (p.pos.mu) as it moves backwards. (After Smith.) amb.oss, ambulacral ossicle; amp, coelom of ampulla; amp.mu, ampullary muscle; circ.f, circular fibres of sheath of connective tissue; c.t, radial strands of dense connective tissue; lev.mu, levator muscles of sucker; long.f, longitudinal fibres of sheath of connective tissue; pos.mu, postural muscle; rad.mu, radial muscle (diaphragm depressor); ret.mu, retractor muscle; suc.cav, cavity of sucker.

A conspicuous calcareous plate lying interradially on the aboral surface is the madreporite (Figs 145, 149 and 154B, mad) perforated by minute canals leading mainly from the stone canal, though some marginal ones lead from the axial sinus (p. 393). Strong ciliary currents sweep its outer surface and its canals have outwardly directed currents. If the madreporite is destroyed it is regenerated rapidly, generally with a single pore, suggesting that this communication with the exterior is vital. The central edge of the plate is incurved to form a flange beneath the perforated area. The small chamber isolated by the flange harbours two vesicles of uncertain function: the madreporic ampulla (Figs 145 and 150B, mad.amp), a diverticulum of the water vascular system, and the dorsal sac (d.s) associated with the haemal system. It is said, but with little apparent justification, that the two vesicles pulsate alternately to maintain a circulation in their respective systems. The stone canal (st.can) leads to the oral surface where it opens to the circumoral ring of the water vascular system (Figs 145 and 150B, w.r) which lies internal to the peristomial skeletal ring. A ridge (Fig. 150C, r), the free edges of which project to form a pair of scroll-like lamellae, juts into the canal from one wall, the cilia over it beating towards the circumoral ring. From this ring a radial canal (Figs 145 and 150B, w.can) passes along the length of each arm above the adductor muscles of the ambulacral ossicles. Between one muscle bundle and the next the canal gives off fine transverse canals to the tube feet. These are alternatively long and short to provide for a larger number of tube feet in a given length. At the junction of canal and podium there is a valve (Fig. 146, v) which can isolate the coelomic fluid on either side and so maintain a differential pressure. The radial canal ends at a sense organ at the tip of the arm (Fig. 149, op). This is an optic cushion, a modified tube foot, conspicuous on account of its red pigment. Between the pigment cells are retinal cells with nerve fibres passing to the radial nerve; the cells are arranged in cups, each overlaid by a lens.

Apart from the eye-spots all other tube feet are similar in structure and subserve tactile, chemosensory and respiratory as well as locomotory functions, and are used to manipulate the prey. The primary function of the water vascular canals would appear to be the maintenance of the appropriate hydrostatic pressure for their successful functioning. The volume of fluid in the channels of the system, excluding the feet, is only 1–2 per cent of the total, so with little or no reserves to draw on immediate replenishment of loss from surrounding spaces is essential. Each tube foot projects from the ambulacral groove and expands distally to a disk-shaped sucker. Proximally it penetrates the body wall and, after receiving a canal from the radial water vascular vessel, passes between two ambulacral ossicles to expand into a thin-walled bladder or ampulla (Figs 145, 146, 148A, 149 and 150B) which projects into the perivisceral coelom. The coelomic epithelium lining the tube foot has two longitudinal tracts of cilia beating in opposite directions and so maintaining a circulation of fluid. The ampulla has coelomic epithelium lining both inner and outer surfaces and, between these, smooth muscle rings lying in the oral–aboral plane (Figs 148A and 150A, amp.mu). The action of these muscles and, more especially, of muscles in the walls of the radial canals is antagonistic to those of the tube foot, into which their contraction forces fluid as its muscles relax, causing it to extend. The wall of the tube foot is considerably thicker than that of the ampulla. It is covered by an epidermis which has gland cells and exceptionally large numbers of sensory cells, especially in the sucker; there is a subepidermal nerve plexus and an

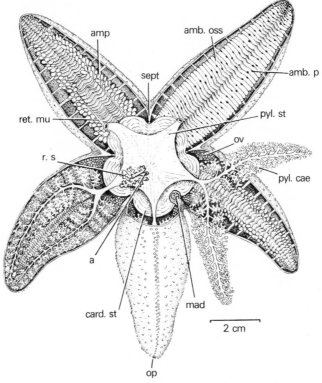

Fig. 149. *Asterias rubens*: dissection from aboral surface. One arm has been left intact, the hepatic caeca and gonads have been cut away from two arms and from one of these the ampullae of the tube feet and retractor muscles of the cardiac stomach have also been removed. a, anus; amb.oss, ambulacral ossicle; amb.p, ambulacral pore; amp, ampulla of tube foot; card.st, cardiac stomach; mad, madreporite; op, optic cushion; ov, ovary; pyl.cae, pyloric caecum; pyl.st, pyloric stomach; r.s, rectal sac; ret.mu, retractor muscle of cardiac stomach; sept, interradial septum.

underlying sheath of dense connective tissue with outer longitudinal (long.f) and inner circular (circ.f) fibres, to which, near the coelomic epithelium, muscles attach along their length or at their extremities. The connective tissue allows extension of the foot when fluid is driven into it, but resists lateral dilatation. Distally it forms an intricate network of dense strands running in the thickness of the sucker, especially near its margin, and inserting at the base of the epithelium (c.t). No muscles are associated with this network.

The muscles of the tube foot comprise a sheath of longitudinal muscles, which act as retractors of the column (ret.mu), a circlet of orientating or postural muscles (pos.mu) around the base of the podium and, distally, the radials (rad.mu) and levators (lev.mu) of the sucker. The action of the longitudinal muscles shortens the foot, forcing coelomic fluid back into the ampulla and radial canal, and differential contraction and relaxation aided by contraction of the postural muscles can bring about flexure. Mucus (Fig. 146, muc.gl) is cf importance in the attachment of the sucker but the principal adhesive force is provided by the cupping of its disk lowering the pressure within the suction cavity. This is done by the levator muscles which exert their force by way of the strong connective tissue fibres, the peripheral distribution of which ensures that the full diameter of the sucker is used. Release from suction is brought about by the antagonistic radial muscles.

It has been shown that the tube feet of each ambulacral groove are responsible for about 10 per cent of the total oxygen uptake and that the remaining 50 per cent is due to exchange across the gills.

In an animal the size of a starfish one would expect to find some system for transporting food, waste and respiratory gases and therefore comparable to the vascular system of other animals: it is one of the more surprising facts about the phylum that such a system is absent. There is no evidence that the water vascular or haemal systems are involved in transport of respiratory gases; this appears to be by way of the perivisceral coelomic fluid. The coelomic fluid is also the important medium for the transport of nutrients which diffuse into it from the large surface area of the pyloric caeca (p. 394) and become readily available to other tissues. It is practically identical with sea water in ionic composition, with such low concentrations of organic compounds that problems of osmotic balance with the external medium are minimized. It contains large numbers of cells (coelomocytes) which can pass freely from one coelomic cavity to another and migrate through the soft tissues as they transport waste. Their site of origin is often alleged to be in nine small glandular bodies, named after their discoverer, Tiedemann, which open to the oral ring of the water vascular system, two in each interradius except the one receiving the stone canal (Figs 145 and 150B, T.b), though there is little evidence to support this. Coelomocytes may also arise in haemal tissue. In the absence of special excretory organs, and with the evidence that these cells carry inorganic particles injected into the starfish to the outside of the body, it is assumed that they play some role in clearing away metabolic waste. They are particularly abundant in the gills and the disks of the tube feet through which they pass and then disintegrate.

There remains to be mentioned one system which has often been regarded as playing the part of a vascular system: this is the haemal system, which comprises a series of intercommunicating channels lying in connective tissue, following the course of the water vascular system and, in addition, branching over the gut and gonads. There is little evidence to suggest that the haemal system is a transport system, or, indeed, what it might do. The oral haemal ring and radial haemal strands (Figs 145 and 150B, haem) are placed immediately beneath the corresponding parts of the water vascular system, and lie in a septum subdividing the surrounding perihaemal coelomic space (Figs 146 and 150A, C3) which separates them from neighbouring tissues. Branches from the radial haemal strand pass to the podia.

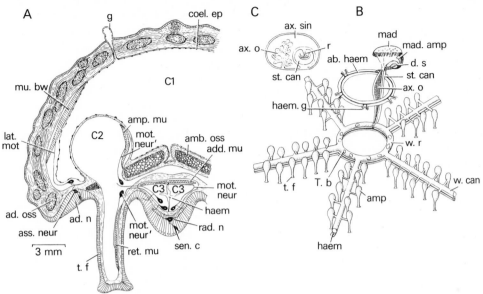

Fig. 150. A, diagram to show the nerve tracts within the arm of a starfish and the innervation of the muscles. The coelomic epithelium is shown only round the perivisceral coelom. (After Smith.) B, diagrammatic representation of the water vascular and haemal coelomic systems of an asteroid. C, Transverse section of axial complex. ab.haem, aboral haemal ring in peri-haemal coelomic canal; add.mu, adductor muscle of ambulacral groove; ad.n, adradial (= marginal) nerve; ad.oss, adambulacral ossicle; amb.oss, ambulacral ossicle; amp, ampulla; amp.mu, ampullary muscle; ass.neur, association (internuncial) neurone; ax.o, axial organ; ax.sin, axial sinus; C1, perivisceral coelom; C2, coelom of water vascular system; C3, peri-haemal coelom; coel.ep, coelomic epithelium of C1; d.s, dorsal sac; g, gill; haem, radial haemal strand in perivisceral coelom; haem.g, haemal strand to gonad; lat.mot, axon of lateral motor neuron; mad, madreporite; mad.amp, madreporic ampulla; mot.neur, primary motor neuron; mot.neur', secondary motor neuron; mu.bw, muscle of body wall; r, ridge projecting into lumen of stone canal; rad.n, association neuron of radial nerve; ret.mu, retractor muscle of foot; sen.c, sense cell; st.can, stone canal; T.b, Tiedemann's body; t.f, tube foot; w.can, radial water vascular canal; w.r, circumoral ring of water vascular system.

The perihaemal coelom and the haemal tissue extend from their circumoral rings aborally, the former as a space known as the axial sinus (Figs 145 and 150C, ax.sin), the latter, within it, as the axial organ (ax.o). The stone canal (st.can), which runs aborally from the circumoral water ring (w.r) also comes to lie within the axial sinus, being attached to its wall by mesenteries, as is the axial organ. The whole complex is intimately associated with the adjacent interbrachial septum (Fig. 145, sept). The sinus communicates with the stone canal aborally, allowing interchange of fluid, and also connects with the outside environment through some of the pores at the margin of the madreporite.

The axial organ connects aborally with two haemal rings: (1) the pyloric haemal ring (pyl.haem) which encircles the stomach near the origin of the intestine and sends branches (pyl.haem.s) to the hepatic caeca in each arm; and (2) the aboral haemal ring (Figs 145 and 150B, ab.haem) around the rectum, from which branches pass to the gonads (haem.g). Only the aboral ring is surrounded by a perihaemal canal and this is isolated from the rest of the perihaemal coelom. A small diverticulum from the axial organ passes to the dorsal sac (d.s) alongside the madreporic ampulla (mad.amp). Some circulation in the haemal system may be brought about by contractions which are said to occur in the haemal strands. The system has direct communication with coelomic cavities, but whether the chemical composition of the contained fluid could in such circumstances differ from that of coelomic fluid is not known. Its coelomocytes wander freely through the coelom.

The alimentary canal and the gonads occupy much of the perivisceral coelom (Figs 145 and 149). The gut is short, passing directly from mouth to anus in the oral–aboral axis, though the digestive glands or pyloric caeca spread into the arms. The mouth (m) opens to a short oesophagus which expands into a capacious chamber, the cardiac stomach (card.st) held in place by five pairs of ligaments (ret.mu), consisting of connective tissue and scattered muscle fibres, which arise on the ambulacral ossicles, one pair in each ray, and are inserted radially on its walls. A small pyloric stomach (pyl.st), constricted from the cardiac, lies immediately above and opens aborally to the intestine; from it five radial ducts (pyl.d) lead to the pyloric caeca (pyl.cae) in the arms. Each caecal duct bifurcates to serve two lobes of the gland which are sacculated and so provide a large area for the uptake of food. Little indigestible matter is taken into the gut and the intestine spans only the short distance between pyloric stomach and anus. Two intestinal diverticula with sacculated walls, the rectal sacs (r.s), lie near the anus (a); their function is unknown.

Asterias rubens is a predacious carnivore and undoubtedly possesses the ability to sense food at some distance, though no special sense organs for this purpose have been detected. Romanes found that he could lead a hungry *Asterias* in any direction by holding a piece of crab meat 2–5 cm away – the distance of perception must depend upon water currents as well as the intensity of the attractant. The principal food is living bivalves, especially *Mytilus*, the shells of which are opened by force; there is no support for the theory that a toxic substance is used. The starfish mounts the bivalve (Fig. 151) and grips the valves with the tube feet at the base of its arms so that the gape of the shell comes close to its mouth. Its body is then arched over the prey and locked into position (by some unknown mechanism, presumably involving muscles and ossicles in the body wall) to resist the tugging action of the tube feet which are first stretched and then, as the shell opens, shortened so that the prey is brought closer to the body. It has been calculated that the combined pull of the tube feet and body muscles on the bivalve's adductor muscles may be as high as 4 000 g, greater than their power of contraction. The pull may be intermittent and continue for several hours, or the shell valves may be parted 0.1 mm in 5–10 min. This minute gap is wide enough for the insertion of a

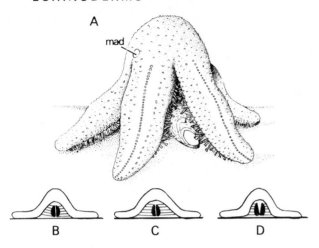

A

mad

B   C   D

Fig. 151. *Asterias rubens*: A, manipulating a bivalve. B, C, D, schematic representation of attack on bivalve. B, with the bivalve orientated with the hinge against the substratum, the tube feet at the base of the arms grip the valves while the remaining feet attach to the substratum. C, the tube feet attached to the shell elongate as the bases of the arms move away from the prey preparatory to D, the pulling phase, when the feet contract and pull the valves apart. (B, C, D, after Christensen.) mad, madreporite.

lobe of the cardiac stomach which is everted by perivisceral coelomic pressure through the mouth. Digestive enzymes from the pyloric caeca are poured over the soft tissues of the prey and the products of digestion together with semidigested fragments are passed into the pyloric stomach and enter the caeca. Digestion is completed there intracellularly, and glycogen and fat stored in the digestive epithelium. The cardiac stomach is returned to its resting position by contraction of the muscles in the ten gastric ligaments. Food does not consist entirely of bivalves, and the same mechanism is employed to feed on tubicolous worms, barnacles and gastropods, from which the operculum is first removed by force applied by the tube feet.

There are ten gonads (Figs 145 and 146, gon; Fig. 149, ov), two in each arm, attached to the body wall close to the proximal end of the ray and lateral to the pyloric caeca. From each gonad sex cells pass to the exterior by a short duct which penetrates the aboral body wall laterally, close to the interbrachial septum, and expands before opening to the surface by 4–7 pores. It has been estimated that one female may give off two and a half million eggs during a spawning period of two hours. One starfish begins to shed its gametes and its neighbours follow. In a number of species it has been shown that this may be due to a neurohormone which passes into the water from the radial nerves, is taken into the tube feet (known to be capable of taking up certain amino acids as well as glucose) and reaches the perivisceral coelom. The substance is present in both males and females. Injection of extracts from the radial nerves into the perivisceral coelom of a ripe starfish (*Asterias forbesi*) triggers the completion of their maturation prior to shedding and releases copious quantities of gametes. The release of the shedding substance under natural conditions appears to be related to an inhibitory substance, also present in the radial nerves, which fluctuates in amount during the year, reaches peak concentration just prior to the period of normal shedding, but disappears during and immediately after spawning. The hordes of liberated sperm swim at random in the water and encounter the ova by chance, sticking to their surfaces.

The nervous system of the starfish is primitive in that it retains the form of a series of nerve nets. The main part, the ectoneural system, lies at the base of the epidermis, is primarily sensory and comprises a plexus, relatively thin, and a radial nerve (Figs 145, 146 and 150A, rad.n) along each ambulacrum connecting with a circumoral ring (n.r) in the peripheral part of the membrane round the mouth. It is differentiated into a superficial zone of randomly arranged fibres, with which the sense cells (sen.c) in the epidermis make synaptic contact, and a deeper zone of aligned fibres which pass around the body wall to the radial nerve and transmit excitations. The tracts are apparently polarized in respect of the direction of conduction. The motor nervous system is a separate plexus developed in the neighbourhood of the muscles and near the coelomic epithelium. It is therefore separated from the ectoneural system by the thickness of the body wall. Lateral to the tube feet, where ossicles are absent, the ectoneural and motor plexuses are closely associated and occasional synapses have been observed – it is presumably here that the system makes the necessary links. Lateral to the tube feet, at the outer ends of the ambulacral ossicles, the subepidermal plexus is thickened into a cord, the marginal nerve cord (Fig. 150A, ad.n). This gives off a series of lateral motor nerves, a pair to each pair of ossicles. Each nerve innervates the muscle connecting ambulacral and adambulacral ossicles and then continues aborally beneath the coelomic epithelium (lat.mot), where it makes contact with the motor nerve plexus which innervates the muscles of the body wall (mu.bw). Above the radial nerve cord there are groups of motor neurons in the floor of the perihaemal canal. Some (mot.neur) innervate the adductor muscles of the ambulacral ossicles (add.mu) and others pass transversely to the tube foot where they connect with secondary motor neurons (mot.neur'). Although the system is fundamentally plexiform, the pseudosegmental organization of the arm seen in the arrangement of ambulacral ossicles and tube feet is impressed on the nervous system giving predetermined nervous linkages.

Movements of pedicellariae and spines scattered over the surface of the body are elicited through the skin plexus: both respond to direct stimulation or stimulation of the adjacent body wall even after it has been isolated from pathways leading to the nerve cord. If the aboral surface of a starfish is stimulated by pressure with a probe different patterns of response result from the progressive spread of impulses through the nervous system. The maximal response, which may involve only the movement of spines and the opening and closing of pedicellariae, is in the vicinity of the point of stimulation and the minimal perhaps 5 mm away, suggesting that the impulses are transmitted through multiple pathways and weaken with a need for facilitation. Further responses involving movement of tube feet can be evoked when a starfish is resting on its aboral surface. If then one lateral surface of an arm is probed the feet of both sides protract and bend towards the point of stimulation. A longitudinal cut

made through the integument between the point of stimulation and the tube feet
inhibits the response, but only over the region of the cut where the pseudosegmental
linear pathways, conducting impulses to the nerve cord, have been severed.

The nerve cords have a similar basic composition to the skin plexus, their greater
thickness being due to the exaggerated depth of the zone of longitudinal fibres. The
fibres run along the arm, some continuing into the circumoral nerve ring and so
down other radial cords. It would appear that such tracts are responsible for
coordinating the activity of the organ systems in widely separate parts of the body
and in particular that of the tube feet. The activity of the tube feet is controlled by
groups of primary motor neurones in the radial nerve cord along each arm which
correspond in number to the tube feet. Their axons extend into each foot where
they connect with secondary motor neurons (Fig. 150A, mot.neur′), the axons of
which pass to the retractor muscles (ret.mu), to the ring of orientating muscles at
the base of the foot and to muscles of the ampulla (amp.mu). Stepping of the feet
depends on excitation within the radial nerve cord, in particular upon the number
of neurons discharging and the extent to which they are activated by stimuli from
the skin plexus, though stepping is maintained if the animal is suspended in water
and subjected to relatively few excitations from the plexus. Sometimes a series of feet
then stop in protraction, but if the arm be stimulated stepping is resumed. Injury to
the radial nerve cord inhibits regular activity of the feet distal to the injury. The
coordination of direction of swing and the pace of the stepping in each arm appears
to depend on groups of neurons at the junction of the radial nerve cords and nerve
ring. An arm cut from a starfish lacking nervous connection with the ring moves
with the cut end forward, but if a segment of the ring is included it moves with the
tip forward. During normal walking one arm may lead, indicating the dominance of
the centre of control of its feet over the four others in the ring; less frequently there
is an interradial direction of movement when two adjacent centres co-dominate.

The coordinated activity of the effector systems of the starfish is maintained by
through-conduction pathways in the nerve cords where rapid transmission of excitation is possible. It is in this respect that the nervous system of the more elaborately
organized echinoderm is in advance of that of coelenterates. In other respects the
nervous systems of these two groups of radially symmetrical animals, in which
movement is dependent on the properties of a hydrostatic skeleton, are similar in
being essentially nerve nets, with small neurons and sense cells invariably bipolar.
The superficial plexus of the starfish has properties similar to those of the coelenterate nerve net: it conducts diffusely and with decrement and the neurons do not
appear to be polarized in the direction of their transmission.

Movement in echinoderms

Echinoderms exhibit a variety of modes of locomotion, some moving over the surface of the sea-bed, some burrowing into rocks or soft substrata and a few being pelagic. Creeping and burrowing are usually carried out by tube feet, aided by the flexible body wall in asteroids and holothurians and by movable spines in echinoids; but some holothurians are secondarily apodous and rely only on the muscles of the body wall, and some echinoids use only their spines. In ophiuroids and crinoids the long narrow arms supported by serially arranged articulating ossicles are the organs of locomotion, levering the body over the substratum with the mouth downwards in ophiuroids and upwards in crinoids. A somewhat similar sinuous flexing of the finely subdivided arms enables feather-stars to swim.

The crinoids (sea-lilies and feather-stars), the first echinoderms to appear in the fossil record, have retained a primitive structure throughout their history. The stalked sea-lilies, inhabitants of deep waters (*Metacrinus, Bathycrinus, Rhizocrinus, Endoxocrinus*), are typically fixed and exhibit only limited bending movements of stalk and arms, though some, such as *Metacrinus*, are said to free themselves from the sea floor and swim in a similar way to *Antedon*, trailing the stalk behind them. *Metacrinus* has rings of cirri on the stem and those near the base are used as grasping organs at reattachment. Thus tube feet appear not to have been the primitive echinoderm means of locomotion and, indeed, in crinoids are directed upwards, away from the sea-floor, and have no ampullae.

When swimming the comatulid *Antedon* (Fig. 152A) rises rapidly from the sea-bed by raising and lowering alternate arms. A group of five arms (one from each ambulacral area) is raised upwards over the disk with the pinnules folded against the axes, then lashed rapidly in the aboral direction with the pinnules extended to give a powerful effective stroke as the second group rises. These movements involve successive contractions of the serially arranged muscles (Figs 152C and 153B, fl.mu) running between the ossicles supporting the arm and indicate a rapid conduction of excitation along the main motor nerve of the arm, the brachial nerve (br.n), which penetrates the ossicles (br.oss).

The brachial nerves arise from a major nerve centre (ab.n.c) lodged in a concavity of the centrodorsal ossicle (cd.oss). Destruction of this centre or section of a brachial nerve stops movement, though section of other parts of the nervous system has no effects. The animal can swim in all directions by varying the power stroke on one side of the body. When it comes to rest it clings to weed or rock with the jointed cirri arising from the base of the disk (cir), capable of only limited movement. It may then crawl by attaching and pulling with the leading arms and pushing with the trailing ones. Sometimes all arms are bent aborally and the animal walks on their tips. Such versatility of movement is associated with a jointed endoskeleton and cannot occur in echinoderms in which the skeleton is reduced or is a rigid case enclosing the whole body.

With locomotion primarily indepedent of tube feet it is not surprising to find that crinoids have no madreporite, hence no direct connection between the water vascular system and the surrounding medium, as in the starfish and sea-urchins which do rely on them for movement. There are, however, ciliated openings (Fig. 153A, B, w.p) in the skin over the oral surface of the disk which lead to the coelom

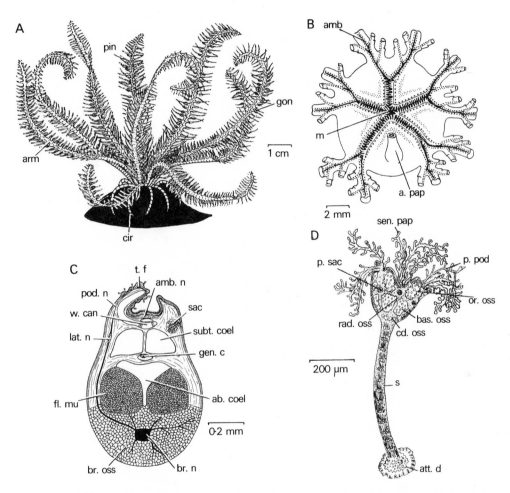

Fig. 152. *Antedon bifida*: A, whole animal temporarily attached. B, oral view of disk and bases of arms. C, transverse section of arm. D, cystidean larva which precedes pentacrinoid larva. ab.coel, aboral coelomic canal; amb, ambulacral groove (= food groove); amb.n, ambulacral nerve; arm, arm with pinnules branching alternately from its sides; a.pap, anal papilla; att.d, attachment disk; bas.oss, basal ossicle; br.n, main brachial nerve; br.oss, brachial ossicle; cd.oss, centrodorsal ossicle; cir, cirrus; fl.mu, flexor muscles; gen.c, genital cord in radial haemal strand; gon, gonad; lat.n, lateral nerve; m, mouth; or.oss, oral ossicle; pin, pinnule; pod.n, podial nerve; p.pod, primary podia; p.sac, primary sacculus; rad.oss, site of radial ossicle—arm grows out as projection from this area; s, larval stalk; sac, sacculus; sen.pap, sensory papilla; subt.coel, subtentacular (= perihaemal) coelomic canal; t.f, tube foot; w.can, radial water vascular canal.

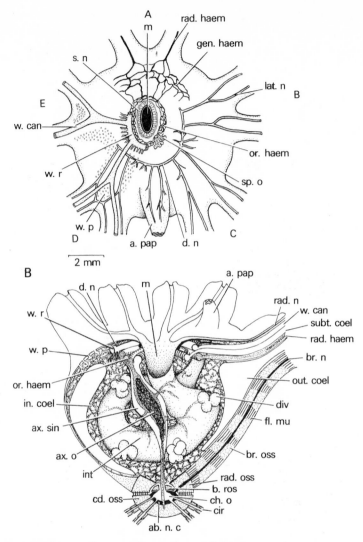

Fig. 153. *Antedon bifida*: A, diagram of oral view of disk, with certain structures seen by transparency—in radius A the haemal system, radii B–D the deep nervous system and radii D and E the water vascular system; central connections in the disk are also displayed. (After Grassé.) B, vertical half of disk, and base of an arm. ab.n.c, aboral nerve centre; a.pap, anal papilla; ax.o, axial organ; ax.sin, axial sinus; br.n, main brachial nerve; b.ros, basal rosette; br.oss, brachial ossicle; cd.oss, centrodorsal ossicle; ch.o, chambered organ; cir, cirrus; div, gut diverticulum; d.n, deep nerve ring; fl.mu, flexor muscles; gen.haem, genital haemal ring; in.coel, inner coelom; int, intestine; lat.n, lateral nerve of arm; m, mouth; or.haem, circumoral haemal ring; out.coel, outer coelom; rad.haem, radial (genital) haemal strand with genital cord arising from genital haemal ring; rad.n, radial nerve leading from superficial nerve ring; rad.oss, radial ossicle; s.n, superficial (epithelial) nerve ring; sp.o, spongy organ; subt.coel, subtentacular (= perihaemal) coelomic canal; w.can, radial water vascular canal; w.p, water pores in upper wall of disk; w.r, water ring with water tubes communicating with inner coelom.

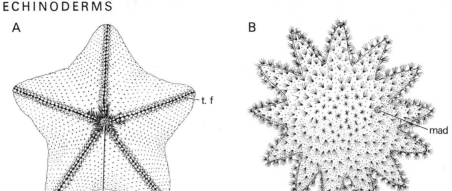

Fig. 154. A, *Anseropoda placenta*: from oral surface to show pentagonal form, tube feet and surface covered with tufts of minute spinelets. The body is wafer thin. B, *Solaster papposus*: from
aboral surface. A multirayed star with broad disk and tapering arms. The aboral spines
resemble paxillae in being arranged in groups borne on a stalk. m, mouth; mad, madreporite;
t.f, tube foot bordering ambulacral groove.

with which the circumoral water ring communicates by way of minute ciliated
ducts (w.r). The protrusion of tube feet is described on p. 422.

The class Asteroidea is represented in the Ordovician, though its place of origin
from the Pelmatozoa is debatable. Moving with the oral surface down and the tube
feet in contact with the substratum these forms have evolved an entirely different
means of locomotion and been able to exploit different ecological niches. The
conditions described in *Asterias* do not do justice to the locomotory versatility of
starfishes. Some if not all Palaeozoic starfish had a burrowing habit and their
structure suggests that the animal dug itself into the sea-bed by flexing the whole
body in much the same way as the Recent *Anseropoda* (=*Palmipes*) (Fig. 154A)
insinuates one side of the body beneath the surface of the shell gravel in which it
lives and pushes with the other, so moving without the use of either spines or tube
feet. This movement is aided by the fact that its body is parchment thin. Ampullae,
which force coelomic fluid into the tube feet, were developed early in the evolution
of starfish and one fossil form (*Chinianaster*) seems to have had suckered tube feet.
However suckers are not present in the primitive living asteroids *Astropecten* (Fig.
155) and *Luidia* which, despite this, can burrow in gravel and fine sand using the
tube feet, and can also walk over the surface. The action of their feet in digging and
walking is coordinated with the musculo-skeletal system of the arms. When
Astropecten walks the amulacral groove (amb.gr) is narrow, the arm broad and the
pointed feet (t.f) hang vertically down; the feet on all the arms are orientated in
the same direction as they carry out their cycle of stepping. During burrowing the
ambulacral groove is widened, the arm narrowed and heightened and the feet, now
far apart, are protracted and bent outwards from the mid-line of the arm and so

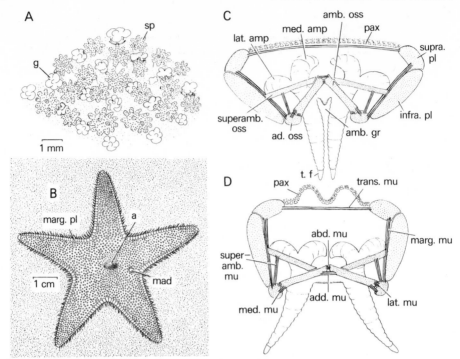

Fig. 155. *Astropecten irregularis*: A, part of aboral surface. B, aboral view of whole animal. C, diagrammatic transverse section of arm in walking position. D, diagrammatic T.S. of arm in digging position. (C, D after Heddle.) a, anus; abd.mu, abductor muscle of ambulacral ossicles; add.mu, adductor muscle of ambulacral ossicles; ad.oss, adambulacral ossicle; amb.gr, ambulacral groove; amb.oss, ambulacral ossicle; g, gill; infra.pl, inframarginal plate; lat.amp, lateral lobe of ampulla; lat.mu, lateral ambulacral-adambulacral muscle; mad, madreporite; marg.mu, marginal muscle; marg.pl, marginal plate; med.amp, median lobe of ampulla; med.mu, medial ambulacral-adambulacral muscle; pax, paxilla; sp, spinelet forming crown of paxilla; superamb.mu, superambulacral muscle; superamb.oss, superambulacral ossicle; t.f, pointed tip of tube foot; trans.mu, pericoelomic transverse muscle.

point in ten different directions, pushing sand away from beneath the animal; they are then withdrawn and the cycle repeated. Meanwhile the dorsal body wall, flexible because it has only small ossicles bearing movable spines or paxillae (pax), is thrown into longitudinal folds along each arm. As the animal burrows deeper the sand heaps up on either side of the arm and eventually falls over its top. Connections between the burrow and surface are maintained at the tips of the arms and sometimes over the disk in the region of the anal papilla. To overcome the resitance of the sand particles high pressures can be built up within the cavity of the tube foot since the ampulla is double (lat.amp, med.amp) and so provides increased surface area per unit volume as compared with the simple ampulla of *Asterias*.

The more elaborate movements of the body during burrowing, which are coordinated with the activity of the feet, involve a more complex series of skeletal elements and muscles along the ventral and lateral surfaces of each arm than are found in *Asterias*. Each ambulacral ossicle (amb.oss), moved on its partner by adductor (add.mu) and abductor (abd.mu) muscles, is linked to the adambulacral (ad.oss) by both a median (med.mu) and a lateral (lat.mu) muscle, not by one as in *Asterias*. Two other muscles attach to each adambulacral ossicle: one (marg.mu) passes up the median surface of the large inframarginal ossicle or plate (infra.pl) to insert on the supramarginal plate (supra.pl), with some fibres passing to the inframarginal; the second (superamb.mu) passes aborally to a rod-shaped ossicle, the superambulacral (superamb.oss) directed transversely across the arm from the ambulacral to the inframarginal ossicle. When the walking posture is assumed the ambulacral groove is narrowed by contraction of the adductor muscles of the ambulacral ossicles, the median ambulacral-adambulacral muscles and the super-ambulacrals, and as a result the dorsal surface of the body is stretched and smooth, with the pericoelomic transverse muscles (trans.mu) relaxed. During burrowing the pericoelomic transverse muscles are contracted along with the abductors of the ambulacral ossicles, the lateral ambulacral-adambulacral muscles and the marginals.

Since *Asterias rubens* has lost the burrowing habit of its progenitors, the ossicles of the arms all tend to be alike and the superambulacrals with the associated muscles are not present.

Asteroids, echinoids and holothurians, unlike crinoids and ophiuroids, mainly rely on tube feet for locomotion and have muscular ampullae projecting into the perivisceral coelom which have probably evolved independently in the three phylogenetic lines. The efficiency of such a tube-foot system is revealed in echinoids where the retracted foot can measure 2–3 mm in length but attain 150 mm when extended. Suckers have also evolved in the three classes, again probably independently, and together with adhesive secretions enable the animals to live on surf-washed shores. It has been seen that in asteroids the radial nerves, the radial parts of the water vascular system and other tubular coelomic systems are outside the skeletal pieces of the ambulacrum. Open ambulacral grooves also occur in crinoids. In ophiuroids, echinoids and holothurians, however, the body wall with its enclosed skeletal elements grows over the grooves, giving protection to the radial structures, with the result that the tube feet emerge from pores in the skeletal plates; the closed groove is called an epineural canal (Figs 156D, 157C and 158B, epin).

The rigidity of the test which encloses the body of an echinoid eliminates any part which it might directly play in locomotion (Fig. 159A). It is composed of meridionally arranged rows of plates extending from the region round the mouth (peristome) to a region round the anus (periproct) which has only small ossicles in it and remains flexible (perip). In the regular echinoid *Echinus* there are twenty rows, two in each ambulacrum (amb.pl) and two in each interambulacram (interamb.pl), all the plates bearing bosses (boss) which carry movable spines. The spines help to anchor the spherical body, counteract its instability, provide protection and, along with the tube feet, are the principal locomotor organs (Fig. 160B). Tube feet, projecting from the broad ambulacral areas, are also stabilizers and locomotor

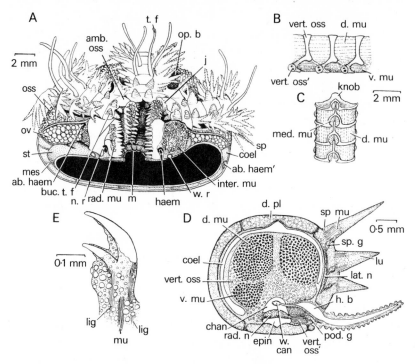

Fig. 156. *Ophiothrix fragilis*: A, half of the disk cut vertically with the intact bases of three arms; gut black; ventral (oral) surface uppermost. The cut on the right passes interradially; on the left it just misses an arm and therefore shows the anatomical arrangement of a radius. B, three vertebral ossicles viewed laterally. These articulate with one another centrally and taper to a narrow flange laterally where the intervertebral muscles attach; proximal surface (direction of mouth) to right. C, four vertebral ossicles and associated muscles viewed dorsally. D, transverse section of arm. E, ambulatory claw from tip of arm. ab.haem, aboral haemal ring in radial position; ab.haem´, aboral haemal ring in interradial position; amb.oss, ambulacral ossicle; buc.t.f, buccal tube foot; chan, median channel of vertebral ossicle through which run radial nerve and water canal; coel, coelom; d.mu, aboral (= dorsal) intervertebral muscles; d.pl, aboral (= dorsal) arm shield; epin, epineural canal; haem, haemal ring in perihaemal canal; h.b, head bulb of podium; inter.mu, external interradial muscle running between half jaws of two adjacent radii; j, interradial jaw (two ambulacral and two toothed adambulacral ossicles); knob, knob-like articulating area on distal surface of vertebra; lat.n, lateral nerve; lig, ligament; lu, lumen of spine filled with connective tissue; m, mouth; med.mu, median dorsal muscles; mes, mesenterial attachment of stomach crossing coelom; mu, muscle moving spine; n.r, nerve ring; op.b, opening of bursa; oss, ossicles of disk; ov, ovary; pod.g, podial ganglion; rad.mu, radially placed muscle linking two halves of jaw; rad.n, radial nerve; sp, spines arising from lateral arm shield; sp.g, ganglion for spine; st, stomach; t.f, tube foot extended; vert.oss, vertebral ossicle; vert.oss´, vertebral ossicle perforated by nerve and water canal; v.mu, oral (= ventral) intervertebral muscles; w.can, radial water canal; w.r, water ring.

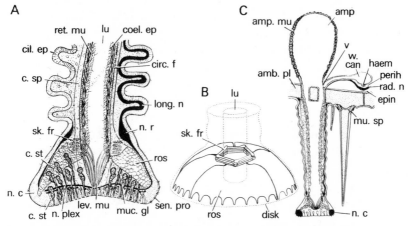

Fig. 157. *Echinus esculentus*: A, diagrammatic longitudinal section of the distal end of a tube foot, passing, on the right, through the longitudinal nerve and one of the indentations in the edge of the skeletal rosette. On the left mucous glands and connective tissue pass through the rosette. Only a few of the levator muscles of the disk are shown. B, perspective drawing to show skeletal elements in the disk of a tube foot. C, diagram of a section across a radius. (A, B after Nichols.) amp, ampulla; amp.mu, muscles of ampulla; amb.pl, ambulacral plate with two pores for tube foot; cil.ep, ciliated external epithelium; circ.f, connective tissue with circular fibres; coel.ep, coelomic epithelium; c.sp, connective tissue with calcareous spicules; c.st, connective tissue strands; disk, edge of disk; epin, epineural canal; haem, radial haemal strand; lev.mu, levators of disk; long.n, longitudinal nerve; lu, lumen of foot; muc.gl, groups of mucous glands; mu.sp, muscles moving primary spine (ring nerve is adjacent); n.c, sensory nerve cells; n.plex, nerve plexus of disk; n r, nerve ring; perih, radial perihaemal canal; rad.n, radial nerve; ret.mu, retractor muscle fibres; ros, rosette; sen.pro, sensory processes; sk.fr, skeletal frame; v, valve; w.can, radial water vascular canal.

organs with a complexity of structure matching their functional efficiency. As the body wall has no involvement in locomotion it has lost all musculature except for what is associated with spines, pedicellariae and the complex feeding apparatus known as Aristotle's lantern.

Regular echinoids, like asteroids, move with the oral surface directed down and with any radius leading, and can reverse direction without turning round. In moving they use tube feet or spines or a combination of both depending on circumstances. Part of the efficiency of the tube feet depends on the fact that they have the most highly developed suckers of any class, with ossicles functionally replacing the connective tissue plate of the asteroid foot. At the distal end of the foot there are a few pentagonal rings of small ossicles, forming a structure known as the frame (Fig. 157, sk.fr) and, spreading out from this, five large ossicles form the rosette (ros) which maintains the shape and width of the sucker. The stem of each tube foot communictes with its ampulla by a double channel recognizable even in dry tests by the rows of double pores in the ambulacral ossicles. The ampulla can be closed off from the radial vessel of the water vascular system by a valve so that on contraction of ampullary muscles fluid passes into the lumen of the tube foot and it is extended. Around the edge of the disk is a sensory ring with processes (sen.pro); when these come into contact with the substratum to which the sucker will attach, an adhesive secretion is poured from multicellular glands discharging on the disk (muc.gl); as contact is secured the centre of the disk is raised by levator muscles (lev.mu). The levators originate on the frame (sk.fr) and, as in asteroids, insert at the centre of the disk. When the sucker is detached the levators relax and the retractor muscles, pulling on one segment of the rosette, lift the edge of the sucker at that point and break the seal. The retractor muscles of the stem of the foot (ret.mu) are antagonistic to the muscles in the wall of the ampulla (amp.mu). They form a layer external to the peritoneal lining of the central cavity of the tube foot and are surrounded externally by layers of connective tissue (c.sp) which allow distension, but resist lateral pressures when the foot is distended.

Activity of the tube feet can be efficient only if the pressures inside and outside the water vascular system are balanced. If the pressure in the system is lowered relative to the surrounding sea water by about 20 cm water, the feet cannot be protruded efficiently, and if there is an equivalent increase in pressure they become stiff tubes. If a sea-urchin is plunged into deep water with the madreporite sealed, the tube feet cannot be properly used; this suggests that changes in pressure are normally registered, if not adjusted, probably at the madreporite. There is no evidence that fluid passes through the madreporite pores during normal activity of the tube feet so long as the external pressure remains constant, and there is negligible loss through them even when violent stimulation causes a large number to contract and force fluid back into the ampullae and channels of the water vascular system. The increased pressure within the test is relieved by a temporary bulging of the only distensible area of the body wall, the peristome, and subsequent expulsion of fluid from the anus.

Movable spines, a characteristic feature of echinoids, arise from tubercles on the plates of the test, the concave end of each articulating with a central boss giving a ball-and-socket joint worked by two conical layers of tissue one inside the other. These arise on a depression surrounding the boss and insert on the spine. If a spine is touched the inner layer, consisting of collagen fibres and fibroblasts, holds it firmly in a powerful defensive position, but if another part of the test is stimulated the spine is moved by the outer layer, which is muscular (mu.sp), and can be orientated to point in any direction. There is a swelling of the subepidermal nerve plexus around the base of each spine, the activity of which must be coordinated with that of other spines and the podia. Spines are also used in other ways: some regular echinoids such as *Paracentrotus* bore mechanically into rock (for protection against wave action) with a rotatory movement of the whole body aided by the abrasive action of the spines and sometimes of the teeth; they may use spines for walking in a coordinated metachronal movement, as in *Diadema*, which moves over

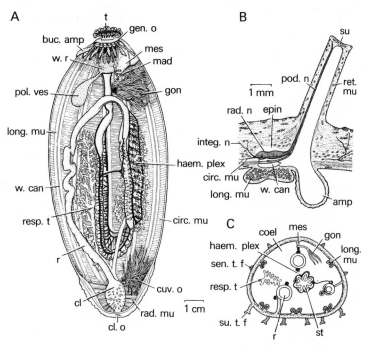

Fig. 158. *Holothuria forskali*: A, body cavity opened by a dorsal longitudinal cut. B, transverse section through a radius. C, diagrammatic transverse section. amp, ampulla of tube foot; buc.amp, buccal ampulla; circ.mu, circular muscles of body wall; cl, cloaca; cl.o, cloacal opening; coel, coelom; cuv.o, cuvierian organ; epin, epineural canal; gen.o, genital opening; gon, gonad; haem.plex, haemal plexus; integ.n, integumentary nerve; long.mu, longitudinal muscles of body wall; mad, madreporic body (stone canal); mes, dorsal mesentery; pod.n, podial nerve; pol.ves, polian vesicle; r, rectum; rad.mu, radial muscles; rad.n, radial nerve; resp.t, respiratory tree; ret.mu, retractor muscles of tube foot; sen.t.f, sensory papillate tube foot; st, stomach; su, sucker; su.t.f, suckered tube foot of three ventral ambulacra (trivium); t, tentacle; w.can, radial water vascular canal; w.r, ring canal of water vascular system.

the substratum at considerable speed, and in *Cidaris* and related genera – all remarkable for their huge spines.

Minute spherical spines or sphaeridia occur in all echinoids except cidaroids, each attached to a tubercle by a muscle and with a nerve ring at the base – a modified area of the subepidermal plexus. Muscle and nerve ring are lacking in clypeastroids. The sphaeridia are covered by ciliated epidermis and are usually lodged in pits in the test, sometimes almost completely enclosed. They may be confined to the ambulacral areas around the mouth or scattered along the entire ambulacra, and are believed to be gravity receptors. The righting reaction of an urchin deprived of them is slow, seemingly from a lack of coordination among the tube feet.

The test of the echinoid offers a solid, inflexible surface with a multitude of crevices in which sediment and larvae of encrusting organisms might settle and hamper the functioning of the spines and the cleansing action of the ciliated epidermis. It is therefore not surprising to find on the test between the spines, on the peristome and more sparingly on the periproct, an array of pedicellariae (Fig. 147D) more varied than described for asteroids – the only other class having these defensive organs. The head of the pedicellariae of echinoids has typically three jaws. It is movable on a stalk which has an internal ossicle (cal.r) either reaching the head, or stopping short and leaving a flexible neck. At the base of the ossicle is a nerve ring and there is also nervous tissue in the stalk and head (n, so). The ossicle is articulated to a minute tubercle of the test by a ball-and-socket joint (jt) encircled by a muscle sheath, and thus bears the same relationship to the test as a spine. Four main types of pedicellariae have been described, each with a number of variants. The tridactyl or tridentate type (b) has a flexible neck bearing three elongated tapering jaws with tooth-like projections. They are the largest and commonest. They snap at objects coming into their reach with a rapid closure of the jaws which have striped fibres in their adductor muscle. In the ophiocephalous (snake-headed) pedicellariae, which occur chiefly on the peristome, the jaws are short, blunt and serrated and have a basal interlocking device coming into action as they close;

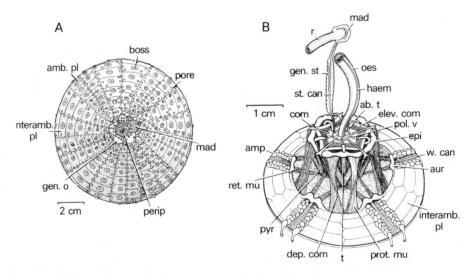

Fig. 159. *Echinus esculentus*: A, diagram of aboral view of dried shell, spines removed. B, Aristotle's lantern and the principal canals of the water vascular system; the coelomic membrane which encloses the lantern is not shown. ab.t, aboral end of tooth; amb.pl, ambulacral plate; amp, ampulla of tube foot; aur, auricle; boss, boss on which spine articulates; com, compass; dep.com, depressor muscle of compass; elev.com, elevator muscle of compass; epi, epiphysis; gen.o, genital pore; gen.st, genital stolon; haem, haemal vessel; interamb.pl, interambulacral plate; mad, madreporite; oes, oesophagus; perip, anus in periproct; pol.v, polian vesicle on water vascular ring; pore, pore, one of pair through which tube foot protrudes; prot.mu, protractor muscle of lantern; pyr, pyramid; r, rectum; ret.mu, retractor muscle of lantern; st. can, stone canal; t, tooth; w.can, radial water vascular canal.

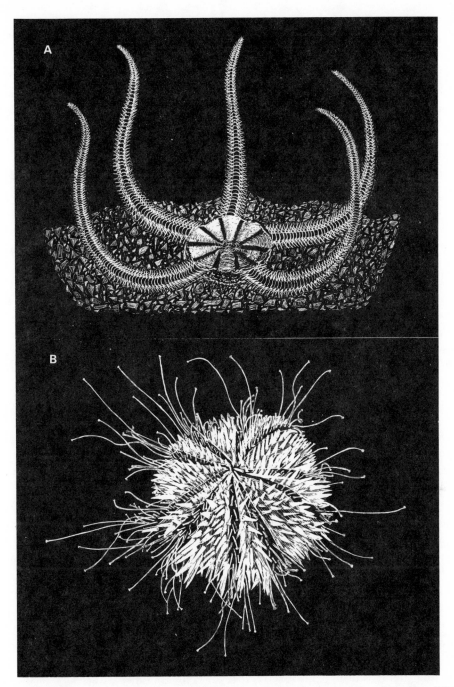

Fig. 160. A, *Ophiocomina nigra*: position assumed when feeding by means of a mucous net. With the base of the arms applied to the substratum and the disk slightly raised, the arms are widespread distally. Mucus which entraps food particles covers the body and especially the spines which are erect and linked to one another by mucous threads. B, *Echinus esculentus*, aboral view, to show extensibility of ambulatory tube feet.

the stalk is flexible and has a wide field of action. The trifoliate (three-leaved) type (a) is small and its broad, short jaws do not meet distally. It searches over the test and bases of the spines for small particles and can break down larger ones. The most specialized pedicellariae are the glandular gemmiform or globiferous ones (c, d). Each of the three jaws is plump at the base, where there is a glandular sac (gl) filled with a toxic secretion, and tapers distally to a sharp tooth or group of teeth (h), bent inwards. The glandular sac may be enveloped in a muscular sheath (mu) which ensures rapid ejection of venom from an opening (op.gl) proximal to the tooth. In some species poison glands also encircle the stalk and have individual pores. The toxic secretion paralyses small animals, will even succeed in driving away a starfish and when injected into man the pain is as severe as from the sting of a bee.

Most echinoids live on hard bottoms, creeping over rocks and exposed to wave action. Specialization of movable spines and tube feet, however, has allowed some urchins to burrow into soft substrata and live a few inches to a foot or more beneath the surface. All irregular urchins burrow, though many are successful only within a limited range of particle size and the departure of their shape from the spherical is linked with their altered mode of life. The spatangoids of heart urchins show the greatest ingenuity in burrowing and living in the sea-bed. Their test is arched above and flattened below, the anus (Fig. 161, a) has migrated from the apical system to an interambulacrum regarded as marking the posterior end, the mouth shows some displacement anteriorly, and the anterior ambulacrum is sunk, giving a characteristic heart shape to the test, while the other four ambulacra form a flower-like figure around the centre of the aboral surface since each is altered to a petaloid shape. Symmetry in these animals is thus bilateral and the body can be regarded as having anterior and posterior ends and dorsal and ventral surfaces. Movement into and through the substratum is brought about by spines, the most effective being broad and spatulate (sp.pl). Numerous smaller spines form protective arches over the ambulacra and in local parts of the body special bands of long spines (a.fasc, in.fasc, sub.fasc, sp.in, sp.sub) with cilia on two sides of the stem produce currents for feeding, respiration and excretion; these bands are termed fascioles and are named according to their location. The functions of the tube feet are similarly varied, but they are not concerned with locomotion. The highly extensible feet of the anterior ambulacrum (fu) build and maintain a respiratory funnel (resp.fu) which allows water to flow from the surface of the sand to the dorsal surface of the urchin, and also grasp surface sediment for food; other feet (sub.t.f) tunnel back from the anus and build a sanitary drain (san); the feet of the dorsal parts of the ambulacra, excluding the first, are respiratory (resp); all ventral podia are sensory (sen) except those in the vicinity of the mouth which manipulate food.

When the sand urchin *Echinocardium cordatum* burrows down from the surface, mounds of sand are heaped up at the front and sides of the body by the activity of the spatulate spines of the plastron (that is, the flattened interambulacral area posterior

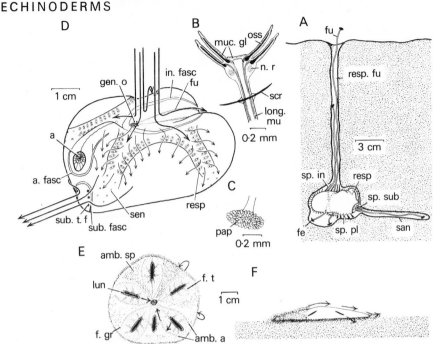

Fig. 161. A, B, C, D, *Echinocardium cordatum*: A, in burrow, lateral view, showing the position of certain types of tube feet. B, diagrammatic longitudinal section of the distal end of a funnel-building tube foot. C, diagram of disk of feeding tube foot which has many papillae covered with mucous glands. D, diagram of dorsolateral view to show the course of the respiratory current; the spines and tube feet are omitted. Water (indicated by arrows) is drawn down a funnel by the ciliation of the body surface augmented by that of the spines of the inner fasciole. It continues down the ambulacra passing out between the respiratory tube feet and is then collected within the subanal fasciole. The cilia on the spines comprising this fasciole direct the current away from the body down the sanitary tube. E, F, *Mellita sexiesperforata*: E, oral view, arrows indicate directions of food transport. F, diagram of individual ploughing through surface sand. Arrows indicate transport of sand. (A – D after Nichols; E, F after Goodbody.) a, anus; a.fasc, anal fasciole; amb.a, ambulacral area; amb.sp, ambulatory spine; fe, feeding tube foot; f.gr, ambulacral food groove; f.t, food tract; fu, position of funnel-building tube foot; gen.o, genital opening; in.fasc, inner fasciole; long.mu, longitudinal muscle; lun, lunule; muc.gl, mucous epithelial glands; n.r, nerve ring; oss, skeletal rod supporting papilla; pap, papillae; resp, position of respiratory tube foot (dorsal part of lateral ambulacra); resp.fu, respiratory funnel; san, sanitary tube; scr, scraper; sen, position of sensory tube foot; sp.in, spine of inner fasciole; sp.pl, spine of plastron; sp.sub, spine of subanal fasciole; sub.fasc, sub-anal fasciole; sub.t.f, position of subanal tube foot, building sanitary tube.

to the mouth), and of those of the lateral ambulacra. The body sinks vertically down and is hidden. As it sinks a tuft of large dorsal spines and the specialized tube feet of the dorsal part of the anterior ambulacrum become active. The disk of each tube foot has a fringe of papillae (for which reason it is called penicillate) supported by calcite rods (oss) and covered by an epithelium of mucous glands (muc.gl); proximal to the disk the foot has a large curved spicule (scr). The mucus from the tube feet is wiped on to the dorsal spines and they, in turn, seem to plaster it on to the wall of a channel they make through the sand, the beginning of a respiratory funnel (resp.fu). As the urchin burrows deeper the spines can no longer maintain contact with the surface and the tube feet, which can be extended a considerable distance beyond the spines, take over the building and maintenance of a comparatively long, narrow channel leading to the surface: the arcuate spicules of the feet scrape away the sand on the walls and mucus from their epithelial glands is applied to it. Currents are drawn down the funnel and over the tube feet on the petaloid areas, which are the animal's respiratory surface, by cilia on the body and on the tall spines which enclose the aboral apex, the inner fasciole (sp.in). In a similar way the tuft of spines forming the subanal fasciole (sp.sub) and the subanal tube feet (sub.t.f), which are penicillate but have no scrapers, mould the short, blind sanitary tube (san). The urchin moves forward in the sand using the spatulate spines of the plastron (sp.pl) and the medium-sized spines of the anterior region of the test to scrape the front wall of the burrow and pass material back under and around the body. Ultimately respiratory and sanitary funnels collapse and new ones are built.

The clypeastroids, sand dollars, or sea biscuits, common on sandy bottoms in tropical and subtropical waters, are regarded as an independent evolution from the regular echinoids. They are flattened in the oral-aboral direction (Fig. 161E, F), have five petaloid aboral ambulacra and the body is covered with a fur of very short spines of various kinds, some ambulatory, others not. There are typically three kinds of tube foot: those in the petaloid areas radiating from the aboral apex are respiratory, but far outnumbered by the small suckered feet which cover much of the test in interambulacral and ambulacral areas; around the mouth are sensory tube feet, also used for feeding. Sand dollars move about just under the surface using spines and tube feet (*Echinocyamus*) or only spines (*Mellita*).

Holothurians or sea-cucumbers lie with the oral-aboral axis parallel to the substratum and the mouth at the anterior end. Three ambulacral and two interambulacral areas are involved in making the underside (trivium); two ambulacral and interambulacral areas comprise the upper side (bivium). Their body wall is soft and leathery resembling that of other coelómates. There is a cuticularized epidermis with glands, a thick dermis and a circular muscle layer (Figs 158A, B and 162, circ.mu) which may form a complete cylinder, but is usually interrupted by five bands of longitudinal muscle (long.mu), lying in the radii. The skeleton has been reduced to microscopic ossicles or spicules, shaped like hooks, rods, crosses or anchors, lying in the peripheral part of the dermis. In a few, ossicles are lacking and in others large flat plates cover certain areas of the body. The body wall is toughened by a layer of collagen fibres beneath the epidermis, which yields to changes in body shape and volume brought about by the muscles acting on the coelomic fluid. The network of fibres stiffens when stimulated mechanically, but how this is brought about is unknown. It is not surprising to find that except in a

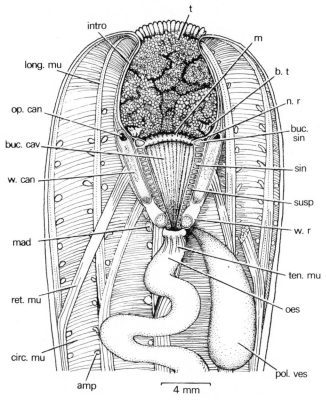

Fig. 162. *Cucumaria saxicola*: Dissection of anterior end with respiratory trees, gonad and associated structures omitted; tentacles retracted. amp, ampulla of locomotory tube foot; b.t, base
of retracted tentacle; buc.cav, buccal cavity; buc.sin, circumoral sinus; circ.mu, circular
muscles of body wall; intro, wall of introvert; long.mu, longitudinal muscle of body wall; m,
lip of mouth; mad, internal madreporite; n.r, nerve ring; oes, oesophagus; op.can, opening of
radial canal to tentacular canal with valve retarding backflow from tentacle; pol.ves, polian
vesicle; ret.mu, retractor muscles of introvert inserting on calcareous ring; sin, perioesophageal
sinus, part of general body cavity; susp, suspensors of buccal wall; t, dendritic tentacles retracted
into cavity of introvert; ten.mu, tensor muscles of oesophagus; w.can, radial canal with calcareous ring seen through wall; w.r, water vascular ring.

few species the adults of these soft-bodied forms have done away with an external
madreporite. In the young the stone canal opens on the dorsal body wall, but
commonly loses contact with it and finally opens to the coelom by a perforated bulb
(mad), the madreporic body, with the cilia lining its pores beating inwards and not
outwards as in external madreporites, since, unlike them, it is in no danger of being
blocked with detritus. Fluid drawn in through the pores replenishes the ampullary
system of the tube feet. Associated with the water vascular ring (w.r) are one or
more polian vesicles (pol.ves), muscular sacs acting as reservoirs of fluid; like the
stone canal they may become more numerous with age.

Freed from the restrictions imposed by a well-developed endoskeleton and in possession of a muscular body wall, the holothurians have exploited a variety of habitats and assumed some bizarre shapes. Some creep over a hard substratum by means of the suckered tube feet of the trivium, perhaps aided by the muscles of the body wall; others have evolved flotation devices and are planktonic; a few can swim and some burrow.

In creeping forms the body shows a dorsoventral differentiation since the podia of the bivium are reduced in number and size and function as sensory papillae. The tube feet of the trivium are the main, and sometimes the sole organs used in creeping. They may persist in three longitudinal rows as in *Stichopus* or be scattered over the whole surface as in *Holothuria*. The disk of their sucker differs from that of asteroids and echinoids. In *Holothuria* it is supported by a single dome-shaped ossicle with fenestrations through which pass connective tissue fibres to insertions on the surface cuticle; there is also a ring of minute spicules at the periphery. The epithelium of the disk is rich in mucous cells and since there are no levator muscles it would appear that their secretion provides the main adhesive force. Retractor muscles, antagonistic to those of the ampulla, may pull on one side of the foot to detach the sucker. Tube feet may be reinforced by body wall muscles in locomotion: the large sea-cucumber *Stichopus panimensis* creeps forwards in caterpillar-like fashion and muscular waves can be seen to be initiated at the anal end and pass anteriorly along the body, synchronized with the activity of the large numbers of tube feet. This use of muscular movements may become the most important factor in creeping and then a marked reduction in the number of podia on the sole is common. *Psolus* (Fig. 163C), found in deep water, is an extreme example: it can creep like a snail on a flat muscular sole which has few tube feet, whilst dorsally the body is arched, covered with scale-like ossicles and has none; both mouth and anus are dorsal. There is thus pronounced bilateral symmetry.

Some pelagic forms, only a few centimetres in length, incorporate papillate podia in their flotation devices. A few are united to form a dorsal transverse membrane, like a sail – in *Peniagone* (Fig. 163B) and in *Pelagothuria* (Fig. 163A) radially arranged papillae (pap) support a web which is immediately behind the oral tube feet (t). The mouth (m) is directed up in *Pelagothuria* and the anus (cl.o) down so that the body resembles a medusa. These pelagic holothurians are not known to the average zoologist; they are inhabitants of ocean waters.

Some holothurians live in burrows in rock crevices, others in sand or mud. They may have their tube feet arranged in five ambulacral bands with little scattering into interambulacral areas as in *Cucumaria* (Fig. 163D) or distributed over the body with little relation to ambulacra as in *Thyone*; other burrowers, like *Synapta*, have lost their locomotor tube feet altogether. A holothurian like *Thyone raphanus* sinks into the sand or mud ventral side down leaving both ends or only the tail protruding. This is accomplished by contraction of the circular and longitudinal

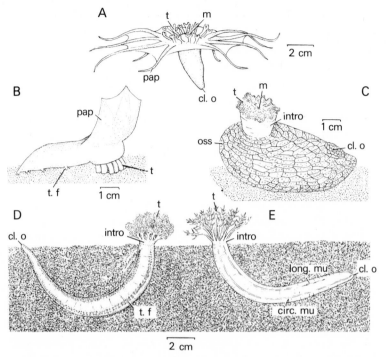

Fig. 163. Holothurians. A, *Pelagothuria ludwigi*. (After Chan.) B, *Peniagone*, with sail. (After Théel.) C, *Psolus fabricii* resting on creeping sole with tentacles contracted. (After Hyman.) D, *Cucumaria elongata* in mud. E, *Leptosynapta inhaerens* in sand. circ.mu, circular muscles; cl. o. cloacal opening; intro, introvert; long.mu, longitudinal muscle band; m, mouth; oss, scale-like ossicles; pap, papillae (modified tube feet supporting web); t, tentacles (oral tube feet); t.f, locomotory tube foot.

muscles of the body wall, aided by the action of the tube feet. A similar position is attained by *Cucumaria elongata*, which penetrates the substratum head first and, using muscles and podia, buries itself, and then each end emerges. The need for the anal end to make contact with the surface reflects its use in respiration (p. 421). The most efficient burrowers are the vermiform, apodous forms in the family Synaptidae. *Leptosynapta inhaerens* (Fig. 163E) lives in sand or mud or amongst *Zostera* and burrows by means of the tentacles around the mouth (t) and the muscular action of the body wall, first loosening the surface of the substratum with the tentacles, then penetrating it and moving round beneath the surface, with only the head emerging occasionally. Respiration does not involve respiratory trees in synaptids so the cloacal opening need not project above the substratum. It can bury itself in five minutes and move through the sand at the rate of 2–3 cm min^{-1}, whereas *Thyone* takes 2–4 h to form a burrow. In burrowing annelids *points d'appui* are provided by chaetae: similar areas where the soil is gripped are provided in *Leptosynapta* by a distinctive type of ossicle (Fig. 147E) shaped like an anchor and associated with an anchor plate. Dilatation and consequent stretching of the body wall brings the anchors, supported against the plates, to the surface of the body, where their two pointed tips project and attach the distended area to the substratum, functionally replacing tube feet. Some apodous forms can stretch themselves to a length of 1–2 m, and similar mobility of the body enables *Leptosynapta inhaerens* and *L. albicans* to be active swimmers.

The structure of the ophiuroid arm indicates its importance in locomotion. It is sharply distinct from and freely movable upon the disk, covered by skeletal plates arranged in an upper, two lateral and an under series and has spines at the sides to give grip (Fig. 156, sp). The under or oral series of plates arches over the ambulacral groove, converting it into an epineural canal (epin). The whole body surface is cuticularized and the dermis, occupied for the most part by the endo-skeleton, rests directly on coelomic epithelium since a muscle layer is missing. The paired ambulacral ossicles have fused to form a series of median vertebrae (vert. oss) which articulate by an arrangement of knobs (knob) and sockets and are moved on one another in various directions by four muscles (d.mu, v.mu). As a consequence the perivisceral coelom which extends into the arm is reduced (coel) and contains no gut caecum. The nerve cord (rad.n) has ganglia related to the muscles between the vertebrae. During locomotion the disk is raised off the ground and any arm or pair of arms may lead, pointing in the direction of movement, and the body is propelled forward in a series of jerks. The arms on either side of the single or pair of leading ones press backwards in a rowing motion, and trailing arms push. In amphiurid brittle stars the leading arm attaches to some object distally and then with a sinuous contraction pulls the body and the other arms forward, or the other arms may push while it pulls. The tube feet (Fig. 164A, t.f), emerge between lateral (lat.pl) and ventral plates (v.pl) and lack both ampullae and suckers. The basal part of each, however, is swollen into a muscular bulb with a nerve ring round its base. Though not sited in the same place as ampullae these bulbs are functionally equivalent, and it may, indeed, have been from the insinking of such a structure into the more capacious coelom of the asteroid that the true ampulla arose. The tube feet are not useless in locomotion but supplement the activity of the whole arm, more effectively in some species than in others. The adhesive properties of mucus secreted by spines and tube feet are effective enough to allow one of the commonest brittle stars of the eastern Atlantic, *Ophiocomina nigra*, to climb vertical walls. In the sand stars of the genus *Ophiura* they are thrust against the substratum to aid the active arms in their power stroke, at the end of which they are suddenly withdrawn by shortening and thrust out again at the start of the next. The principal protractive force comes from the muscular bulb at the head of the tube foot (Figs 156 and 165, h.b; cont.b). The capacity of the bulb is severely limited by the adjacent vertebra but this does not determine the degree of extension of the foot since fluid can be pressed out of the radial canal and returned there when the foot retracts. In some

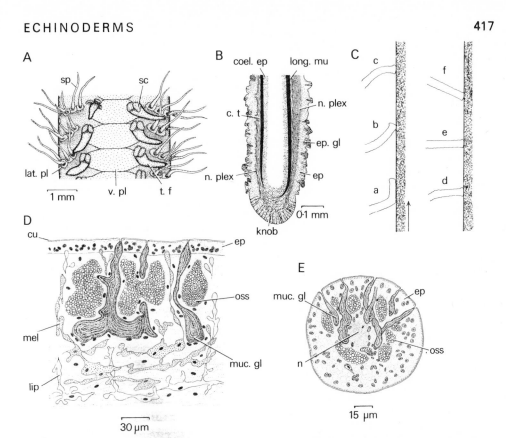

Fig. 164. *Ophiocomina nigra*: A, part of oral surface of arm. B, longitudinal section of distal part of tube foot. C, successive phases of movement (a–f) of tube foot in climbing a vertical surface. D, transverse section of upper arm plate to show calcareous and pigment layers in relation to mucous glands. E, T.S. arm spine to show basal ends of multicellular glands clustered around the central nerve. (A, B, C after Smith; D,E after Fontaine.) c.t, connective tissue; coel.ep, coelomic epithelium; cu, cuticle; ep, epidermis; ep.gl, epidermal gland; knob, terminal knob of tube foot; lat.pl, lateral arm plate; lip, lipocyte; long.mu, longitudinal muscles; mel, melanocyte; muc.gl, multicellular mucous gland; n, central nerve; n.plex, neurofibrillar plexus; oss, ossicle; sc, tentacle scale; sp, arm spine; t.f, tube foot; v.pl, ventral arm plate.

species there is an additional nonmuscular storage sac arising from the radial canal at the junction of each pair of transverse canals (exp.ves). The flow of fluid is regulated by a special valve at the entrance to the head bulb (v) which, unlike corresponding valves of other echinoderms is worked by extrinsic muscle fibres (mu.v) arising on the wall of the bulb and crossing its lumen to insert on the two flaps which comprise the valve. The force of the muscles pulls the flaps apart even if the hydraulic pressure in the foot is greater than in the radial canal. The canal can be constricted at intervals between the origins of successive pairs of feet (const) to isolate lengths which can form expansion chambers for fluid when they retract, an arrangement reminiscent of crinoids (p. 422). The connective tissue (Fig. 164B, c.t) of

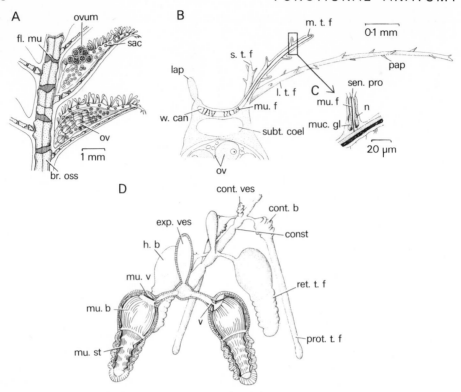

Fig. 165. A, B, C, *Antedon bifida*: A, two pinnules with ripe ovaries, the eggs being discharged from the upper one. (After Grassé.) B, transverse section of pinnule to show group of three tube feet on one side and a lappet on the other; one foot is cut longitudinally. C, longitudinal section of papilla considerably enlarged. D, generalized ophiuroid: diagram of part of the radial water vascular system showing three pairs of tube feet, their head bulbs and accessory vesicles. One pair has been cut longitudinally to show the longitudinal muscle layer differentiated into two systems. (B, C, D, after Nichols.) br.oss, brachial ossicle; const, constriction in radial water vascular canal; cont.b, contracted head bulb; cont.ves, contracted vesicle; exp.ves, expanded vesicle; fl.mu, flexor muscles; h. b, head bulb of tube foot; lap, lappet bordering food groove; l.t.f, long tube foot; m.t.f, medium tube foot; mu.b, stout longitudinal muscles of bulb; muc.gl, mucous gland; mu.f, muscle fibres; mu.st, longitudinal muscles of stem; mu.v, extrinsic muscles of valve passing to wall of bulb; n, nerve cell; ov, ovary; ovum, ovum being discharged; pap, papilla; prot.t. f, protracted tube foot; ret.t. f, retracted tube foot; sac, sacculus; sen.pro, sensory process; s.t. f, small tube foot; subt.coel, subtentacular (= perihaemal) coelomic canal; v, valve; w.can, water vascular canal.

the tube foot comprises an outer layer of longitudinal fibres and an inner envelope, most conspicuous in the bulb, of a deformable matrix with a double spiral of inextensible fibres similar to that described for nemerteans and related to the control of fluid pressure. The envelope and fibre systems are well developed in the podia of *Ophiura* and the burrowing amphiurids, which are distinguished by their speed and precision of action, but the envelope is slight and the layer of longitudinal fibres relatively thick in the slower moving podia of *Ophiothrix fragilis* and *Ophiopholis aculeata*.

The stone canal of ophiuroids is connected to the outside, usually by a single pore though in some species there may be up to twelve. The hydropore is not aboral as is the madreporite of asteroids and echinoids, but located in an interradial oral shield. Thus the stone canal and axial gland, both enclosed in the axial sinus, run oralwards from their respective rings just above the jaw apparatus. The water ring gives off five polian vesicles, in addition to the five radial canals, and also branches to the buccal tube feet.

The most advanced ophiuroids of the order Euryalae dwell mostly in deeper waters and are known from dredged material. They are covered by a naked or granulated skin and mostly lack scales. The disk may be up to 10 cm in diameter and the long arms may be simple or greatly branched as in *Gorgonocephalus* and *Astrophyton*. The arms are more flexible than in other ophiuroids and can coil around objects and roll up in the vertical plane because of the shape of the articulation of the vertebral ossicles.

Respiration of echinoderms

Whether primarily evolved for this activity or not tube feet provide a respiratory surface of paramount importance: they are thin-walled sacs in which a circulation of coelomic fluid is maintained, and are in constant motion in an active animal, often extending some distance through the water. The general surface of the body is permeable to oxygen; and movement of water over it is sustained by general (asteroids, echinoids) or partial (crinoids, ophiuroids) ciliation as well as the action of appendages – spines, pedicellariae, tube feet. The crinoids, with their finely divided arms, have no respiratory areas other than the body surface and the tube feet, whereas in other classes accessory structures have developed. Thin respiratory evaginations of the body wall emerging from between the ossicles, have already been described in *Asterias* (Figs 146 and 150). In other starfish gills are confined to the aboral surface, either grouped in small clusters (*Pectinaster*) or scattered (*Astropecten*, Fig. 155, g; *Luidia*). The two burrowing forms need to protect such delicate structures: the aboral surface carries small plates each with an erect column bearing at its summit a crown of movable spinelets – the entire piece is called a paxilla (sp). The spinelets can be raised vertically to expose the gills or lowered to form a chamber protecting and enclosing them. The spinelets have gland cells secreting mucus which forms a sheet protecting the gills from sediment as the starfish burrows. In *Astropecten* and *Luidia* a groove runs along the edges of each arm between the large marginal plates; in the grooves are tracts of small spines, each club-shaped, with two longitudinal bands of cilia, one on either side, which produce a strong orally directed current out of the spaces between the paxillae in which the gills lie.

In echinoids the most significant respiratory area is that provided by the tube feet, and since their ampullae project freely into the body cavity they give maximal surface area for the diffusion of oxygen into the coelomic fluid. The fact that each ampulla (Fig. 157) is connected to the foot by two canals and that a water circuit is said to be maintained through them may be of respiratory significance. In regular echinoids (except cidaroids) they are locally supplemented by ten bushy tufts on the periphery of the soft peristome, one pair in each interambulacral area. They are out-pouchings of a special part of the coelom enclosing Aristotle's lantern which is closed off from the main perivisceral cavity by a delicate membrane (see p. 424). The volume of fluid within these gills is controlled by two sets of muscles associated with the lantern, one acting to increase the volume of the lantern coelom and draw fluid out of the gills and the second to decrease the volume and force fluid into them. The diffusion of oxygen into this part of the coelom by way of respiratory tufts is associated with the activity of the lantern; it is not significant for the animal as a whole. Respiratory movements can be evoked by stimulation of the nerve ring; under normal conditions they are not rhythmical, but become so when increased carbon dioxide acts as a stimulant.

In cidaroids the rows of tube feet continue across the peristome to the mouth and there are no gills. In urchins belonging to this group the lantern coelom has five large bushy sacs known as Stewart's organs, projecting into the perivisceral coelom which facilitate interchange of gases.

In the irregular echinoids, clypeastroids and spatangoids, gills are wanting and the tube feet (Fig. 161, resp) of the petaloid ambulacra are specialized for respiration and have no other function. They are suckerless and lobulated or leaflike to increase their surface area. A water flow maintained over the surface of the test by cilia which beat away from the apical area passes down the respiratory ambulacra and flows around the podia which divert small streams out of the petaloid areas; the subanal fasciole (sub.fasc) is largely responsible for directing the current down the sanitary tube. *Spatangus purpureus* burrows in gravel to a depth of about 5 cm and if no connection with the surface is maintained by spines the respiratory current, brought about by the ciliation of the body, flows through the interstices of the particles. As in all clypeastroids, which are shallow burrowers, there are no tube feet for building a burrow; this contrasts with *Echinocardium cordatum* which burrows in sand to about three times the depth and has special tube feet to maintain a respiratory funnel.

Ophiuroids and holothurians have neither papulae nor gills. The brittle stars have bursae (Fig. 156, op.b), sac-like invaginations of the oral wall of the disk at the bases of the arms, through which water is circulated; their walls are thin, even thinner than those of the podia, as might be expected in an internal situation.

Commonly ten bursae project into the interior of the disk, occupying the spaces between stomach pouches; there may be double this number (*Ophiothrix fragilis*) or they may be lacking. Each opens by a slit or occasionally a double pore (inlet and outlet in *Ophioderma*) and the sac may be simple or have diverticula passing to the adjacent interradial muscles of the jaw apparatus (*Ophiothrix*, *Ophiocomina*). The circulation of water is maintained by cilia or the pumping action of the aboral muscles of the disk. Despite the presence of such spaces tube feet provide a substantial respiratory surface and exchange of fluid between the radial canals of the water vascular system and the tube feet relays oxygen which has diffused into them from the surrounding water into the canal systems. Other pathways of diffusion in the arms are restricted by the vertebrae.

In the holothurians, with the exception of those (such as synaptids) in the order Apoda, the intestine ends in a cloaca from which arise a pair of diverticula known as respiratory trees (Fig. 158, resp.t). They are branching tubules projecting into the large perivisceral coelom, thin-walled, muscular and with their ultimate branches ending in vesicles. Water is rhythmically pumped into them and then expelled. Oxygen and carbon dioxide exchange takes place across the membrane of the vesicles in response to a diffusion gradient; the oxygen tension in the coelomic fluid is consistently lower than that of water leaving the respiratory trees in species which have been studied. In *Holothuria tubulosa* over half the oxygen uptake is by exchange through the trees and the rest through the tube feet, including the feeding tentacles, and the integument. Pumping of water into and out of the trees by the cloacal walls is cyclic. There may be 6–10 successive inflows per minute to fill the expanding tubules and then one violent expiration. The anus is closed during inflow, the cloacal sphincter and circular muscles of the cloacal wall relaxed and the radial muscles of the cloaca (rad.mu), arising from the body wall, contracted; the water flows in under its own pressure. Then the radial muscles relax, the cloacal sphincter closes and contraction of cloacal muscles forces water through the relaxed trees. Meanwhile, relaxation of the muscles of the body wall ensures that hydrostatic pressure in the coelom does not counter inflow pressure. Outflow is brought about by the contraction of muscles of the body wall and respiratory trees with the anus open, the radial muscles contracted and muscles of the cloaca relaxed. The control and coordination of muscles of the respiratory pump involves the nerve ring and radial nerves.

The Apoda, as the name implies, have no tube feet, except for 10–12 modified feeding tentacles, nor have they respiratory trees. They are vermiform in shape with a large surface area and the skin is comparatively thin. It is roughened by sense buds, each encircled by gland cells, and these compensate for the lack of podia. The skin and the tentacles, which arise from the ring canal and have no ampullae, provide the respiratory surfaces.

Feeding and functioning of the gut

Echinoderms exploit many kinds of food and collect it by methods which reflect the diversity of their organization. The crinoids, the most primitive, are the only ones which are exclusively ciliary mucous feeders, their small, active tube feet producing and manipulating the food-trap. In *Antedon bifida* the epithelium of the tube feet is papillated (Fig. 165, pap), each papilla having sensory processes (sen.pro) at its tip which arise from basal nerve cells (n) and some mucous glands (muc.gl) surrounding a central muscle fibre (mu.f). Detritus touching the processes is thought to cause the muscle to contract reflexly and the glands to secrete mucus which is manipulated by the feet to form a net. The tube feet of the arms and pinnules are arranged in groups of three. Whereas those of the arm are all approximately the same size, those of the pinnules can be graded as long, medium and short. The short stand erect at the sides of the ambulacral groove, the medium are directed outwards at an angle of about 45°, while the largest spread further to form the main catchment area and periodically flick towards the food groove to transfer their catch. This movement (the only one they can perform) is associated with the development of retractor muscles only on the median surface of the foot: extension must be hydraulic. The food string is picked up by the medium-sized feet bending outwards and then inwards towards the groove; they have retractors on median and lateral surfaces. The smallest feet manipulate the food into the groove and the cylindrical arrangement of their retractors facilitates their universal movement. Cilia in the food groove direct the catch to the mouth, and a series of lappets (lap) keeps the food within the groove.

When *Antedon* is feeding two groups of tube feet, one on either side of the arm or pinnule, are protruded simultaneously by hydraulic pressure. This pressure is not built up by a head bulb or ampulla as in ophiuroids and asteroids, but by the adjacent part of the broad radial canal (w.can) which has muscle fibres (mu.f) spanning its lumen. On contraction the fibres constrict the canal at the base of the two groups of podia and so isolate a chamber which is then also constricted and fluid is forced into the feet. The action of these muscles leaves the central part of the canal open as a through channel. In the pinnule the need for a through channel is reduced, though it is represented by a central area of the radial canal devoid of muscles.

Stalked crinoids have received little attention, but it is probable that a similar mechanism controls their tube feet. It may be argued that contraction of radial canals is the ancestral method of extending tube feet and that muscular bulbs and ampullae are devices which evolved later as tube feet became transformed into locomotor organs and greater forces had to be dealt with.

Suspension feeding is also carried out by ophiuroids, but this is only one of several methods of obtaining food. A single species may have diverse feeding mechanisms so

ensuring that a variety of food sources in the environment is available to it. When a
brittle star like *Ophiocomina nigra* (Fig. 160A) finds itself in a water current it assumes
a special feeding position with the disk, and therefore the mouth, raised slightly
from the substratum, the arms lifted into the water and widespread, their tips
curved inwards and their spines diverging. The arms swing continuously and gently
from side to side through an angle of about 40°, scouring the water. The body is
coated with mucus from unicellular glands and a tangle of mucous threads links
the spines. Small particles accumulating on the net are continuously removed as the
mucus is carried across the aboral surface and drawn towards the oral surface of the
arm by cilia. Meanwhile the tube feet, covered with adhesive secretion from
unicellular glands, lick material from the spines. They scrape themselves clean on
scales, modified spines at their base, and then mould boluses of food and mucus and
pass them oralwards from tip to tip. The same method of collecting food enables the
common brittle star, *Ophiothrix fragilis*, to thrive in congested aggregations on the
sea bottom, up to 340 m^{-2}. Species which burrow in soft substrata may feed in a
similar way. For example *Amphiura chiajei*, a common sand-dwelling ophiuroid of
European coasts, which lives in a burrow maintained by mucus, projects the arms
into the sea to catch food particles on secretions covering the podia; the podia
transport the catch to the mouth. This species can also grasp large pieces of food by
a loop in the arm and convey them to the mouth by arm retraction, a method
commonly employed by ophiuroids when feeding on chunks of detritus and even
active prey.

In holothurians food is collected by a circlet of tentacles (Figs 155, 158, 162 and
163, t) which are highly modified buccal tube feet, very muscular and well supplied
with glands and sense cells. Their shape varies and is the basis on which the class
may be subdivided into orders: in Dendrochirota they are long and arborescent, in
Aspidochirota they form shield-shaped tufts and in the order Apoda they are
pinnate. The shape of the tentacles is related to the different ways in which food is

Most ophiuroids are predominantly feeders on detrital material, roughly selected
for edibility by the buccal tube feet (Fig. 156, buc.t.f) which appear to have
gustatory discrimination and can sense food without contact. This may account for
the inclusion of much material of no food value which is gathered along with small
polychaetes, molluscs, young echinoderms, algal cells and protozoans. The possession of jaws, however, also allows many like *Ophiothrix fragilis* to browse on sessile
algae. The jaws (j) are interradial, their bases broad and they taper centrally so that
the mouth is five-angled. Each is composed of ossicles from two adjacent arm bases
and is therefore formed of two halves which are not fused but joined by muscle
(rad.mu), and so can move on one another; other muscles (inter.mu) join the base
of one jaw with the next. The ossicles involved are the first pair of ambulacrals
(amb.oss) and the first pair of adambulacrals, the latter with teeth formed from
movable spines.

gathered. The dendrochirote holothurians (*Cucumaria, Thyone, Psolus*) spread their ten tentacles through the water to form a plankton trap and, provided there is a current, are able to get enough nourishment without moving about. *Psolus* has carried this trend furthest and with its limpet-like form, dorsal tentacular crown and almost immobile habit is well adapted for such a mode of life. *Cucumaria* and *Thyone* also use their tentacles to disturb the substratum and put more food particles into suspension. In these genera, the tentacular crown and surrounding peristome are part of an introvert which can be retracted into the body for protection if the animal is disturbed. The muscles effecting retraction (ret.mu) are anterior branches from the longitudinal muscle of the body wall (long.mu) which cross the coelom to insert on the radial members of a ring of ten ossicles (five radial, five interradial) which encircles the initial part of the gut and to the five radial components of which the longitudinal muscles of the body wall are also attached. The calcareous ring encloses a coelomic space (sin) across which strands of connective tissue and muscle fibres pass to the gut wall. The water ring (w.r) encircles the gut posterior to the ossicles; from it five radial canals (w.can) pass forward to the buccal tube feet (tentacles) and their ampullae, and then curve back inside the body wall to the posterior end of the body (Fig. 158, w.can). When a dendrochirote holothurian is feeding the tentacles on which the food adheres are cleaned off in the fore-gut one by one since they have no ciliary tracts to move food. The tentacle bends through the mouth, the lips closing around its base, and it is wiped clean as it is pulled out.

The shield-shaped tentacles of aspidochirotes, such as *Holothuria*, which has twenty, have each a short stem terminating in a large number of short horizontal branches. They are powerful organs used to shovel organic debris from the sea-floor into the mouth. There is no introvert, but when the tentacles contract the adjacent body wall closes over them. In contrast the long pinnate tentacles of the Apoda are used to sweep surfaces and gather up detrital matter, the method used by *Leptosynapta* (Fig. 163E, t) in gathering microorganisms and detritus from algal growths.

Echinoids have a skeletal masticatory apparatus called from its discoverer the lantern of Aristotle. It is fully developed only in regular echinoids and reduced or wanting in the irregular ones, except clypeastroids. Most regular urchins will eat anything but, equipped with jaws and teeth which can be protruded through the mouth, some, like *Echinus*, *Diadema* and *Strongylocentrotus*, are able to graze on kelp, or like the small sea urchin *Psammechinus miliaris* scavenge the epifauna, including barnacles, crabs and shelled molluscs, the hard exoskeletons of which are fragmented with precision and force. When feeding the buccal tube feet actively hold food which is being fragmented, and mucus secreted from the oral region lubricates the action of the teeth. The muscular lips surrounding the mouth and the flexibility of the broad peristomial membrane are adaptations necessary for the wide gape required for the action of the lantern. The membrane extends to the edge of the test where ambulacral and interambulacral plates (Fig. 159B, interamb.pl) are modified to give attachment areas for muscles of the lantern and form a structure known as the perignathic girdle. Each pair of ambulacral plates carries an arch-like auricle (aur) and each interambulacral pair a solid ridge. They mark the attachment of a peritoneal fold which encloses the entire lantern and its musculature, so separating a lantern coelom, with which the gills communicate, from the perivisceral coelom.

The lantern is formed of five main interradial pieces (pyr) called pyramids because of their shape; their apices lie towards the mouth. Each is made up of two closely approximated halves meeting at the apex and is joined to the next by short transverse muscle fibres by which they can be rocked on one another. This movement is transferred to the tooth (t) which each pyramid supports. At their upper or aboral end an arched transverse bar, the epiphysis (epi) rests on the two pieces of the pyramid and between one epiphysis and the next a slender radial piece, the compass (com), passes outwards from the vicinity of the oesophagus (oes). Beneath each compass is a stouter piece, the rotule (not shown in Figure). Each tooth is embraced by ridges from the pyramid near its lower end, and near its upper end is gripped by the epiphysis. Its oral end, formed of especially hard calcareous material, projects into the buccal cavity and is protrusible through the mouth. Its aboral end (ab.t), the growing area which compensates for the considerable wear, is soft. The whole apparatus is operated by a complex series of muscles. The protractors of the lantern (prot.mu) are flat bands extending from the epiphyses to the paragnathic girdle at the interambulacra. Their contraction pulls the lantern outwards and the teeth are exposed. Retractor muscles (ret.mu), which pull the lantern back partially or wholly and also open the teeth, originate on the radially situated auricles (aur) of the paragnathic girdle and insert on the lower ends of the pyramids. Internal and external rotular muscles connect the epiphyses with the corresponding rotule and as the epiphyses are articulated to the half pyramids, their movements, brought about by these muscles, are transmitted to the teeth. The compasses and their muscles are not connected with feeding, but are part of the respiratory apparatus. A flat pentagonal muscle (elev.com) in the coelomic epithelium covering the aboral surface of the lantern attaches to the compasses. It is the elevator of the compasses and on contraction raises them and so increases the volume of the lantern coelom and draws fluid out of the gills. Two slender muscles (dep.com), which insert on the outer end of each compass, run, external to the lantern protractors, diverging to their origins on adjacent interambulacra. These are depressors of the compass and on contraction force fluid into the gills. Oxygen diffuses through the thin peritoneal layer separating the lantern and perivisceral coeloms.

Irregular echinoids, clypeastroids and spatangoids, which are typically burrowers, feed on detrital material and microorganisms. Amongst clypeastroids the lantern is low and broad and the peristomial membrane narrow, so the teeth are scarcely protrusible. In the small *Echinocyamus pusillus*, the only British member of the group, the buccal tube feet scour the gravel and coarse sand in which it lives for organic particles and these are collected by the teeth. In most clypeastroids ambulacral feeding grooves lead to the mouth and as the animal moves through the sand small particles which fall between the dorsal spines are trapped in mucus and directed orally to the grooves. In some sand dollars the test is perforated with elongated holes or lunules. Their function has been demonstrated in the keyhole urchin *Mellita sexiesperforata*, common in Jamaica (Fig. 161, E,F). As it ploughs horizontally through the substratum by means of ambulatory spines on the oral surface, sand is pushed on to the aboral surface of the test. Large particles are transported posteriorly on the tips of club-shaped spines, the larger of the two types on the aboral surfaces, and deposited in the sand, leaving the body at the posterior end or by way of the lunules (lun). Fine particles, selected only on size, fall between the spines which are ciliated at the base, where currents carry them to food tracts (f.t) on the oral surface, either over the margin of the test or through the lunules. The food tracts lead to the ambulacral grooves (f.gr). Food in the grooves is loosely aggregated in mucus which may have come from the second type of aboral spine; it is carried to the mouth by podia. Other podia of the oral surface probe between the sand grains for food.

The spatangoids gather organic particles with oral tube feet, up to 40 in number. The feet are prehensile and pass the particles towards the mouth where they are scraped off against spines surrounding the peristome. Organic matter falling on the aboral surface is trapped in mucus and transported orally by cilia, the main stream passing down the anterior ambulacrum to the vicinity of the buccal tube feet. *Echinocardium cordatum* also collects surface organic deposits with the funnel-building podia (p. 411), which transfer their catch to the anterior ambulacrum.

Asteroids are highly efficient predators, eating any slow-moving or sessile animals; they are also carrion feeders and a few collect particles in suspension. The feeding process of most starfish resembles that of *Asterias* (Fig. 151) which has been described on p. 394. Sometimes the prey is held directly against the mouth and if small enough may enter the stomach whole. A large starfish, *Pycnopodia helianthoides*, which has 24 arms with 15 000 tube feet, may reach a diameter of a metre and can lift and swallow whole animals such as hermit crabs and sea-urchins. The force of the tube feet can remove such animals as chitons and limpets from rocks and the sucking action of the stomach can draw worms from their tubes and barnacles from their parapet of plates. It is presumably by this means that the crown of thorns starfish (*Acanthaster planci*) strips living polyps from the underlying coral skeletons so destroying vast tracts of reef. Some different or special feeding methods occur among asteroids: the burrowing *Astropecten* cannot feed as *Asterias* does because of its stiff arms and suckerless tube feet; it uses the tube feet around the mouth to push prey rapidly through the extended gape. It feeds selectively mainly on juvenile bivalves, sensing and locating them with precision and preferring species with a low resistance to anaerobic conditions which will open soon after being ingested and expose the soft tissues. The stomach is everted only to eject the empty shells.

The scarlet starfish *Henricia*, living sublittorally on rough ground where there is a high phytoplankton concentration, feeds on particles in suspension. In its feeding position (Fig. 166) the mouth remains open, the central part of the body is raised slightly from the substratum and one or more rays extend through the water, their tips upturned, while the suckered tube feet of the others grip the substratum. Strong ciliary currents along the ambulacral areas direct particles, which may be trapped in a mucous string, towards the mouth.

Whereas in crinoids, holothurians and echinoids the gut is a coiled tube leading from mouth to anus, and glands can be incorporated in its length, in asteroids it has such a direct passage through the short oral–aboral axis that the glandular tissue has to be constricted from it and spread through the arms; in ophiuroids it is sac-like, restricted to the disk, with neither intestine nor anus.

The coils of the crinoid gut lie entirely within the disk. From the centrally placed mouth of *Antedon* (Fig. 153, m) the alimentary tract makes a complete turn in a clockwise direction, as viewed from the oral surface, then leads to the anus on a muscular papilla (a.pap) which can be extended and directed away from the disk to void the faeces. A sphincter surrounds oesophagus and rectum, but muscles are feebly developed and cilia which beat towards the anus are the important agencies in transporting food. The gut gives off a number of caeca (div) the last two long and branched, and digestive enzymes are produced mainly if not entirely by their epithelium; food does not enter them and ciliary currents drive their secretions into the main lumen of the gut. Digestion appears to be wholly extracellular. The faecal

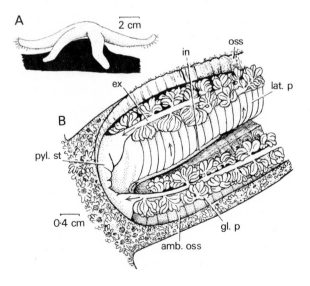

Fig. 166. *Henricia sanguinolenta*: A, in feeding position with the central part of the disk raised and two arms directed through the water. B, base of an arm opened aborally to show the pyloric caeca and their enlarged ducts. Arrows show directions of ciliary currents. amb.oss, ambulacral ossicle; ex, excurrent channel: gl.p, glandular pouch of pyloric caecum; in, incurrent channel; lat.p, lateral pouch on duct; oss, ossicles of body wall; pyl.st, pyloric stomach.

matter is embedded in a gelatinous substance derived, at least in part, from the wall of the rectum.

Correlated with the high proportion of indigestible matter which holuthurians ingest is the fact that the gut traverses the long oral–aboral axis three times (Fig. 158) between mouth and anus in *Holothuria* and may be even more extensive in others (*Thyone*). The digestive tract, supported in the perivisceral coelom by mesenteries, is divisible into buccal cavity, oesophagus, stomach (a short muscular region clearly defined in *Thyone, Cucumaria* and synaptids, but scarcely demarcated in *Holothuria* and *Stichopus*), and intestine which makes up the major length. There are no diverticula.

The food contains many abrasive particles and the oesophagus and stomach are cuticularized as a protection against them; much mucus is also secreted. The tract is ciliated in parts but cilia are not powerful enough to transport the heavy and bulky food, passage of which is effected by peristaltic waves in the gut musculature. In *Synapta* these are visible through the body wall and occur every two seconds.

Much of the haemal tissue of the holothurian is associated with the gut with two main branches, one on either side. The more conspicuous branch, the dorsal sinus, is connected to the ascending limb of the intestine by numerous tufts forming a rete mirabile or "wondrous network" (haemoplex). Haemal tissue from this plexus permeates the wall of the gut and contains an unusually large number of amoebocytes, many of which escape and are also found in the coelomic fluid. The various enzymes in the yellow digestive fluid liberated into the gut of an individual which is feeding are probably derived from the epithelium, though there is much to be learned about their sources and identities. Digestion takes place within the lumen and although digestive products are picked up by the amoebocytes and carried into the haemal system, this is probably not the whole story, as once thought, for the results of more recent investigations suggest that nutrients in solution pass directly to the coelomic fluid. Studies in *Leptosynapta, Holothuria* and *Thyone* show that there is active uptake of glucose.

When excessively irritated many holothurians undergo evisceration. In large aspidochirotes (*Holothuria, Stichopus*) the cloaca ruptures and on contraction of the body wall the whole digestive tract is emitted through the rupture along with one or both respiratory trees, and usually the gonad. Eviscerated individuals survive and regenerate the organs they have lost: the body wall must be the source of energy for this.

In a few genera of holothurians such as *Holothuria* and *Actinopyga* less excessive irritation causes the animal to curve its posterior end towards the source of irritation and, with a general contraction of the body wall, emit from the cloacal region long, sticky threads which entangle even such active predators as lobsters and render them incapable of movement; meanwhile the holothurian creeps off. The threads are derived from organs which branch from the base of the respiratory trees as a few or a tuft of long blind tubules called Cuvierian after their discovery (cuv.o). Under a peritoneal covering the tubules have an outer layer of specialized gland cells (perhaps also of peritoneal origin) resting on a layer of collagen fibres packed into a tight spiral formation and held thus by muscle fibres; centrally is a minute canal continuous with the cavity of the respiratory trees and cloaca. When the holothurian is irritated it seems that water in the respiratory trees is forced into the tubules which then elongate within the coelom. The cloacal wall ruptures, probably at a preformed weak spot, the tubules emerge from the rupture, peritoneal wall outwards, washed out by water from the gut. After discharge the muscle layers no longer hold the connective tissue in spiral shape so they elongate into threads several feet long, their surface cells rupture, freeing an extremely sticky substance to entangle possible predators. Holothurians with this habit are sometimes called "cotton spinners".

The gut of the echinoid (Fig. 167) not only agrees with that of the holothurian in being long and coiled, but also in having extensive networks of haemal lacunae branching from two longitudinal sinuses associated with stomach and intestine. It is, however, enclosed in an inflexible test and some peculiar features of its organization relate to this. The gut is longest and best differentiated in regular urchins, macrophagous feeders fragmenting plant and animal material with the lantern teeth. The sand dollars, detritus feeders with a reduced lantern, have a shorter, broader gut, though irregular urchins, which also gather food with tube feet and ciliary currents but have no lantern, retain the long gut. The alimentary tube is divisible into buccal cavity, oesophagus, stomach, intestine and rectum. The buccal cavity is that part of the gut passing through the lantern. The oesophagus (oes), emerging from the lantern in the regular urchin, passes vertically, then descends orally, to a junction with the stomach, generally marked by a blind pouch or caecum. The stomach and the intestine (int), except in sand dollars, are typically festooned from the inner surface of the test by mesenteries (mes), the festooning being related to the pentamery of the body. The rectum (r) ascends vertically to the anus (a). In the heart urchins the stomach is characterized by a large caecum in its proximal part.

In the primitive urchins known as cidarids and diadematids a groove runs the length of the stomach. In more advanced echinoids this is constricted off as a narrow tube or siphon (si) through which a current of water is maintained from the oesophageal end backwards. The water is drained from the food in the oesophagus, which is thereby concentrated in the stomach, and after passing through the siphon is squirted on to the intestinal contents as a kind of enema. The siphon may also be implicated in the control of the gut volume which, in an animal encased in a rigid test filled with incompressible fluid, must remain more or less constant, and which is complicated by the fact that the gut is capacious and the volume of food in it varies considerably over short periods of time. In regular urchins it has been shown that

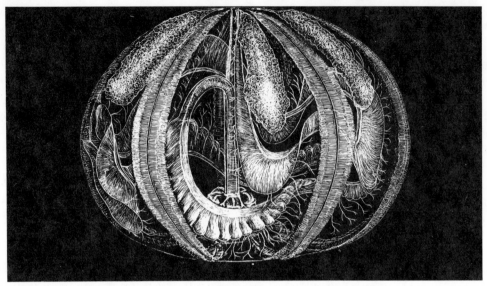

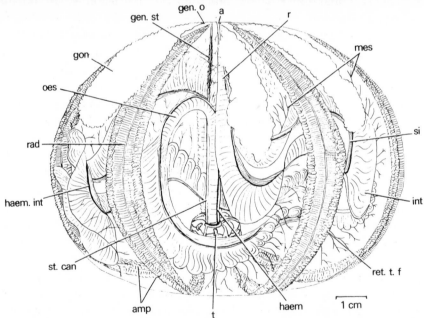

Fig. 167. *Echinus esculentus*, test removed. a, anus; amp, ampullae of tube feet; gen.o, genital opening; gen.st, genital stolon; gon, gonad; haem, haemal vessel arising from oral haemal ring; haem.int, haemal vessel accompanying intestine; int, intestine; mes, mesenterial supports of intestine; oes, oesophagus; r, rectum; rad, radius along which run nerve, perihaemal canal, haemal strand and water vascular canal; ret.t.f, retracted tube foot; si, siphon; st.can, stone canal arising from water vascular ring; t, tooth bending over aboral surface of Aristotle's lantern.

the volume of fluid varies inversely with the quantity of food; the ingestion of food is compensated for by ejection of fluid from the anus and defecation by the uptake of fluid at the mouth. To permit this and allow the flow into the siphon it is necessary for water to move easily through the gut contents and this is achieved by a special treatment of the ingested material. The food in the buccal cavity is compacted into subspherical pellets of regular size which are covered in tough mucus not destroyed by, but permeable to enzymes and sufficiently permanent for the pellets to retain their identity even after defecation. The end of the rectum is attached to the edge of the periproct by a coelomic membrane enclosing a periproctal sinus, and around the anus is a separate perianal sinus. The walls of both sinuses have muscle fibres which, by altering the pressure in the cavities, assist in expulsion of faeces.

The epithelium lining the gut is ciliated, except in the oesophagus, and has gland cells. Mucous glands are abundant in the oesophagus and zymogen glands in the stomach of regular echinoids, except for one species, the common long-spined urchin of the Caribbean, *Diadema antillarum*, where enzymes are said to be secreted by the caecum; this is also the site of their production in irregular urchins. In the regular urchin the food pellets are moved along the oesophagus by peristalsis; this is the only region of the gut with separate circular and longitudinal muscle layers.

Assimilation efficiency is high and this may be associated with the long time food remains in the gut. In the purple urchin, *Stronglocentrotus purpuratus*, feeding on the brown weed *Macrocystis*, the digestive efficiency has been recorded as 80 per cent. When this alga was labelled with ^{14}C the easily absorbed, sweet alcohol, mannitol, its chief storage product, was detectable in the coelomic fluid soon after feeding commenced and rose to a peak 4–6 h later. Starved individuals fed on the red weed *Iridaea flaccidum* labelled with ^{11}C showed over 90 per cent of the activity removed from the gut contents during the first day, as the alga passed through the oesophagus and stomach. The activity which disappeared from the gut contents appeared in the coelomic fluid, reaching a peak level a few hours after feeding started. Labelled galactoside, hydrolysed by the gastric enzymes, was distributed to body tissues by this course whereas lipids, less readily transferable, were stored in the gut wall, the primary storage organ for nutritive reserves throughout the phylum.

The significance of the network of haemal lacunae (Fig. 167, haem.int) in the gut wall remains unknown. It is unlikely to function as a true circulatory system though there is active movement and possibly restricted circulation in the vessels of the stomach. It has been suggested that this may be concerned with the extraction of nutrients from the gut and their passage to the coelomic fluid.

The functional regions of the gut are precisely defined in asteroids, as has been described for *Asterias rubens*. The tract is lined by a tall columnar epithelium of ciliated and mucous cells, and cells secreting enzymes are limited to the pyloric caeca in macrophagous starfish. Ciliary currents flow from the cardiac to the pyloric stomach, radially across the floor of the pyloric stomach, into the pyloric ducts, and so to the caeca; they leave the ducts aborally. Enzymatic secretion is directed out of the pyloric caeca and along ciliated channels in the wall of the cardiac stomach which, whether everted through the mouth or remaining *in situ*, smears the secretion over the food. It has been shown for *Asterias forbesi* that the channels follow the branching of connective tissue fibres in the gastric ligaments and whereas the cilia over the rest of the gastric epithelium sweep aborally, those in the channels sweep orally. At least in some asteroids the digestive fluid contains a complete proteolytic system comparable with that in the vertebrate pancreas. In *Henricia* (Fig. 166), a microphagous feeder which has been studied in some detail, enzyme-secreting cells are also scattered in the epithelium of the cardiac and pyloric stomachs and the enlarged ducts of the pyloric caeca. In all asteroids the caeca, serving as the chief organs of absorption, present a large surface area to the perivisceral coelom through which soluble nutrients can diffuse to the coelomic fluid for distribution. The haemal system associated with the gut is poorly developed as compared with holothurians and echinoids and its significance is again unknown.

The gut of the microphagous feeder must provide strong ciliary currents to circulate the particulate food and mix it with digestive juices; in microphagous asteroids the ducts of the pyloric caeca have been greatly enlarged for this purpose, and the cardiac stomach, so voluminous in predatory starfish, is quite small. In *Henricia* the arm is deep, with a stiff wall composed of ossicles (oss) which send up through its thickness erect columns tipped with a bundle of spinelets. This encasement allows only limited flexibility and so ensures the successful functioning of a ciliary mechanism within the ducts of the pyloric caeca which extend through the length and depth of the arm. Ten ducts make up the greater volume of the digestive tract and open widely to the large pyloric stomach (pyl.st). Each duct is deep in the oral–aboral direction with a narrow channel on its oral edge, whilst the glandular parts of the caecum form a series of pouches opening from a similar channel on the aboral edge. A centrifugal current from the cardiac stomach extends along the ventral channel (in) while the lateral walls carry a series of grooves or pouches (lat.p) in which the effective beat of the cilia is aboral and leads to the openings of the glandular pouches (gl.p). The pouches are the focal point of the circulation, and form series on either side of the dorsal channel of the duct, each pouch with an opening deep enough to separate a ventral inhalant current from a dorsal exhalant one leading to the dorsal channel (ex) and thence to the aboral part of the pyloric stomach. Particles entering the pouches are circulated in the digestive juice for some minutes before entering the exhalant stream. Even then they may not be rejected to the intestine but recirculated within the pyloric stomach. The digestive system is similarly modified in other particulate feeders (*Porania, Linckia, Echinaster*) where the diverticula in the arms may be even more elaborate.

The rectal caeca, characteristic of asteroids, are particularly large though their function remains obscure.

The digestive system of ophiuroids is anatomically the simplest in the phylum yet information concerning its function is scanty. The mouth (Fig. 156, m) surrounded by the peristomial membrane, is on the aboral surface of the jaw apparatus (j) which is therefore never hidden from view as in echinoids. There is a short oesophagus and a sac-like stomach (st) filling the available space within the disk, to the wall of which it is anchored by mesenteries (mes). The gastric wall is folded peripherally to form ten pouches fitting between the bursae. Only in one Indo-Pacific species (*Ophiocanops fugiens*) does the stomach extend from the disk into the arms. The epithelium of the stomach has flagellate cells, mucous cells and, especially within the pouches, cells filled with secretory granules which are presumably enzymatic. A strong protease, acting in both acid and alkaline media, has been reported for *Ophiura texturata*, also an amylase, and there is probably a lipase.

Functioning of the haemal system

The significance of the haemal system in echinoderm organization has intrigued many investigators. The axial gland or organ, a spongy mass of tissue transversed by a haemal channel, has been regarded as its centre in asteroids, ophiuroids and echinoids and pulsations of low magnitude and frequency, observed in some parts of the axial complex in young individuals, initiated the idea that the axial organ might be a heart and the focal point of a circulatory mechanism. In asteroids and ophiuroids at least, the axial sinus (of coelomic origin) could then be regarded as a pericardial cavity. In the axial haemal complex of the echinoid *Strongylocentrotus pupuratus*, vessels have been described which are said to connect in a specific way with the haemal system, the perihaemal coelom and the ambulatory system, and have been described as a kind of heart which brings about a one-way circulation within the haemal strands for the transference of fluid to coelomic spaces. The histology of the axial organ in most echinoids, however, suggests that its chief significance is secretory rather than circulatory and its association with the haemal system is rather to allow dispersal of its secretion to the perivisceral coelom by way of the haemal strands. The organ has also been assigned an excretory function in echinoids since it takes up foreign matter directly, or indirectly from coelomocytes. Endocrine functions have also been suggested for the axial gland of crinoids (Fig. 153, ax.o) which consists of tubules of glandular cells embedded in connective tissue and covered externally by the coelomic epithelium of the axial sinus (ax.sin); it is closely associated with a plexus of haemal tissue orally, known as the spongy organ (sp.o). This multiplicity of supposed functions reflects our genuine ignorance of the real role of the haemal system in the life of the echinoderm. In general its strands follow the course of the water vascular system and branch over the gut and gonads; they consist of connective tissue with a ramifying system of intercommunicating spaces lacking a definite lining, which are better called lacunae than vessels. The connective tissue is covered externally by coelomic endothelium and most of the system is enclosed in tubular parts of the coelom, the perihaemal system.

No haemal system is likely to function as a true circulatory system. In summary it can be suggested that pulsatile and contractile parts of the system promote the exchange of materials between coelomic fluid and various organ systems, especially gut and gonad, and perhaps disperse secretion from the axial organ. In animals the size of echinoderms a circulatory system of some kind is essential for transporting respiratory gases, nutrients and waste, and these functions are primarily fulfilled, as indicated earlier, by the fluid of the perivisceral coelom, kept in circulation by cilia and the movement of organs. The fluid contains numerous corpuscles of various types, which wander freely through the body. They seem to be involved in a variety of transport processes.

1. In some holothurians they contain haemoglobin and probably carry oxygen. It is significant that these cells are abundant in such burrowing forms as *Thyone*, *Cucumaria* and *Synapta*, where respiratory difficulties might be expected.

2. The abundance of amoebocytes related to the gut of echinoids and holothurians has suggested a role in the transport of food; in holothurians it seems certain that they do this and it has even been suggested that they carry digestive enzymes manufactured in the rete mirabile of the haemal system into the intestine for release.

3. Certain amoeboid cells have been shown to pick up waste, migrate to the exterior by a variety of routes and disintegrate.

4. They act as part of the mechanism by which coelomic fluid clots when exposed to injured tissue.

Reproduction of echinoderms

With few exceptions the sexes are separate but cannot be distinguished externally. Sperm and ova leave the body directly from the gonad and fertilization is external. Hence there are neither copulatory organs, nor accessory glands, nor storage sacs for sperm or ova associated with the reproductive system, and the enormous size of the ripe gonad is indicative of the numbers of gametes produced. The primary gonocytes do not arise in the gonad, but are presumably of peritoneal origin, and migrate there along a circumscribed route which is closely associated with the haemal system and forms a cord known as the genital rachis. It arises near the axial organ, the peritoneum over which may produce the germ cells, connects with an aboral ring round the gut from which branches run into the arms where their tips dilate to form the gonads. In a few species sexuality is unstable and hermaphrodites occur and in a few others hermaphroditism is normal.

Primitive echinoderms probably had a single gonad opening by a single pore, yet among extant forms only the holothurians retain this arrangement. Their gonad is in the anterior part of the perivisceral coelom and comprises a variable number of tubules radiating from an attachment to the interradial mesentery of the bivium through which the gonoduct passes to the gonopore; it becomes very voluminous at maturity. In contrast crinoids have numerous gonads located in the arms. In *Antedon* (Figs 152 and 165) there is one in each pinnule with the exception of those adjacent to the disk. The animals are gregarious and spawning in a group is more or less simultaneous. As the ova leave the gonad they are fertilized and adhere to the pinnule, perhaps up to thirty on each, until the free larval stage has developed. A number of antarctic comatulids form a brood pouch as an invagination of the body wall adjacent to the gonad, and the free larva is more or less suppressed; the eggs are large and their numbers reduced.

Asteroids, ophiuroids and echinoids have the gonads pentamerously arranged in the interradii and projecting into the perivisceral coelom; they open interradially. The gonads of asteroids consist of bunches of tubules attached on each side of the interbrachial septum, usually two in each arm. Each discharges by one or a group of pores, though in a few species in the families Luidiidae and Astropectinidae gonads are serially arranged along the arm, each with its own gonopore. Colour differences between the ripe testes and ovaries visible through the body wall allow the sexes to be distinguished. A female *Asterias* has been said to liberate 2.5 million eggs in two hours. Typically the eggs and larvae are planktonic. Some species have special breeding habits which may be related to anatomical differences. *Asterina gibbosa*, for example, the common pincushion star of western Europe, and *A. exigua* of Australia, attach their eggs to the under surface of stones, and to make this possible have gonads and gonopores on the oral side. The large eggs, 0.5 mm in diameter, adhere by their membranes and even after hatching the young remain in contact with the membranes for a time. Some brood the young, usually, like *Henricia sanguinolenta*, by arching their arms over the eggs. If the arms are stiff this is impossible and the brood chamber must then be aboral, the eggs being protected among paxillae and, as the embryos increase in size and push the paxillae aside, the body wall of the parent is stretched and the young starfish come to lie in a depression. In *Leptasterias groenlandica* the few eggs develop to young starfish in pouches of the cardiac stomach; meanwhile the female does not feed.

The gonads of ophiuroids (Fig. 156, ov) are confined to the disk, are attached to the coelomic wall of each bursa, and discharged by a temporary opening into it. Fertilization is either external (*Ophiura, Amphiura, Ophiocomina, Ophiothrix*) and the egg develops to a free larva, or in the bursa where the juveniles develop. Developing young may be found within the bursa of the small littoral brittle star *Amphipholis* throughout the year. Each is initially secured by a stalk which grows out from the bursal wall, and it has been suggested that the numerous haemal sinuses there bring nourishment to the embryos since, unlike other viviparous ophiuroids, they have little yolk.

Regular echinoids have five gonads (Fig. 167, gon) suspended by mesenterial strands and appearing more or less fused at maturity, while the irregular ones have four, since the retreat of the periproct from the centrally located apical plate system in the young urchin destroys the fifth. Each gonad tapers to a short gonoduct aborally which opens by the gonopore in the corresponding genital plate (Fig. 159, gen.o). In a few species the gonopores are mounted on papillae in males (*Psammechinus miliaris, Echinocyamus pusillus*) or in both sexes. Species which brood their young, chiefly cidaroids and spatangoids from antarctic waters, show sexual dimorphism. The large eggs of cidaroids are carried on the peristome or around the periproct and the area may be sunken, while in spatangoids it is the deepened petaloid areas which form the brood chambers. In both special spines keep the young in place.

Classification of echinoderms

Echinodermata
   Pelmatozoa (= Crinozoa)
      Crinoidea
      [†Cystoidea]
      [†Blastoidea]
      [†Heterostelea]
      [†Edrioasteroidea]
   Eleutherozoa
      Asteroidea
         [Somasteroidea]
         Euasteroidea
            Phanerozonia
            Spinulosa
            Forcipulata
      Ophiuroidea
         [†Stenurida]
         [†Auluroidea]
         [†Ophiurida]
         Ophiurae
         Euryalea
      [†Ophiocystoidea]
      Echinoidea

Perischoechinoidea
   [†Bothriocidaroida]
   [†Echinocystitoida]
   [†Palaechinoida]
   Cidaroida
Euechinoidea
   Diadematacea
   Echinacea
   Gnathostomata
   Atelostomata (with Spatangoida as one section)
Holothuroidea
  Dendrochirota
  Aspidochirota
  Elasipoda
  [Molpadonia]
  Apoda

14

Minor phyla

THE BIG NAMES in the animal kingdom – protozoans, echinoderms, arthropods and the like – are well known to the zoologist no matter how superficial his knowledge of the subject. In addition to these large groups, however, there is a collection of smaller phyla often lumped together as the "minor" phyla which are unfamiliar to all but serious students, although some of them are amongst the commonest of animals. The reasons for their neglect are various – they are often small and insignificant (unsuccessful animals often are); they sometimes occupy habitats not easily accessible (unsuccessful animals are often driven into poorer habitats); but principally, they are taken to be so far from the main evolutionary stream that, however interesting in form and function in themselves, they are unconnected with the more important story that the zoologist has to tell.

The minor phyla appear to have originated as parallel or diverging branches of the main phyletic stems, and so occur at all grades of evolution. The sponges have arisen as an alternative way of forming a multicellular body to that presented by coelenterates; ctenophores show a different way of adapting the body of a proto-coelenterate for pelagic life from that followed by jelly-fishes; however successful in itself it has proved less adaptable and the ctenophores remain a small group. At a more complex level of organization two collections of minor phyla may be distinguished, one at the acoelomate and one at the coelomate grade. It should be pointed out, however, that many zoologists would regard at least some of these animals as belonging to aberrant classes of the nearest major phylum or perhaps attached as appendices to them, rather than as independent phyla. It is likely, indeed, that this treatment reflects with some truth the origin of the groups as early offshoots from main phyletic lines and explains their possession of some of the characteristics of their more successful relatives.

The minor acoelomate phyla include the nemertines, associated with platyhelminths, alongside which they have been dealt with (p. 140), and a series of other groups which all seem to have some relationship with nematodes. These include the Acanthocephala and the Rotifera, the former a group of parasitic worms, the latter the wheel animalcules abundant in fresh waters, whose organization is simplified by smallness. A further group may be mentioned, the Endoprocta (= Entoprocta) or Kamptozoa. These are small animals superficially like the bryozoans mentioned in the next paragraph and originally regarded as only a separate class of that phylum. Like bryozoans they are attached, they have a circlet of ciliated tentacles setting up a feeding current and a U-shaped gut with small nerve centres round the oesophagus. Unlike bryozoans they have no coelomic cavity, the anus lies within the ring of tentacles and not outside it, the feeding current enters the cone of tentacles from outside and escapes away from the animal along the axis of the cone whereas the flow is the reverse in bryozoans, and they have a pair of protonephridia which bryozoans lack. These differences are so fundamental as to make nonsense of any classification which unites the two groups and to emphasize how much of the superficial form of an animal may be imposed upon it by its adaptations for a particular mode of life.

The minor coelomate phyla fall into two groups according to whether they ally themselves with the echinoderm-chordate evolutionary line (the deuterostomes) or to the annelid-arthropod-mollusc line (the protostomes). The former includes the phyla Pogonophora and Hemichordata, the first of which is dealt with on page 440. The second contains four phyla: (1) Chaetognatha, or arrow-worms, represented mainly by the genus *Sagitta*, abundant animals in the marine plankton where they live on small copepods and many kinds of larval forms, including fish; (2) Brachiopoda, or lamp shells; (3) Bryozoa, the sea-mats, moss animals or polyzoans; and (4) Phoronida, a group of small marine worm-like animals which inhabit tubes often aggregated into clumps. Of these four phyla the chaetognaths stand apart from the rest and, indeed, from most groups of the animal kingdom and their relationships are enigmatic. The other three, however, show clear resemblances and form a natural superphyletic grouping. The most conspicuous features which they share are a sessile mode of life, a U-shaped gut, a body divided by partitions into three sections of which the most anterior forms a small lobe, known as an epistome, which overhangs the mouth dorsally, the second forms a ring-shaped section lying around the first part of the gut and the third is the main part of the body and contains the viscera. These are not formed as annelid segments are and the body is not segmented. The middle section contains the nerve ganglia and carries a series of tentacles, primitively set in a horseshoe shape around the mouth and epistome and its coelomic space sends a canal along each tentacle. The tentacle, together with the base from which they spring, are collectively known as a lophophore and constitute the main food-gathering organ of the animals. Because of the universal occurrence of a lophophore, animals in the phyla Phoronida, Brachiopoda and Bryozoa are sometimes referred to as belonging to a group Lophophorata. Although primitively a simple crescentic ring round the mouth as in the freshwater group of bryozoans known as Phylactolaemata (Fig. 177), the lophophore changes shape in more advanced animals. The nature of the change depends on whether the animal remains solitary, like *Phoronis* or a brachiopod, or whether it bceomes colonial, as in all bryozoans. In the former case the lophophore is greatly elongated to increase its food-catching capability and has then, for reasons of economy of space, to be rolled into spirals or other complicated shapes; in the latter, the horseshoe simplifies to a circle, possibly because, as has been said of other colonial organisms, the combined effect of many zooids increases the catching power of each individual and so allows a simplification of its structure.

The Brachiopoda and Bryozoa are relatively large by comparison with some other minor phyla and were of considerable importance in earlier geological epochs. For these reasons they are briefly dealt with in the following pages although nothing further will be said about chaetognaths or *Phoronis*.

Pogonophora

At intervals the sea has proved to contain animals thought to be long extinct, such as the crossopterygian fish *Latimeria* and the monoplacophoran mollusc *Neopilina*; indeed, there are still those who hope optimistically that a persistent plesiosaur or ichthyosaur may emerge somewhere from its vast expanses. Not often, however, have animals of a group both new to science and unrepresented in the fossil record been discovered. This was, nevertheless, what happened during the course of the Dutch *Siboga* expedition in the early part of the century when, off the coasts of the East Indian islands, there were dredged at considerable depths (500–2 000 m) numerous tubes containing elongated wormlike animals of almost linear dimensions. The first description of these creatures, called *Siboglinum* from the name of the vessel and the shape of the body, was published by Caullery in 1914, though he was uncertain of their zoological position as the material was not sufficiently well preserved to allow proper investigation. Since then this and related animals have turned up in quantities wherever conditions are right: they are apparently amongst the commonest of deep-water animals. Dr D. B. Carlisle, translator from the Russian of Ivanov's memoir on the group Pogonophora to which they have been assigned, records a remark by Sir Alister Hardy, referring to the cruises of R.R.S. *Discovery II* in the Antarctic, that many tons of them "must have been shovelled overboard by some of the leading marine biologists of the day" because they were not recognized as biological material – and this doubtless also happened on many earlier oceanographical collecting expeditions.

Pogonophorans are tube-dwelling animals, their tubes being plunged into the soft substratum which they inhabit to such depths that it was some time before the hind end of the animal was known, the collecting dredge shearing it off. For this reason, and because the animals possess no alimentary tract – not even during development – Caullery was uncertain whether the animals were individuals or whether they might not prove to be nonfeeding zooids of a colonial organism, the connection with the colony having been destroyed. Further collections, however, have never shown a colonial habit so the absence of a gut must remain as an unusual characteristic of the group and pose the question of how the animals are nourished. Experimental investigation of this and other topics has been greatly hampered by the almost complete inactivity shown by pogonophorans after collection; this may well be due to damage or the changed conditions of the laboratory compared with the natural surroundings.

Body form

The general dimensions of the pogonophoran body are extraordinary; its length is never less than one hundred times the breadth and may be six times that. Many pogonophorans are less than 1 mm in diameter though they may be many centimetres long like *Zenkevitchiana longissima* (36 cm long, 0.8 mm broad); some, like *Spirobrachia grandis* are stouter (25 cm long, 2.5 mm broad) but small ones like *Siboglinum minutum*, may be excessively thin (5.5 cm long, 0.1 mm broad). The body is divisible into a forepart (Fig. 168A, f.part) which carries one or more tentacles (t) and a much longer hind part known as the trunk (tr) on which are set papillae and other projections. Though in places these are regular in arrangement the body does not exhibit a true metamerism. The general function of these structures is to give the animals a grip on the inner wall of its tube and allow it to clamber within it. The most prominent adhesive devices are:

1. The bridle, or frenulum (br), a darkly pigmented, V-shaped cuticular thickening, incomplete dorsally, held in a groove between raised epidermal ridges.
2. Serially arranged, usually called metameric papillae (Fig. 168A, C, m.pap) on the most anterior part of the trunk, containing glands (gl) and tipped, in most genera, with small scales of cuticle known as plaques (pl) to help in gripping the inside of the tube.
3. Irregularly arranged papillae of similar construction occur behind the point where the metameric arrangement stops.
4. The region carrying papillae ends with a number of annuli (ann), usually two; these are ridges running round the body, each containing two or three rows of small cuticular rods, toothed at their free edges, resembling annelid uncini (p. 192), and acting as fixing devices, like them.
5. In the postannular region further papillae are found, often regularly spaced, and like the preannular ones, glandular and provided with cuticular plaques and some with setae (s).
6. At the posterior end of the body, there is a slightly dilated anchor (anc) with regular rows of setae projecting from its surface.

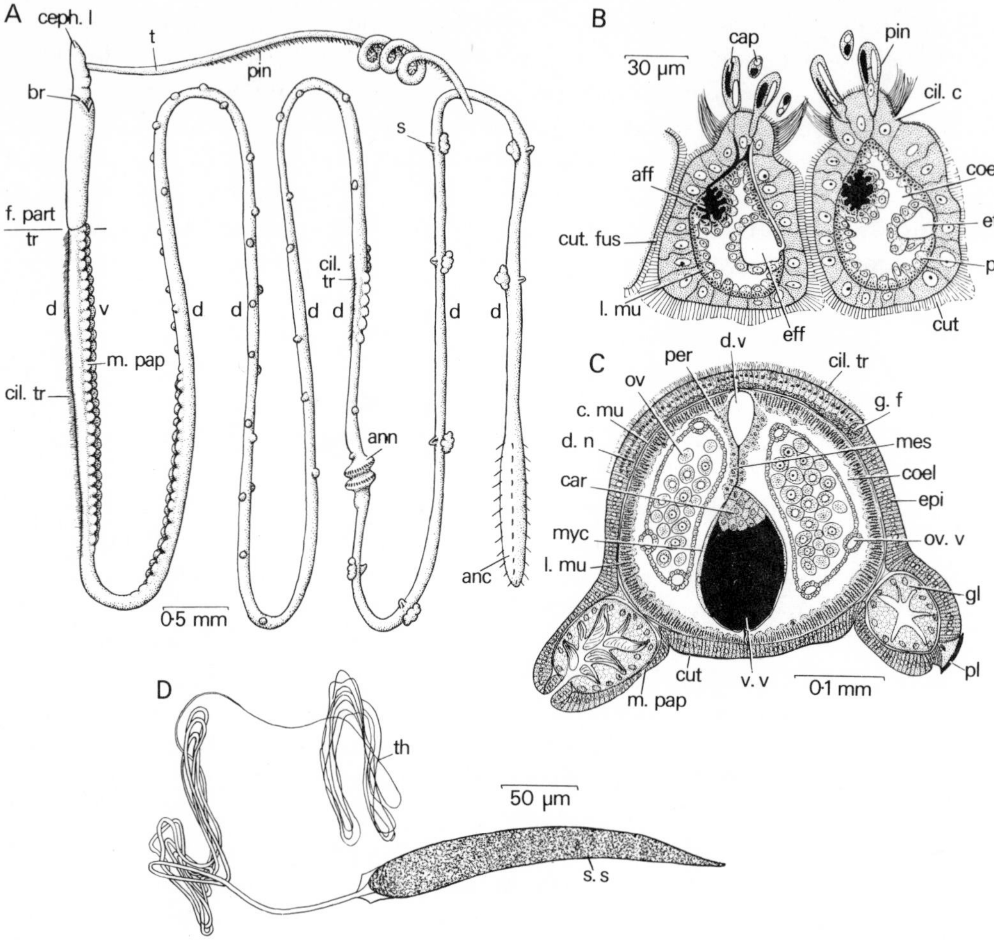

Fig. 168. A, Right lateral view of whole pogonophoran, based on *Siboglinum atlanticum*; B, *Siphonobrachia ilyophora*, transverse section of two tentacles with pinnules; C, *Polybrachia annulata*, transverse section of anterior part of trunk; D, *Oligobrachia dogieli*, spermatophore. (A, after Southward and Southward; B, after Gupta and Little; C and D, after Ivanov.) aff, afferent branchial vessel; anc, anchor; ann, annulus; br, bridle; cap, capillary in pinnule; car, heart body; ceph.l, cephalic lobe; cil.c, ciliated cell; cil.tr, dorsal ciliated tract; c.mu, circular muscle; coel, coelom; cut, cuticle; cut.fus, fusion of cuticle between tentacles; d, dorsal; d.n, dorsal nerve; d.v, dorsal vessel; eff, efferent branchial vessel; epi, epidermis; f.part, posterior limit of forepart; g.f, giant fibre; gl, gland; l.mu, longitudinal muscle; mes, mesentery; m.pap, metameric papilla; myc, myocyte; ov, ovum in ovary; ov.v, ovarian blood vessel; per, peritoneal cell; pin, pinnule; pl, cuticular plaque; s, seta in papilla; s.s, sperm sac; t, tentacle; th, threadlike part of spermatophore; tr, anterior limit of trunk; v, ventral; v.v, ventral vessel.

At the front end of the body, set on what (in ignorance of the true relationships) is conventionally accepted as the probable ventral side, and just behind the anterior tip of the cephalic lobe (ceph.l), is a series of tentacles, very variable in number; the presence of this tuft of tentacles, reminiscent of a beard, gives the name Pogonophora (beard-carrier) to the group. The genus first described, *Siboglinum*, has only a single tentacle, but most have more – *Nereilinum*, 2; *Oligobrachia*, 6–9; *Cyclobrachia*, 9; *Lamellisabella*, 28–31; *Galathealinum*, 268 – to name a few of the better worked genera. Presumably the tentacles can emerge from the tube into the water like the fan of a polychaete worm, but it does not seem likely that, when numerous, they are able to separate from one another since the sides of neighbouring tentacles are adherent, the cuticle over the surface of one stuck to that on the other (Fig. 168B, cut.fus). While most epidermal cells on the tentacles have a cuticle (cut) made of interwoven layers of fibres of some collagen-like substance, interpenetrated by microvilli, there are also ciliated cells (cil.c) and pinnule cells (pin). The last two types form longitudinal tracts facing the central concavity of the bunch of tentacles. The pinnules are set in a double row and each is formed of a single cell, the base of which is partly embedded in the epidermis whilst the rest extends into the water as a finger-shaped projection perhaps 750 μm long. Each pinnule contains an intracellular capillary loop (cap) connecting with afferent (aff) and efferent (eff) longitudinal vessels lying in the tentacle, so that they appear to be primarily respiratory devices ventilated by a current created by the ciliated cells. The pinnules are almost certainly not the sole respiratory surface: there is a broad ciliated tract (C, cil.tr) on the dorsal side of the first part of the trunk and sometimes also further back which may be responsible for moving water within the tube. The animals expose such an enormous surface area in relation to their diameter that diffusion of respiratory gases should not be difficult.

The body wall of the trunk consists of an epidermis (epi) similar to that over the tentacles, with occasional mucous gland cells and others with lipid or protein contents. The cells rest on a basal lamina, internal to which lie circular (c.mu) and longitudinal (l.mu) muscle fibres, a peritoneal epithelium (per) and a coelomic space (coel) split into right and left halves by a mesentery (mes). Some doubt exists as to whether the longitudinal muscle fibres belong to separate cells; they are probably lodged in the bases of the peritoneal cells which would, in these circumstances, be myoepithelial. In the forepart of the body the general arrangement of the body wall is similar but the plan of the coelom is different. The posterior section of the forepart resembles the trunk in that the body wall bounds coelomic cavities divided by a mesentery, though separated from the trunk cavities by a transverse diaphragm; in its anterior section, however, the coelomic space is not paired but median, small and not in contact with the body wall. From this anterior coelomic space a canal extends up the centre of each tentacle.

Respiration and excretion

The distribution of whatever oxygen may be picked up at the respiratory surfaces to the deeply buried posterior part of a pogonophoran must be due to the blood vascular system. This consists of two longitudinal vessels, one dorsal (d.v), one ventral (v.v), running the length of the body in the mesentery internal to the muscular layer of the body wall. The dorsal vessel sends three branches forward into the cephalic lobe and, in the trunk, gives off numerous lateral branches which, at least in some parts, connect with the ventral one. Anteriorly at the base of the tentacles, the ventral vessel becomes much more muscular and forms a heart, placed as a branchial heart where its pulsations drive the blood through the small vessels of the tentacles and pinnules to the dorsal vessel and its branches. In some pogonophorans (*Siboglinum*, *Oligobrachia*, placed in an order Athecanephria) the heart has a small cavity lying partly around it, suggesting a pericardial cavity; but in most (order Thecanephria) there is no such space. The walls of the main vessels have an external coating of contractile cells (myc) and so presumably aid in circulating the blood. The blood is red with haemoglobin.

The vascular system also establishes close relations with the excretory organs. These have the form of a pair of coelomoducts located in the forepart, to the anterior coelomic cavity of which they open internally and on to the ventral surface of which they discharge externally. Each duct is U-shaped, has a small bladder-like swelling in the thickness of the body wall and is connected to its partner of the opposite side by an anastomosing duct. Most excretory activity occurs in the U-shaped section and is facilitated by the intimate contact established between it and the vascular system. In the group Athecanephria the coelomoducts rest closely on the lateral branches of the dorsal vessel to the cephalic lobe; in the group Thecanephria, however, the ventral wall of the dorsal vessel is dilated to form a renal sac, filled with blood, within which the coelomoducts lie.

The activities of the pogonophorans are controlled by a so-called brain, a concentration of nerve cells and fibres located within the epidermal layer over the dorsal surface of the cephalic lobe and anterior surface of the forepart. From here a nerve trunk (d.n) runs posteriorly in a dorsal position while others run along the tentacles. It is not possible to say anything about the functioning of this system save that a giant fibre (diameter up to 12 μm) extends along the trunk as far as the annuli (g.f) and is presumably responsible for the abrupt "panic" withdrawal into the tube which pogonophorans, like all tubicolous animals, exhibit when disturbed.

Reproduction

Pogonophorans are gonochoristic, that is, have separate sexes. The gonads lie in the trunk and discharge to the exterior by a pair of modified coelomoducts; in males their openings are placed laterally immediately behind the septum separating forepart and trunk, in females they lie some way further back. In males the paired testes lie dorsally in the coelom one on either side of the mesentery and run throughout the hinder part of the trunk. Anteriorly each joins the coelomoduct whose glandular walls pack the sperm into spermatophores (Fig. 168D) with well-defined specific characters but always provided with a long filament at one end. In females the paired ovaries (C, ov) lie in the anterior part of the trunk coelom and each discharges posteriorly into the mouth of the coelomoduct, though not fusing with it as the testis does with the male duct. The eggs are led to the genital pore along the coelomoduct.

The formation of spermatophores makes it highly unlikely that the gametes are simply broadcast to meet by chance, a method that would offer no sensible explanation of the highly elaborate structure of the spermatophore and its filament. Such larvae as have been described, too, have been found within the tube of the female which acts as a brood pouch. On these grounds it makes better sense to suppose that a transfer of spermatophores from male to female takes place in which it would seem that the tentacles must be involved and which is facilitated by the density of the populations in which these animals occur – over 100 in 0.5 m^2.

The outstanding problem in pogonophoran functioning is the means by which they obtain and digest food in the absence of an alimentary canal. There are several ways in which this could be achieved: by trapping food and exposing it to a process of external digestion followed by phagocytosis of particles; by pinocytosis of the soluble products of an external digestion; by pinocytosis or direct uptake of organic matter dissolved in the sea water. Historically, the first of these mechanisms was the one supposed to operate and Ivanov proposed that the tentacles at the anterior end of a pogonophoran acted like those of a sabellid or serpulid worm and captured small organisms or detrital particles from the water, but that they then curled round them and secreted enzymes over them prior to absorbing the products of this digestive process since there was no mouth to which they could be sent. In animals with many tentacles it was easy to imagine the central cavity of the tentacular crown acting as a digestive chamber and the pinnules functioning as intestinal villi; in those, like *Siboglinum*, with only a single tentacle it was supposed to corkscrew round the food. Later investigation has failed to provide support for this idea: the tentacles do not separate from one another to provide an extensive filtering area; at the ultrastructural level there is little evidence for the secretion of enzymes; autoradiographic studies on the uptake of labelled foodstuffs show no indication that it occurs predominantly on the tentacles. We are therefore driven back on the idea of a direct uptake of organic molecules in solution in the water, a suggestion first put forward by Pütter many years ago but largely rejected. Modern techniques involving the use of radioactive tracers allow the path followed by such molecules to be accurately traced and there is now a considerable body of evidence to suggest that the direct absorption of such matter in solution is a prominent part of the energy sources of many invertebrates. Entry of amino acids into epidermal cells has been recorded in members of most invertebrate phyla except arthropods whose thick cuticle probably prevents it. Indeed, the uptake of amino acids by the polychaete worm *Clymenella torquata* has been shown to be at a level in short-term experiments which would, if maintained over long periods, require more than the measured oxygen consumption of the worm for its respiratory oxidation. Investigation has also shown that there is the necessary amount of dissolved amino acids available in the water in which this worm lives (74.17 μmol l^{-1}). (There is no suggestion that *Clymenella* does, in fact, obtain all its amino acids in this way; it does feed, but the work shows that such uptake is possible and that the material is available.)

Similar work has been done with *Siboglinum* and *Oligobrachia*, revealing an uptake in the forepart of the body of ^{14}Cphenylalanine and ^{3}Hglycine. When the animals were within their tubes the rate was 0.000 23 μg mg^{-1}h^{-1}, rising to 0.001 37 μg mg^{-1}h^{-1} when they were removed from their tubes, quantities which were readily available in the water over the sediments in which the pogonophores were embedded (calculated at about 8 000 μg l^{-1}). The only indication that there might be an external digestive process occurring was the fact that labelled carbon from a protein food was also absorbed, though much more slowly than were amino acids. Pogonophorans would therefore seem to have become totally or nearly totally dependent for food on a process of direct uptake, a process common enough in other animals but never in them more than supplementary to absorption by way of the gut. To this activity their habitat adapts them in view of the richness in organic solutes of the water over sediments and it is likely that the unusual proportions of the body, giving a relatively immense surface, contribute to the same end.

Pogonophorans are undoubtedly highly modified for their particular mode of life and such a degree of adaptation is apt to obscure relationships. These were originally thought to lie clearly with the deuterostomes or echinoderm-chordate line, to which the animals investigated by Ivanov appeared to show resemblances in the triple division of their coelom, with its most anterior part unpaired and connected to the exterior, in the presence of a dorsal nerve cord and in a superficial likeness to such creatures as *Balanoglossus* and the other hemichordates. Since then the discovery of the terminal anchor or opisthosoma, which was not known to Ivanov with its rows of setae and numerous internal septa, which confer a degree of metamerism on the animals, has substantially contradicted this view. Modern investigators prefer to regard the group as representing an aberrant offshoot from the annelid stem.

Brachiopoda

Brachiopods, like lamellibranch molluscs (p. 515), have the soft parts of the body enclosed in a bivalved, calcareous shell secreted by the mantle; the valves are

opened and closed by muscles, and food particles in suspension are collected from a
water current which is maintained by cilia on an elaborate structure contained
within the mantle cavity; as in molluscs, too, waste products and sex cells are shed
into this cavity. Here the superficial resemblance to the lamellibranch ends.
Brachiopods have an extensive coelom. All are attached and the orientation of the
body is at right angles to that of a bivalve mollusc since the hinge line of the shell is
posterior, not dorsal, and the gape directed anteriorly; the shell valves are dorsal
and ventral (Fig. 169C, d.v, v.v). The feeding current which passes through the
shell gape is set up and maintained by cilia on numerous filaments (D, fil) on
two elaborate arms or brachia extensions of the lip (lip) – this is the lopho-

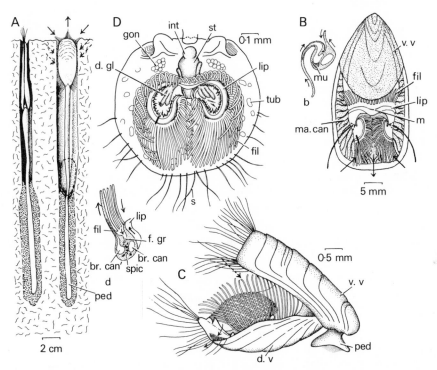

Fig. 169. *Lingula unguis*. A, in burrow, in side view (left) and dorsal view (right); broken line
shows position of animal on contraction; above this point the burrow is wide enough to permit
movement; below, the pedicle is adherent to the substratum. B, part of ventral valve removed
to show lophophore; the pedicle is removed and the posterior end of the body directed upward
to correspond with D. Thick line across lophophore indicates position of section b. *Pumilus
antiquatus*. C, lateral view whilst feeding. D, ventral valve removed; thick line through lophophore indicates position of thick section d in which parts corresponding to those in b are
labelled. Arrows indicate water and ciliary currents. (A, after François; C, D, after Atkins.)
br.can, great brachial canal; br.can', small brachial canal; d.gl, digestive gland; d.v, dorsal
valve; f.gr, food groove; fil, filament of lophophore; gon, gonad; int, intestine; lip, lip of
lophophore; m, mouth; ma.can, mantle canal; mu, muscle; ped, pedicle; s, seta; spic, spicule;
st, stomach; tub, tubercle of shell overlaid by mantle; v.v, ventral valve.

phore which is equivalent to the gills of a bivalve in providing a feeding and respiratory area. It was once thought that the arms could be protruded from the shell and used as a foot, hence the term Brachiopoda or arm-foot. Fossils date back to Cambrian times. Indeed existing genera, about 70, represent a remnant of a more numerous and diversified group which populated Palaeozoic and Mesozoic seas until superseded by the more efficient lamellibranchs. About 1 600 extinct genera have been described, so the group has proved to be a treasure trove for palaeontologists; it has been much neglected by neontologists. One well-known brachiopod, *Lingula*, is a persistent form since the Cambrian and, judging from the hard parts, it has remained unchanged.

Shell and movement

Recent brachiopods are all marine, permanently attached to the substratum by a stalk, the pedicle (Fig. 169A, C, ped), or, comparatively few, by cementation of the ventral valve (the inarticulate *Crania* and the articulate *Lacazella* – see below).

The larva selects the site of attachment which cannot be changed throughout the life of the animal, nor can it re-establish itself if uprooted, though some fossil species were not firmly attached and could be rolled from place to place by water currents. The valves of the shell are bilaterally symmetrical and in certain respects dissimilar. The pedicle is associated with the ventral or pedicle valve which is typically larger than the opposing dorsal valve, also called the brachial on account of its relationship with the lophophore. Most brachiopods attach with the ventral valve uppermost for which reason (as has become customary) the figures show them in this orientation. The valves, frequently ovoid or circular in outline and sometimes elongate, are typically convexly curved, the ventral more than the dorsal, which in some species is flat. The popular name lamp-shell refers to the resemblance of the ventral valve with a posterior upturned spout to the oil lamps of the Romans. In one group, the oldest, to which *Lingula* belongs, the shell valves are held together by muscles only, whereas in the second group they are hinged posteriorly by a tooth-and-socket arrangement, the two teeth being on the ventral valve and the sockets on the dorsal. This second group, the articulate brachiopods, is evidently derived from the former, the inarticulate, an evolution which may have taken place more than once.

The shell is covered externally by a thin organic layer, the periostracum (Fig. 170A, C, per), and chitin forms part of the calcareous layers; here the inorganic constituent is calcium phosphate in inarticulates (except *Crania*) and calcium carbonate in *Crania* and articulates. The periostracum is secreted by cells lining a periostracal groove which, as in bivalves, runs along the inner surface of the mantle skirt not far from its free edge, and the rest of the shell is secreted by the outer epithelium of the mantle. Increase in size is recorded by lines or ridges that mark the external surface and represent earlier positions of the mantle edge and the shell

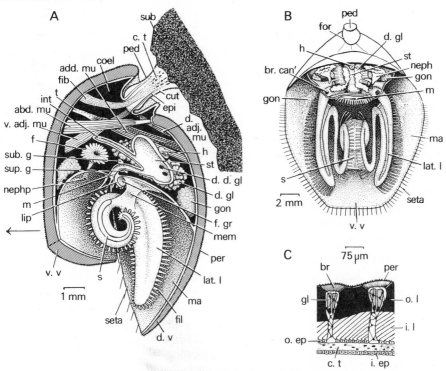

Fig. 170. Terebratulid organization. A, sagittal half of generalized terebratulid; the distal half of the ventral valve has been removed and the lateral lobe of the lophophore displaced dorsally. The coelomic cavity is black. Arrow indicates level of section of Fig. 171A. B, *Magellania*: dorsal view after removal of most of the dorsal valve and mantle; the body cavity is opened dorsally. C, vertical section through part of shell and mantle to show caeca. abd.mu, abductor muscle; add.mu, adductor muscle; br, brush; br.can', small brachial canal; coel, coelom; c.t, connective tissue; cut, cuticle; d.adj.mu, dorsal adjustor muscle; d.d.gl, duct of digestive gland; d.gl, digestive gland; d.v, dorsal valve; epi, epidermis; f, funnel of nephridium; f.gr, food groove; fib, fibres; fil, filament; for, foramen in shell; gl, gland cell; gon, gonad; h, heart; i.ep, inner mantle epithelium; i.l, inner layer of shell; int, intestine; lat.l, lateral loop of lophophore; lip, lip of lophophore; m, mouth; ma, mantle; mem, membrane supporting spiral part of lophophore; neph, nephridium; nephp, opening of nephridium to mantle cavity; o.ep, outer mantle epithelium; o.l, outer layer of shell; ped, pedicle; per, periostracum; s, septum connecting spiral parts of lophophore; st, stomach; sub, substratum; sub.g, suboesophageal ganglion; sup.g, supra-oesophageal ganglion; t, tendinous tissue; v.adj.mu, ventral adjustor muscle; v.v, ventral valve.

is often longitudinally ridged. Along this edge the epithelium may form papillae (Fig. 170C) which become embedded in the shell and elongated as its thickness increases; in dead shells these points of entry of the papillae appear like dots or puncta. The papillae may be extremely fine in inarticulates (*Lingula*, *Discinisca*) or large and branching distally (*Crania*). In punctate articulate shells the cells at the distal end of a papilla, where it is embedded in the outer calcareous layer (o.l) secreted at the mantle edge, become glandular (gl) and adhere to the periostracum by brush connections (br). These structures recall the aesthetes of chitons (p. 546), but appear to be concerned with the production of mucopolysaccharide for the repair of the periostracum and the maintenance of the shell.

The pedicle, issuing posteriorly from the shell, is short in most living species so that the shell is closely attached. Its distal end adheres to the substratum by root-like extensions: the adhesion is strong and the animal cannot be dislodged without injury. The pedicle supports the weight of the shell holding it in a position relative to the substratum which for most species is a hard material such as rock, shell or coral. However, some attach to algae, others to ascidian tests, while lingulids (Fig. 169A), living in sand or mud, anchor the exceptionally long and flexible pedicle by a sticky secretion exuding from the surface of its distal end. The pedicle of inarticulates differs in origin and structure from that of articulates. In the former it arises as an outgrowth from the posterior end of the ventral part of the mantle skirt, has an evagination of the coelom within it and contains muscles allowing rotation, extension and retraction of the shell. In the latter it is developed from the caudal region of the larva (absent in inarticulates) and is devoid of muscle and coelom; under the thick cuticle (Fig. 170A, cut) is a one-layered epidermis (epi) and the whole of the interior is occupied by connective tissue (c.t), with a central core of dense longitudinal fibres (fib). The short pedicle of the articulate emerges from a foramen (for) on the back of the ventral valve and movement of the shell on the pedicle is restricted: the muscles which effect movement do not penetrate the pedicle, but attach to its external surface (v.adj.mu, d.adj.mu). The pedicle of lingulids emerges from between the two slightly notched valves, and is attached to the ventral one.

The sole protective reaction of which the brachiopod is capable is a rapid and firm closure of the valves by contraction of adductor muscles; at the same time muscles associated with the pedicle may rotate the shell into a different orientation, draw it nearer to the substratum or, in *Lingula*, within it. The snap closure of the valves, essentially the movement of the brachial valve relative to the pedicle valve, contrasts with the gradual opening brought about by the relaxation of the adductors and contraction of abductors, or openers. In inarticulates the arrangement of the muscles is complicated by the fact that they must act to hold the valves together as well as to open and close them. In articulates two adductors (Fig. 170A, add.mu) originate side-by-side near the posterior end of the ventral valve (v.v) and pass dorsalwards, each forking to two insertions on the brachial valve. Fibres comprising the smaller posterior fork are striated, while those of the anterior fork are smooth; the former snap the shell shut and the latter hold it closed tightly for long periods. This is a mechanism reminiscent of the quick and catch muscles of bivalves (p. 517). The origins of the paired abductors (abd.mu) are anterior to those of the adductors and the muscles pass obliquely backwards to insert on a process of the dorsal valve projecting ventrally between the two dental sockets: by pulling on this cardinal process they depress the posterior end of the dorsal valve and raise its anterior margin. These muscles are smooth. They have some attachment to the dense connective tissue of the pedicle and on contraction elongate the pedicle and so erect the shell. There are also adjustor muscles used to move the shell sideways or up and down on the pedicle. Paired dorsal adjustors (d.adj.mu) run from the ventrolateral areas of the surface of the pedicle to the hinge areas of the dorsal valve, and ventral adjustors (v.adj.mu) from its dorsolateral areas to the ventral valve lateral to the adductors. All these muscles pass through the coelom (coel). Most articulates have the contractile length of the adductors and abductors shortened by the introduction of tendinous tissue (t) and this has the effect of reducing their effective length and increasing power. The contractile fibres are confined to the dorsal end of the adductors and the ventral end of the abductors.

Since in inarticulates both valves are movable the system of muscles is more complex: there are two pairs of so-called adductors and a variable number of oblique muscles, three pairs in *Lingula*, which allow the valves to rotate on one another and to shear laterally; the shearing action in *Lingula* is used in the maintenance of the burrow. There are no abductors. The valves gape only a few degrees, a movement which can be effected by the adductors. The posterior adductors are near the posterior end of the shell and when acting alone cause the shell to gape; the anterior adductors, considerably further forwards, are differentiated into two portions, perhaps quick and catch. In inarticulates the muscles are not shortened by tendinous tissue.

A rapid closure of the shell is evoked in response to tactile stimulation of the shell surface, of chitinous setae projecting from the mantle edge and of the mantle edge itself. The mantle edge is sensitive to light, touch and chemical stimuli, though no receptor cells have yet been located. It is supplied by nerves from the cerebral ganglia through which there is a simple reflex circuit to the adductors. The setae (seta), present in most brachiopods, extend the sensitivity outwards from the edge; they arise from follicles occurring at regular intervals just within the mantle edge posterior to the periostracal groove, and, like annelid chaetae, may be moved by muscles. In some species long setae form a trellis over the mouth of the shell straining the inhalant current of larger particles.

Feeding and digestion

As in other suspension feeders the food-collecting organs are large and elaborate. Indeed, the lophophore of the brachiopod spreads throughout the mantle cavity which occupies two-thirds the volume within the shell. It is an outgrowth of the anterior body wall in the oral region, bearing ciliated filaments, and extends on either side of the mouth (Fig. 169B, m). In small species (*Pumilus*) it is attached to the body wall throughout its length. In larger species the right and left halves are free, away from the oral region, forming two arms or brachia which are supported by a variety of structures. In these species the brachia pass through extensive changes during ontogeny. They elongate to increase their food-collecting capacity and become coiled or lobed in a variety of shapes to accommodate themselves within the mantle cavity. The mouth is medially placed in a transverse groove, the food groove (Fig. 169d, f.gr), which extends along each brachium and is bounded dorsally by a continuous lip (B, D, d, lip) and ventrally by a ridge bearing the filaments (fil). The brachial axis is supported throughout its length by a blind coelomic canal (large brachial canal, d, br.can) with muscle fibres in the wall which control the position of the axis, acting against the incompressible coelomic fluid. The canal is embedded in a firm collagenous connective tissue, which, in some articulates, such as terebratulids, contains calcareous spicules. In most there is, in addition, an internal brachial skeleton (Fig. 171A, sk) in the connective tissue on the side of the lophophore away from that bearing the filaments. This is the brachidium, secreted by an invagination of the same epidermis as secretes the dorsal valve. It may be merely a pair of prongs, called crura (sing.crus), projecting anteriorly into the lophophore from the hinge area and strengthening its base, but usually the crura are extended forwards into the lophophore as a loop, which may be short as in *Terebratulina*, or may extend the whole length of the arms as in *Magellania*. Movements of the lophophore, as distinct from those of the filaments, are limited. Its positioning as a whole is controlled by muscles attached to the valves and running to its base. Each filament has a central coelomic canal which connects with a longitudinal brachial canal (small brachial canal) (Figs 169A and 171A, br.can', op), near its base and this in turn opens to the perivisceral coelom. Bands of muscle fibres, mainly longitudinal in direction, run along the frontal and abfrontal boundaries of the filamentar canal and are responsible for shortening the filament and posturing it during feeding. The muscles on the frontal surface are the better developed and in a number of articulates are known to be striped, a feature related to the flicking movements of which the filaments are capable. Hydrostatic pressure in the coelomic canals of the filaments provides a skeleton against which the muscles work – it decreases as the filaments are extended and increases as the longitudinal muscles contract. Extension of the filaments may be a resting state produced by the pressure of the coelomic fluid and momentarily disturbed by contraction of the muscles in the wall as the filaments move and shorten.

The filaments of the lophophore are ciliated. As in bivalve molluscs, lateral cilia maintain a water current into and out of the mantle cavity (Figs 169 and 171) which passes between the filaments in a frontal-abfrontal direction. Suspended particles trapped between the filaments or impinging on their sticky frontal surfaces are directed towards the food groove by a broad band of frontal cilia. Particles below a certain size, which includes such important items as diatoms and dinoflagel-

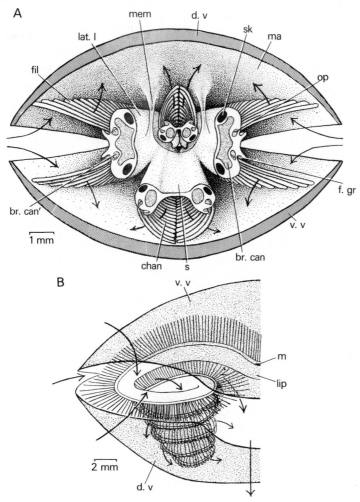

Fig. 171. A, terebratulid cut transversely at the level of the middle of the mantle cavity indicated by the arrow in Fig. 170A and viewed anteriorly, to show lophophoral structure and relationships. B, left half of brachiopod such as *Tegulorhynchia* to show the left half of a spirolophous lophophore in the mantle cavity. Arrows indicate water currents. (After Rudwick.) br.can, great brachial canal; br.can′, small brachial canal; chan, channel formed by overarching filaments; d.v, dorsal valve; f.gr, food groove; fil, filament; lat.l, lateral lobe of lophophore; lip, dorsal lip; m, mouth; ma, mantle; mem, suspending membrane; op, opening from small brachial canal to filamentar canal; s, septum linking arms of spiral part of lophophore of terebratulid; sk, skeleton (= brachidium); v.v, ventral valve.

lates, enter the groove and are passed by cilia to the mouth; there is no qualitative sorting mechanism. Large and heavy particles drawn in with the water interrupt the feeding mechanism. The occasional one results in the flicking of the filaments it encounters, and they contract to allow free passage of the particle to the exhalant stream. When large numbers enter the lateral cilia stop beating, the frontals reverse their beat and the unwanted particles are trapped in copious mucus secreted by the filaments, carried to their tips and to the mantle edge. A violent expulsion of water from the mantle cavity, with the sudden closure of the shell, clears away all unwanted matter.

In small brachiopods with a simple type of lophophore (Fig. 169C, D) the two arms spread anteriorly and, when the shell gapes, the circlet of filaments curves forwards towards the pedicle valve: the ridge bearing the filaments lies along the edge of the outspread lophophore and the food groove (Fig. 169d, f.gr) is median to this. Water is drawn into the centre of the tentacular ring (so comes into direct contact with the frontal cilia) and passes out between the filaments leaving the mantle cavity on each side. The increased length of the arms with subsequent evolution of the lophophore is accommodated either by the looping of each arm, as in *Lacazella*, or by the free end of each arm (the growing end) spiralling; there is a combination of these two methods in terebratulids (Fig. 170), where each arm of the lophophore forms a laterally placed loop (lat.l) which grows forwards free of the body wall and then describes a spiral median to the loop; the two arms of the spiral are united across the median plane by a septum (s) and suspended from the dorsal body wall (mem). In the loop, which follows the curvature of the shell, the two limbs are united with one another and their blind coelomic canals (great brachial canals) fuse (Fig. 171A, br.can). The rows of filaments on the loop are held in such a way that the dorsal series makes contact with the inside of the dorsal mantle (ma) and the ventral with the ventral mantle, so enclosing a space which communicates with the outside world through the gape of the shell. Because of the direction of beat of the lateral cilia water enters this space. The septum linking the right and left spirals is functionally equivalent to the fusion of the two arms of the loop in that it forms the floor of a channel, the side walls of which are formed by the filaments held with the tips of the right and left series approximated (chan). The spiral so formed on each side of the lophophore is continuous with the lateral loop. Some of the water which enters the lateral channels passes between the filaments to the exhalant part of the mantle cavity, and the rest continues into the spirals to be filtered by the filaments there. Water can escape from the mantle cavity only at the extreme anterior end where there is a gap between the two arms. In the type of lophophore known as a spirolophe (Fig. 171B) the tips of the brachia diverge from one another and from the mantle surface and each brachium spirals, commonly dorsally, as in *Tegulorhynchia* and *Lingula*. The tips of the filaments of the basal coil of *Tegulorhynchia* touch the mantle of the dorsal valve and filaments on other turns make contact with those on the next more ventral coil. In this way the space within the spiral connects with the ventral part of the mantle cavity and the gape of the shell, but is separated by a screen of filaments from the dorsal part of the mantle cavity. Water enters the ventral part of the mantle cavity, then the central part of the spirals and escapes into the dorsal part by being filtered through the filaments. In *Discinisca* (and possibly other forms) the lophophore spirals ventrally and its central space therefore communicates with the dorsal exhalant spaces of the mantle cavity. Figure 169B,b shows that in *Lingula* the filaments are related to the food groove in such a way that their frontal surfaces face laterally and not medially; the whorls of the spirals lie in a vertical plane and not horizontal as in *Tegulorhynchia*.

Particles entering the mouth (Figs 169B, 170A, m) from the food groove are entangled in secretion from the wall of the gut and directed through the oesophagus into the stomach (Fig. 170A, st) by cilia and the peristaltic action of the oesophageal wall. A sphincter guards the entrance to the intestine (int). When food has accumulated in the stomach there are rhythmic pulsations of the lobes of the digestive gland (d.gl) which fills much of the coelom and opens into the stomach by broad ducts (d.d.gl), usually one to three pairs. Food is sucked into the lumen of the gland, retained for a second or so then expelled. This activity results in particles being brought into contact with the digestive cells of the gland and also the expulsion of secretion and waste from the gland. There are secreting cells in the digestive gland epithelium and they may be responsible for some digestion in the lumen of the stomach, but intracellular digestion is more important. Small particles are taken up by the digestive cells where they are digested in vacuoles and the waste expelled. Phagocytes may also be involved in the uptake of particles. Waste from the stomach is released into the intestine, rotated by cilia and compacted with secretion to form pellets. In inarticulates the pellets are ejected from the anus which opens to the mantle cavity, generally on the right lateral body wall. In articulates the intestine is blind (int) and antiperistaltic contraction of the gut wall ejects the pellets through the mouth. Faecal matter is transported to the mantle edge by cilia and flushed clear of the body by a sudden snapping of the shell valves. This occurs at fairly regular intervals, the rhythm being interrupted when silt accumulates in the mantle cavity and pseudofaeces are ejected. The valves reopen immediately and feeding is resumed.

Other activities

Brachiopods have two fluid systems, coelomic and circulatory. The former is spacious and the latter relatively poorly developed. The perivisceral coelom (coel) is divided by dorsal and ventral longitudinal mesenteries suspending the gut, and by transverse mesenteries, but these must not be regarded as septa indicating segmentation. Muscles related to the movement of the valves pass through the coelom posteriorly and the gonads (gon), usually four and derived from localized areas of peritoneum, discharge the sex cells into it; these are freed to the mantle cavity by way of the large funnels of one or two pairs of excretory organs, usually taken to be metanephridia (f). The peripheral parts of the body of a brachiopod, the lophophore and its filaments and the mantle skirt, are pervaded by coelomic canals branching from the perivisceral coelom. The ramifications of the canals in the mantle (Fig. 162B, ma.can) vary from genus to genus and may be of taxonomic value, while the canals in the lophophore have a constant arrangement with the large brachial canal, supporting the lophophoral axis, cut off from the perivisceral coelom in the adult in all species. The coelomic epithelium is ciliated and cilia maintain a circulation of fluid in the canal systems. The fluid is coagulable and contains a variety of cells, some phagocytic and carrying waste to the metanephridia. The ramifications of the coelom in the peripheral parts of the body over which water circulates suggest that the fluid has a respiratory function. In *Lingula* there are coelomocytes containing haemerythrin with limited ability to carry oxygen which may be related to the infaunal habit of this brachiopod.

The circulatory system comprises a dorsal channel supported by the dorsal mesentery with a contractile vesicle or heart (Fig. 170, h) in the region of the stomach. Its fluid has no respiratory pigment. In *Terebratulina* the vesicle contracts once every 30–40 s. The channel forks anteriorly and enters the arms of the lophophore, sending a branch to each filament; posteriorly it serves the gonads, metanephridia and mantle lobes, branching through the mantle canals and following the surface adjacent to the mantle cavity. This branching system thus duplicates the course of the coelomic canals; in each the channels end blindly giving a to-and-fro circulation efficient enough for animals with a low metabolic rate.

The nervous system of the brachiopod is little known and most information comes from work done in the last century. The most complete accounts concern one articulate, *Gryphus vitreus*, and three inarticulates, *Crania*, *Discinisca* and *Lingula*. In both groups the nervous tissue is close to the epidermis and in the inarticulates located within its base. Ganglia are few. As in other invertebrates there is a circumoesophageal nerve ring. From the dorsal part of this ring, nerves pass to the dorsal lip, one arising from each of two small supraoesophageal ganglia (sup.g) in articulates; in inarticulates no such ganglia are defined. The rest of the lophophore is innervated by nerves from the circumoesophageal connectives which pass ventrally to large suboesophageal ganglia (sub.g). It is from these ganglia that all other parts of the body are innervated.

The sexes are separate in the majority of brachiopods and the colour of the gonad is their only distinguishing feature. In a few genera, including *Argyrotheca*, *Pumilus* and *Platidia*, hermaphroditism occurs. As the gonads mature they bulge into the perivisceral coelom and may spread into adjacent areas of the mantle canals. The animals are gregarious, spawning of a large number of individuals occurs simultaneously and fertilization is typically external. The sex cells are liberated into the coelom and directed by ciliary currents through the metanephridia to the mantle cavity where they come under the influence of the exhalant water current. Large numbers of eggs are liberated at a single spawning; for *Glottia* this has been estimated at about 10 000. A few species brood their young up to the free-swimming larval stage common to all brachiopods. They produce fewer eggs. The brood pouches of the small terebratulid *Argyrotheca* are enlargements of the metanephridia, while in *Lacazella* there is a single median pouch in the ventral mantle lobe fitting into a depression in the underlying valve, and the eggs are attached to two lophophoral filaments which hang into it. These habits indicate that fertilization is either within the mantle cavity or in the metanephridia. This implies that sperm carried in with the inhalant water current must pass the lophophoral filter, though the mechanism by which this is achieved is unknown.

To many zoologists brachiopods are little-known animals, the inconsiderable vestiges of a group which had its heyday in Palaeozoic times and has since been supplanted by more efficient competitors, particularly by bivalved molluscs. Bivalves, it is true, show features which allow them to work more efficiently than brachiopods, such as the fusion of filaments to give the eulamellibranch gill which, with the muscular pumping of water through the mantle cavity, permits filtration of a greater volume than can be done by the unjoined filaments of a brachiopod; they also can exploit ways of life closed to brachiopods by virtue of their foot, by which they can move from place to place, and their partially closed mantle cavity and siphons, which let them live within a substratum whilst retaining contact with the overlying water. Despite the attractiveness of this idea it is too simple. Brachiopods underwent their greatest radiation in Palaeozoic times but were drastically reduced in numbers at the beginning of the Mesozoic by some catastrophic change affecting marine habitats which occurred at that time. This may have taken the form of a lowering of sea level or fall in salinity but, whatever its nature, it either directly destroyed brachiopods or severely limited the habitats in which they could survive. Later, as better conditions appeared, brachiopods failed to re-establish themselves as a major component of the population and the bivalves took their place, successfully invading habitats now empty of the brachiopods which had previously excluded them. It is still possible, today, to find sublittoral areas dominated by brachiopods with few competing bivalves.

Bryozoa

Bryozoa is the name given to a phylum also known as Polyzoa and Ectoprocta; it may be translated "moss animals" and refers to the way in which its members, all colonial and nearly all sessile, form moss-like growths on a variety of substrata, which may be rocks or shells but are frequently the surface of algae. One class of the phylum, known as the Phylactolaemata, contains exclusively freshwater animals; the remaining bryozoans, placed in the two classes Stenolaemata and Gymnolaemata, are predominantly marine. Their study by neontologists, like that of brachiopods and other groups with hard parts that have fossilized well, has not been made easy by the development of an extensive vocabulary of special terms invented by palaeontologists to describe minutiae of skeletal structure, and by a failure of communication between palaeontologists and those working on the living animal. The state of affairs in relation to bryozoans is made worse by the persistence of terms used in the past to describe their anatomy by reference to an interpretation of their organization based on a total misunderstanding. As in most modern accounts the use of these terms will be minimized, but it is neither possible nor, to permit understanding of more detailed texts, is it sensible to avoid them altogether.

Organization of a simple bryozoan

In order to make the anatomy of bryozoans intelligible the structure of a simple and abundant form, *Membranipora membranacea*, will be described (Fig. 172). Comparative study of bryozoans suggests that the simplicity of this genus may not be primitive, but it serves the purpose admirably. The animal forms a delicate growth over the fronds of *Laminaria* and is, indeed, partial to the agitated conditions of the water in which this weed grows, avoiding less exposed sites. It is colonial and the growth is composed of a large number of individuals, each more or less a

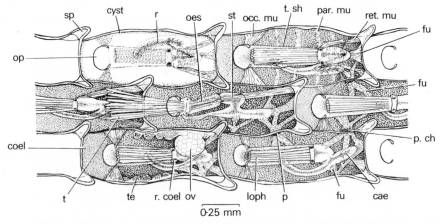

Fig. 172. *Membranipora membranacea*: several zooids of colony in ventral view, anterior ends to the left. cae, caecum of gut; coel, coelom; cyst, cystid; fu, funiculus; loph, lophophore; occ.mu, occlusor muscle of operculum; oes, oesophagus; op, operculum closing orifice; ov, ovary; p, pore; par.mu, parietal muscle; p.ch, pore chamber formed by abortive bud; r, rectum opening to tentacle sheath under retracted lophophore; r.coel, ring coelom; ret.mu, retractor muscle; sp. spine; st, stomach; t, tentacle; te, testis; t.sh, tentacle sheath.

parallelepiped, lying in a single layer on the frond. Each zooid is small, measuring about a millimetre in length, one-third of that from side to side and less in thickness. A large colony may contain thousands of zooids; the largest described was said to contain 2.3×10^6, though almost certainly not all were alive. The lower surface of each zooid is morphologically dorsal and lies in contact with the frond, the upper surface (the ventral) is exposed to the sea water, whilst the other four, right, left, anterior and posterior, are pressed against those of their neighbours. The zooids are arranged in lines with the individuals in one line alternating with those in the next so that the pattern of contact is quincuncial, that is, one zooid lies at each corner of a square and a fifth at its centre (like bricks in a wall or trees in an orchard). Towards the anterior end of the upper surface lies the only opening, known as the orifice, oval in shape, and closable by a membranous lid or operculum (op) hinged on its posterior lip. When conditions are right the anterior end of the zooid may be protruded for feeding or defecation, but at the least disturbance it is rapidly retracted and the operculum shut. What emerges from the orifice is a short length of cylindrical body carrying at its summit a lophophore, that is, a circular base supporting a circlet of about sixteen tentacles (t). At the centre of the lophophore lies the mouth whilst the rectum (r) opens just external to the lophophore on the dorsal side. The name Ectoprocta for the phylum is derived from the situation of the anus outside the circlet of tentacles.

It is clear from the position of mouth and anus that the gut must be U-shaped, a shape common in sessile animals. It lies suspended in a coelomic cavity (coel) by a number of strands each called a funiculus and comparable to a mesentery; one major strand (often the only one, to which the name is then restricted) attaches the apex of the U to the posterior wall of the zooid (fu). The coelom is bounded by the external wall of the zooid (cyst), which is therefore the equivalent of the body wall or somatopleure of other coelomates, although it tends to be called the cystid by students of bryozoans whilst the gut and lophophore are known as the polypide. Like any other somatopleure it is composed of an outer epidermis and an inner peritoneal lining. The epidermis secretes a dead exoskeleton (known as a zooecium) which immobilizes the body wall; correlated with this muscles are not usually developed between epidermis and peritoneum. The exoskeleton is composed of a superficial cuticle with underlying calcareous material; the cuticle contains chitin and because of its similarity to the layer over the calcareous shells of brachiopods and molluscs is sometimes called a periostracum.

A moment's reflection shows that it is impossible for all six sides of the zooid to be rigid for inward and outward movement of the lophophore could not then occur. In *Membranipora* the whole ventral body wall, known as the frontal membrane (Fig. 175A, f.mem), is pliable, except for a narrow, marginal, calcified ledge called the gymnocyst (gym), and has a number of transverse bundles of muscles (par.mu) embedded in it, which run from an origin on the side walls of the zooecium. When the muscles (the parietal muscles) contract the frontal membrane is pulled down-wards, raising the pressure within the coelomic cavity; it is this which, first, opens the operculum and then forces the lophophore through the orifice. Retraction of the lophophore is brought about by relaxation of the parietal muscles and contraction of a pair of retractor muscles (Fig. 172, ret.mu) inserted on the right and left sides of the lophophoral base and originating on the posterior parts of the body wall. An occlusor muscle (occ.mu) runs from each side of the operculum to the lateral wall of the cystid; together they close the operculum. Retractor and opercular muscles contract vigorously so that withdrawal of the zooid is an abrupt movement like that of many sessile animals at the threat of danger; eversion, by contrast, is a tentative, exploratory action brought about by steady increase in coelomic pressure. The part of the body wall which runs between the orifice and the base of the lophophore when the zooid is extended becomes the tentacle sheath (t.sh) the lining of a cavity in which the tentacles lie, when the zooid is retracted.

The rest of the anatomy of a zooid is very briefly dealt with. There is neither vascular nor excretory system. The great area of the lophophore and frontal membrane, together with the short distance from these to the deepest part of the zooid permit diffusion of gases to supply all the animal's respiratory needs, helped by the movement of coelomic fluid as a zooid expands or withdraws. A pair of ganglia (Fig. 173A, d.g) lies dorsally at the base of the lophophore, perhaps formed by the separation into right and left halves of the single ganglion more commonly found in bryozoans. A delicate circumoesophageal nerve ring (p.n.r) links the ganglia and also makes connections with a more substantial nerve ring in the base of the lophophore (a.n.r) from which both sensory and motor nerves pass into the tentacles (t.n). The cerebral ganglia innervate the tentacle sheath, the gut, the retractor muscles and the body wall, where there lies a nerve net. The ganglia are not placed in the main perivisceral coelom, but in a ring-shaped section (r.coel) lying in the basal part of the lophophore and separated from the main coelom by a diaphragm (dia). The ring coelom sends an extension into each tentacle (t.coel). This arrange-ment allows the coelomic fluid, which acts as skeletal support for the tentacles, to be little influenced by pressure changes in the main coelom and is to be correlated with their ability to flick radially though not to shorten. Each zooid is hermaph-rodite, the gonads (Fig. 172, ov, te) developing from the peritoneal lining of the lateral walls of the zooid. The gametes escape into the coelom and make their way to the external medium by a pore at the summit of the intertentacular organ, a short tubular projection from the mid-dorsal line of the tentacle sheath distal to the anus, which develops as the sex cells ripen. Precisely where fertilization occurs is not known. The possibilities which are open are (1) self-fertilization within the coelom as happens in *Bugula*; (2) cross-fertilization within the coelom by sperm from another zooid which have entered by way of the intertentacular organ – this

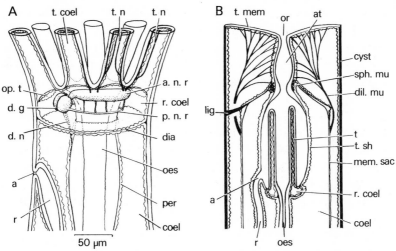

Fig. 173. A, anterior end of an extended zooid of *Membranipora membranacea* with the distal parts of the tentacles cut off. B, diagram of a longitudinal section through the anterior end of a retracted stenolaematous zooid; on the left of the diagram the section passes through one of the ligaments which anchor the anterior end of the membranous sac; on the right it passes between ligaments. a, anus; a.n.r, anterior nerve ring; at, atrium; coel, perivisceral coelom; cyst, cystid; d.g, dorsal ganglion; dia, diaphragm; dil.mu, dilator muscle of atrium; d.n, dorsal nerve; lig, ligament; mem.sac, membranous sac; oes, oesophagus; op.t, opening from ring coelom to tentacular coelom; or, orifice; per, peritoneum; p.n.r, posterior nerve ring; r, rectum; r. coel, ring coelom; sph.mu, atrial sphincter muscle; t, tentacle; t.coel, tentacular coelom; t.mem, terminal membrane; t.n, tentacular nerve; t.sh, tentacle sheath.

probably happens where zooids are of single sex as in *Flustra*; (3) external fertilization, which seems the most likely in *Membranipora membranacea*.

Membranipora, like all bryozoans, uses its lophophoral tentacles to create a feeding current. Each tentacle carries cilia on its lateral and frontal walls which beat so as to suck water into the cone of the lophophore from above and drive it outwards between the tentacles. Suspended particles may be caught and transferred to the mouth by frontal cilia, but the animals are better described as impact feeders in that what is mostly ingested is the stream of particles travelling in the central axis of the lophophore and impinging on the neighbourhood of the mouth, which are then sucked into the gut by movements of the wall of the fore-gut. The cells of the first part of this, near the mouth, are ciliated, but the greater part is covered with highly vacuolated cells with lateral walls strengthened by thickenings which were at one time mistaken for the striations of muscle fibrils. The epithelium rests on a well-developed layer of circular muscles; it is this section which sucks food into the gut when the mouth is open and passes it to the stomach when the mouth is closed. A valve separating fore-gut from stomach prevents backflow. The stomach is Y-shaped, the two arms connecting with fore-gut and hind-gut respectively, the base forming a caecum held at its apex by the funiculus.

The lining of the stomach seems to be the main source of digestive enzymes and also the main site of uptake of food; both extracellular and intracellular digestion appears to take place. The hind-gut receives the indigestible residues of the food in the form of a mucous string produced by the rotation of the food particles in the stomach. This is then segmented into pellets which escape when the animal is extended.

Although there seems to be no link between the nervous system of one zooid and those of its neighbours, so that each behaves as an independent individual, there are ways in which food and oxygen can probably be exchanged. This is by way of a series of pores (p) located in the adjacent walls of each zooid; though normally plugged by cells belonging to the body wall communication between zooids can take place here. Each pore, as seen below, represents an abortive zooid of the colony.

Colony formation, polymorphism and protective devices

Membranipora membranacea grows so as to form a layer one zooid thick over the surface of *Laminaria* or *Fucus*, all zooids facing in the same direction. The colony grows towards the base of the weed so that in early summer, when the previous year's growth is cast off, at least part has colonized the new shoot. Most bryozoans with a calcified cystid tend to the same encrusting growth as *Membranipora*, like *Flustrallidra* (formerly *Flustrella*) on *Fucus* and *Umbonula* and *Mucronella* on stones and shells. Colonies may also be erect, attached only by the base. If calcified they then mimic corals in their mode of growth (*Cellaria*, *Pentapora*): if not calcified they look more like seaweeds, though usually lacking their characteristic colours. *Flustra*, often cast upon beaches, resembles a piece of *Fucus*, but it is straw-coloured; *Bugula*, a genus commonly used as a bryozoan type, looks like a finely divided red weed which has lost its colour after death, each branch composed of two lines of alternating zooids. *Alcyonidium* forms gelatinous growths attached to rocks or weeds which resemble weeds in colour and texture: they may be so abundant as to form a distinct zone on the beach. The simplest colonies consist of stolons from which zooids arise at intervals. The stolen may adhere to a substratum as in the genus *Aetea*, or be erect and branching, with zooids set regularly along the main axes or grouped in bunches (*Bowerbankia*). The stolon consists of zooids composed of cystid only, without any polypide.

The sessile habit exposes *Membranipora* and other bryozoans to the direct attack of predators. Not many animals like bryozoans to eat but echinoids often take them and nudibranch molluscs of the family Limaciidae feed exclusively on them. It is not surprising, therefore, that a relatively unprotected frontal surface like that of *Membranipora* is uncommon and that devices to protect it have been evolved. In *Membranipora* a short spine (sp) lies at the posterolateral corners of each zooid and gives the colony a

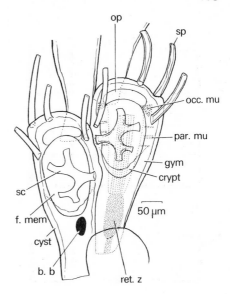

Fig. 174. *Scrupocellaria*. b.b, brown body; crypt, cryptocyst; cyst, cystid; f.mem, frontal membrane; gym, gymnocyst; occ.mu, occlusor muscle; op, operculum; par.mu, parietal muscle; ret.z, retracted zooid; sc, scutum, a large spine protecting the frontal membrane; sp, spine.

slightly spiky surface, but this bryozoan gains most protection from predators by the agitation of the water in which it lives and the way in which the waving of the *Laminaria* fronds against one another tends to wash off unattached animals. In other genera, the development of spines is much greater. They tend to grow round the orifice and at the edge of the frontal membrane. In *Scrupocellaria* (Fig. 174) a single, massive, branched spine (sc) lies over this delicate surface and in *Beania* a series of spines arches over it. In one gymnolaematous group (the cribriform bryozoans) spines grow centrally over the frontal membrane from the lateral margins and fuse in the mid-line (Fig. 175C); they also fuse with their neighbours but leave a series of pores through which water passes as the frontal membrane moves up and down. In another series of genera (Fig. 175B) of which *Cellaria* may be mentioned a calcareous shelf known as a cryptocyst (crypt) grows medially from the posterior and lateral edges of the cystid, close to the ventral surface, so as to subdivide the coelom into a major part, containing the viscera, which lies below (dorsal to) the cryptocyst and a minor part, known as the hypostegal coelom which lies above it and below the frontal membrane. The parietal muscles (and so perhaps the movable part of the membrane) are now restricted to the area near the orifice where they pass through gaps in the cryptocyst (par.mu). There are other pore-like structures in the calcareous matter of the cryptocyst but these are plugged with papillae of tissue. The development of a cryptocyst is presumably of advantage in protecting the polypide but the naked frontal membrane is still external to it. In one group of gymnolaematous bryozoans, the Ascophora, this situation has been improved upon (Fig. 175D) in that an invagination of the original frontal membrane undercuts the cryptocyst and so creates under its cover a sac communicating with the outside

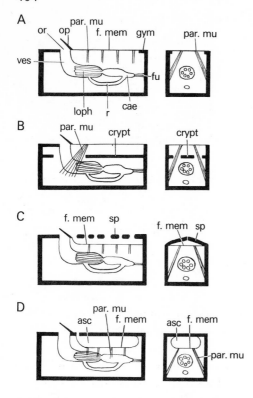

Fig. 175. Diagrams to show different methods used by gymnolaematans for the extrusion of the lophophore. The diagrams on the left represent longitudinal sections through zooids, those on the right, transverse sections. A, malacostegoid zooid with naked frontal membrane, no cryptocyst and scattered parietal muscles; B, coelostegoid zooid with naked frontal membrane, well-developed cryptocyst and consequent anterior concentration of parietal muscles; C, cribrimorph zooid with frontal membrane protected by spines; D, ascophoran zooid with ascus and completely hard ventral surface. asc, ascus opening by ascopore behind orifice; cae, caecum; crypt, cryptocyst; f.mem, frontal membrane; fu, funiculus; gym, gymnocyst; loph, lophophore retracted into tentacle sheath; op, operculum; or, orifice; par.mu, parietal muscles; r, rectum; sp, spine; ves, vestibule.

world, which has the parietal muscles running from its floor to the zooecium. This is the ascus (asc), a compensation sac, the floor of which behaves like the frontal membrane of *Membranipora*. The ascus develops in this way in *Schizoporella* but in *Umbonula* it seems to be created by growth of the gymnocyst so as to give a new ventral surface with calcareous skeleton, external to the original one which becomes the floor of the ascus: evolution of a compensation sac, therefore, may have occurred more than once.

A different method of extrusion of zooids is found in the Stenolaemata (Fig. 173B). Here the orifice (or) lies at the anterior end of a tubular body wall (cyst). When the zooid is withdrawn, the orifice is closed by a flexible membrane, the terminal membrane (t.mem), with a central pore leading through an atrium (at) to a tentacle sheath (t.sh); elsewhere the body wall is immobile. On extension the zooid emerges through the pore and the tentacle sheath, vestibular walls and terminal membrane then form a column on the summit of which the lophophore is carried. Retraction is due to retractor muscles as in the gymnolaematous forms, but the mechanism for extrusion is different. To understand how it works three special anatomical features must be noted:

1. From the base of the vestibule a membrane, the diaphragm, runs to the outer wall of the zooid and is attached there at eight points (lig); gaps between these points let the coelomic space around the vestibule and under the terminal membrane communicate with the posterior coelom.

2. From the attachments of the diaphragm a structure known as the membranous sac (mem.sac) extends posteriorly like a test-tube subdividing the perivisceral coelom into an inner part around the gut and an outer part in contact with the body wall. It is with this outer part that the anterior coelom connects.

3. Muscles (dil.mu) run from the terminal membrane and atrial wall through the gaps in the diaphragm to attachments on the outer wall of the membranous sac. When these muscles contract the atrial walls are pulled outwards, the terminal membrane inwards; both actions raise the pressure on the coelom and this is transmitted backwards to compress the membranous sac and squeeze the lophophore outwards. This is the explanation offered by Borg, who first discovered the membranous sac; perhaps, however, the full significance is not understood since the rise in coelomic pressure would evert the lophophore if the muscles attached to the body wall and no sac were present. In these circumstances the terminal membrane would act like a frontal membrane, with which, indeed, it is probably homologous.

Accompanying colonial organization in many groups is a polymorphism which allows specialization of individuals for different functions. In Gymnolaemata most zooids are capable of both feeding and reproduction (autozooids) although in Stenolaemata ordinary zooids are sterile and there are special reproductive gonozooids. In the Gymnolaemata zooids are modified for protection and communication. The first of these take the forms known as avicularium and vibracularium. In a genus like *Flustra* (Fig. 176A) an avicularium can be seen to be easily derived from an ordinary zooid by reduction of the polypide (vp), leaving the cystid (cyst), its orifice (or) leading to a blind vestibule (ves), and the operculum (op) and muscles which work it. The operculum and edge of the orifice form a grasping device which can hold debris or small epizoic organisms. As in an autozooid the operculum is opened indirectly by parietal muscles (par.mu) acting on the frontal membrane (f.mem) but shut by the direct pull of occlusor muscles (occ.mu). To improve its efficiency the base of an avicularium commonly narrows to a stalk, the orifice lip and operculum become pointed and the whole structure projects from the surface of the colony. In many genera, like *Bugula* (Fig. 176B) it becomes remarkably like a bird's head in shape and carries out snapping movements; these features suggested the names avicularium for the zooid and mandible for its operculum. A vibracularium is developed from an avicularium by great elongation of the mandible and regression of the remainder, producing a long whip-like structure which moves over

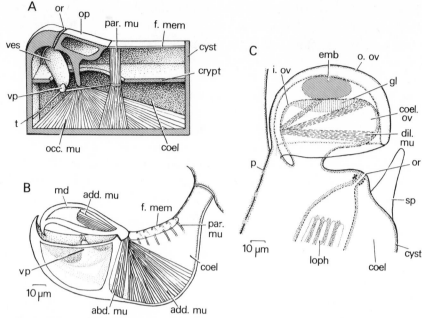

Fig. 176. Avicularia and ovicells. A, avicularium of *Flustra* to show resemblance to ordinary zooid; B, highly modified avicularium of *Bugula* drawn with the same orientation as A; C, ovicell of *Bugula*. abd.mu, abductor muscle of mandible; add.mu, adductor muscle of mandible; coel, coelom; coel.ov, coelomic cavity of ovicell, an extension of that of the zooid; crypt, crypto-cyst; cyst, cystid; dil.mu, dilator muscles of ovicell which open its cavity to the exterior; emb, developing embryo in cavity of ovicell; f.mem, frontal membrane; gl, glandular floor of cavity of ovicell; i.ov, inner wall of ovicell; loph, lophophore, retracted within tentacle sheath; md, mandible, modified operculum; occ.mu, occlusor muscle of operculum; o.ov, outer wall of ovicell; op, operculum; or, orifice; p, pore; par.mu, parietal muscles; sp, spine; t, tendon anchoring polypide; ves, vestibule; vp, vestige of polypide.

the surface of the colony and probably disturbs epizoic organisms about to settle there.

Reproduction

In *Membranipora* and some other genera the whole development occurs in the sea; most bryozoans, however, brood their eggs until they have developed into self-supporting larvae. Seemingly most primitive is the condition in which the eggs are fertilized (? self-fertilized) in the coelom, where they develop until they finally rupture the body wall and escape (*Nolella dilatata*); more commonly they enter an incubatory pouch formed by an invagination of the body wall enclosing an external space. The common condition (Fig. 176C) is where the incubation chamber known as an ovicell resembles a ball and socket joint. One outgrowth from an autozooid forms a socket (o.ov) and surrounds a second globular outgrowth (i.ov). Between the two lies an external space. The fertilized egg (emb) is passed into this space from the perivisceral coelom through an opening at the base of the lophophore mid-dorsally. This is the supraneural pore, a reduced version of the intertentacular organ of *Membranipora*. It is likely that some of the cells lining the brood chamber (gl) produce a histotrophe since embryos removed from it die.

The embryo develops inside the ovicell to a free-swimming larval stage which has different shapes in different species but is always unlike the kind of larva, known as a cyphonautes, which develops from the unbrooded egg of a animal like *Membranipora*. Ultimately all settle as the first member (ancestrula) of a colony which arises by a repeated process of budding, the pattern of which largely determines the final shape of the colony. In its simplest form budding occurs at the anterior end of a zooid. Here a bulge of the body wall grows out to form the body wall of a new individual from which, in turn, the new polypide is budded. If only terminal budding occurs a linear colony arises. Commonly, however, lateral as well as terminal buds form so that a more fan-like growth is produced. In *Membranipora* it is terminal budding at the edge of the colony that causes its expansion over the substratum; new rows of zooids are formed as the colony expands, however, by lateral buds. Even where rows are in contact the formation of lateral buds is not totally inhibited: it is they which break through into neighbouring zooids and form the pores setting one zooid in communication with the next.

It will be recalled that bryozoans possess no definite excretory organs. There is little doubt that much waste matter such as respiratory gases and ammonia can diffuse through the epidermis and gut wall. This does not, however, appear to be adequate for sustained health since bryozoans exhibit a cycle of degeneration and regeneration which seems partly to solve excretory problems though also acting as a perennating device. As a zooid ages the walls of the stomach come to contain greater and greater quantities of brown granules presumably excretory in nature, to such an extent as to interfere with the digestion and absorption of food. Ultimately there occurs a degeneration of the polypide, during which its remains condense to a spherule coloured brown by the gut pigment and known as a brown body (Fig. 174, b.b). The tissues of the cystid remain alive and may in due course bud a new polypide by the same process as produces a polypide in the growth of a colony. The fate of the brown body varies with the kind of bryozoan. In some it remains encysted within the coelomic cavity of the new polypide – by counting the brown bodies one can tell how many times that zooid has regenerated; in most, however, it is incorporated into the new polypide and discharged.

Freshwater bryozoans

Nothing has been said in the above descriptions specifically about the freshwater bryozoans placed in the class Phylactolaemata, yet nearly all is applicable to them.

There are, however, certain respects in which their organization differs and which indicate that this group is, in general, more primitive, although it has also certain features adaptive for life in fresh water. They are, indeed, relatively common in temperate areas in summer, fixed to the underside of leaves or to submerged wood. The first obvious difference lies in the fact there is no polymorphism in phylacto-laematous colonies – all zooids are autozooids. A second is in the shape of the lophophore, a double horseshoe-shaped array of tentacles set so that the open end of the horseshoe is dorsal and its arms right and left (Fig. 177A). The mouth (m) lies between the two sets of tentacles on the vertical side, under a dorsal lip known as an epistome (epi). These are primitive features and the lobing of the lophophore is reminiscent of brachiopods. The lophophore functions as in other bryozoans, directing a stream of particles into the ever-open mouth, but its shape allows it to carry more tentacles – up to about one hundred in a genus such as *Cristatella* – and this may compensate for the lesser quantities of food available in fresh water.

The cystid of a phylactolaematan is soft and histologically more complex than that of other bryozoans. The epidermis secretes a cuticle, sometimes thin and chitinous, sometimes thick and gelatinous; between it and the peritoneal epithelium lie circular and longitudinal muscle fibres. These are particularly well developed in the basal wall of the colony in the genus *Cristatella* which is able to creep slowly like a slug; smaller colonies of *Lophopus* and *Pectinatella* have also been observed to move. In all the rate is very slow, never more than about half a millimetre in an hour. Protrusion of the lophophore is due to contraction of the somatopleuric muscles raising the pressure within the coelom, a method possible only because of the lack of calcification of the body wall. Retraction is due to retractor muscles like those of *Membranipora*. The relationships of lophophore, tentacle sheath and orifice are also similar, save for the absence of operculum and for the persistence of a small section of the vestibule as a collar-like fold round the base of the zooid even when it is fully protruded (dup). Only in the genus *Fredericella* do lateral cystid walls persist to isolate zooid from zooid: in all others they vanish. The colony then has a single, continuous, common coelomic cavity (coel) into which the individual polypides hang and to the walls of which they are moored, each by its own funiculus (fu). The peritoneal lining is ciliated and maintains a steady current along a fixed path which undoubtedly aids respiration, the distribution of food and excretion. In *Cristatella* cells laden with waste collect, under the influence of this circulation, in an excretory vesicle (ex.ves) near the epistome from which they burst out at intervals. Although some osmoregulatory organ, such as a nephridium, might have been expected to occur in freshwater animals such as these, nothing of the sort has been described and their osmoregulatory control remains unknown.

Spermatozoa develop on the funiculus of each zooid (te) and eggs (ov) at a spot on the ventral body wall below the vestibular fold. How the egg is fertilized is unknown, although it seems likely to be brought about while it is still in the ovary

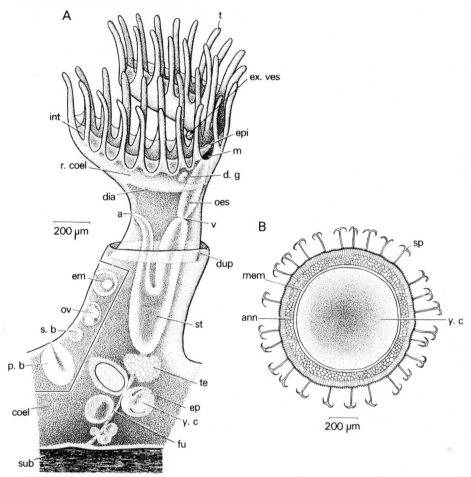

Fig. 177. Phylactolaematan structure. A, zooid of *Cristatella* (the area bounded by the black line belongs to the ventral surface); B, statoblast of *Cristatella*. a, anus; ann, annulus, float of air-filled chambers; coel, perivisceral coelom; d.g, dorsal ganglion; dia, diaphragm separating ring coelom from perivisceral coelom; dup, fold in cystid; em, embryo in embryo sac; ep, part of statoblast derived from epidermis; epi, epistome; ex.ves, excretory vesicle; fu, funiculus; int, intertentacular membrane; m, mouth; mem, membrane; oes, oesophagus; ov, ovary; p.b, primary bud; r.coel, ring coelom with extensions into tentacles; s.b, secondary bud; sp, spine; st, stomach; sub, substratum; t, tentacle; te, testis; v, valve; y.c, yolky cells of statoblast.

by sperm from the same zooid or others in the same colony. It then enters an invagination of the body wall called the embryo sac or ooecium (em), where it develops into the first zooid of a new colony. Proximal to the ovary on each zooid is a budding zone (p.b, s.b) from which new zooids are added to the colony. A parent zooid can normally give rise to only a limited number of daughter zooids before dying; colonies, therefore, can grow so as to contain only a certain number of zooids, usually many fewer than in gymnolaematans.

Like many freshwater animals the bryozoans flourish in summer and die down in winter. Overwintering is accomplished by means of structures known as stato-blasts (Fig. 177B) which have a limited capacity to withstand drying but are well adapted to resist extremes of temperature; they are thus reminiscent of the gemmules of freshwater sponges in function – and even resemble them in appear-ance. Statoblasts are produced in enormous numbers – one estimate suggested that the colonies of *Plumatella repens* on 1 m^2 of water plants gave rise to 8×10^5 statoblasts as they disintegrate at the onset of winter. They grow on the funiculus of each zooid, especially in autumn, though they also appear earlier in the season, escape and germinate to produce new colonies from which it is probable that most over-wintering statoblasts arise. A number of epidermal cells from the cystid (Fig. 177A, ep) invades the funiculus where they come to surround a group of cells laden with food granules and derived from the peritoneum covering the funiculus (y.c). The epidermal cells split to give the two layers which grow round the central mass of yolky cells. These, together with the inner epidermal layer, provide the rudiment from which the first zooid of a new colony will grow; the outer epidermal layer, on the other hand, secretes a thick chitinous cuticle (Fig. 177B, mem) as a protective device which later splits into two closely fitting halves. In some genera, such as *Plumatella*, a peripheral ring of air-filled cells, the annulus (ann), is added to the equator of the shell and acts as a float; in others, like *Cristatella*, hooked spines (sp) project radially from the annulus and anchor the statoblast to a substratum on which it may find the right conditions for germination. Still others, like those of *Fredericella*, have neither flotation nor attachment devices and sink directly to the bottom.

Classification of lophophorates

[Chaetognatha]
Brachiopoda
  Inarticulata
  Articulata
[Phoronida]
Bryozoa (=Polyzoa Ectoprocta)
  Phylactolaemata
  Stenolaemata
    Cyclostomata
    [†Cryptoporata]
    [†Trepostomata]
    [†Cryptostomata]

Gymnolaemata
  Ctenostomata
  Cheilostomata
    Anasca
     Cribrimorpha
    Ascophora
Pogonophora
  Athecanephria
  Thecanephria

15
Molluscs

THE MOLLUSCS constitute a group of animals of extraordinary diverse form which have become adapted to a larger number of different modes of life, ranging from the completely attached oyster at the one extreme to the agile, powerful and intelligent squid at the other. The way in which the molluscan body – fundamentally identical whatever the mode of life – has been modified to allow this range of habit is one of the most interesting chapters of functional anatomy.

Organization of a primitive mollusc

The best known are probably the kind known as snails, which the zoologist places in the class Gastropoda. These animals, however, have undergone a very peculiar reorganization of their basic anatomy and it will be easier to indicate the essential features of a mollusc and the way in which they work by describing a more primitive animal in the first place, a chiton or coat-of-mail shell.

Chitons are common intertidal and sublittoral animals found clinging to the underside of stones at low water. Their body is depressed to the half-streamlined shape common to all creatures that have to retain their hold on a substratum in a current of water. In dorsal view all that is visible is the skin of the dorsal part of the body, known as the mantle or pallium, and a centrally placed calcareous shell, made of eight articulating pieces or valves. Should the chiton be washed off its rock the articulated shell allows it to curl up like a hedgehog and protect its underside, but this is a specialized feature of chitons and not a characteristic molluscan feature, though the possession of a shell is. The margin of the dorsal surface, peripheral to the shell, and known as the girdle, is protected by large numbers of calcareous spicules. If the ventral side of the mollusc is examined it is noticed that the greater part is occupied by a broad, flat foot by which the animal both anchors itself and moves. Anterior to this is the head, little more than a protuberance carrying the mouth. The body of the chiton is therefore composed of a head, a foot and a third part called the visceral hump, which contains the viscera, is covered by the mantle and roofed by the shell. This triple division of the body is characteristic of all molluscs.

The ventral surface (Fig. 178A) reveals another basic molluscan feature: the mantle extends from the edge of the visceral mass to form the roof of a cavity which lies around head and foot and which contains a number of structures. This cavity is the mantle (or pallial) cavity and its roof is the mantle skirt (m.sk). The mantle cavity contains, in the mid-line posteriorly, the anus (a); on either side of this is placed a sensory hillock testing the nature of the water within the cavity and known as an osphradium (os). Anterior to this lie gills (ct), varying in number according to the species, and hanging downwards from the mantle skirt. Excretory and genital openings discharge into the mantle cavity medial to the gills. This series of structures and apertures, collectively known as the pallial complex, is regularly found within the molluscan mantle cavity.

Each gill (Fig. 178B, C) consists of an axis (ct.ax) bearing a series of semicircular leaflets (ct.l) on its anterior and posterior sides. The axis contains the main afferent and efferent branchial vessels (aff.br.v, eff.br.v) interlinked by capillaries in the leaflets. This type of gill structure is, again, one of the features of molluscan organization and its particular pattern is indicated by the special name ctenidium. In a living chiton it can easily be shown that the mantle skirt is raised from the substratum at a point near the anterior end where a current of water enters the mantle cavity (I), whilst a similar raising at the central posterior point (E) allows the water to leave. This current ventilates the mantle cavity, bringing in oxygen to the gills and, as it passes over the excretory, genital and anal openings, taking away waste and, in the appropriate season, gametes. The current is created by cilia lying on the flat surfaces of the leaflets of the ctenidia, known as lateral cilia (lat.c). They beat so as to drive water from the more lateral part of the mantle cavity, between the ctenidia and the mantle skirt, to the more central part lying between the ctenidia and the foot. Since the tips of the leaflets on neighbouring ctenidia touch the series of gills forms a perforated screen subdividing the mantle cavity into a lateral inhalant part and a median exhalant section into which the openings of kidney and gonad open. In the gill axis the afferent branchial vessel is situated in the exhalant side and the efferent vessel on the inhalant side. This arrangement shows that the molluscan gill, like most gills, operates on the counter-current principle, blood and water moving in opposite directions so that the former leaves the respiratory surface in equilibrium with water of maximal oxygen content.

Chitons are browsers scraping algae and detrital material from the rocks on which they live. The mouth is situated on the apex of a short down-turned snout (m) and leads into a buccal cavity. On the floor of this there lies a mobile, tongue-shaped elevation, the odontophore, covered with a cuticle. Over its mid-dorsal line runs a ribbon of chitinous teeth which are secreted at the inner end of a caecum, the radular sac, projecting backwards from the posterior end of the buccal

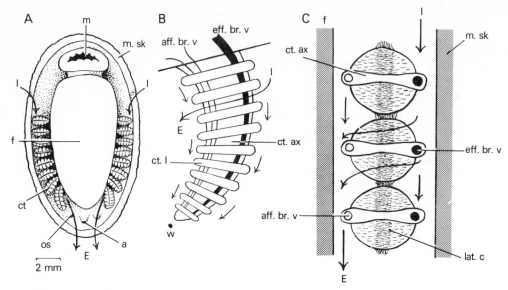

Fig. 178. A, a chiton seen in ventral view; B, one gill, seen in anterior view, hanging from body into mantle cavity, the foot to the left, the mantle skirt to the right; C, a horizontal section through part of the mantle cavity containing three gills, the anterior end to the top of the diagram. Large arrows show direction of water currents in the mantle cavity, small arrows of ciliary currents on the gills. (Based on Yonge.) a, anus; aff.br.v, afferent branchial vessel; ct, ctenidium; ct.ax, ctenidial axis; ct.l, ctenidial leaflet; E, exhalant water currents; eff.br.v, efferent branchial vessel; f, foot; I, inhalant water currents; lat.c, lateral cilia on broad surfaces of ctenidial leaflets, creating water currents; m, mouth; m.sk, mantle skirt or girdle; os, osphradium; w, waste collected by ciliary currents indicated and dropped off tip of gill.

cavity. The details of this will be considered later (p. 484), but the presence of a radula is another basic characteristic of the molluscs.

The features which have emerged from this description of a chiton as essentially molluscan are thus: the triple nature of the body; the shell secreted by the mantle; the mantle cavity with its complex of organs; the structural and functional pattern of the gill; the use of the radula as a feeding organ. In addition it may be added that molluscs are coelomate, though their main body cavity is a haemocoel.

Organization of a winkle

The gastropod body in such an example as the edible winkle (*Littorina littorea*), shows a clear division into two parts, the head-foot on the one hand, the visceral hump on the other (Fig. 179). The former is concerned with locomotion and sensation and bears the mouth; the latter deals with the visceral activities of respiration, digestion, excretion and reproduction. The visceral hump is connected to

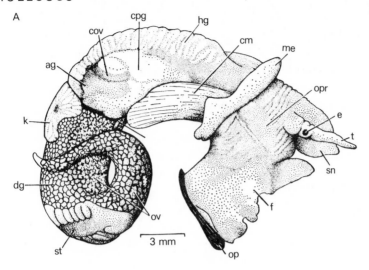

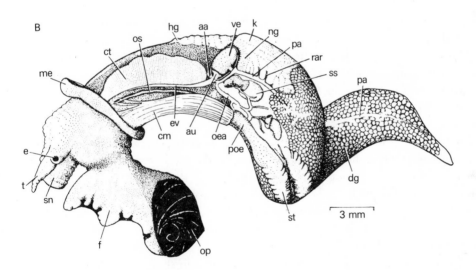

Fig. 179. *Littorina littorea*: female removed from the shell and seen A, from the right; B, from the left. Some of the organs are seen by transparency. aa, anterior aorta; ag, albumen gland; au, auricle; cm, columellar muscle; cov, covering gland; cpg, capsule gland; ct, ctenidium; dg, digestive gland; e, eye on eye stalk; ev, efferent branchial vessel; f, foot; hg, hypobranchial gland; k, kidney; me, mantle edge; ng, nephridial gland; oea, oesophageal artery; op, operculum; opr, ovipositor; os, osphradium; ov, ovary; pa, posterior aorta; poe, posterior oesophagus; rar, renal artery; sn, snout; ss, style sac region of stomach leading to intestine; st, stomach; t, tentacle; ve, ventricle.

the head-foot by way of a relatively narrow stalk through which run all the systems – nervous, alimentary, vascular – which lie in both parts.

Since the head-foot is a part of the body organized for movement with the head going first, it shows a bilateral symmetry like that of other free-living animals; the visceral hump on the other hand lies permanently within the shell, its shape unaffected by movement since this is usually very slow, and for reasons to become clear later, coiled in a spiral, like the shell which covers it.

The body wall of the visceral hump, the mantle, or pallium, secretes the shell. Since the visceral mass never emerges from its shelter the mantle is a delicate tissue almost devoid of muscles and covered by a squamous epithelium. The head-foot can also be withdrawn into the shelter of the shell if the situation requires, but must at other times emerge; its body wall is therefore thick, tough and muscular and often equipped with gland cells to keep the surface moist and lubricate movement. The cells tend to be sunk below the epidermis and gathered into distinct glandular masses.

Retraction of the head and foot into the shell is brought about by the contraction of a muscle, the columellar muscle (Fig. 179, cm), which runs from the central pillar of the shell, or columella, branching like a fan, into the foot, the head and the several parts of these, so that when it contracts each is pulled into the shelter of the shell. The last part to reach this – and therefore that nearest the mouth of the shell when the animal is withdrawn – is the posterior end of the foot. This bears a thick cuticular secretion, the operculum (op), which effectively stops the entrance to the shell against predators or adverse conditions. The columellar muscle fibres are not striped, are capable of extreme shortening on contraction and can maintain themselves in the contracted state for very long periods of time; whether this is managed by the same physiological machinery as is employed by the adductor muscles of bivalves is not known, though likely. It is also uncertain how a winkle emerges from the shelter of its shell when the columellar muscle relaxes. There are no muscles directly antagonistic to it which could protract the animal; presumably the head and foot are inflated by blood pressure as part of the extrusion mechanism, but this is certainly not the whole story.

If a winkle can withdraw the head-foot into the shell when danger threatens, it follows that the visceral hump cannot occupy the full volume of the shell; there must be some other space into which the head-foot passes, and this must act as a kind of compensation sac, filling with water when the head-foot emerges, emptying again as it is withdrawn. This space is the mantle or pallial cavity which is excavated out of the anterior face of the visceral hump and lies directly above and behind the head, the posterior part of which forms its floor. Its roof and its side walls are formed by the mantle skirt, on the inner surface of which is placed the set of structures known as the pallial complex. It is easy to show that the cavity is ventilated; carmine or other minute suspended particles in the water will be seen to enter it on the left and to emerge on the right, and it will be noted that the current is maintained through the narrow chink-like openings which persist when the animal is nearly fully withdrawn and the operculum all but blocks the mouth of the shell.

Locomotion is brought about by activity of the foot, which is provided with a flattened ventral surface or sole on which the animal creeps, the movement being lubricated by the secretion of mucus and similar material from a collection of gland cells lying within and under the epidermis of the sole (the sole gland) and aggregated into a mass at the anterior end (the anterior pedal gland) and distributed to some extent by cilia on the epidermal cells. The actual movement is due to the passage of waves of muscular contraction along the pedal sole from the anterior tip to the posterior tip. There are two sets of these, out of phase with one another, one on the right half of the foot, the other on the left (Fig. 180, wv). As a wave passes a given point the foot there is raised and lowered on to the substratum again in advance of its original position. Forward creeping thus consists of a series of small steps taken alternately by the two sides of the foot. These are accomplished by contraction of the pedal musculature which (apart from the bundles of columellar

Fig. 180. Ventral views of three prosobranchs to show locomotor waves on the sole of the foot; A, *Littorina saxatilis*; B, *Gibbula cineraria*; C, *Pomatias elegans*. All the animals are creeping towards the top of the page but in A and B the waves are travelling in the reverse direction. *Pomatias* steps alternately with the two halves of the foot and the left half is shown raised above the substratum as it moves forward. gv, transverse groove on foot, characteristic of trochid group to which *Gibbula* belongs; lsf, left half of foot raised; m, mouth; t, tentacle; wv, wave travelling over sole of foot.

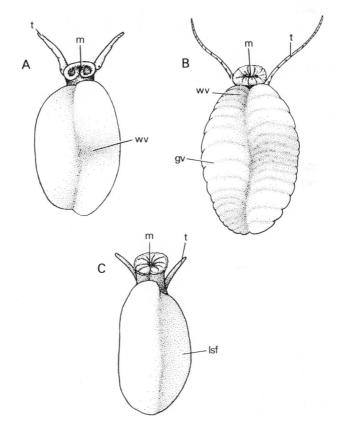

A

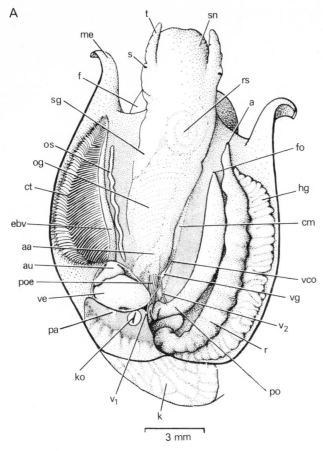

B

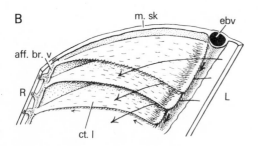

Fig. 181. *Littorina littorea*: A, animal removed from shell and mantle skirt cut mid-dorsally to show contents of mantle cavity; B, part of mantle skirt and ctenidium. Large arrows show water currents, small, ciliary currents. a, anus; aa, anterior aorta; aff.br.v, afferent branchial vessel running longitudinally in mantle skirt at right side of gill, receiving branches from hypobranchial gland; au, auricle; cm, columellar muscle; ct, ctenidium; ct.l, ctenidial leaflet; ebv, efferent branchial vessel in ctenidial axis; f, foot; fo, female opening; me, mantle edge; m.sk, mantle skirt; og, oesophageal gland; os, osphradium; pa, posterior aorta; po, pallial oviduct; poe, posterior oesophagus; r, rectum; rs, radular sac; s, eye; sg, salivary gland; sn, snout; t, tentacle; v_1, nerve to heart and kidney; v_2, genital nerve; vco, visceral connective; ve, ventricle; vg, visceral ganglion.

fibres) takes the form of a complex web of fibres, some longitudinal, some transverse, some oblique from side to side and dorsal to ventral and all under the control of nerves from the pedal ganglia. Although all the fibres seem to contract on the passage of a wave along the foot the responsibility of a group of them for a particular phase of the locomotor activity has not yet been worked out.

Respiratory exchange probably occurs at a number of places on the body of a winkle, wherever the skin is sufficiently delicate, but is mainly sited on the surface of the ctenidium (Fig. 181). This lies on the left side of the mantle skirt and consists of an elongate axis running almost the entire length of the mantle cavity and carrying a large number of triangular lamellae (ct, ct.l) which are attached to the ctenidial axis by one angle and have one side lying along the mantle skirt (m.sk) to the right of the axis. Blood enters each leaflet at its right end (aff.br.v), passes through a capillary network within it and escapes into an efferent ctenidial vessel (ebv) which runs along the ctenidial axis to the heart which is placed in the visceral hump at its innermost end. Respiratory exchange occurs between the blood in the leaflets and the current of water which ventilates the mantle cavity.

The water current is maintained by cilia on the gill leaflets (ct.l). The epithelium on their flat sides is largely ciliated, the direction of beat being from the lower left to the upper right edge. Water is therefore sucked by these lateral cilia, from the left half of the mantle cavity and pushed into the right half, and it is this which is responsible for the respiratory current.

Since winkles creep over the substratum and are littoral in habit, the water which is sucked into the mantle cavity is liable to contain suspended matter; this could damage the delicate tissue of the gill, clog the channels between leaflets and conceivably block the whole mantle cavity. Ciliary currents, however, run over the epithelium of the walls in such a way as to clear the cavity of detrital material. This may follow one of three different pathways depending upon the size of the particles (see Fig. 186D). Since the opening leading into the mantle cavity is rather narrow the velocity of the water passing through it is high; as soon as the opening is passed, however, the diameter of the mantle cavity enlarges rapidly and the rate at which the water moves drops correspondingly. The fall in velocity has the effect of making the largest particles (A) settle on the walls, and especially on the floor, near the mouth of the cavity. Here they come under the influence of ciliary currents which rapidly carry them to the exterior, where they fall off the body or are washed away.

Smaller particles (B) are carried deeper into the mantle cavity. Since the dimensions of this are increasing, however, the velocity of the water still tends to drop and sooner or later many particles fall on to the floor of the cavity near the mid-line of the body or even on the right side. These, too, come under the influence of ciliary currents but are borne towards the exhalant opening on the right. The very finest particles (C) may never settle and can, therefore, traverse the mantle cavity from side to side in the water current. Many, however, collide with the gill leaflets, or hit the mantle skirt or other surfaces and, were they not removed, would ultimately block the cavity. Ciliary currents on the gill leaflets and on the walls of the mantle cavity deal with these. Along the left and right edges of the gill leaflets cilia (frontal and abfrontal respectively) drive such particles to the tip of the leaflet, where they fall on to the floor and join other particulate material there for ejection by the exhalant aperture. Their passage is made easy and their return to the general circulation of water prevented by secretion of mucus from goblet cells placed in the gill and pallial epithelia. On the roof of the mantle cavity, in the mantle skirt to the right of the ctenidium, there is developed a particularly large collection of gland cells the secretion of which ensnares particles which land in this area. This is the hypobranchial or pallial mucous gland (Fig. 181, hg), often with its epithelium flung into folds and ridges so as to give greater secreting surface within a given area.

Since most of the suspended particles which are introduced by the water ventilating the mantle cavity are extruded on the right, this is clearly the most advantageous point at which to add other waste material to the outgoing stream. It is not surprising, therefore, to find the anus (a) discharging faecal matter to the mantle cavity on the right, close to the edge of the mantle skirt, and the kidney (ko) opening to the same current, though at a deeper level in the mantle cavity. Eggs also leave the body of the female winkle (fo) by nearly the same route.

The vascular system of the winkle, like that of almost all molluscs, is partly open, partly closed; that is, the blood sometimes runs within vessels with walls of their own and sometimes within unwalled spaces in direct contact with muscles, nerves and viscera. The heart lies in a pericardial cavity placed in the visceral hump, on the left, at the base of the ctenidium, whence it receives the greater part of its blood. A little blood comes, however, direct from the kidney, where it has escaped the obvious oxygenation to which the rest has been exposed, though it probably does not depress the oxygen content of the blood in the heart to any serious extent. The heart is a systemic heart, primarily concerned in pumping blood to the remainder of the body, which is reached by way of two major blood vessels, an anterior (or cephalic) aorta supplying the head-foot and a posterior (or visceral) aorta taking blood to the visceral mass. In these areas the main vessels break into small, and these into smaller ones, but ultimately the blood enters cephalic, pedal or visceral haemocoelic spaces and bathes the organs directly. From these sources blood returns to the heart, predominantly by way of a renal portal system and the capillaries of the ctenidium, though a small amount may by-pass the kidney and a further small amount by-pass the gill. It should be realized that the haemocoelic spaces are not spacious cavities around the viscera, but are clefts and interstices in a web of tissue, connective, muscular and nervous, which lies around and between the organs of the body in much the same way as the parenchyma of flatworms. The dimensions of the spaces are often not very different from those of capillaries. They therefore offer the same advantages as a capillary bed for the exchange of materials between blood and tissues, and are likely to have the same effects on the blood pressure. In some situations, nevertheless, larger sinuses do occur.

Since it has already been indicated that blood may be used to dilate and make turgid various parts of the body, it would seem likely that valves exist within the vascular system controlling the flow or distribution of blood within the winkle.

Valves do occur in the heart, directing flow through it, but it is not easy to find them elsewhere in *Littorina littorea*. At one point, indeed, where the main vein from the head-foot (Fig. 182B, ocp) and the main efferent from the visceral haemocoel join (vv), prior to entering the kidney, strands of connective tissue (cn) traverse the lumen of the vessel and suggest a possible control of its diameter, but elsewhere any effect of this kind must be due to strangulation of a vessel during the contraction of surrounding muscles. In *Littorina* one sheet of muscle and connective tissue completely separates the visceral from the cephalopedal haemocoels, and in other gastropods other sheets are also to be found, all of which must exert some control over the flow and distribution of blood within the animal's body.

The heart is two-chambered, being made up of an auricle and ventricle (Fig. 179B, au, ve), and lies in a pericardial cavity of coelomic origin. The walls are muscular, the ventricle more so than the auricle; the muscle fibres are striated. As in vertebrates the heart is largely autonomous and contains its own pacemaker, possibly located in the ventricle, the rhythm of which is affected by cardiac nerves. There is considerable argument about the effect of stimulation by these nerves (which originate in the visceral ganglia), and nothing is known about them in the winkle, but it is likely that, as in vertebrates, some fibres are inhibitory and some excitatory; the effect of the former seems to be commonly the more pronounced in experimental investigation.

Some uncertainty exists as to how the auricle becomes filled with blood since it seems improbable that there is sufficient residual arterial pressure after the blood has traversed the equivalent of three successive capillary beds (body, kidney and ctenidium) to drive it along the vein leading from the ctenidium to the auricle. One theory of heart action invokes the fluid-filled pericardial cavity as an aid towards auricular expansion. When the auricle contracts its contents are propelled into an empty ventricle; the volume of fluid in the heart remains unaltered. When the ventricle contracts, however, its contents pass into the vessels, and this would leave the heart empty. The decrease in pressure which accompanies ventricular systole, however, is conveyed via the pericardial fluid to the auricle, which is thereby dilated and filled with blood sucked in from the efferent branchial vessel. This theory involves an accurate alternation of auricular and ventricular contraction, which is observed in many molluscs.

The excretory system has already been seen to be intimately connected with the vascular system. It consists of a single kidney sac (Figs 179 and 181, k) of coelomic origin, connected to the pericardial cavity by a minute renopericardial canal, and discharging to the inner end of the mantle cavity (ko). The pressure changes in the pericardial cavity which accompany ventricular contraction must not only dilate the auricle but also tend to suck fluid from the kidney into the pericardial cavity. Movement of fluid is unlikely to occur since the bore of the canal is extremely fine and the pressure required to force liquid through it correspondingly high, and since

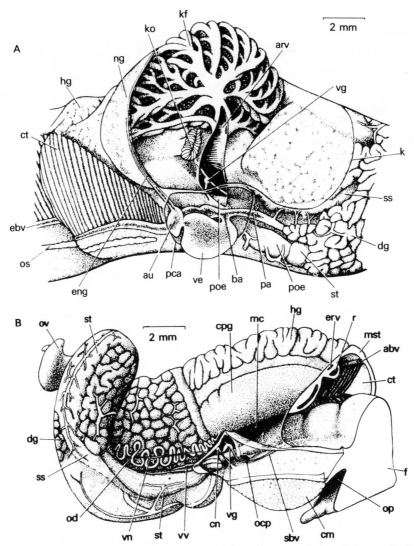

Fig. 182. *Littorina littorea*: A, dissection from the left to show vessels in relation to the kidney. The pericardial cavity has been opened, exposing the heart; the kidney has also been opened and its afferent vein slit from the point ventrally near the visceral ganglion where it receives blood from head-foot and viscera, to the point where it branches into the kidney folds. B, dissection of approximately the same area from the right with the base of the afferent renal vein opened as well as some of the spaces which drain into it. The anterior end of the animal has been cut away. abv, afferent branchial vein; arv, afferent renal vein; au, auricle; ba, bulbus aortae; cm, columellar muscle; cn, connective tissue strut; cpg, capsule gland; ct, ctenidium; dg, digestive gland; ebv, efferent branchial vein; eng, efferent vessel from nephridial gland; erv, efferent renal vein; f, foot; hg, hypobranchial gland; k, cut edge of kidney sac; kf, folds on wall of kidney; ko, kidney opening to mantle cavity; mc, mantle cavity; mst, cut edge of mantle skirt; ng, nephridial gland; ocp, opening of vessel from head-foot to afferent renal vessel; od, oviduct; op, operculum; os, osphradium; ov, ovary; pa, posterior aorta; pca, cut edge of pericardium; poe, posterior oesophagus; r, rectum; sbv, suboesophageal part of visceral loop; ss, style sac region of stomach; st, stomach; ve, ventricle; vg, visceral ganglion in blood vessel; vn, visceral nerve; vv, visceral vein.

it is lined by cilia which beat from pericardial cavity to kidney. It is more likely that there may be a loss of liquid from the cavity of the heart to the pericardial cavity and that this is drained away to the kidney and ultimately excreted. Were the blood volume to rise through osmotic uptake this process might also increase and so provide part of an osmotic regulation. The fluid within the pericardial cavity seeps through the heart wall from the blood within and recent work on the ultrastructure of the heart wall shows a system of blood channels reaching very close to the epicardial layer of cells.

There appear to be two major sites of excretory activity within the kidney; one of these is a series of folds placed on the right and dorsal walls of the kidney chamber (Fig. 182A, kf), the second is a structure called the nephridial gland (Figs 179B and 182A, ng) which lies on the left wall, abutting against the pericardial chamber. Blood passes from head, foot and visceral hump into a large afferent renal vessel (Fig. 182A, arv) which branches repeatedly within the folds just referred to. The blood is brought into minute vessels lying under the epithelium lining the kidney sac, allowing the cells there to extract excretory material from it. Granules of excretory material may often be seen in them; they are presumably shed and escape to the mantle cavity. Most of the blood is then collected into an efferent renal vessel and conveyed to the gill; some, however, is led to the nephridial gland, where there is an intimate intermingling of fine capillaries and kidney cells lining tubular crypts opening to the kidney sac. Ultimately the blood from this gland is collected into a blood vessel (eng) which takes it directly to the auricle (au) so that it does not pass through the gill. Precisely what kind of activity goes on in the nephridial gland is unknown; for a number of reasons it seems more likely to be involved in ionic regulation rather than in elimination of nitrogenous waste.

Feeding in winkles is carried out by means of the radula, placed over the dorsal surface of a tongue-like protuberance from the floor of the buccal cavity, the odontophore. It is covered with a thick cuticle, to which the radular ribbon is fused, and contains a number of skeletal structures called cartilages because of their histological resemblance to vertebrate cartilage. These provide surfaces for the origin and insertion of a complex array of muscles involved in the feeding process, the whole structure being known as the buccal mass. More detailed study of the radula shows a continuous cuticular base of chitin and protein on which is set a regular series of transverse rows of teeth, each row precisely repeating that in front as regards the number, shape and size of the teeth it contains. Seven teeth lie in each row, differentiated to a certain extent in accordance with the work they do. There is a central or rachidian tooth occupying the middle of each row and flanked, on each side, by three rather similar teeth, the two more median of which are known as lateral teeth and the one furthest from the mid-line as a marginal tooth. All are secreted by special cells (odontoblasts) located at the innermost end of the radular sac, and fused on to the ribbon, which is secreted from the cells forming the floor of the radular sac immediately in front of the odontoblasts. As row after row of teeth is produced the whole apparatus migrates forwards along the radular sac until it emerges on to the dorsal surface of the odontophore in the buccal cavity. During this migration, the teeth are closely applied to cells forming the roof of the radular sac by the activity of which they are tanned and impregnated with inorganic salts (mainly iron and silicon) so that when they are fully formed they are hard and tough. In use they ultimately get worn and broken but are replaced by new ones emerging from the radula sac. In gastropods which give their radula hard wear – browsing on minute algae on rocks, for example – replacement is frequent, and the radular sac is long, often several times as long as the body of the animal, whereas in carnivores or those that feed on soft food it is short. This must reflect the rate at which the radular ribbon is moving; if it moves rapidly, a long radular sac has to be provided for the process of tooth formation, if not, a short one suffices. In both cases the time spent in the radular sac is comparable.

When feeding, a winkle opens the mouth, which lies on the down-turned end of a short snout, and brings the buccal mass forward from its normal position of rest in the buccal cavity until the tip of the odontophore projects through the mouth. This is achieved by buccal protractor muscles which run from body wall to buccal mass, the former being immobilized by blood pressure within the cephalic haemocoel. As protraction occurs the cuticle covering the buccal mass and supporting the radula is pulled forwards over the extruded tip of the odontophore by the contraction of a second set of muscles running from the underside of the cuticle to the cartilages. This has the effect of pulling a certain number of the most anterior rows of radular teeth over the pointed tip of the odontophore. When this happens the teeth behave in characteristic fashion (Fig. 183). Within the buccal cavity they all lie flat on the surface of the odontophore (od) with their denticulated tips pointing inwards, and usually partially folded into a groove; this protects the buccal tissues from possible damage. When pulled outwards, however, the teeth erect themselves as they pass the odontophoral tip and the lateral (l_1, l_2) and marginal (m) teeth also rotate laterally so that each row forms a fan of upstanding and radiating teeth. The groove on the odontophore is abolished at the tip and the cuticle is made taut. If the outward movement of the cuticle and radula is arrested and reversed the teeth fold inwards and lie flat once more. The area over which the change in the disposition of the radular teeth occurs is known as the bending plane (b.p), and every time teeth move over it outwards they erect themselves and rotate sideways; every time they are pulled over it inwards they lie down and fold themselves away. This is the essential feeding action of the radula. It is important to realize that the movements of the teeth are brought about by the mechanical properties of the cuticle, of the attachment of the teeth to it, and their reaction to changing tensions put upon them; the teeth are not erected by the direct pull of muscular strands on them.

After the odontophore has been protruded and the radula pulled outwards over

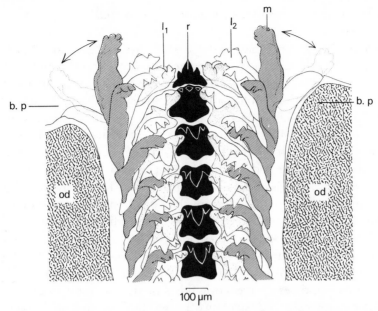

Fig. 183. *Littorina littorea*, part of the radula seen from above. The radula lies over the dorsal surface of the odontophore at the bottom of the diagram. At the top it bends over the tip of the odontophore, where the bending plane lies, on to its lower surface. At the bending plane the teeth erect and swing sideways (as shown by the arrows and dotted lines) as they move on to the ventral side of the odontophore; they move centrally and fold downwards as they are returned to the dorsal side. (Based on Ankel.) b.p, line of the bending plane; l_1, l_2, first and second lateral teeth; m, marginal tooth; od, odontophore; r, median or rachidian tooth.

its tip so as to erect and splay out the teeth it is applied to the substratum, or to the surface of weed or other food, and the radular teeth pulled inwards over the bending plane. When this happens their tips rake the surface of the object to which they have been applied and loose diatoms, detrital particles and the like, or shreds of material torn from the thallus of the weed or body of the prey are caught and pulled into the buccal cavity. This inward motion is accentuated by simultaneous or subsequent withdrawal of the odontophore into the buccal cavity. These two actions are due to the contraction of further sets of muscles, those for the former running from the radular cuticle to the cartilages, those for the latter being slips of the columellar muscle which run from the buccal mass to the shell. A variation of the feeding movements may involve the repeated scratching of the substratum or food plant by the radular teeth to free particles which may then be raked up into the buccal cavity. This is brought about by a to-and-fro movement of the radula over the odontophore tip, or by a movement of the odontophore under the radula.

The winkle, unlike some gastropods, has no jaw embedded in the roof of the buccal cavity to aid the radula in feeding. Salivary glands, rather solid masses of glandular material connected to the roof of the buccal cavity by a duct on each side, secrete a saliva which contains mucus, lubricating the radular movements, and an amylolytic enzyme.

Opening from the posterior part of the buccal cavity, dorsal to the radular sac, is the oesophagus, which leads to the stomach in the visceral hump. It is clearly divisible into an inflated, spindle-shaped, anterior half and a narrow, tubular, posterior part. The anterior half has a complex structure. On its dorsal wall run two tall longitudinal folds with a deep channel between; all three structures extend forwards on to the roof of the buccal cavity, where the food which has been collected by the radula is transferred to them. Here it comes under the influence of a strong, backwardly directed ciliary current which transports it towards the stomach, aided by peristalsis of the oesophageal walls. The lateral and ventral walls of the oesophagus are quite different from the dorsal and bear a series of transverse folds which run from the lateral margin of one longitudinal dorsal fold to the lateral margin of the other. On the sides of these folds ciliary currents beat towards the free edge and along this currents run towards the dorsal food channel. The folds secrete enzymes, and the mechanism just described ensures that a mixture of food and enzymes of salivary and oesophageal origin is passed back to the stomach.

One further complexity – of structure rather than of function – must be mentioned here: the food channel and dorsal folds lie dorsally at the anterior end of the oesophagus and the glandular portions of the oesophagus are ventral. If these structures are followed posteriorly, however, it will be found that the food channel slowly curves round the left side of the oesophagus until it comes to lie ventrally, whilst the glandular part slowly curves up the right side until it has become dorsal. It will also be noted that the anterior aorta, passing forwards from the heart to the head, and various nerves running from the head to the visceral hump are similarly involved in a course which leads them to cross from one side of the body to the other, all those that lie over the gut passing from right to left as they are traced backwards, all those that lie under the gut passing from left to right when traced backwards. This state of affairs results from a process known as torsion which will be dealt with below (p. 490); it is enough for the present to note that it seems to affect all gastropods at least temporarily during their life history.

The posterior part of the oesophagus, a relatively simple tube with a few longitudinal folds running along its walls, passes through the narrow neck linking head-foot to visceral hump and enters the stomach, a long, narrow organ running up the visceral hump with the oesophageal aperture half way along its right side. The part of the stomach into which the oesophagus discharges is separated internally from the rest by a tall fold stretching to the uppermost tip of the organ. Food, with admixture of salivary and oesophageal enzymes, is thus led to the end of what is, functionally, an extension of the oesophagus within the stomach, in which digestion continues. At the same level as the oesophageal opening, but discharging to the other side of the fold, are the ducts of the digestive gland, a voluminous structure occupying the bulk of the visceral mass; these bring in further digestive secretions. In the same area, much of the gastric wall is covered by a cuticle, elevated into a boss known as the gastric shield (Fig. 184B, g.sh). Against this – at least to a certain extent – the contents of the stomach, with their triple set of enzymes, are squeezed by muscle in the gastric wall. Digested food escapes and enters the digestive gland, to be absorbed by its cells. Ciliary currents on the gastric wall act on particles that work loose from the main mass, and it may be that they get swept into the digestive gland and ingested. The indigestible residue of the meal

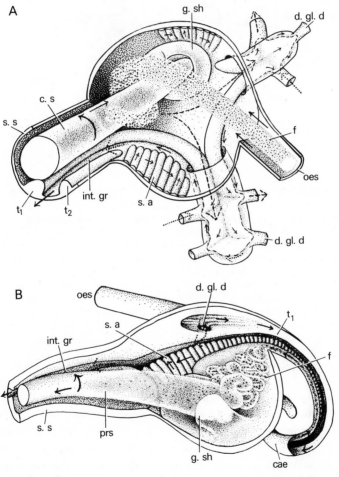

Fig. 184. The stomach of A, a generalized bivalve, B, a generalized prosobranch gastropod. For further explanation see text. cae, spiral caecum; c.s, crystalline style; d.gl.d, duct of digestive gland; f, string of mucus with embedded food particles; g.sh, gastric shield; int.gr, intestinal groove; oes, oesophagus; prs, protostyle; s.a, sorting area; s.s, style sac; t_1, major typhlosole; t_2, minor typhlosole.

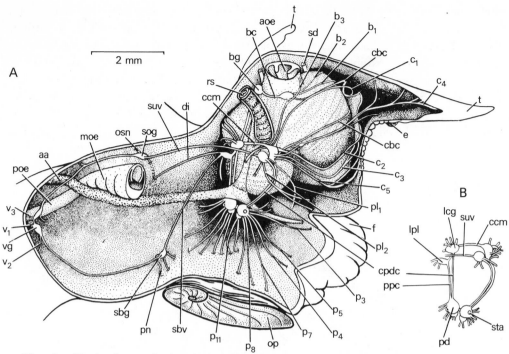

Fig. 185. *Littorina littorea*: A, dorsolateral dissection to show the nervous system. The gut has been cut behind the buccal mass and it and the radular sac pulled forwards through the nerve ring and turned dorsally; most of the radular sac and the salivary glands have been removed. B, diagram of nerve ring. aa, anterior aorta, dividing anteriorly to supply head and foot; aoe, anterior part of oesophagus; b_1–b_3, nerves from buccal ganglia; bc, buccal commissure; bg, buccal ganglion; c_1–c_5, nerves from cerebral ganglia; cbc, cerebrobuccal connective; ccm, cerebral commissure; cpdc, cerebropedal connective; di, dialyneury linking supra-oesophageal and left pleural ganglion; e, eye; f, foot; lcg, left cerebral ganglion; lpl, left pleural ganglion; moe, mid-oesophagus; op, operculum; osn, osphradial and branchial nerve; p_3–p_{11}, nerves from pedal ganglia; pd, left pedal ganglion; pl_1–pl_2, nerves from pleural ganglia; pn, pallial nerves; poe, posterior oesophagus; ppc, pleuropedal connective; rs, radular sac in radular sinus; sbg, suboesophageal ganglion; sbv, suboesophageal part of visceral loop; sd, salivary duct (cut); sog, supra-oesophageal ganglion; sta, statocyst; suv, supra-oesophageal part of visceral loop; t, tentacle; v_1–v_3, nerves from visceral ganglia; vg, visceral ganglion.

gets carried, partly by muscular, partly by ciliary means, into the distal end of the stomach, called the style sac (s.s), which lies lower in the visceral hump. Here it is mixed with quantities of locally secreted mucus and forms the structure known as the protostyle (prs), which is rotated within the style sac and gradually moved towards the intestine by ciliary currents; within this it becomes segmented into faecal pellets which are ultimately discharged to the mantle cavity by way of the anus.

This major constituent of the faeces consists of material which has never passed anywhere but through the stomach. It is therefore called the "stomach string".

Some particles from the stomach get swept into the digestive gland and may be rejected again from that If they are phagocytosed, parts of them may resist digestion and, again, be rejected. In addition, the digestive gland has an excretory function, and discharges to the stomach material extracted from the blood by which it is bathed. There thus emerges from the digestive gland ducts waste matter with three possible origins, to be added to the stomach string. This faecal and excretory component constitutes the "liver string" and it is carried through the style sac along a special groove, the intestinal groove (int.gr), bordered by conspicuous ridges, the typhlosoles, into the intestine (dotted arrows in Fig. 184A), where it becomes plastered on to the surface of the stomach string and the whole structure is consolidated. In winkles it is not easy to separate the two elements in the final faecal pellet, but it can easily be done in such gastropods as top-shells.

The activity of the winkle is governed primarily by the nervous system, hormonal control (so far as is known) being limited in gastropods and confined to a few features of the reproductive system.

The nervous system (Fig. 185) consists of three pairs of ganglia located in the head-foot, two pairs dorsal to the oesophagus (cerebral and pleural ganglia), one pair ventral, buried in the anterior part of the foot (pedal ganglia), and all interconnected so as to produce a nerve ring round the gut. From the pedal pair numerous nerves (p_3–p_{11}) pass to innervate the muscles of the foot, from the cerebrals nerves (c_1–c_5) pass to the tentacles, the eyes and the skin of the head as well as to a pair of ganglia, the buccal ganglia (bg), which provide special centres for the innervation of the buccal mass and oesophagus. The pleural ganglia give off only a small number of nerves ($pl_{1,2}$), mainly to the edge of the mantle skirt, but from them originates a nervous loop which passes back to visceral ganglia (vg) in the base of the visceral hump from which nerves (v_1–v_3) run to all the viscera. It is not possible to say much more about the nervous system of the winkle than this, since hardly any functional analysis has been made.

The reproductive system is male or female in pattern, the gonad lying in the upper parts of the visceral mass and connected by a duct to an opening in the mantle cavity. From there a ciliated pathway carries the genital products to a penis placed on the right side of the head in males or to an ovipositor, placed in a corresponding position and presumably homologous, in females.

There is nothing more than a simple duct connecting testis to mantle cavity in males. Across the mantle cavity, however, the ciliated pathway for the sperm is elaborated into a prostate gland and the sperm reach the penis in prostatic secretion. The seminal groove extends to the tip of the penis which can be greatly enlarged by inflow of blood so that it can enter the mantle cavity and reproductive aperture of the female. A row of penial glands set along the side opposite to the seminal groove may help to secure the penis to the tissues of the female.

Sperm passed into the female duct fertilize eggs which are enclosed within capsules prior to being set free to develop in the plankton. The female duct is therefore necessarily more complex than the male in view of the larger number of functions which it has to fulfil. Its initial stretch, a conduit for eggs running down the visceral hump from the gonad, is simple, but at the base of the mantle cavity it expands into a voluminous mass of glands which lie in and thicken the mantle skirt at its extreme right. The innermost of this collection of structures is the receptaculum seminis in which lie the sperm which have been received from the male in copulation. These are initially discharged into a caecum opening from the main female duct just within the external aperture and known as the copulatory bursa and make their way from this to the receptaculum along a groove, the ventral channel, running along the mid-ventral line of the intervening stretch of duct. Elsewhere, this is elaborately glandular and extended into lobes. The uppermost lobe secretes albumen, a food for the developing embryo; a lower one secretes a kind of shell known as the egg covering and the remainder a capsule within which 1–5 eggs are ultimately enclosed.

The eggs pass down the duct from the ovary, are fertilized by sperm from the receptaculum seminis and then pass into the glandular duct. Here each is embedded in a mass of albumen and enclosed in an egg covering. Then, usually in twos or threes, they are enclosed in a capsule, a disk-shaped structure about 1 mm in diameter with a central swelling where the eggs lie within a gelatinous fluid. When completed this structure passes through the female aperture and is carried forwards and out of the mantle cavity by the ovipositor. The capsules are pelagic and from them free-swimming veliger larvae escape.

Both ovipositor and penis are seasonal in their development, reaching maximal size during the breeding season and being reduced to insignificance at other times. There is evidence that this is controlled by hormones of gonadial origin.

Functioning of the gastropod mantle cavity

Gastropods are not animals that can be dismissed lightly in any account of functional anatomy: they exhibit adaptations of structure which fit them for an astonishing number of different activities. Their original mode of life was probably not significantly different from that exhibited by a winkle, but from such a starting point evolutionary change has led to the production of a vast array of different forms. Before these are described, however, it is necessary to comment on one feature of gastropod structure, already mentioned, which distinguishes them from all other molluscs; this is torsion, a process which takes place during the development of all gastropods and the effect of which persists either for their whole lifetime or for a period within that. Torsion consists of a rotation of the visceral hump upon the head-foot so that what starts as its posterior surface, facing backwards over the hinder margin of the foot, is twisted forwards to become its anterior surface, facing forwards over the head. This has the effect of bringing the original left side to the right and the original right side to the left and of introducing a spiral turn into the narrow neck of tissue which connects the base of the visceral hump to the dorsal surface of the head-foot. This particular twist has nothing whatsoever to do with the spiral coiling of the gastropod shell, and is shown by all structures running from the one part of the body to the other in this region. From the relationship of the structures involved to one another, and from observation of its occurrence during development, it can be said that the direction of rotation of the visceral hump on the head is always counter-clockwise when the animal is viewed from above with the head anteriorly.

Torsion has brought the mantle cavity with its contained pallial complex from the posterior end of the body, where it is found in all other classes of the phylum, to an anterior one; what has been the advantage of this shift? And has it been the sole benefit conferred? Two answers may be made to the first question. The first is that given by Garstang, who pointed out that so long as the mantle cavity was posteriorly placed the first part of the body that was pulled into it for protection under the shell was the foot and the last part was the biologically more valuable head. After torsion the situation is reversed and it is the foot which is withdrawn last and therefore exposed to the attack of predators; damage to this perhaps matters less than damage to the head and can, at any rate, be minimized by the development of the operculum, and torsion is therefore advantageous. The second answer which may be quoted is that of Morton, who pointed out that the mantle cavity is much more than an actual or potential shelter for the head – it is a respiratory chamber in which the ctenidia lie and to which, oddly, the anus and excretory organs discharge. In its original situation on the posterior face of the visceral hump the efficiency with which these activities could be carried on was considerably lowered by the fact that the cavity faced backwards, so that its connection to the outside tended to be constricted by a backward toppling of the shell due to forward movement of the animal, which, though facilitating the removal of excretory matter, made its ventilation difficult, especially since the motive power for the respiratory current was the weak suction of cilia rather than muscle. After torsion, the tendency of the shell to topple backwards opened the mouth of the cavity, and forward movement facilitated the entry of water and so improved ventilation. So far as the second question goes, it is generally conceded that, apart from the gains due to the new position of the mantle cavity, the other consequences of torsion appear disadvantageous. Its persistence in spite of them therefore emphasizes the relative immensity of the increased efficiency of the organs in the mantle cavity.

The question of the working of the mantle cavity in its new situation may now be discussed since this appears to have been a major functional factor in the evolution of gastropods. The original gastropods, as may be argued from the fossil record and the most primitive existing forms, differ from the winkle in being provided with two sets of many of the components of the pallial complex – two gills, two osphradia, two kidneys – and of some other related structures, such as the auricles of the heart. The living forms (*Pleurotomaria, Scissurella*) are not bilaterally symmetrical since the visceral hump and shell are coiled in a spiral just as are those of the winkle, but, presumably, they originated in a bilaterally symmetrical group, perhaps the extinct Bellerophontacea, and became asymmetrical later. Two points should be borne in mind: (1) no wholly convincing functional reason has ever been proposed to explain why the visceral mass became spirally coiled; and (2) traces of spiral coiling have been found in *Neopilina galatheae*, the most primitive known mollusc, suggesting that it is a very early feature of the phylum. Coiling may represent an attempt to allow expansion in the volume of the visceral hump without exaggerated increase in its length and producing a mechanically unmanageable body form. However, it introduces a factor which has played a significant part in the further evolution of the class in allowing ample room for the development of organs on the left while constricting those to the right of the mid-line.

The earliest gastropods had two gills in the mantle cavity (Fig. 186A). Respiratory water therefore entered it (I) from both right and left sides, washed over the gills and converged upon the median area of the mantle cavity (cc), into which the anus (a) and the two excretory organs (lko) discharged. The effluent stream therefore escaped, not on the right side as in the winkle, but in the mid-line. It was directed to as dorsal a position as possible, but it may well have been that in the new, anterior situation of the mantle cavity some of the current was deflected downwards as the animal crept forwards, resulting in a wash of waste material over the head and its sense organs. This may not be a particularly serious matter, may, indeed, be much less important to an aquatic than a terrestrial creature, but, nevertheless, seems to have been disadvantageous since no living gastropod retains the arrangement and all have modified the circulation of water and avoided the situation. The modifications have been achieved in one of three ways, two of which are less important than the third.

In the first group (Fig. 186B) the dorsally directed stream of effluents from the mantle cavity may be thought of as retarding the growth of that part of the mantle skirt against which it impinges, and hence of the overlying shell. It thus emerges (E) through a bay or slit in both mantle skirt and shell (ao) and the former may extend outwards to form a short tube or siphon which directs the current away from the head. This is found in the slit-limpets *Emarginula* and *Puncturella*, and in *Scissurella* and *Pleurotomaria*. In a further series of these animals, such as the keyhole limpet *Diodora* and the ormer *Haliotis*, the lower edges of the slit, once it has been formed, re-join and so convert it into a hole. In *Haliotis* a series of holes forms as the animal grows. By these devices this group of gastropods successfully directs the stream of water away from the head. To this extent the change is advantageous, but in so far

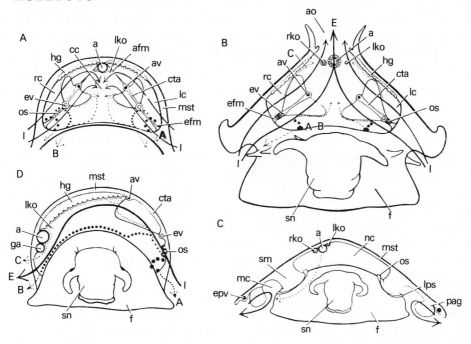

Fig. 186. Diagrammatic transverse sections to show the arrangement of water and ciliary currents in the mantle cavity of A, a hypothetical primitive gastropod; B, *Diodora*; C, *Patella*; D, a mono-tocardian. Continuous arrows show water currents; broken arrows, ciliary currents A, B and C and the particles carried by them are indicated by the size of the dots. a, anus; afm, membrane attached to side of ctenidial axis carrying afferent branchial vein; ao, apical opening of mantle cavity; av, afferent branchial vein; cc, central part of mantle cavity; cta, ctenidial axis; E, exhalant; efm, membrane attached to side of ctenidial axis carrying efferent branchial vessel; epv, efferent vessel from pallial gills; ev, efferent branchial vessel; f, foot; ga, genital aperture; hg, hypobranchial gland; I, inhalant; lc, left part of mantle cavity; lko, left kidney opening; lps, lateral pallial streak, sensory organ; mc, mantle cavity; mst, mantle skirt; nc, nuchal cavity; os, osphradium; pag, pallial gill; rc, right part of mantle cavity; rko, right kidney opening; sm, shell muscle; sn, snout.

as it breaks open the shell and allows possible desiccation of the mantle cavity if the snail is exposed at low tide, it is not, and it is probably significant that gastropods showing this adaptation are a small group of mainly sublittoral forms not contributing much to the fauna of any area.

The second group is also a small one as regards the number of species which it contains, but in numbers of individual animals it must be regarded as one of the most successful of all molluscan groups; these are the animals known as limpets, most of which have become adapted for clinging to rocks in the intertidal region. Here (Fig. 186C) the ctenidia have been lost (*Patella, Patina*; one remains in *Acmaea*), or, some would say, reduced along with the osphradia (os) to a vestige which may have some sensory, but certainly has no respiratory significance. The mantle cavity therefore becomes essentially a cloacal cavity, receiving the anus (a) and the outlets of the kidney and reproductive organs (lko, rko). Respiratory exchange is effected through a series of new structures, pallial gills (pag), which lie in part of the mantle cavity, but not in that part to which anus and kidneys open and in which the ctenidia lay; instead, they are located under the edge of the mantle skirt round the entire periphery of the animal or in its lateral and posterior parts, depending on the species. Cilia on their surface maintain a flow of water sucked in from outside. Faecal matter is discharged in well compacted rods carried posteriorly by ciliary currents for discharge. This adaptation frees the animal from the possibility of waste matter contaminating the head and does not destroy the completeness of the shell, with the result that many limpets, as is well known, are able to withstand long intertidal periods clinging to the surface of a rock without danger of drying up. It will be noted at a later stage (pp. 546, 547) that this arrangement is very similar to what is found in the chitons and *Neopilina*. The patellacean limpets have returned by a long evolutionary pathway to a state similar to that from which their remote ancestors started.

Neither of the solutions of the problem of the sanitation of the mantle cavity yet mentioned, however, is the one most productive of successful forms. This involves a much more radical alteration of the organization of the pallial complex and rests upon the fact that there is present in the gastropods, from their start, a tendency for the right side to be smaller, less well developed, than the left, as indicated by the spiral coiling of the shell. If this tendency be exaggerated (Fig. 186D), with the organs of the left half becoming predominant, and those of the right half decreasing in size and importance, the mid-line (and so the exhalant current) will be transferred towards the right side and finally, with the disappearance of the right gill and osphradium, the anus (a) and outgoing current (E) come to lie completely at the right margin of the mantle cavity, as already described for the winkle. The right kidney disappears as an excretory organ though its opening is retained as an outlet for the gametes (ga), the left kidney (lko) being the sole excretory organ.

One further modification occurs. In the primitive gastropod, and in those with two ctenidia, each ctenidium consists of an axis bearing two rows of leaflets, one on each side. In those under discussion now the left ctenidium, the sole remaining one, loses the leaflets attached to the left side of the axis (cta), which becomes fused by its left side to the mantle skirt and carries only the single right row of leaflets projecting across the mantle cavity to the right.

The gastropods so far mentioned are thus divisible, on the basis of the structure and functioning of the mantle cavity, into two major grades: a more primitive, less adaptable, group with a double set of structures (gills, osphradia, kidneys, auricles), and a more advanced group, which has proved extremely adaptable, with only one. The former group is called the Diotocardia (alternative terms are Archaeogastropoda and Aspidobranchia), the latter the Monotocardia (alternatives

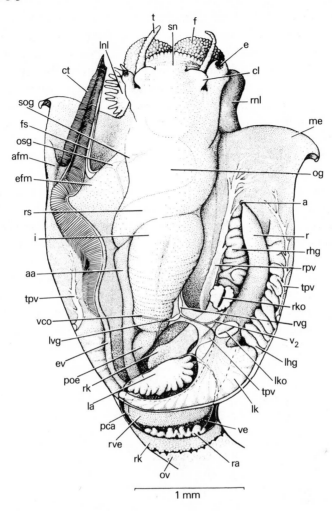

Fig. 187. *Monodonta lineata*: dissected to show the contents of the mantle cavity. a, anus; aa, anterior aorta; afm, afferent branchial membrane; cl, cephalic lappet; ct, ctenidium; e, eye on eye stalk; efm, efferent membrane of ctenidium; ev, efferent branchial vessel; f, foot; fs, supporting tissue of free part of ctenidium; i, intestine seen through body wall; la, left auricle; lhg, left hypobranchial gland; lk, left kidney or papillary sac; lko, left kidney opening; lnl, left neck lobe; lvg, left visceral ganglion; me, mantle edge; og, oesophageal gland, by transparency; osg, osphradial ganglion, by transparency; ov, ovary; pca, pericardial cavity; poe, posterior oesophagus, seen through body wall; r, rectum; ra, right auricle; rhg, right hypobranchial gland; rk, right kidney; rko, right kidney opening, also genital opening; rnl, right neck lobe; rpv, right pallial vein; rs, radular sac, by transparency; rve, rectum passing through ventricle; rvg, right visceral ganglion; sn, snout; sog, supra-oesophageal ganglion; t, tentacle; tpv, transverse pallial vein leading blood from body to mantle skirt on way to gill; v_2, genital nerve; vco, visceral loop; ve, ventricle.

are Mesogastropoda + Neogastropoda, and Pectinibranchia). Intermediate stages between the two grades are to be found amongst the top-shells (Fig. 187), which have only one gill (ct) and one osphradium, but have two kidneys (lk, rk) and two auricles (la, ra) and in the Neritacea (e.g. the nerite of British rivers, *Theodoxus fluviatilis*, Fig. 188). It is almost certain that the monotocardian grade of organization has been reached by more than one stock.

Both groups exhibit the full effects of torsion. Since this, while undoubtedly beneficial, has also brought new problems, it is not surprising that some gastropods have reversed the process and end their development with untwisted visceral hump. The gastropods can therefore be divided into the Prosobranchia, including both Diotocardia and Monotocardia, and the Opisthobranchia, the former showing torsion throughout their lives, the latter, though usually unable to exclude all traces of it from their development, later undergoing a process of detorsion which partly or wholly undoes its effects. There is also a third group, the Pulmonata, which retain a disposition of the body indicating a partial torsion but have undergone so many other modifications of structure (notably the development of a lung) that they are usually separated from the other two groups. Many opisthobranchs and pulmonates take a short cut during development and do not undergo more than the partial torsion they exhibit as adults. Their ancestors, however, had undergone full torsion not a partial torsion, which they exhibit. The process of detorsion in

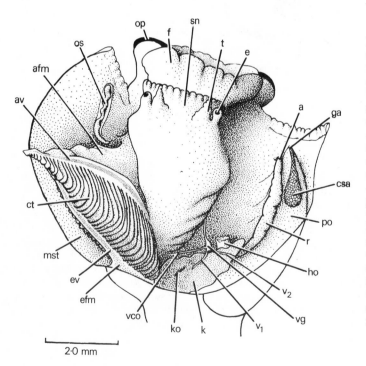

Fig. 188. *Theodoxus fluviatilis*: dissected to show the contents of the mantle cavity. a, anus; afm, afferent membrane of ctenidium; av, afferent branchial vessel; csa, crystal sac on oviduct, a store of grit used to strengthen wall of capsule in which eggs are laid; ct, ctenidium; e, eye; ev, efferent branchial vessel; f, foot; ga, genital aperture; ho, vestigial right ctenidium; k, kidney; ko, kidney opening; mst, mantle skirt; op, operculum; os, osphradium; po, oviduct in mantle skirt; r, rectum; sn, snout; t, tentacle; v_1, nerve to heart and kidney; v_2, genital nerve; vco, visceral loop; vg, visceral ganglion.

opisthobranchs, by which the mantle cavity and its contents are returned to the back end of the body is possibly not itself advantageous, since natural selection originally favoured those gastropods which carried out the reverse change, but can be so in relation to some other change happening simultaneously. This seems to be primarily the loss of the shell, so as to produce a naked slug. Before this can occur the viscera have to be provided with shelter compensating for absence of the shell and this is achieved by a gradual uncoiling and sinking downwards of the visceral mass (*Umbraculum*, Fig. 190A) until it is incorporated in a head-foot swollen to accommodate it. As this occurs the mantle cavity becomes gradually shallower until it vanishes, leaving the three apertures which it contains (anal, excretory and reproductive) on the general surface of the body. The ctenidium is lost and may be replaced by secondary gills if the animal is large, though cutaneous respiration may be adequate if it is small. Numerous stages in the evolution of the slug type may be seen. Thus in the opisthobranch barrel shell, *Acteon*, the visceral hump is still enclosed within a spirally coiled shell large enough to allow the animal to retract completely, and the mantle cavity faces forwards; in the bubble-shell, *Haminea* (Fig. 190B), and to a much greater extent in the boat or canoe shell, *Scaphander*, the shell, though large, can no longer contain the soft parts, and the mantle cavity has become shallow and lies on the right side; in the sea hare (Fig. 190C), *Aplysia*, the mantle cavity is similar, but the viscera are largely withdrawn into the expanded haemocoel of the head-foot and the shell is hardly calcified and almost wholly overgrown by the mantle. Finally, in the nudibranchs (Fig. 190E, F), recession of the visceral hump into the head-foot, disappearance of the shell and mantle cavity and detorsion are all complete. The dorids still retain a greatly altered pallial complex in the combination of anus, kidney opening and modified ctenidium (ct), but in the eolids the presence of cerata (cer) has abolished the need for the retention of the gill.

The changes described must obviously have repercussions on the mode of life; no longer can the animal expose itself as boldly as a limpet within the intertidal zone or invade brackish or fresh water with desiccation and osmotic upset minimized by an impermeable shell. Instead the animals are limited to the marine habitat and tend to be sublittoral, coming ashore only at the breeding season to copulate and deposit their eggs. Deprived of the protection given by the shell against possible predators many come to rely on protective coloration (opisthobranchs and dorids), on warning coloration (eolids with cnidosacs) and on repugnatorial glands embedded in the mantle producing a discharge which may range from mere unpleasantness to possible toxicity.

Among those gastropods which retain a mantle cavity two main innovations may be mentioned. These are (1) its conversion into a lung, and (2) the invention of a ciliary method of collecting food.

The invention of a lung is almost wholly confined to the group known as the

Pulmonata, but has also been managed by a limited number of prosobranchs of which one (*Pomatias elegans*) is not uncommon in chalk or limestone districts of Britain. In these prosobranchs the ctenidium and osphradium have been lost and uptake of oxygen occurs by increased vascularization of the roof of the mantle cavity, partly achieved by an extension of the kidney into the mantle skirt. A similar trend is noticeable in two species of winkle which inhabit the topmost areas of the shore, *Littorina saxatilis* and *L. neritoides*; the former has a reduced gill, the latter none, and since these animals are related to *Pomatias* it seems likely that *Pomatias* has become terrestrial by migration on to land from an ancestral marine habitat rather than by way of estuaries and freshwater, which is by far the commoner route. *Acicula fusca* is another terrestrial prosobranch without a gill, but the animal is so small that it inhabits the water films found among masses of decaying leaves (usually beech on calcareous ground) and is, in a sense, aquatic, respiring by keeping at most a bubble of air within the mantle cavity.

In pulmonates, which do not undergo more than about 90° torsion, the mantle cavity never develops an extensive opening and communicates with the outside only by a restricted opening on the right (the pneumostome) which can be opened and shut. This gives it the characteristics of a lung and when the pneumostome is open, movement of the floor forces air out and sucks more in. Nevertheless, it has been calculated that in ordinary circumstances diffusion through the open pneumostome is adequate to satisfy respiratory needs. Like many respiratory surfaces, however, that of the pulmonate allows loss of water and, as the animals are sensitive to this, the pneumostome can remain open for long periods only when the air is very humid; at other times it must close, and it is in these circumstances that forced ventilation is of advantage. The bulk of the gaseous exchange within the lung occurs over the richly vascular surface of the mantle skirt where vessels project into the cavity, covered only by a delicate epithelium (Fig. 189C); this ridging increases the surface area two or threefold. The pulmonate lung is unusual among respiratory organs in using increased pressure to augment gas uptake; with the pneumostome open, its floor is lowered, sucking air in; the pneumostome then closes and the floor raised so increasing the internal pressure momentarily before the pneumostome opens once more. Some respiratory exchange also occurs through the general surface of the body. The rectum elongates so as to discharge faecal matter close to the pneumostome and the kidney is drawn out into a long ureter which opens alongside the anus. These arrangements minimize contamination of the lung and reduction of respiratory efficiency.

Many pulmonates are aquatic and live in freshwater, having probably evolved from a stock of marsh-dwelling snails; because these animals rise to the surface of a shallow aquarium to fill the lung with air it has been assumed that they always do so in nature. Recent work, however, suggests that this is unlikely and that sometimes the bubble of air in the mantle cavity acts as a physical gill and allows

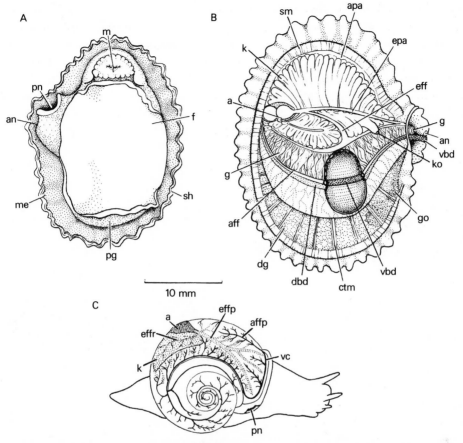

Fig. 189. *Siphonaria*: A, in ventral view; B, in dorsal view after removal from shell. An oval window has been cut in the mantle skirt and gill to expose the floor of the mantle cavity. C, *Helix*: diagram of the circulation, mainly in relation to the mantle skirt. a, auricle; aff, afferent vessel to gill; affp, afferent vessel to lung; an, anus; apa, afferent pallial vessel taking blood to kidney; ctm, bands of muscle and connective tissue; dbd, ciliated band on surface of mantle skirt helping to maintain water flow over gill; dg, digestive gland; eff, efferent vessel from gill to heart; effp, efferent vessel from lung to heart; effr, efferent vessel from kidney to heart; epa, efferent pallial vessel; f, foot; g, gill; go, gonad; k, kidney; ko, kidney opening; m, mouth; me, mantle edge; pg, repugnatorial gland; pn, pneumostome; sh, shell; sm, shell muscle; vbd, ciliated band on floor of mantle cavity helping to ventilate gill; vc, venous circle in pallial edge.

exchange of gas with the surrounding water, as with many aquatic insects. This permits the snails to live without drowning at depths, or at distances from the shore at which surfacing would not be possible. Even close to the shore it has been shown that many pulmonates surface only when the temperature rises and drives oxygen out of solution. All young aquatic pulmonates seem to have the mantle cavity full of water, and it becomes air-filled, partly for respiratory, partly for hydrostatic reasons, only when they become adult. This suggests that cutaneous respiration is adequate when the body surface is relatively large but requires supplementation as the animal grows bigger; cutaneous respiration is likely to be adequate also in waters which contain much oxygen. In two families of aquatic pulmonates, the Planorbidae and the Ancylidae, cutaneous respiration is made more efficient by the development of secondary gills – in the ram's horn snail, *Planorbarius corneus*, a quadrilateral flap with its own complete vascular supply, and possibly other folds within the mantle cavity; in the freshwater limpets, *Ancylus* and *Acroloxus*, folds lying

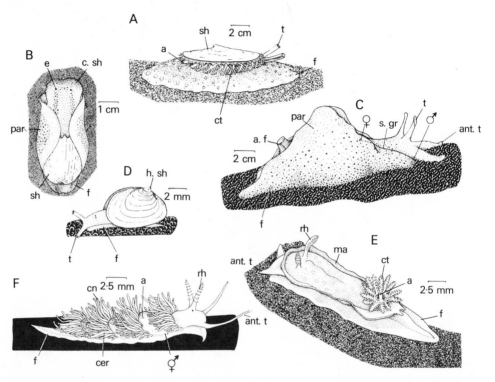

Fig. 190. Opisthobranchs. A, *Umbraculum mediterraneum*, a tectibranch found on hard bottoms; B, *Haminea navicula*, a tectibranch found on mud; the animal drawn is just beginning to burrow; C, *Aplysia californica*, a tectibranch; D, *Berthelina limax*, a sacoglossan with bivalved shell crawling on the weed *Caulerpa*; E, *Goniodoris nodosa*, a dorid nudibranch; F, *Facelina auriculata*, an eolid nudibranch. a, anus; a.f, anal funnel, an exhalant pallial tube on which the anus opens; ant.t, anterior tentacle; cer, ceras, containing digestive gland; cn, cnidosac; c.sh, cephalic shield; ct, ctenidium; e, eye; f, foot; h.sh, helicoid beginning of shell which remains attached to one valve; ma, mantle; par, parapodium or lateral expansion of the foot which bends upwards as a protective cover over the mantle cavity and shell; also used in flapping swimming; rh, rhinophore, the name for the special shape of tentacle found in nudibranchs; s.gr, sperm groove from genital aperture in mantle cavity to penis in head; sh, shell; t, tentacle; ♀, position of female opening in mantle cavity; ♂, opening of penis pouch; ☿, position of male and female openings.

outside the reduced mantle cavity with less well-developed vessels. Other pulmonates use other parts of the body – tentacles in the pond snails (*Lymnaea* spp.), fringing
processes of the mantle edge in the bladder snail *Physa* – but all must be external to
the mantle cavity since that has lost the power to develop the ciliated epithelium
which could maintain a water current through it. This must have occurred when
the pulmonate stock first adapted for land life and it has not proved possible to
re-acquire. One pulmonate genus, the marine *Siphonaria* (Fig. 189), to which belong
many species of limpet-like animals found on beaches throughout the tropics, has
succeeded in evolving secondary, ciliated gills within a mantle cavity through which
a current of water is maintained. This is an obvious adaptation to the fluctuating
conditions of an intertidal habitat whereas the state of permanent immersion in
which the freshwater pulmonates live has not generated a sufficiently intense
selective pressure to require it, and has allowed exposed gills to develop. The mantle
cavity itself, however, has had to be retained, since it acts as a compensation sac as
the snail moves in and out of the shell; only in freshwater limpets where the need for
this is least has it been reduced.

The uptake and transport of oxygen in aquatic pulmonates is also improved by
the occasional presence of the pigment haemocyanin in lymnaeids and the regular
occurrence of haemoglobin in planorbids. Both of these allow the snails to absorb
more oxygen, particularly from water with little oxygen in it, but haemoglobin is
distinctly better than haemocyanin in this respect. As in diving animals in general,
gastropods are relatively insensitive to the concentration of carbon dioxide in their
bodies.

Feeding in gastropods

No pulmonate has become a ciliary feeder since on adaptation to land life ciliated
epithelium was lost from the mantle cavity and it has proved beyond the powers of
secondarily aquatic forms to regain it. Some opisthobranchs, the gymnosomatous
pteropods or sea-butterflies, have evolved a ciliary food-collecting mechanism. They
are planktonic and feed on diatoms and similar minute phytoplanktonic organisms. As might be expected, the mantle cavity is not involved in this, and the animals
make use of expanded outgrowths of the foot with which they swim (the butterfly's
wings) as food-collecting surfaces; from these ciliary currents carry the food to the
mouth. Most gastropods which use ciliary means of collecting food, however, are
prosobranchs and since the evolution of the habit may be followed in this group it
allows us to see how the high degree of adaptation exhibited by the ciliary feeding
bivalves has probably come into being.

It will be recalled that particulate matter is inevitably drawn into the prosobranch mantle cavity with the respiratory water stream and is rejected, according to
size, either on the left or right. Some prosobranchs have, so to speak, learned that

this material has nutritive value and collect it as food, the less advanced types using it as a supplement to what they gather in the ordinary way with the radula, the more advanced relying on it as their sole source. An example of the first type is *Bithynia tentaculata*, a common freshwater prosobranch. In this animal large particles are still expelled from the mantle cavity on the left, but those which are gathered on the right are kept apart from the faecal matter and travel along a groove on the floor of the mantle cavity to a point under the right tentacles, embedded in mucus from glands on the ctenidium, the hypobranchial area and the pallial floor. At intervals the mixture of mucus and particles seems to be raked into the gut by the radula, though some authors deny this and others say that it happens only in snails from certain localities. Advance on this state of affairs is shown by the river shell *Viviparus* and the auger or screw shell *Turritella*, in which the volume of water entering the mantle cavity has been increased by elongating the cavity and allowing room for a longer ctenidium, by elaboration of the groove on the pallial floor and the provision of greater quantities of mucus. Maximal modification is exhibited by

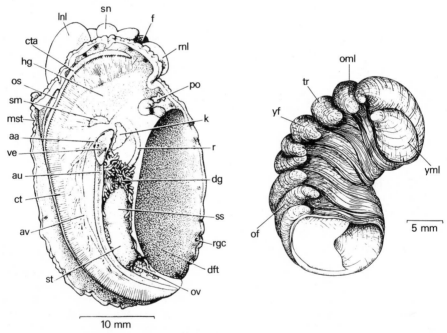

Fig. 191. *Crepidula fornicata*: left figure, dorsal view of animal removed from shell; right figure, chain of snails. aa, anterior aorta; au, auricle; av, afferent branchial vessel; ct, ctenidium; cta, ctenidial axis; dft, dorsal surface of foot; dg, digestive gland; f, foot; hg, hypobranchial gland; k, kidney; lnl, left neck lobe; mst, mantle skirt; of, old female; oml, old male; os, osphradium; ov, ovary; po, oviduct; r, rectum; rgc, repugnatorial gland; rnl, right neck lobe; sm, shell muscle; sn, snout; ss, style sac; st, stomach; tr, transitional form; ve, ventricle; yf, young female; ym, young male.

the limpet-like *Calyptraea chinensis* (Chinaman's hat shell), and *Crepidula fornicata* (slipper limpet) in which the mantle cavity and ctenidium (Fig. 191, ct) are as long as the whole animal, the leaflets of the ctenidium are elongated into filaments which touch the pallial floor at their tips (Fig. 192) and so subdivide it completely into an infrabranchial inhalant chamber on the left (ibr) and a suprabranchial exhalant one on the right (sbr). Particles which enter the mantle cavity are caught

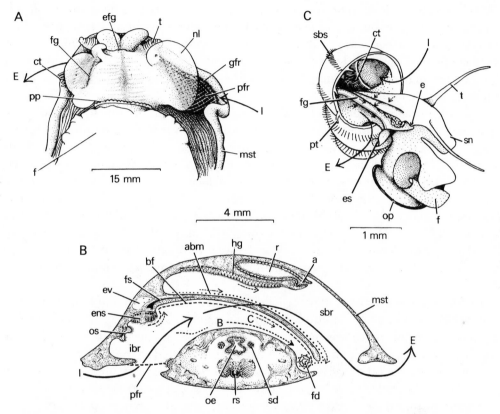

Fig. 192. A, B, *Crepidula fornicata*: A, ventral view of anterior part of body; B, transverse section to show currents in mantle cavity; C, young specimen of *Viviparus contectus*. Continuous arrows show water currents, broken arrows ciliary currents. a, anus; abm, mucous stream on abfrontal surface of gill filament; B, path followed by medium-sized particles; bf, sheet of mucus on frontal surface of gill trapping fine particles; C, path followed by minute particles; ct, ctenidium; E, exhalant; e, eye; efg, anterior end of food groove; ens, endostyle; es, exhalant siphon; ev, efferent branchial vessel; f, foot; fd, food and mucus rod in food groove; fg, food groove; fs, skeletal support of gill filament; gfr, groove from pallial filter to food pouch; hg, hypobranchial gland; I, inhalant; ibr, infrabranchial part of mantle cavity; mst, mantle skirt; nl, neck lobe; oe, oesophagus; op, operculum; os, osphradium; pfr, pallial mucous filter collecting coarse particles; pp, front part of foot; pt, pallial tentacle; r, rectum; rs, radular sac; sbr, suprabranchial part of mantle cavity; sbs, periostracal bristles on shell; sd, salivary duct; sn, snout; t, tentacle.

on the ctenidium and transported to the mouth as before on mucus coming predominantly from a special gland on the ctenidial axis, the endostyle (ens), so called because its function parallels exactly that of the endostyle of the tunicates. What most distinguishes these two genera from the others, however, is the development of a second filter stretched across the mouth of the mantle cavity on the left which traps large particles about to enter. Though *Turritella* is equipped with tentacles which also do this, it rejects the particles and uses the tentacles solely as a device to reduce the amount of particulate matter coming in – necessarily, since it lives in a muddy habitat, and excess would soon clog the gill and spoil the entire feeding mechanism. The calyptraeids, however, spin a mucous net (pfr) which is gradually rolled up with the particles which have been caught and transported to a sac on the edge of the mantle skirt over the head whence it can be collected at intervals by the radula.

The adaptations exhibited by these ciliary-feeding prosobranchs include: the increased size of mantle cavity and ctenidium; increased secretion of mucus; filamentous gill leaflets; reduction of radula, since it is used only to pull mucus with entangled food particles into the gut; reduction of salivary glands, since much mucus comes from other sources; reduction of locomotor ability, since the ciliary water currents bring food even to a stationary animal; reduction of the snout, since the animal has only a short distance to reach for its food. All these are features which will be seen later to be characteristic of the lamellibranchs, most of which are wholly dependent on ciliary feeding, and this gastropodan evolutionary sequence suggests how the highly adapted lamellibranch condition has probably come into being.

Reduction in locomotor ability reaches a climax in the vermetids which are very successful inhabitants of the littoral zone of warmer seas, where they may form a conspicuous band along the shore. The shell, coiled in the young stages, is later partly or wholly uncoiled and is embedded in or cemented to the substratum, which may be the shell of others of the same species; in some the operculum is reduced or lost and the animal can retreat far into the shell. Although some vermetids are predominantly ciliary feeders (*Dendropoma maxima*) and exhibit such associated adaptations as a long mantle cavity and a gill with elongated filaments, they all have a special pedal gland producing a viscous secretion which is elaborated into a network of cords on which plankton is trapped. Mucus-trap feeding (Fig. 196A) has become of increasing importance within the group and is related to hypertrophy of the pedal gland and the development of two extensive, pedal tentacles to weave the mucus; the gland opens at the anterior edge of the metapodium which is large while the rest of the foot is small. When laden with plankton the trap is hauled in by the radula assisted by the pedal tentacles. *Serpulorbis gigas*, occurring in the Mediterranean, casts out threads up to 30 cm long away from the shell, whilst in some social species such as *S. squamigerus*, occurring in California, a communal net is formed. In these species the gill filaments are triangular and appear to have no function in feeding, whilst the pedal gland reaches its greatest development.

The ordinary gastropod collects its food by means of the radula and, like the dentition of mammals, this shows adaptation to the type of food eaten (Fig. 193). In the most primitive type, the rhipidoglossan, which is found in such animals as keyhole limpets (*Diodora*) and top-shells (*Gibbula, Calliostoma*) the important feature is a large number of fine marginal teeth (mrt) in each row. As the radula is pulled in over the bending plane these act as a brush, sweeping small particles which are lying on the surface of the feeding ground into the mid-line, where they are caught on the recurved parts of the central teeth and so drawn into the buccal cavity. These animals are thus primarily microphagous and live on diatoms and detritus,

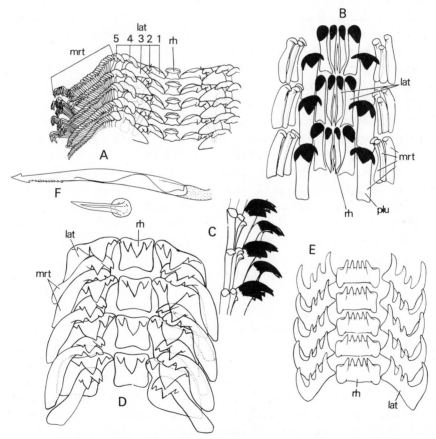

Fig. 193. Radulae. A, rhipidoglossan (*Haliotis*), dorsal view; B, docoglossan (*Patella*), dorsal view and C, side view; D, taenioglossan (*Littorina*), dorsal view; E, rachiglossan (*Buccinum*), dorsal view; F, single teeth of a toxoglossan radula, upper from *Conus*, lower from *Mangelia*. lat, lateral tooth; mrt, marginal tooth; plu, pluricuspid tooth; rh, rachidian or median tooth.

mainly of vegetable origin, which are gathered from the surfaces on which they browse. Their teeth are not adapted to rasp and their buccal mass is provided with many tensor muscles which allow its position to be adjusted so that the marginal teeth just brush the surface.

From this kind of beginning may be traced the taenioglossan radula, which is that found in *Littorina* and most lower monotocardian prosobranchs. Here the number of marginals is reduced and their mechanical power increased and, though the radula can still function in the same way as the rhipidoglossan, it can also rasp, when the teeth are brought into firmer contact with rocks or weeds. The same evolutionary trend is continued in the highest monotocardians (whelks and the like), which possess a rachiglossan radula, devoid of marginals and so incapable of brushing, and used solely for rasping, plucking or pulling fibres out of the food mass; these animals are all either carnivores or carrion feeders. Some other carnivores do not gnaw their prey as the rachiglossans do, but swallow it whole; these fall into a group known as the ptenoglossans, in which the radula forms a spiny covering to the buccal mass, with all the spines recurved. The prey is seized by the extruded buccal mass and pulled into the gut entire (*Ianthina*) (Fig. 194A). Another group includes the cones, which have vastly modified the radula so that it is composed of a series of isolated teeth which become loaded with poison and may be discharged into the prey, usually a fish, a worm or another gastropod, which is then ingested whole.

In opisthobranchs and pulmonates the use of the radula is different again. In many tectibranchs (*Philine, Scaphander*) the radular teeth are used to grasp small prey which is ingested whole; in these the teeth are not operated, as in proso-branchs, by being pulled over the bending plane; instead, they are made to erect by changing tensions set up in the membrane over the buccal mass to which they are attached by the contraction of muscles. Thus they may erect and lie down at the same position, without any other movement being involved. In nudibranchs, which feed on coelenterates, bryozoans and sponges, the radula is a rasping organ, but in one group, the sacoglossans, it is used to puncture the cellulose wall of filamentous algae and the contents of the cells are then sucked out through the hole thus made. In the pulmonates the slug *Testacella* (Fig. 194B), which is carnivorous and feeds on earthworms, has a buccal mass and spiny radula which function like those of ptenoglossans. The typical pulmonate, however, is a vegetarian and rasps algal growths, toadstools or larger plants. The radula is no longer a ribbon, but a broad band with a large number of teeth in each row, the central ones little different in size and shape from the laterals. The whole structure forms a rasp which is placed on the substratum and moved bodily over it as one might use a piece of sandpaper. The same method is employed by one group of prosobranchs, the diotocardian limpets (*Patella*), which possess what is known as a docoglossan radula (Fig. 193B, C), a narrow ribbon but, like the pulmonate radula, one in which little movement of the teeth on the underlying cuticle takes place. The buccal mass carrying the ribbon is passed outwards on to the substratum (rock with diatoms, or weed) and then driven forwards over that as one might use a plane. At the same time the radular teeth are drawn inwards over the advancing tip of the buccal mass, pulling in whatever they may have gathered. This use calls for great muscular strength so that the buccal mass of a limpet is large and powerful; it also wears out teeth rapidly so that the radular sac is very long.

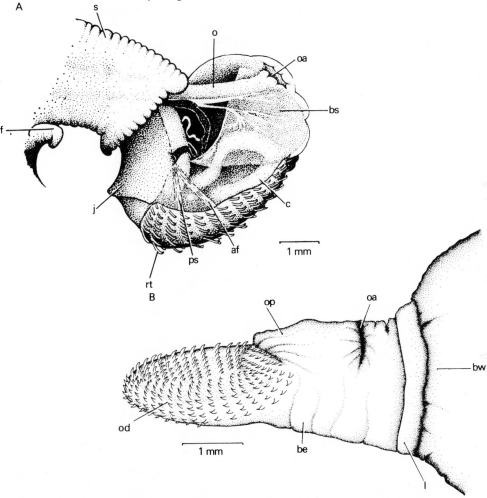

Fig. 194. A, *Ianthina janthina*: anterior end of animal from the right, showing the extroverted buccal walls forming a grasping apparatus for feeding. B, *Testacella maugei*: anterior end from the left showing a similar buccal eversion. (After Crampton.) af, flexor muscle in buccal mass; be, epithelium of buccal wall; bs, septum within buccal mass controlling blood so as to keep it turgid; bw, body wall of head; c, cartilage of odontophore; f, anterior edge of foot; j, jaw; l, lip; o, oesophagus with underlying retractor muscle; oa, opening of oesophagus; od, odontophore carrying radular teeth; op, oesophageal pouch; ps, muscle of odontophore; rt, radular tooth; s, snout.

In the above paragraphs mention has been made of carnivorous gastropods; these can use only certain kinds of animal as prey, since they are themselves so slow of movement. Their usual food is sponges, bryozoans, hydroids or tunicates – food which may be grazed as if it were a plant. A few gastropods have learned to deal with other kinds of prey, again animals which do not run away when attacked, but retire within a shell or tube, such as other molluscs or worms. Gastropods in the genera *Natica, Polinices, Nucella, Urosalpinx, Ocenebra* and related genera can bore through the calcareous shells of their prey in order to rasp the flesh; this they do partly mechanically, using the radula, holding the shell with the foot (Fig. 195A). The mechanical activity of the radula is completed by a more important chemical action, since all possess a glandular structure known as the accessory boring organ (ABO). In the naticaceans it lies on the lower lip; in the muricaceans it lies on the sole of the foot and is retractable into a pouch there. The ABO has been seen to be placed into the borehole when the snail rests after a period of active radular rasping. The secretion of the gland is not strongly acid (pH 4) and its precise action is still

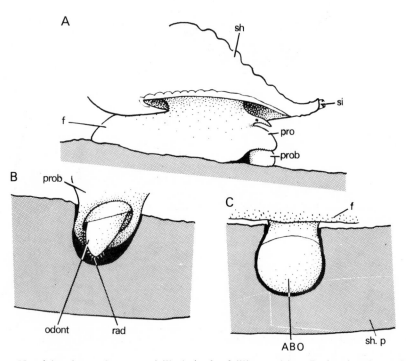

Fig. 195. *Urosalpinx cinerea*, the oyster drill; A, in the drilling position; B, showing the proboscis tip and radula in the borehole; C, showing the accessory boring organ (ABO) in the borehole. (Based on photographs by Carriker.) f, foot; odont, odontophore; pro, propodium arched over the proboscis to make it secure; prob, proboscis; rad, radula; sh, shell; sh.p, shell of prey; si, siphon.

not known, but it certainly affects the shelly material so that its mechanical removal by the radula is made easy.

A further method of feeding is that used by the pyramidellids (*Odostomia*, *Chrysallida*) which suck blood or other fluid from other animals on which they prey as ectoparasites (Fig. 196B). This they are able to do because the jaw has been rolled up to form a tube tapering at its tip like a hypodermic needle. This can be plunged into the body of the prey from the mouth, the lips of which have been modified to form a sucker at the tip of a long, mobile proboscis. The blood is sucked into the oesophagus by the action of a pump created out of part of the buccal cavity. These animals have no trace of radula.

It was mentioned above that sacoglossans feed on filamentous algae, slitting open the cells with their lance-like radular teeth and sucking the fluid contents into the gut by means of a muscular pump attached to the buccal cavity. The fate of the chloroplasts in this food is unusual; they are not digested at once (though this may happen ultimately), but retained within the cells of the digestive gland as active organelles still functioning as they did in the plant from which they came. Since light is essential for this the sacoglossan digestive gland is not the solid mass of tubules which it is in other gastropods but a loose mass ramifying throughout the superficial parts of the body. Most of its cells contain chloroplasts and this imparts a general green hue to the animals, making them both beautiful and difficult to find

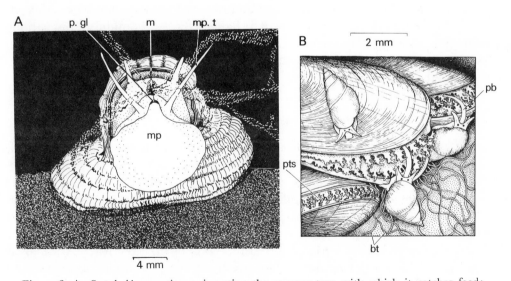

Fig. 196. A, *Serpulorbis squamigerus*, ingesting the mucous trap with which it catches food; B, *Odostomia scalaris*, feeding on small mussels. bt, byssus threads of mussel; m, mouth; mp, metapodium; mp.t, metapodial tentacle; pb, proboscis of gastropod; p.gl, opening of pedal gland; pts, pallial tentacle of mussel.

against the fronds of weed on which they live. The physiology of the captive chloroplasts may resemble that of symbiotic algae (p. 87).

A somewhat similar exploitation of an ingredient of their food is found in eolid nudibranchs (Fig. 190F). These animals feed on anemones or hydroids, clamping parts of the column wall or whole polyp heads between powerful jaws and then rasping them with their comb-shaped radular teeth. Somehow, in a way not fully understood, the nematocysts of the anemone are prevented from discharging and they pass intact into the nudibranch gut. The digestive gland is confined to the processes (cerata, sing. ceras) set on the dorsal surface of an eolid (cer), which are also richly vascularized and involved in respiration. The single, lobed tubule of the digestive gland of each ceras is divisible into three sections: (1) a basal part occupying most of the length of the ceras, lined by digestive cells and the seat of the secretion of enzymes and the uptake of food; (2) connected to the distal end of this, a narrow ciliated canal which leads to (3) the cnidosac (cn), an ovoid sac at the tip of the ceras where it opens to the outside. Nematocysts are passed through the digestive gland, along the ciliated canal and are finally ingested by the cells lining the cnidosac. Despite their journeyings they are still capable of exploding, though they may no longer react to the kind of change which brings that about in coelenterates and which it may not be in the mollusc's power to reproduce. They still react however, like all nematocysts, to mechanical pressure and this can be created by contraction of a thick layer of muscle running round the cnidosac. The behaviour of an eolid is such as to make the tips of the cerata its most obvious part – when attacked the processes erect and radiate from the animal like a hedgehog's spines so that if a predator bites it is liable to get a mouthful of discharging nematocysts escaping from the cnidosacs. The colouring of eolids is often vivid with bands of contrasting hue encircling the apices of the cerata: it is tempting to see in this a warning coloration.

Functioning of the gastropod gut

Once food is ingested, it passes along the oesophagus to the stomach. Salivary glands are primarily lubricating, though enzymes occasionally occur in their secretion. In prosobranchs the oesophagus is the seat of extensive secretion of digestive enzymes from a series of oesophageal glands. This arrangement, usually similar to that described for *Littorina*, becomes profoundly modified in the rachiglossans, in relation to the development of their proboscis, which allows the animals to probe into crevices in search of carrion, or into the body of the prey itself. A proboscis (Fig. 197, pb) is an elongation of the snout with the mouth (m) at its tip and containing the buccal mass and radula, sometimes as long as the rest of the animal. When not in use it is retracted within a proboscis sac (ps). Its presence demands elongation of the gut as well as of the snout, and this affects the oesophagus (aoe),

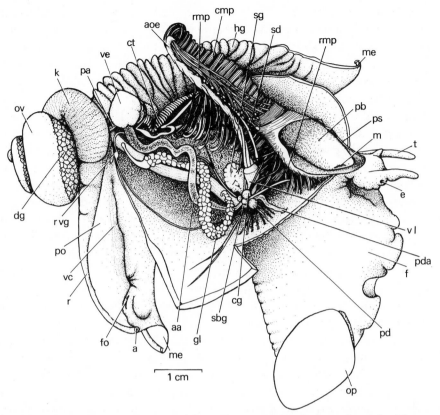

Fig. 197. *Buccinum undatum*: dissection of anterior part of body from the right. The mantle skirt has been cut in the mid-line and deflected above and below; the proboscis has been exposed by opening the proboscis sheath. a, anus; aa, anterior aorta; aoe, anterior œsophagus; cg, right cerebral ganglion; cmp, circular muscle of proboscis sheath; ct, ctenidium; dg, digestive gland; e, eye; f, foot; fo, female opening; gl, gland of Leiblein; hg, hypobranchial gland; k, kidney; m, mouth; me, mantle edge; op, operculum; ov, ovary; pa, posterior aorta; pb, proboscis lying in proboscis sheath; pd, right pedal ganglion; pda, artery to foot; po, oviduct in mantle skirt; ps, wall of proboscis sheath, cut; r, rectum; rmp, retractor muscles of proboscis; rvg, right visceral ganglion; sbg, suboesophageal ganglion; sd, right salivary duct, cut; sg, left salivary gland; t, tentacle; vc, ventral channel in capsule gland, seen by transparency; ve, ventricle; vl, valve of Leiblein.

round which, it will be recalled, runs the nerve ring (cg), a structure which has not changed its situation within the body. The tendency for the oesophagus, with its expanded glandular portion, to be pulled forward through the nerve ring as the proboscis elongated in evolution has provoked a drastic modification of structure: the glandular part has been stripped away from the dorsal food channel so as to form an apparently separate gland (the so-called gland of Leiblein) connected to the oesophagus by a duct (gl). In many rachiglossans (*Nucella, Buccinum*) the line along which the separation has taken place is marked by a modified piece of epithelium. In these animals the gland secretes a strong proteolytic enzyme, whereas in vegetarians its secretion contains carbohydrase.

In opisthobranchs and pulmonates no trace of oesophageal gland persists; instead the oesophagus often enlarges to form a crop with a gizzard, within which the food may be mixed with enzymes usually of digestive gland origin, and broken up.

The stomach shows comparable evolutionary change. The most primitive type is found in top-shells (Fig. 184B) and consists of two portions, a morphologically anterior part, roughly globular and receiving the oesophagus (oes) and ducts from the digestive gland (d.gl.d), and a posterior part, tubular, leading to the intestine (s.s). The globular part also connects with a spiral caecum (cae). Food from the oesophagus is led past the opening of the ducts of the digestive gland and is presumably mixed with enzymes from that source; it then passes along the spiral caecum and emerges from that into the globular part of the stomach, in the form of a mucous thread. Here the thread is coiled into an irregular mass which is squeezed against the cuticular gastric shield (g.sh), and the expressed, dissolved foodstuff passed to the cells of the digestive gland for absorption. The indigestible residue, mixed with much mucus, enters the tubular part of the stomach, the style sac, and is there compacted into the beginnings of a faecal rod which later passes along the intestine and, broken into lengths, is discharged at the anus. In higher forms of prosobranchs, as microphagy changes to macrophagy and a vegetarian diet to a carnivorous one, the spiral caecum is lost and the globular part of the stomach becomes a simpler bag-like structure within which digestion occurs, though the style sac persists; in opisthobranchs and pulmonates the oesophagus, expanded into a crop, largely replaces the stomach as the site of digestion and the latter is marked only by the entrance of the ducts of the digestive gland. The style sac becomes unrecognizable as the initial part of the intestine. In animals which discharge faeces to the mantle cavity the intestine is long and primarily concerned with elaborating pellets or rods which will not disintegrate in the cavity if discharged during an intertidal period, though some, perhaps even considerable, uptake of dissolved food also occurs there. With the gradual suppression of the mantle cavity which marks gastropod evolution the activity of the intestine as a mould for faecal pellets becomes less important and in many opisthobranchs it is an exceedingly short tube.

Reproduction of gastropods

The primitive diotocardians broadcast eggs and sperm and do not copulate: other gastropods, with few exceptions, do. The monotocardians have, normally, separate sexes and the males are distinguishable by their generally smaller size and by their penis. Females have always the more elaborate reproductive system since they have to make provision for the reception and storage of sperm, the provision of food for the eggs and their encasement within protective covers; in addition, many have devices for fastening the spawn to an appropriate substratum. The general arrangement of the ducts is as in a winkle, though the final product is more usually an egg capsule which is fastened to weed or rock than pelagic. In lower forms this is brought about by simple pressure of the foot; in some higher forms (*Lamellaria, Trivia*) which embed their egg capsules in ascidian colonies (Fig. 198), a hole is eroded by the radula and the egg case is then pushed into the hole by a protrusible papilla on the middle of the sole of the foot. In rachiglossans this is converted into a glandular structure. The egg capsule, still soft, escapes from the oviducal aperture in the mantle cavity and is transported round the right side of the head-foot to the sole, where it enters a special pedal gland found in females only. This grasps it and presses it against the substratum so as to convert its base into an expanded plate (Fig. 199B, b) which fits tightly against the rock and fastens it firmly; it also gives a characteristic shape to the rest of the capsular wall so that all capsules laid by a particular female have the same shape and markings. In the opisthobranchs and pulmonates the eggs are laid in jelly masses, arranged in ribbons in the former and the aquatic pulmonates, but often laid singly in humid areas by terrestrial snails and slugs. The largest land snails, such as *Achatina*, the giant snail of the tropical Old World, may have the egg enclosed in a calcareous shell. Opisthobranchs and pulmonates are hermaphrodite and cross-fertilization is the rule, each animal acting

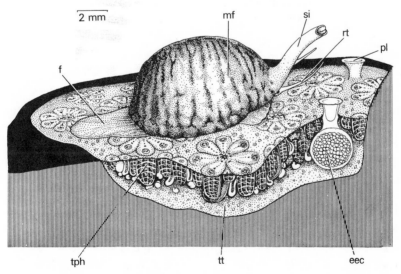

Fig. 198. *Trivia monacha*: cowrie crawling on a colony of the tunicate *Botryllus schlosseri* which has been cut vertically in the foreground to show the zooids and an egg case of the mollusc embedded in it. A second capsule projects from the tunicate in the background, right. eec, eggs in capsule; f, foot; mf, mantle fold over shell; pl, plug of mucus in egg capsule; rt, right tentacle; si, inhalant siphon; tph, pharynx of zooid of tunicate; tt, test of tunicate.

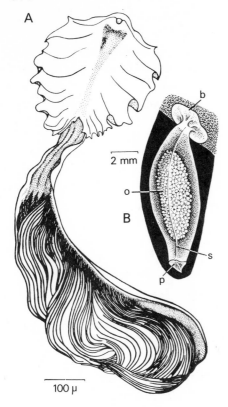

2 mm

100 μ

Fig. 199. A, spermatozeugma of *Cerithiopsis tubercularis* showing longitudinal rows of eupyrene spermatozoa fastened to the tail of a giant apyrene spermatozoon. B, *Nucella lapillus*, egg capsule. b, base, attached to rock by ventral pedal gland; o, mass of developing eggs; p, plug of mucus in mouth of capsule through which young escape; s, suture along which right and left halves of capsule (secreted by right and left walls of the capsule gland in the female reproductive duct) are joined.

simultaneously as both male and female, giving and receiving sperm. This is advantageous in that both animals are fertilized by the one act and, since snails are slow moving and very dependent on climatic conditions for activity, compensates for the fact that meetings between animals may not be frequent. These hermaphrodites are commonly autosterile but a few (some lymnaeids, physids and planorbids) can fertilize their own eggs with their own sperm. A few prosobranchs are also hermaphrodite. Of these the best known are *Crepidula* and *Calyptraea*. Whereas the latter lives in isolation *Crepidula* lives in chains (Fig. 191), each chain consisting of a group of animals, the oldest at the bottom clinging to an empty shell, the youngest at the top, the chain being a more or less permanent grouping of individuals which progress from one end to the other as they age. The youngest are male, the oldest female and centrally there are one or two in the process of changing sex and remodelling their genital ducts to carry out new processes. Metamorphosing larvae are attracted to such chains and settle on them, becoming male; if they settle in isolation they become female and may attract a settling larva to act as male. It has been shown that the development of the male cells in the gonad depends upon the secretion of a male hormone in the head; when this ceases, or if the gonad be removed from its influence, the male phase is lost.

A few prosobranch species and some pulmonates are characterized by the reduction or total absence of the penis. In pulmonates this may mean that the animal can act only as a female, though it is often able to evert the terminal part of the male duct to make an intromittent organ of a sort. In prosobranchs the lack of penis appears to have been overcome by the use of structures known as sperma- tozeugmata. Many species of prosobranch and some pulmonates produce two types of sperm, one, the normal one, called eupyrene, which fertilizes the eggs, the other, with modified nuclear condition, known as dyspyrene or apyrene, of doubtful importance in the sexual biology of the species. In aphallic species such as *Cerithiopsis tubercularis*, *Clathrus communis* or *Ianthina janthina* the apyrene spermatozoa become very large and possess many flagella; the normal sperm become attached to their surface and the composite structure is known as a spermatozeugma (Fig. 199A). They are liberated and swim into the genital ducts of females of the same species so that cross-fertilization is assured.

The primitive gastropod was marine and probably a microphagous type of diotocardian prosobranch living on a hard substratum. From this evolved the sole successful surviving types of diotocardian, the patellacean limpets, on the one hand, adapted for rock-clinging and capable of resisting intertidal periods in inhospitable surroundings, and the top-shells on the other. The latter are well on the way towards the body pattern of the monotocardians, which have become adapted for a vast variety of modes of life: herbivorous, carnivorous or parasitic; creeping, free-swimming, or floating. In floating forms (the Violet snail, *Ianthina*) the position of the animal at the surface of the sea is maintained by the formation of a float by the foot (Fig. 200, mfl); this is done by enclosing bubbles of air in mucus secreted by the pedal gland and weaving a mat of these to which, in some species, the capsules containing the eggs are attached. Free-swimming forms such as the heteropods *Carinaria*, *Pterotrachea* (Fig. 200B, C) and the pteropods use the foot, lobed in pteropods, for swimming, lighten themselves by reducing the size of the visceral mass and so of the shell, which also loses much of its calcareous matter; in heteropods, which are carnivores, the eyes become greatly enlarged, allowing the animals to capture their prey by sight.

Organization of bivalves

The bivalved molluscs, as the name suggests, are those in which the shell is composed of two parts, placed one on each side of the body. It is secreted, as in gastropods, by the epithelium of the outer face of the visceral hump and mantle skirt but calcifies from two centres instead of one. Mid-dorsally there lies between the valves an area of mantle known as the isthmus, which secretes conchiolin – the

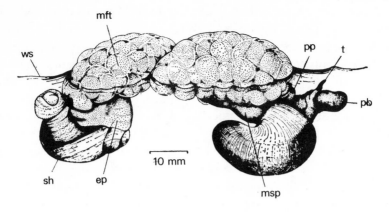

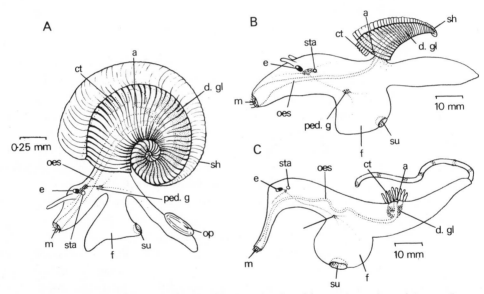

Fig. 200. Uppermost figure: *Ianthina janthina* in normal position at the surface of the sea. Lower figures: three heteropod gastropods to show gradual departure from the typical prosobranch form. Though drawn in the normal position (dorsal side up) these animals swim upside down. A, *Atlanta*; B, *Carinaria*; C, *Pterotrachea*, which has lost the shell. (Based on Ankel.) a, anus; ct, ctenidium; d.gl, digestive gland and gonad; e, eye; ep, epipodium; f, foot; m, mouth; mfl, mucous float; msp, mesopodium; oes, oesophagus; op, operculum; pb, proboscis; ped.g, pedal ganglion; pp, propodium; sh, shell; sta, statocyst; su, sucker on foot (characteristic of males only); t, tentacle; ws, surface of water.

organic matrix within which the calcareous salts of the shell are deposited – but where no calcification takes place. This gives rise to a structure called the ligament, the elastic properties of which force the valves apart. Originally in line with the two valves which flank it, the ligament tends to move so as to lie either external to the shell or internal to it when it may be called a resilium. In the former case, approximation of the valves stretches the ligament, in the latter it compresses it; in both cases the elasticity of the ligament forces the valves apart when the deforming force is removed, and the shell gapes. The deforming force is normally applied by two powerful muscles, the adductors, one placed anteriorly and the other posteriorly; when a bivalve dies these muscles can no longer contract and the ligament forces the shell open. A dead bivalve has, therefore, always a gaping shell.

Closure of the shell is a protective action against predators or adverse environmental conditions, which calls for two component activities on the part of the adductors. The first is the ability to contract rapidly so as to get the shell closed against a predator, the second is the ability to maintain a prolonged closure, over an intertidal period, for example. Muscles adapted for the one kind of contraction are not normally adapted for the other; it is not surprising, therefore, to find that adductor muscles usually contain two types of muscle fibre, one called the quick muscle, able to contract rapidly in a twitch but unable to sustain the contracted state, the other known as the catch muscle, not able to contract rapidly but when contracted, able to stay in this state for extended periods with minimal expenditure of energy. These two types of fibre, histologically distinct, are usually intermingled in the adductors so that the muscle appears homogeneous; in some bivalves, however, like *Nucula* (Fig. 202) and *Pecten* (Fig. 210) the two types of fibre are segregated and two visibly different parts are distinguishable.

The shell acts as a skeleton for the origin of muscles, as a protection against predators and in burrowing forms it also helps to keep mud or sand out of the mantle cavity and to anchor the animal. Boring bivalves, however, run little risk from these sources and to them incomplete closure of the shell is not serious; with this slackening in the force of natural selection the ligament tends to become less well formed and may, as in the group Adesmacea (the piddock *Pholas*, the shipworms *Teredo*, *Xylophaga*), be vestigial or lost.

The original bivalves probably arose from an animal comparable in many ways to the ancestral gastropod, with a conical or dome-shaped shell and a mantle skirt hanging downwards over head and foot like a curtain. It crept over the substratum on the sole of the foot. From the beginning the bivalve line has relied on a microphagy not based on the use of a radula, with the same adaptations becoming obvious as have already been listed for ciliary-feeding gastropods. A second evolutionary trend was towards a burrowing habit in adaptation to which the foot lost the original creeping sole and became axe-shaped, forming a narrow, wedge-shaped structure which could be insinuated into the substratum, swollen at its tip to form an anchor up to which the body of the bivalve could then be pulled. With the adoption of such a mode of life and dependence on microphagy modification of the mantle cavity was essential to ensure that its contents (upon which the ciliary food collecting depends) are neither damaged nor clogged by incoming particles of substratum. This has been achieved by extending the mantle skirt ventrally on each side of the body so that it wraps over it completely and hides it from view. As part of the adaptation for ciliary feeding the head failed to develop and the foot and such later developments as siphons are the only parts of the body which can be thrust out from between the two halves of the mantle skirt and from the shelter of the shell.

The edge of the mantle skirt is marked by three parallel ridges, an inner, a middle and an outer (Fig. 201). The outer ridge (o.m.f) secretes the horn-like, periostracal external covering (per) as well as the outer calcareous part of the shell (o.sh). The middle ridge (m.m.f) is primarily sensory and, in the absence of sense organs such as those on the gastropod head, has become important; it is frequently drawn out into short tentacles to acquire fuller exposure to the environment. The innermost ridge (i.m.f) forms a structure comparable to the velum at the periphery of the hydrozoan medusa and, like it, concerned with modifying the movement of water. In the primitive bivalve the margins of the mantle skirt run, wholly separate from one another, from the anterior end of the ligament dorsally to the posterior end of the ligament dorsally, with only a narrow gap between them through which the foot may be protruded and through which water enters and leaves for feeding and respiratory functions. Initially (Fig. 202) the inhalant current (I) comes in anteriorly and the exhalant current (E) emerges posteriorly after washing over the gills (ct) and the various apertures which discharge to the mantle cavity (the nut-shell *Nucula*).

This arrangement, however, makes functional sense only so long as the bivalve

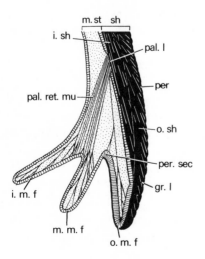

Fig. 201. Section across margin of mantle skirt and shell of a bivalve. gr.l, growth line; i.m.f, inner mantle fold or velum; i.sh, inner layer of shell; m.m.f, middle mantle fold, sensory; m.st, mantle skirt; o.m.f, outer mantle fold, secreting periostracum; o.sh, outer layer of shell; pal.l, pallial line, where pallial retractor muscles (pal.ret.mu) attach to shell; per, periostracum over outer surface of shell; per.sec, point where periostracum is secreted; sh, shell.

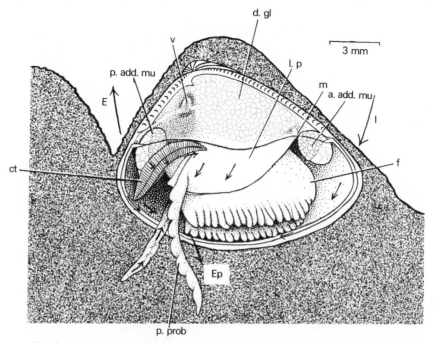

Fig. 202. *Nucula*, seen from the right, in its natural position in the substratum after removal of the right valve and the right half of the mantle skirt. Arrows show the situation of the exhalant (E) and inhalant (I) water currents, and the direction of ciliary currents within the mantle cavity. The arrow marked Ep indicates where pseudofaeces are ejected. (Based on Yonge.) a.add.mu, anterior adductor muscle, its anterior part quick, its posterior part a catch muscle; ct, ctenidium; d.gl, digestive gland; f, foot, showing right and left halves with sole between; l.p, right outer labial palp; m, position of mouth; p.add.mu, posterior adductor muscle, its posterior part quick, its anterior part a catch muscle; p.prob, left palp proboscis, an extension of outer labial palp; v, ventricle of heart, seen by transparency.

moves over the surface of the substratum or is only slightly buried. Should the mode of life involve deeper penetration of the substratum an anterior inhalant current would be disadvantageous in introducing too much particulate matter into the mantle cavity. In adaptation to this and to the posterior location of the ctenidia in the mantle cavity the inhalant stream comes to lie posterior (Fig. 203A); it still (as in gastropods) is ventral, and the exhalant stream (E) dorsal, but the two are often contiguous with only the posterior end of the gill between (the mussels *Mytilus*, *Anodonta*). Further modification of the mantle edge occurs with further adaptation for burrowing and boring. In the former way of life the danger of contamination of the mantle cavity by excess silt is ever present and there is a trend towards fusion of the right and left edges of the mantle skirt to minimize it; obviously, at least three openings must be retained, the two posterior ones for the entry and exit of the respiratory and feeding stream of water, and an anteroventral gape through which

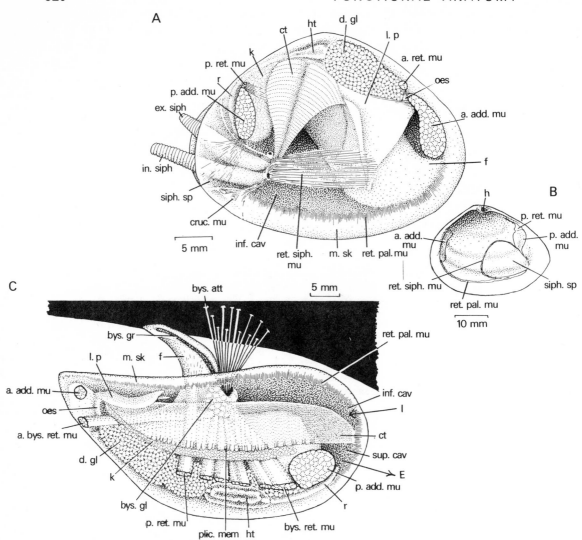

Fig. 203. A, *Scrobicularia plana*, removed from shell and seen from right side; B, *Scrobicularia plana*, right shell valve, internal view; C, *Mytilus edulis*, removed from shell and seen from right side. Arrows show the situation of the exhalant (E) and inhalant (I) currents to the mantle cavity. a.add.mu, anterior adductor muscle and its attachment to shell; a.bys.ret.mu, anterior byssus retractor muscle; a.ret.mu, anterior pedal retractor muscle and its attachment to shell; bys.att, attachment of byssus threads to substratum; bys.gl, byssus gland in foot; bys.gr, byssus groove leading byssus secretion along foot; bys.ret.mu, byssus retractor muscle (only the most posterior of a series is labelled); cruc.mu, cruciform muscle; ct, ctenidium; d.gl, digestive gland; ex.siph, exhalant siphon; f, foot; h, hinge; ht, heart; inf.cav, infrabranchial part of mantle cavity; in.siph, inhalant siphon; k, kidney; l.p, outer labial palp; m.sk, mantle skirt; oes, oesophagus; p.add.mu, posterior adductor muscle and its attachment to shell; plic.mem, plicate membrane (small folds where outer gill meets mantle skirt, said to be main site of oxygenation of blood); p.ret.mu, posterior pedal retractor muscle and its attachment to shell;

the foot may be thrust for locomotion. In boring animals little grit may be present in the burrow to enter the mantle cavity but fusion is often as complete as in burrowing animals (see Figs 205 and 206). Neither type can get sufficient food from the water in the burrow or in the interstitial spaces of sand or mud: they must have more or less continual access to open water. If the animals are no more adapted for burrowing than those just described they will obviously not be able to do more than hide under the surface of sand or mud with the posterior end of the mantle and shell breaking the surface to allow entry and exit of water; only if at least temporary interruption of this is not critical may the animal move deeper. The more highly adapted burrowing and boring bivalves, however, have evolved siphons.

These are tubular prolongations of the mantle edges around the inhalant and exhalant areas at the posterior end (Figs 203A and 208, ex.siph, in.siph). They may be formed from gutter-shaped extensions of (1) the inner ridge only; (2) the inner and middle ridges; or (3) the whole mantle edge. In the first case union of the right and left lobes may be partly by ciliary junctions (*Yoldiella*) or by tissue fusion (*Yoldia limatula*); in the others the tissues of the right and left component are always fused. When the middle ridge of the mantle skirt is involved in the formation of siphons it may produce sensory tentacles at their tips (the cockle *Cardium*, the gaper *Mya*), and when the outer ridge contributes it secretes periostracum over them as it does elsewhere over the shell (*Mya*). When only the inner ridge is involved neither tentacles nor periostracal covering can be present (the tellins *Scrobicularia*, *Tellina*). With the help of siphons the mollusc may burrow or bore to a considerable depth so as to be safe from the attention of most predators and yet retain the connection with the external medium necessary for feeding, respiration and the discharge of excretory and genital products.

Within limits, the siphons are expendable in that they may be regenerated if lost, though during the period of re-growth the bivalve is at some disadvantage. Danger of loss is minimized in that the siphons are commonly retractile within the shelter of the shell by retractor muscles (ret.siph.mu). These are local exaggerations of a muscle (ret.pal.mu) which is present along the entire length of the mantle edge, from which fibres run to a nearby origin on the inner surface of the valves. On their contraction, the edge is pulled inwards and not trapped between the valves as they close. The attachment of the fibres to the substance of the shell interferes with the process of calcification so as to make the line along which it occurs, the pallial line (Fig. 203B, ret.pal mu), visible on the inner side of a cleaned shell, running roughly parallel to the edge. If siphons are developed the pallial line bends anteriorly from

r, rectum; ret.pal.mu, pallial retractor muscle and its attachment to shell; ret.siph.mu, siphonal retractor muscle and its attachment to shell; siph.sp, siphonal space (space between edges of mantle skirt but external to mantle cavity into which siphons may be withdrawn) marking the pallial sinus on the shell; sup.cav, suprabranchial part of mantle cavity.

the shell edge near the posterior end to form a bay, the pallial sinus (siph.sp). This marks the area within which the siphons are accommodated on retraction and its depth is roughly parallel to the extent to which they are developed. It is thus possible for a palaeontologist to tell, by inspection of a fossil valve, whether the animal was siphonate or not, how long the siphons were likely to have been, sometimes the angle at which they were extended, and hence to make a reasonable deduction about its mode of life. The two adductor muscles (a.add.mu, p.add.mu) may also be regarded as expansions of the pallial muscle. All siphonate forms have a pallial sinus but not all are capable of totally retracting the siphons and closing the shell over them; in some the siphons merely shorten very much and have the greater part of their length within the shell, the rest projecting through a gap. This is particularly obvious in animals which have large siphons covered by periostracum as in *Mya* (called a "gaper" for this reason), and which live deeply buried so that even a slight retraction removes the siphons from the grasp of a potential predator.

Feeding of bivalves

Though all bivalves are microphagous they do not all collect their food in the same way. There are three fundamental methods: (1) by the labial palps, employed by about a dozen genera of primitive bivalves known as the protobranchs; (2) using highly modified ctenidia to trap small animals, found in a second small group of about ten genera known as the septibranchs; (3) using the cilia on the ctenidia to create a water current which is then strained of minute suspended particles, found in the groups Filibranchia and Eulamellibranchia, which constitute by far the major part of the class (about 400 genera). Some evolutionary advance, adapting the ctenidia for more efficient food gathering, is obvious in this last group. Before considering it, however, it is necessary to say something about the method employed by the protobranchiate bivalves.

These animals live near the surface of soft marine substrata and maintain contact with the overlying water directly (*Nucula*, Fig. 202) or by means of siphons (*Nuculana*, *Yoldia*); in the former water enters at the anterior end of the mantle cavity and leaves posteriorly, whereas in the siphonate forms it enters by the ventral and leaves by the dorsal siphon, both placed posteriorly. In both types the ctenidia (ct) are comparatively small structures and conform to the primitive molluscan pattern of an elongated axis carrying a double row of triangular leaflets. The axis is attached to the body in the posterior part of the mantle cavity with the leaflets projecting into the cavity. It usually runs obliquely from an anterodorsal to a posteroventral position, the angle with the vertical being least in *Nucula* and greatest in the siphonate forms. The flat, apposed sides of the leaflets bear lateral cilia sucking water from the anteroventral inhalant part of the mantle cavity into the posterodorsal exhalant part. As in gastropods, cilia set along the inhalant edge of the leaflets (frontals) and others along the exhalant edge (abfrontals) clear particulate matter from the gill and deposit it in the hypobranchial (inhalant) part of the mantle cavity, where it forms masses of material held together by mucus from goblet cells on the ctenidia. These look like faeces, for which reason they are called pseudofaeces. They fall off the gills on to the mantle skirt where ciliary currents convey them to ventral or posterior edges; at intervals the animal contracts the adductor muscles and the force of this blows the pseudofaeces away (Ep).

The gills of protobranchs have little to do with feeding, though they do trap some particles. Feeding is achieved primarily by means of the labial palps (l.p), structures derived from the exaggerated ciliated ring (known as the velum) by which the larva swims and collects its food. There are two on each side connected across the mid-line to the corresponding ones on the other side of the body; the connection is anterior to the foot (f), but on each side the two palps lie alongside that structure. The mouth (m) lies between the anterior and posterior pair medially and anteriorly. Each outer palp carries, near its posterior end, a tentaculiform outgrowth, the palp proboscis (p.prob) – which can be stretched beyond the valves, into or on to the substratum, where its ciliated and glandular surface picks up particles. These travel down the proboscis and pass on to the apposed inner faces of the palps, which are ridged and grooved to form a sorting apparatus; if the particles are small they gradually move over the palp towards the mouth and are ingested; if they are larger, however, they will be carried to the posterior tip of the palp off which they fall to join the pseudofaeces and be ejected from the mantle cavity. This sorting seems to be based on size alone and bears no relation to the edibility of the particles. The protobranchs, therefore, though microphagous, depend predominantly on the palp proboscides for the collection of food.

The filibranch and eulamellibranch bivalves (sometimes grouped together as the polysyringian bivalves) collect their food in the same way as the ciliary-feeding gastropods, by straining the water passing through the ctenidia. To augment both the amount of water exposed to filtration and the area over which this occurs a radical alteration has been made in the form of the gills, whilst subsidiary changes have been made to increase the efficiency of the filtration. The primary alteration in the ctenidium converts each originally triangular plate into a V-shaped filament (Fig. 204A) attached by the apex of one limb to the ctenidial axis (ct.ax) and anchored by ciliary contacts at the apex of the other to the mantle skirt, the side of the foot or the visceral mass. Each axis with its double row of leaflets therefore forms a W-shaped structure in cross-section; the space (sup.cav) between the arms of the V's is suprabranchial, exhalant, and connected to the exhalant area of the mantle edge; the space ventral to the W (inf.cav) is infrabranchial, inhalant, and connected to the inhalant area of the mantle edge. In more primitive bivalves of this type the neighbouring filaments are linked to one another only by interlocking clumps of cilia (cil.j) and the descending (des) and ascending (as) halves of one and the same

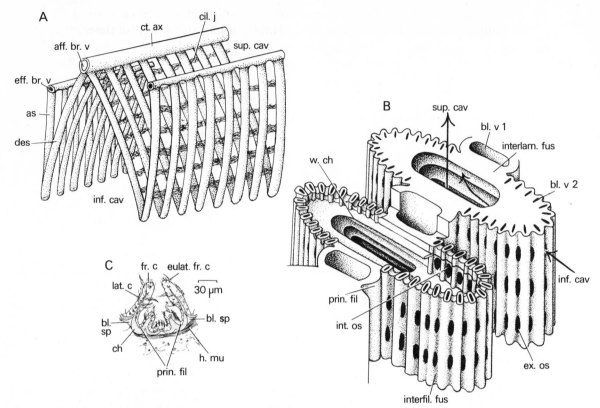

Fig. 204. Stereograms to show the structure of A, filibranch gill and B, eulamellibranch gill. In the former a short length of ctenidial axis is shown with some of the double series of V-shaped filaments arising from it. The number of ciliary junctions linking neighbouring filaments has been reduced to simplify the drawing. In the eulamellibranch a block of tissue from an outer gill has been isolated by cuts parallel to the ctenidial axis; the nearer face therefore comprises the ascending parts of the filaments and the more remote face their descending parts. The gill is folded and the filaments at the base of the folds enlarged so that it is heterorhabdic. Neighbouring filaments fuse at regular intervals, leaving ostia through which water passes. Asending and descending filaments also fuse at intervals making horizontal partitions partly blocking the suprabranchial cavity. The nearer fold is cut at a level between these partitions, the further fold along one. Arrows show the direction of water currents. C, *Solen*. A principal filament, flanked by two ordinary filaments, cut in transverse section to show arrangement of cilia on the gill. aff.br.v, afferent branchial vessel in ctenidial axis; as, ascending filament; bl.sp, blood spaces; bl.vi, blood vessel connecting with afferent branchial; bl.v2, blood vessel connecting with efferent branchial; ch, chitinous support for filaments; cil.j, ciliary junction; ct.ax, ctenidial axis; des, descending filament; eff.br.v, efferent branchial vessel; eulat.fr.c, eulaterofrontal cirri, beating towards apex of filament; ex.os, external opening of ostium to infrabranchial chamber of mantle cavity; fr.c, frontal cilia, beating along filament (through the paper); h.mu, horizontal muscle; inf.cav, infrabranchial part of mantle cavity; interfil.fus, interfilamentary fusion; interlam.fus, interlamellar fusion; int.os, internal opening of ostium to suprabranchial chamber of mantle cavity; lat.c, lateral cilia, beating towards base of filament (in plane of paper) to create water current; prin.fil, principal filament; sup.cav, suprabranchial part of mantle cavity; w.ch, water channel within gill.

filament are not connected at all save at the point of inflexion; this gives rise to a gill of delicate construction known as the filibranch type and seen in the edible mussel *Mytilus*. In more advanced bivalves (Fig. 204B) neighbouring filaments fuse with one another at definite levels (interfil.fus), so reducing the interfilamentar spaces, which are long, narrow slits in the filibranch type, to rows of apertures like port-holes (ex.os, int.os). At the same time the ascending and descending halves of a certain number of filaments unite with one another across the piece of suprabranchial cavity which lies within the gill (interlam.fus). This gives greater complexity of structure but much greater firmness: it is known as the eulamellibranch type. Within the grade further advance may lead to the plication of the gill surface and this is usually accompanied, for mechanical reasons, by an enlargement of the filaments lying at the base of the groove between folds (prin.fil). In addition to these changes the ctenidial axis comes to lie more horizontally and so to increase in length.

The presence of folds not only increases the surface area of the gill and so the current-producing power but also allows a coarse filtration of particles to be achieved by the folds, in addition to the fine one due to the filaments. For this reason folding is most commonly found in bivalves inhabiting a coarse substratum in which the entry of large particles is most probable.

Filibranch and eulamellibranch gills filter food in similar fashion. Water is driven from the inhalant to the exhalant parts of the mantle cavity by lateral cilia (Fig. 204C, lat.c), set along the sides of the filaments in filibranchs, set in the pore-like openings (ostia) of the eulamellibranchs. The water is filtered by rows of laterofrontal cirri (eulat.fr.c) set along the length of the filaments, which flick particles from the water on to the surface of the filament facing the current, that is, facing the inhalant part of the mantle cavity. These cilia are compound cirri and when examined with a scanning electron microscope are found to have a pinnate structure which greatly increases their catching power. On the surface of the filaments the particles come under the influence of frontal cilia (fr.c) which are beating either ventrally (towards the apex of the V's of each gill) or dorsally (towards the ctenidial axis, the mantle skirt or foot) according to the species being examined. Mucus must play some part in trapping the particles and keeping them close to the surface of the gill; its precise role is uncertain but it does not seem to act by forming a continuous sheet over the surface as in tunicates and Amphioxus and in the prosobranchs *Crepidula* and *Calyptraea*. Particles collected by the gills are forwarded to the labial palps whose rigid faces bear a complex array of ciliary currents which must exert some control over what reaches the mouth and what is rejected. At one time both gills and palps were thought to exercise a strict selection of particles but recent work has made this role less certain. Single large particles or groups of small particles clumped in mucus (which tend to be treated as one large particle) fall off the gill edge, join material rejected by the palps and contribute to the pseudofaeces.

The filtration of water by bivalves is very efficient: American oysters (*Crassostrea virginica*) can filter up to 37 l h^{-1} at 24°C and may strain particles down to 1 μm in size.

The last group of bivalves to be considered is one known as the septibranchs which includes animals grouped into a small number of genera and living in deeper waters; they are carnivorous. A carnivorous bivalve may sound a highly improbable combination and the group cannot be regarded as successful; like several other rather inefficient animal stocks, however, the animals manage to retain their way of life by living in impoverished habitats where competition is reduced because more labile stocks have successfully adapted for life in more strenuously competitive situations. The septibranch mantle cavity is subdivided into anteroventral and posteroventral parts by an oblique partition lying between the foot medially and the two halves of the mantle skirt laterally, though the separation is not complete since the septum is perforated by pores (*Cuspidaria*) or branchial sieves (*Poromya*). Pores are simple, ciliated apertures; branchial sieves are openings crossed by about half-a-dozen ciliated filaments, almost certainly the vestiges of a ctenidium. The septum is muscular, the muscles being attached to the shell so that it can be raised or lowered within the mantle cavity. When this happens there are obvious changes in the relative volumes of the lower inhalant half and the upper exhalant half, the former being increased in volume when the septum is raised and decreased when it is lowered. The pores and sieves allow the septum to behave as if it were a complete partition when it moves dorsally, but as a leaky one when it moves ventrally. This has a double cause: the presence of dorsally directed valves on the pores, and the fact that the septum moves dorsally abruptly but is lowered slowly and gently. On lowering, water passes dorsally through the pores into the exhalant chamber; on raising, however, water is sucked into the mantle cavity by way of the inhalant siphon in a moderately forceful manner and an equal volume is simultaneously expelled from the exhalant part. This constitutes the feeding mechanism since the inward current is sufficiently powerful to suck into the mantle cavity any small enough animal (particularly if moribund or dead) in the immediate neighbourhood of the mouth of the inhalant siphon. To facilitate this, the opening is relatively larger than in a ciliary-feeding bivalve. Once within the mantle cavity the food is grasped by the labial palps, which are more muscular than is usual in bivalves, and thrust into the oesophagus.

Functioning of the bivalve gut

Once food has been gathered by a bivalve, whether by palp proboscis or ctenidium, it is sorted by the labial palps into large particles which are rejected and small ones which are passed to the mouth where they enter the oesophagus, along which they are propelled by cilia to the stomach. This method of moving food and faecal matter applies to the entire length of the alimentary canal, which is markedly deficient in muscle, and it may be correlated with the minute size of the food particles.

The successful functioning of the bivalve gut depends upon the stomach, which possesses the same general structure as in a gastropod; it is connected to a digestive gland occupying much of the visceral mass and extending into the dorsal parts of the haemocoel within the foot. The stomach is usually globular, with the oesophageal opening anteriorly, ducts from the digestive gland opening to the right and left sides, an extension (the dorsal hood) dorsally and the style sac, from which the intestine arises, originating ventrally at the posterior end. Anteroventrally on the right there is a caecum, primitively spirally coiled, and associated with some ducts from the digestive gland, which corresponds to the similar structure of the primitive prosobranch gastropods.

The typical bivalve (all save protobranchs) possesses a structure known as a crystalline style (Fig. 184A, c.s); this lies in the style sac (s.s) and projects forwards and dorsally into the stomach where it ends with its tip thrust against an uprising ridge of cuticle, the gastric shield (g.sh), placed on the left side. Cilia on the wall of the style sac rotate the style and drive it slowly out of the sac against the gastric shield; here it gradually dissolves in the stomach contents. To renew the style, secretion continues within the style sac, where two ridges, the major and minor typhlosoles (t_1, t_2), run lengthwise, separating the chamber in which the style lies from the intestinal groove running along it from stomach to intestine (int.gr). These ridges are the site of secretion of the style substance, largely mucus with adsorbed diastatic enzymes, and also of the ciliary currents driving the style towards the stomach. Its solution there liberates the enzymes associated with it. The style has another role to fill in the functioning of the stomach: food from the oesophagus enters the stomach in the form of a mucous string with embedded particles (f); this becomes wrapped round the head of the style which acts as a capstan, winding the food around it. The food is thus drawn into the stomach and simultaneously associated with diastatic enzymes. Occasionally, as in the freshwater zebra mussels, *Dreissena*, and the wedge shells, *Donax*, the style sac is completely separate from the intestine.

The further treatment of the food involves it being sent into the ducts of the digestive gland, where the bulk of it is ingested by cells and digested intracellularly. The elaborate sorting of gills and palps, ensuring selection of the smallest particles only, is preparatory to this, and a further sorting takes place on the walls of the stomach, though the mechanism by which the selected particles are forwarded to the ducts is still not wholly obvious. In animals such as the tellins, *Tellina*, or the cockles, *Cardium* – and in most others – one of the typhlosoles separating style sac from intestinal groove extends across the floor of the stomach on the right side and thence curves across the anterior wall to the left. On both sides it is related to ducts of the digestive gland (d.gl.d) and, particularly on the right, sends a mobile tongue-like extension into the duct. This is richly ciliated, the direction of the ciliary beat being into the ducts (pecked arrows in Figure); the intestinal groove, which lies alongside the typhlosole, carries a strong ciliary current directed towards the intestine. The apparent duct into which the tongue of typhlosole just described enters, is a vestibule from which lead the ducts proper of the digestive diverticula; these have outwardly beating cilia (dotted arrows in Figure), so that inward movement of food particles is perhaps due to a compensation current. It has been suggested that the absorptive activities of the cells lining the diverticula also aid suction of material into the ducts from the stomach. However achieved, it is clear from feeding experiments that (except in protobranchs) fine particulate matter, sieved from the current in the mantle cavity by the gills and food material in solution are ultimately ingested by the cells which form by far the commoner of two types lining the tubules of the digestive diverticula. The particles are taken into vacuoles and digested there. Soluble matter is probably the sole uptake in protobranch bivalves.

The diverticula contain a second type of cell, usually located in the crypts of the tubules and devoid of vacuoles, though often ciliated. The function of these remains problematical. When first discussed they were regarded as immature cells which would later replace the absorptive type. More recent work has shown that the digestive gland is the source of enzymes which pass to the stomach and bring about digestion there. This may be a phasic activity of the absorptive cells or may be the function of the cells in the crypts. Still a third activity of the digestive diverticula of bivalves, in which these cells might be concerned, is excretion; for this the cells are adapted, like corresponding ones in the gastropod digestive gland, by their broad base lying against the intertubular haemocoelic spaces from which excretory matter might be absorbed.

Feeding and digestion have recently been re-examined in bivalves and shown to be linked, in a number of animals, with environmental cycles. In many living in intertidal situations the animal exhibits a feeding cycle during high tide periods when a series of adductions of the valves pumps into the mantle cavity water which can then be strained by the gills. Later this material is passed to the stomach where it undergoes some digestion and travels to the digestive gland and is absorbed and the digestive processes are completed. At the end of this phase of activity the cells in the digestive gland degenerate and are later regenerated. There are thus phases of secretion, absorption, disintegration and regeneration. In some genera these follow one another precisely in relationship with pumping and an external rhythm, in others they may partly overlap, in still others they may all be going on simultaneously.

Particles which have not entered the ducts of the digestive diverticula and waste matter from that source ultimately enter the intestinal groove and pass into the intestine, along which they are carried by the beat of cilia on the epithelium. These particles are not necessarily lost as food, because the lumen of the intestine contains amoeboid cells which may phagocytose them, digest them intracellularly and transport them round the body. Vast numbers of amoebocytes also lie in the connective tissue around the intestinal wall. Ultimately the indigestible residue is discharged from the anus and leaves the mantle cavity by the exhalant opening. In many bivalves the intestine is flung into extensive coils which lie in the visceral section of the foot; the function of these is not known, unless it be merely to provide a long enough passage for the amoebocytes to make an effective attack on the food passing along.

In some genera, however (the file shell *Lima, Ostrea, Scrobicularia, Mya*), enzymes are secreted from the wall and some extracellular digestion occurs along with

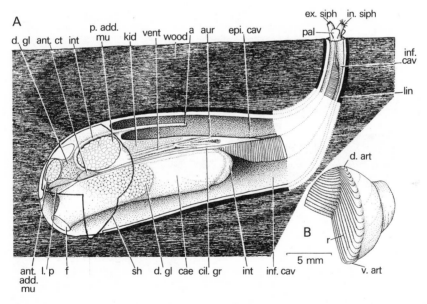

Fig. 205. A, Diagrammatic representation of a shipworm (*Teredo*) lying in its burrow in a piece of wood. Dotted lines indicate a length of the animal not drawn. B, left shell valve, outer surface. a, anus; ant.add.mu, anterior adductor muscle; ant.ct, anterior part of ctenidium linked to the main posterior part by a ciliated groove; aur, auricle of heart; cae, caecum of stomach seen through body wall; cil.gr, ciliated groove linking two parts of ctenidium; d.art, dorsal point of articulation between shell valves; d.gl, digestive gland; epi.cav, epibranchial or exhalant part of mantle cavity; ex.siph, exhalant siphon; f, foot; inf.cav, infrabranchial or inhalant part of mantle cavity; in.siph, inhalant siphon; int, intestine seen by transparency; kid, kidney; lin, calcareous lining of burrow; l.p, labial palp; p.add.mu, posterior adductor muscle; pal, pallet; r, ridge on shell used in making burrow; sh, outline of left valve of shell; v.art, ventral point of articulation between shell valves; vent, ventricle of heart.

uptake of the digested food. Whether this is the explanation of the coiling of the intestine, whether mid-gut digestion is a widespread phenomenon and whether other functions such as osmoregulation and excretion are carried out here are all points which still have to be made clear.

The food of most bivalves is bacteria, dinoflagellates, diatoms and similar phytoplanktonic organisms, and probably also includes dead organic particles of similar size. This may be collected from suspension in the water (*Mya, Ensis, Mytilus*) or sucked up by a mobile inhalant siphon moved over the surface of the substratum (*Tellina, Scrobicularia*). There are one or two species, however, which have more special ways of feeding. One of these is the shipworm (*Teredo, Xylophaga*) which bores as a larva into wooden piers, floating trees, or wooden ships and persists in this habitat for the remainder of its life (Fig. 205A); since the hole leading into the burrow is made by the entering larva and is not thereafter enlarged, the surface

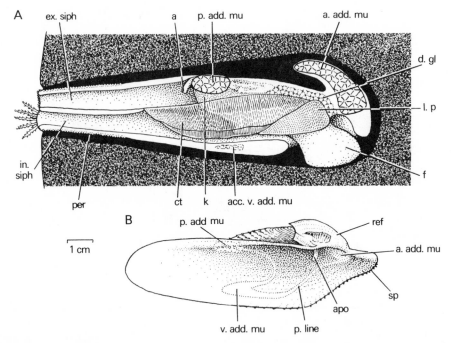

Fig. 206. *Pholas dactylus*. A, whole animal *in situ* in its burrow, seen from the right; the shell is not shown. B, the left valve, showing its inner surface. It is placed so that it would be correctly situated if superimposed on A. a, anus; a.add.mu, anterior adductor muscle and its scar on the shell; acc.v.add.mu, accessory ventral adductor muscle; apo, apophysis; ct, ctenidium; d.gl, digestive gland; ex.siph, exhalant siphon; f, foot; in.siph, inhalant siphon; k, kidney; l.p, labial palp; p.add.mu, posterior adductor muscle and its scar on the shell; per, periostracum on siphonal tube; p.line, pallial line; ref, reflected part of shell; sp, spines on outer surface of valve used in boring; v.add.mu, scar of ventral adductor on shell.

of the wood may appear sound, though the deeper parts are riddled by a mass of burrows. Neighbouring burrows never join. Boring is done by sharp-edged ridges on the anterior face of each valve (Fig. 205B, r); the bivalve grips the wall of its tube with the small, sucker-like foot (f) and then, by appropriate movement of the muscles which run from foot to shell rocks the valves upwards and downwards in a rotary movement during which the edges scrape particles from the wood and so slowly enlarge the burrow. The valves are small and cover only the extreme anterior end of the animal which thus pushes slowly through the wood: the much greater posterior part of the body is not in direct contact with the wood but is covered by a thin secretion of calcareous material which adheres to the wood and forms a lining to the tube (lin). The wood is ingested and stored in a special caecum (cae) attached to the stomach and is passed to the digestive diverticula for ingestion and digestion. Though wood may form the principal intake of the shipworm it can never be its total food since it is almost certainly deficient in nitrogen and vitamins. Just as wood-eating insects supplement their diet to supply these deficiencies by keeping protozoans or a mycetome (a group of special cells containing symbiotic micro-organisms) within the body, the shipworm still takes phytoplankton in limited quantities for the same reason.

There are other bivalves like the piddock (*Pholas*, Fig. 206) and the date mussel (*Lithophaga*) which can bore into rocks. It is usually assumed that the same mechanism as the shipworm uses to make its burrow in wood allows these animals to excavate rock, but there is evidence that in *Lithophaga* an acid secretion helps to dissolve the calcareous rocks into which the animals burrow. All rock borers are inevitably dependent on outside sources of food, and perhaps for this reason have wide openings to the burrows in which they lie.

A second group of eulamellibranch bivalves with unusual feeding devices is that of the giant clams of coral reefs (*Hippopus*, *Tridacna*). These are related to the ordinary edible cockles (*Cardium*) of European shores, but are immobilized, sometimes by their immense size, sometimes by being fastened to the coral rock on which they lie by a tuft of byssus threads (Fig. 207, bys) secreted by a special gland on the foot and similar to those by means of which the common mussel (*Mytilus*) is anchored. Their body has undergone a curious distortion, so that although the foot (f), with its byssus, and the viscera retain their normal situations in relation to one another, the hinge of the shell (h) has moved ventrally so that it lies near the byssus and the siphons (ex.siph, in.siph) have moved dorsally so as to occupy the position where the hinge would normally be. This migration has the effect of bringing the siphons to the uppermost part of the animal, where they lie between the edges of the mantle skirt (m.e) expanded into great masses of flesh, exposed freely between the valves. Here, they present a great area to the light and house many symbiotic algal cells which are thus enabled to carry on an active photosynthesis and reproduction, presumably using carbon dioxide, phosphates and nitrates supplied by the bivalve

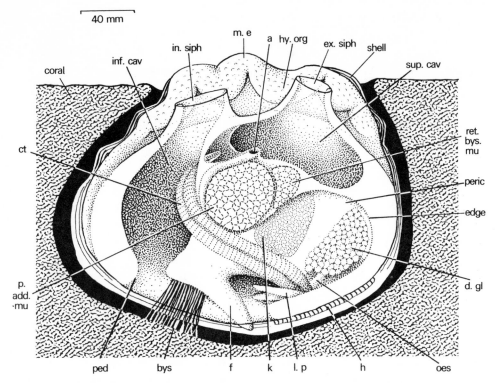

Fig. 207. *Tridacna*, seen *in situ* in its burrow in coral, from the right after removal of the right valve. (Based on Yonge and Purchon.) a, anus; bys, byssus threads attaching to coral; ct, ctenidium; d.gl, digestive gland; edge, dorsal limit of infrabranchial part of mantle cavity; ex.siph, exhalant siphon; f, foot; h, hinge; hy.org, hyaline organs in which symbiotic algae are cultured; inf.cav, infrabranchial part of mantle cavity; in.siph, inhalant siphon; k, kidney; l.p, outer labial palp; m.e, thickened mantle edge; oes, oesophagus, seen by transparency; p.add.mu, posterior adductor muscle, the only one present; ped, posterior edge of pedal gape through which foot may be extended; peric, pericardial cavity; ret.bys.mu, byssus retractor muscle; sup.cav, suprabranchial part of mantle cavity.

(see p. 89). Indeed, what in the common cockle is a cuticular development perhaps subserving a lens-like action in relation to photosensitive organs on the siphons has, in these great clams, been transformed into a light-collecting device (hy.org) to force the plants in the siphonal greenhouses, the tissues of the mollusc being themselves protected against the tropical light by the development of dense pigments in the surface layers. The algal cells thus bred are apparently also devoured by amoebocytes which distribute them round the body. Giant clams also feed in the typical bivalve way.

Functioning of the vascular system of bivalves

The vascular system in bivalves is like that of gastropods in including considerable

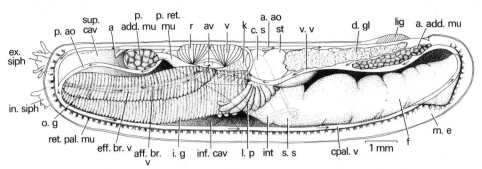

Fig. 208. *Ensis*, a young specimen, seen from the right as a transparent object. Arrows show the course of the blood flow in the vascular system. a, anus; a.add.mu, anterior adductor muscle; a.ao, anterior aorta; aff.br.v, afferent branchial vessel; av, opening of auricle into ventricle; cpal.v, circumpallial vessel in edge of mantle skirt; c.s, crystalline style within stomach; d.gl, digestive gland; eff.br.v, efferent branchial vessel; ex.siph, exhalant siphon; f, foot; i.g, inner gill; inf.cav, infrabranchial part of mantle cavity; in.siph, inhalant siphon; int, intestine in foot; k. kidney; lig, ligament of shell; l.p, labial palp; m.e, mantle edge; o.g, outer gill; p.add.mu, posterior adductor muscle; p.ao, posterior aorta; p.ret.mu, posterior retractor muscle of foot; r, rectum; ret.pal.mu, pallial retractor muscle; s.s, style sac embedded in base of foot; st, stomach; sup.cav, suprabranchial part of mantle cavity; v, ventricle with rectum running through it; v.v, visceral vein drawing blood from anterior part of body, including circumpallial vein, and running to kidney.

haemocoelic spaces around the viscera and in the foot, though their dimensions are larger. The heart consists of a ventricle (Fig. 208, v) which comes to lie around the rectum (r) during development and into which two auricles (av) discharge. In primitive bivalves these arise from a narrow base set along the efferent branchial vessel (eff.br.v) of each side, but in more advanced forms with the changed position of the gills the attachment to the vessel is extended and the auricle becomes triangular in outline. Blood is discharged both anteriorly (a.ao) and posteriorly (p.ao) from the ventricle, but the vessels (anterior and posterior aortae) cannot be homologous with the two aortae of gastropods. The anterior passes forward and supplies blood to the head, the foot and the visceral mass, while the posterior goes backwards and takes blood to the mantle and siphons. Blood is returned from the anterior part of the mantle, the viscera and the foot into a vessel (v.v) which takes it to the kidney (k), thence to the gills and back to the heart.

Movement in bivalves; attached bivalves

It used to be thought that much of the movement of a bivalve, such as protrusion of the foot and elongation of the siphons, was due to an influx of blood. However true this may have been of the ancestral mollusc, the pressure with which the blood is despatched from the heart is totally inadequate in modern bivalves to distend these organs directly at the rate at which they are seen to move; pressure other than arterial pressure is necessary, on this view, to bring about movement. Blood is, indeed, involved in the manipulation of these parts of the body but only in so far as it provides a fluid skeleton on which muscles may act, as in coelenterates or annelids. In such operations there must be some mechanism for retaining the blood within the required area, as do the septa of an earthworm, so that it will not just be squeezed away when muscles contract. In the foot this is provided by a valve, the valve of Keber, placed across the main exit from the pedal haemocoel into the vessel leading to the kidneys. When this is closed the foot is a self-contained unit and contraction of the appropriate muscles causes change in its shape. It has similarly been shown that in the mud-dwelling bivalve, *Scrobicularia plana*, the walls of the siphons behave as if they were short lengths of a worm's body, retaining a constant volume, though varying in length and breadth. This is due to the antagonistic action of longitudinal and radial muscle fibres (circular muscles are absent) on the blood contained within the siphonal wall. No sphincter muscle at the base, which would isolate the fluid, has been described, though a comparable effect must somehow be produced. The water contained within the siphon is not part of the machinery since the opening at the siphonal tip is never shut except when the siphon is fully retracted. In the deeply burrowing animals of the genus *Mya*, however, the water contained within the siphon and mantle cavity is part of the mechanism by which extension is brought about. This occurs by the adductor muscles contracting so as to approximate the two valves, thereby increasing the pressure on the water within the mantle cavity; this is transmitted to the water within the siphons and brings about a change in the dimensions of the siphonal walls. This mechanism operates because the animal is able to keep the siphonal apertures and the pedal gap closed so that the pallial cavity is watertight, and because the strength of the adductor muscles is greater than that of the siphonal musculature, though the latter is presumably inhibited by nervous activity during extension. Withdrawal of the siphons is achieved by the activity of the siphonal musculature, the adductors being passive.

With the shrouding of the head by mantle folds the tentacles and eyes are usually said to be lost. It is more probable, however, that a true head has never evolved in bivalves. In primitive molluscs such as *Neopilina* and chitons the head does not amount to more than a protuberant area around the mouth; only in gastropods and in cephalopods, where the head lies in a part of the body which can be projected far out of the mantle cavity and where the mode of feeding has required sensory information and the development of a radula and muscles to move it, has the head become a prominent part of the organization of the mollusc. In bivalves the head, as site of the major sense organs, is functionally replaced by the edges of the mantle skirt, which are the only parts of the body, along with the foot, to come into direct contact with the environment. In many burrowing bivalves this contact is further limited to the siphons and they are often rich in tentacles, eye-spots and chemoreceptors. The eye-spots are usually simple groups of sense cells, sometimes with an overlying thickening of cuticle acting as lens, but in the scallops the eyes set around the edge of the mantle skirt are complex structures with an inverted retina as in vertebrates; the relatively minute nervous system of the bivalve, with its small number of nerve cells, makes one wonder what the animal gains from eyes of such complexity, though a simple detection of a shadow cast by a moving predator may be all the message that is necessary. Correlated with the epifaunal life of scallops eyes are set around about 340° of their total circumference. In some relatives of the scallops, the thorny oysters *Spondylus*, the sensory processes of the mantle edge are extended, for better functioning, on long processes of the shell which later, when they are no longer at the margin of the shell, acquire the secondary function of making attack by predators more difficult. With the division of the pallial cavity into infra- and suprabranchial parts by elaboration of the ctenidia, the osphradium (which retains its primitive position on the ctenidial axis) becomes located in the exhalant, suprabranchial section of the mantle cavity and so loses its value to the animal as a chemo- or mechanoreceptor placed in the path of the incoming water. It, too, is therefore functionally replaced by pallial sense organs, usually the tentacles on the inhalant aperture.

Although the foot, the siphons and perhaps the mantle edge may be capable of altering shape by means of intrinsic muscles acting upon contained blood or other fluid as a hydrostatic skeleton, movement of any one of these parts in relation to others calls for the contraction of muscles which connect them to the valve of the shell. Some of these have already been mentioned like the siphonal and pallial retractors, which pull in the siphons and the edge of the mantle skirt when the animal is disturbed and the shell closes. The foot is likewise provided with muscles running to the shell and there may be four different groups. Two retractor muscles on each side are of nearly constant occurrence; one runs posteriorly (Figs 203A and 208, p.ret.mu) and attaches to the shell alongside the posterior adductor, the other passes anteriorly (a.ret.mu) alongside the anterior adductor. Close to this, but less constant in occurrence, a muscle on each side runs to the posterior end of the foot and, on contraction, pulls this forward and so aids in its protraction. Still less commonly a pair of muscles run from a central position on the shell to the foot and act as pedal elevators. The digging cycle of burrowing bivalves (Fig. 209) involves shell muscles and both the blood in the body and the water in the mantle cavity. As in burrowing polychaete worms the cycle starts with a phase in which the animal holds its position in the substratum whilst it probes for the next downward step; the probe is then anchored and the rest of the body pulled down towards it. The first anchor is provided by the two shell valves which press against the substratum because of the ligament while the adductor muscles are relaxed. During this stage contraction of the circular and transverse muscles of the foot acts on the contained blood so that the foot penetrates the substratum. At the end of the pedal elongation

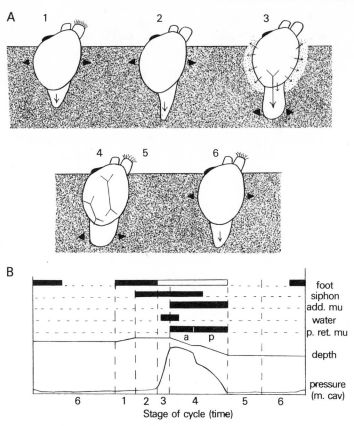

Fig. 209. A, 1–6, stages in the digging cycle of a bivalve. Single-headed arrows show movement of blood and water; double-headed arrows contraction of muscles. 1, valves press against substratum to anchor shell, siphons open, foot extends; 2, siphons shut; 3, adductor muscles contract, forcing water out between valves to loosen substratum, and blood into foot; 4, foot expanded, anchored to substratum whilst contraction of pedal retractor muscles pulls shell downward-siphons open; 6, return to stage 1. B, the events of the digging cycle plotted against time. Black indicates the periods of contraction of adductor muscles (add.mu), pedal retractor muscles (p.ret.mu, a, anterior; p, posterior), of closure of the siphons, of ejection of water from the mantle cavity and of downward elongation of the foot. The white part of the foot record is the period during which it acts as anchor. The depth at which the animal lies and the pressure within the mantle cavity (m.cav) are also shown. (After Trueman.)

the adductor muscles contract, whilst the siphons are closed. This contraction has three effects: (1) it frees the shell anchor from the sand; (2) it blows water out of the mantle cavity and so loosens the surrounding sand; and (3) it distends the tip of the foot with blood squeezed out of the body, converting it into a penetration anchor. Contraction of the anterior and posterior pedal retractor muscles then pulls the body towards the foot, the adductors keeping the valves shut so as to exclude sand particles and the bivalve is poised for the start of another cycle. Extension of the foot, being due to local muscles, involves blood pressures of only about 10 cm of water; when the valves are adducted at the next phase of the cycle pressures may rise to ten times that value. The cycle may be readily observed in such animals as *Scrobicularia*, *Tellina* or (particularly rapidly – at a rate of about 90 probes per minute) in the razor shell *Ensis*, which can quickly burrow to a depth of 1–2 feet. Ordinary locomotion of bivalves like the freshwater mussel is a slower version of the same movements.

Though the typical bivalve can move thus, some have become attached and largely immobilized on the substratum. This applies to the common edible mussel, *Mytilus*, which is fastened (Fig. 203C, bys.att) by a series of radiating fibres of material secreted from a gland, the byssus gland (bys.gl), in the base of the foot (f). This organ has changed its function so as to act as a manipulator of byssus threads, is small and pointed and has a groove (bys.gr) along its ventral side leading to the apex from the opening of the byssus gland at the base. When about to attach itself the mussel extends the foot between the anterior ends of the valves and rests the tip against a chosen substratum, usually rock or the shell of another mussel, but sometimes weed if the animal be small. At this stage the groove is closed by approximation of its edges, and becomes filled with byssus secretion, which spreads into a little disk at the tip of the foot. The groove is now opened so as to expose the secretion to sea water which hardens it rapidly into a tough, horny fibre. After a moment to let this happen, the mussel then repeats the process by applying its foot elsewhere, keeping the first secreted thread taut the meanwhile. In this way the bivalve anchors itself by a conical bundle of fibres. The shell tapers markedly to the anterior end, whilst bulging posteriorly and this allows the animals, which are gregarious, to arrange themselves in clusters over relatively small areas of attachment. Though not spontaneously breaking free, they can re-attach themselves if the byssus threads are accidentally torn.

The final stage in the trend towards attachment is exhibited by oysters (*Ostrea*, *Crassostrea*) and by thorny osters (*Spondylus*) which fasten themselves to appropriate hard substrata at the end of their larval life by secretion of byssus material applied by the foot between the underside of one valve (the left in *Ostrea* and *Crassotrea*, the right in *Spondylus*) and the substratum. Once this has set they are fixed irrevocably. The same end is reached in a different way by the saddle-oysters (*Anomia*) which similarly settle at the end of larval life and attach by means of byssus threads which become calcified; these are applied to the rock or other substratum at the edge of the larval shell. Later, however, the shell grows round the byssus, which then appears to emerge from a hole in the under valve.

These animals are sessile. To them, therefore, apply the same rules of symmetry as apply to other sessile animals, and for the same reason – that all directions are potentially alike. Oysters, therefore, become radially symmetrical, or as nearly so as bivalve organization allows, and the ctenidium extends over an arc of c. 150° in

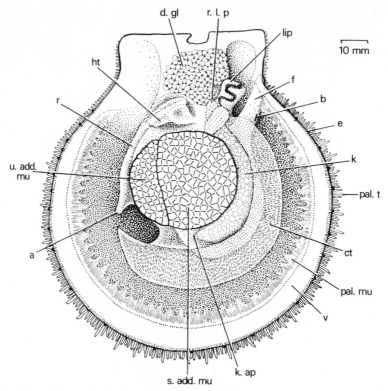

Fig. 210. *Pecten maximus*. The animal has been removed from the shell and is lying on its left side. The mantle flaps are separate except dorsally, where they are notched to allow passage of the ligament from valve to valve. a, anus; b, aperture of byssus gland on foot; ct, ctenidium; d.gl, digestive gland; e, eye on mantle edge; f, foot; ht, heart; k, kidney; k.ap, opening of kidney; lip, lip apparatus; pal.mu, pallial muscle; pal.t, pallial tentacle on edge of mantle flap; r, rectum; r.l.p, right inner labial palp; s.add.mu, striated, quick, part of adductor muscle; u.add.mu, unstriated, slow part of adductor muscle, catch muscle; v, velum.

Ostrea, 180° in *Anomia* and 210° in *Spondylus* within the approximately circular shell, which is closed by the activity of a single, centrally placed adductor muscle; this corresponds to the posterior one in bivalves of more conventional shape.

Oddly enough, the same approach to a circular outline has been reached by the scallops (Fig. 210), which far from being fixed, are active, swimming bivalves. They have the same central adductor, the same extended ctenidium (160° in *Pecten*) but are able to swim vigorously by ejecting water from the mantle cavity by way of two channels lying one on either side of the ligament. Escape of water from other parts of the cavity when the adductor contracts is prevented by the development of extensive flaps which act as valves from the innermost fold of the edge of the mantle skirt. The animal therefore progresses as if it were taking bites out of the water. The windowpane shell (*Placuna placenta*) is also circular in outline and so thin from side to side as to make one wonder how the entire organization of a bivalve can be packed within it. It is a relative of *Anomia*, but lies free on the bottom. In view of this and the situation in scallops it seems that several bivalve stocks have become cemented to the substratum but that some of their members must have later reverted to an unattached mode of life.

Organization of cephalopods

In complete contrast to the two groups which have been discussed so far are the members of the third major class of the Mollusca, the Cephalopoda. These animals have successfully – and surprisingly – transformed the typical molluscan organization into the equipment which an active predator requires: acute sense organs, fast neuro-muscular machinery and an intelligent brain to link the two. In addition, effective ancillary systems for respiration and circulation have been brought into existence so that the rest of the body may function at this new high level of activity.

The living cephalopods exhibit several grades of organization. At an evolutionary level much lower than that exhibited by most members of the class is the archaic *Nautilus*, the pearly nautilus of the western Pacific, which retains the traditional external molluscan shell. It has modified it greatly, however, since the shell (Fig. 212A, sh) is divided by a number of septa (sep) into a corresponding number of chambers, only the last, most recently formed and largest of which is occupied by

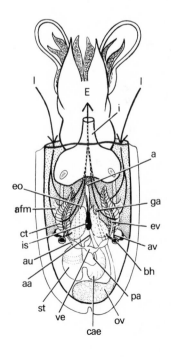

Fig. 211. Diagram of a coleoid cephalopod to show water currents in the mantle cavity and some internal organs. a, anus; aa, anterior aorta; afm, membrane attaching ctenidium to body; au, auricle; av, afferent vessel to gill; bh, branchial heart; cae, caecum of stomach; ct, ctenidium; E, exhalant; eo, opening of kidney; ev, efferent vessel from gill to heart; ga, genital aperture; I, inhalant; i, funnel; is, ink sac, opening to rectum; ov, ovary; pa, posterior aorta; st, stomach receiving oesophagus and with muscle band; ve, ventricle.

the animal. From the apex of its visceral hump, at the innermost end of the chamber, a tubular, vascular extension of the body, known as the siphuncle (siph), runs through all the other chambers of the shell and is able to secrete gas into them. In this way the heavy shell, which might be expected to weigh the animal down and restrict it to a bottom-dwelling mode of life, has been transformed into a buoyancy device by means of which *Nautilus* may float or swim at the surface of the sea. The amount of lift given by the shell is probably under the control of the animal. In addition to its primitive type of shell *Nautilus* also differs from other Recent cephalopods in showing a doubling in number of many pallial organs: thus there are four gills, four auricles and four kidneys. The locomotor organs, sense organs and brain are still at a low level of organization.

All other living cephalopods (known collectively as Coleoidea) have converted the shell into an internal structure and only in one genus (*Spirula*) does it show a chambered arrangement in any way reminiscent of the nautiloid shell. In other genera some trace of chambers may persist in a shell of soft, chalky consistency as in the cuttlefish *Sepia* (Fig. 212B), but frequently all the calcareous matter has been lost, leaving a plate of periostracal material of horny appearance and consistency, like the "pen" of the squid *Loligo*, or the shell may vanish altogether, as in the octopuses. Where it has any substance, the shell still retains, to varying degrees, its original molluscan function as a skeleton, giving origin to muscles. It has also retained its original cephalopod role of flotation device, though the way in which it now acts is markedly more sophisticated than in nautiloids and the control over the amount of lift obtained is much more pronounced.

In an animal like *Sepia* a section through the shell or cuttlebone shows that it is made up of a series of layers separated by chambered spaces less than 1 mm in height. The calcareous layers are added to during the life of the animal by secretion from the pallial epithelium lining the space in which the shell is lodged. The spaces are filled partly with gas, partly with liquid, the whole having a specific gravity of about 0.6. The lift which this provides just counterbalances the weight of the cuttlefish in the water so as to make it effectively weightless. In addition, the animal can vary the relative amounts of gas and liquid in the chambers and so alter the amount of lift which it gets. This is done by an active movement of the liquid component of the mixture. The variation in lift tends to occur in a circadian rhythm, the animals being denser by day and less dense at night; this is thought to be associated with their natural cycle of behaviour, since they tend to lie half buried in sandy bottoms by day and to emerge to hunt for food at night.

The body of the cephalopod exhibits a clear division into head-foot and visceral mass; the subdivision of the former into head and foot is less noticeable than in the other molluscan classes. Since cephalopods are largely free-swimming animals the foot is not required for creeping or burrowing and is available for adaptation for

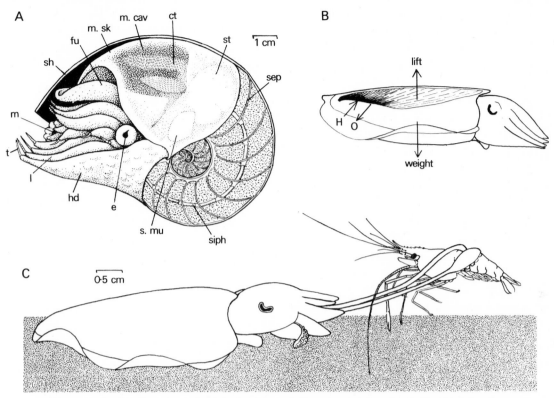

Fig. 212. A, *Nautilus*: the animal is shown partially retracted into its shell, which has been bisected sagittally to show the chambers and siphuncle. B, diagram of the factors involved in the maintenance of buoyancy by a cuttlefish. Liquid within the chambers of the cuttlefish bone is shown black, gas is shown white; the former is concentrated in the oldest, most posterior chambers, the latter in the younger, anterior ones. The hydrostatic pressure of the sea water (H) is balanced by an osmotic pressure (O) between the blood and the liquid in the chambers. The cuttlefish bone has a density of 0.6 and so gives a lift which balances the tendency of the mollusc to sink. C, a cuttlefish captures a prawn, shooting out two long tentacles bearing suckers at their tips. The prey is retracted within the other eight arms, bitten by the jaws and poisoned by saliva. (B, after Denton and Gilpin-Brown; C, after Wilson.) ct, ctenidium, one of four; e, eye; fu, funnel, scroll-like; hd, hood, acting as an operculum; l, lobes on head bearing tentacles; m, mouth, surrounded by lips; m.cav, mantle cavity; m.sk, mantle skirt; sep, septum of shell; sh, shell; siph, siphuncle; s.mu, muscles by which animal grips shell; st, stomach; t, tentacle, ringed but without suckers.

other functions. Part has retained connection with locomotion and has been transformed into a structure known as the infundibulum or funnel, an exhalant siphon leading out of the mantle cavity. As all cephalopods are bilaterally symmetrical in relation to their active mode of life, this siphon is placed in the mid-line of the pallial aperture like the exhalant siphons of such bilaterally symmetrical gastropods as *Emarginula* and *Diodora*, with the ingoing current entering the mantle cavity on each side. In *Nautilus* the siphon (fu) has a scroll-like shape, but in other living cephalopods the edges of the scroll have fused to produce a tube with an internal valve ensuring that no water can enter the mantle cavity by that route. At one time it was believed that the rest of the foot had been transformed into a series of tentacles or arms lying around the head – whence the name Cephalopoda – but it is now thought that they are cephalic in origin. They carry suckers and are the means by which the animals grasp objects as prey, or as part of a locomotor process, or as part of a tactile exploration of their environment.

The arms form the basis of the classification of the Recent cephalopods other than *Nautilus*. *Nautilus* is itself placed in a subclass called Nautiloidea, all other members of which are extinct; some of its principal characteristics have already been incidentally mentioned. Other living genera fall into a single subclass called the Coleoidea (Dibranchia is an alternative name often met with) with the contrasting characters already indicated as differentiating them from the nautiloids. The coleoids may be further divided on the arrangement of the arms into three orders:

1. A basic group, the decapods, with eight short and two long and retractile tentacle-like arms.

2. A more advanced group, Octopoda, presumably of decapod origin, with only eight arms. These animals have no shell and the eight arms are often interconnected by a web of flesh.

3. The sub-order Vampyromorpha, which includes one genus of deep-sea cephalopods with eight arms and two others which seem not to be homologous with the tentacles of the decapods.

The orientation of the cephalopod body in space is different from that of other molluscs (Fig. 212): the animal normally swims or rests on the bottom with the head-foot in front and the visceral mass behind, and with the mantle cavity on the ventral side of the visceral mass. These topographical dispositions are the basis for normal descriptions of cephalopod anatomy though on morphological grounds it is clear that cephalopod ventral is molluscan posterior, dorsal equals anterior, anterior equals ventral and posterior dorsal.

Decapods, such as the squids and cuttlefishes, are powerful swimmers, rivalling the fishes in their mastery of this skill. It is accomplished in two ways. Quiet swimming is achieved by means of waves of contraction passing along fins placed horizontally along the side of the body ($=$ visceral hump). Active pursuit of prey, or avoidance of predators, however, is brought about by a different mechanism. Because of the reduction or virtual absence of a shell the mantle skirt has been liberated and has become mobile. Its musculature is hypertrophied and it becomes a powerful means of ventilating the mantle cavity (Fig. 211). Water is sucked inwards at right and left and is then expelled medially by way of the funnel, locking arrangements between body and mantle skirt ensuring that it does not escape the way it came in. The expulsion may be very forceful since there is an elaborate arrangement of differentially graded, giant nerve fibres running to the pallial musculature, the smallest and shortest to the more anterior areas, the largest and longest to the more posterior. Since their conduction rate is directly related to their diameter this ensures that all the pallial exhalant musculature is set into nearly simultaneous contraction, giving rise to a powerful jet of water forced through the funnel. The animal is thus jet-propelled, and, as the funnel is mobile and flexible and may be pointed in many directions, the cephalopod can shoot forwards towards prey, backwards, up or down to avoid danger to itself. In the latter case it can add to the confusion of the predator attacking it by expelling along with the jet a mass of ink from the ink sac (is), which is a modified anal gland lying alongside the rectum. The ink is not so much a smoke-screen as a distraction to capture the attention of an attacker whilst the cephalopod makes off.

Such vigorous movement of water through the mantle cavity, though primarily for locomotor purposes, is also important for respiration and allows a much more efficient ventilation of the ctenidia than would have taken place had the cephalopods, like other molluscs, relied on cilia for this purpose. Indeed, the two activities must be interlinked and the potentiality for rapid motion is useless unless an increased respiratory rate can be maintained. The ctenidia have no cilia and their leaflets contain an elaborate capillary bed for the oxygenation of the blood. The blood flow is such that it does not follow the counter-current principle used by other molluscs, but the ventilation rate is more than adequate to compensate for this. The blood contains the pigment haemocyanin, capable of carrying considerable quantities of oxygen (about 3.8–5.0 volumes per cent): this is greater than corresponding figures for other invertebrates, but appreciably less than what is taken up by the haemoglobin of a fish (about 10–20 volumes per cent). The cephalopods suffer other disadvantages by comparison with fishes in their respiratory physiology: vertebrate venous blood always contains a lot of oxygen, that of squids hardly any (0.37 volumes per cent); vertebrate muscle usually contains myoglobin with an associated store of oxygen, cephalopod muscles have no comparable reserve; because of this almost complete withdrawal of oxygen from the circulating blood and the absence of a store the cephalopod is extremely sensitive to oxygen lack. This may be one of the reasons for their frequent inability to survive capture.

The blood is pumped round the body in an elaborate vascular system with true capillaries linking the arterial and venous sides. There are three hearts, a systemic one (au, ve) centrally placed in a capacious pericardial space in the visceral mass, which receives oxygenated blood from the ctenidia (ct) and dispatches it to the body, and two branchial hearts (bh), one at the base of each gill and also in the pericardial cavity, which receive the deoxygenated blood from the system and pump it through the gills. Circulation is further aided by the contractility of the walls of many main vessels. As in other molluscs the kidneys are placed so that blood on its way from body to gills has to pass through them. In cephalopods the main veins to the branchial hearts pass through the kidneys and excretory tissue is developed on their walls, an arrangement which looks less efficient than that found in gastropods and bivalves. There is no renal portal system as in those groups.

Cephalopods are all carnivores, feeding mainly on crustaceans or fish. *Sepia* catches prawns (Fig. 212C) by shooting out the two long retractile tentacles, armed with suckers at their tip, and then pulling the prey to the mouth. Squids behave similarly. In octopods, however, the capture of food is different: the animal creeps or gently swims close to its intended prey, then jumps on to it, trapping it under the arms and smothering its attempts to escape by means of the web. In all, the later events are similar. The animal captured is passed to the mouth which lies at the centre of the radiating arms, and which is provided with a strong pair of jaws shaped like the beak of a parrot. The body of the prey is pierced with these and saliva, which contains a neurotoxic agent tyramine, is injected to kill it. The radula is then used to scrape out food, or, in some cases, digestive enzymes may be vomited outwards to bring about an external digestion of the prey. In either case the particles or digested material are later ingested, passed down a long oesophagus to the stomach located near the apex of the visceral mass. The stomach is split into proximal and distal parts linked by a narrow connection, provided with a sphincter, where open ducts form the digestive gland. The proximal part (Fig. 211, st), to which the oesophagus opens, forms a kind of crop or gizzard. It frequently possesses a cuticular lining which can triturate the food in the presence of enzymes from the digestive gland. The distal part, from which the intestine arises, is ciliated, often spirally wound and extended into a caecum (cae), and provided with internally projecting leaflets. Here further digestion in the presence of a different series of enzymes occurs and some absorption of the products of digestion takes place. Some uptake also occurs in the intestine. The cilia on the leaflets direct indigestible material into the intestine and to this is added indigestible matter from the gizzard which has by-passed the caecum. The secretion which enters the gizzard comes from a different lobe of the digestive gland from that which goes to the caecum. This lobe is often referred to as the pancreas and the main one as the liver; these terms facilitate reference but should not be understood in the vertebrate sense.

The male cephalopod uses one of its arms (different in different species) to transfer spermatophores to a seminal receptacle placed either on the buccal membrane (*Sepia*) or in the mantle cavity (*Loligo, Sepiola*) or into the female duct directly. The spermatophores are made within the male duct and transferred through the funnel to the arm used as a copulatory organ or hectocotylus. In most cephalopods this is thrust, often after some courtship behaviour, so as to release the spermatophores in the appropriate place. The sperm released from the spermatophores may then have to migrate, presumably along a chemical gradient, so as to reach the female duct or eggs. After fertilization these are laid in cases of jelly strings. The eggs are fairly large with quantities of yolk and the young hatch as miniatures of the adult, all trace of a veliger larval stage being suppressed. In some cephalopods – the best known, perhaps, being the paper nautilus *Argonauta* – the hectocotylus, after being placed in the mantle cavity of the female, is broken off and left there to liberate the spermatophores which it bears. When early investigators found this *in situ* they regarded it as a parasitic worm and placed it in a genus *Hectocotylus*, whence the name of the modified male arm.

The last point to which attention should be directed in any summary of cephalopod structure is the complexity of the sense organs and brain. The major sense organ is undoubtedly the eye, which has often been compared to that of vertebrates, although the basis of construction is not the same. Eyelids, cornea, iris, lens and retina all occur, but the lens is cuticular and the retinal photosensitive cells look outward; the fibres of the optic nerve run to the brain from the base of the retina so that there is no blind spot. Focusing is effected by moving the lens backwards and forwards as some fish do, not by altering the strength as in most vertebrates. There is an area equivalent to the vertebrate fovea where vision is most acute and *Sepia* has about 10^5 mm^{-2} retinal cells in this area. The eye has been shown to form good images. In addition to eyes cephalopods possess statocysts which lie close to the ventral side of the brain with more than one collection of sensory cells (cristae) in each; there are also olfactory pits placed on the side of the head near the entrance to the mantle cavity where they are probably able to test the water entering. They may, therefore, be regarded as functional replacements of the lost osphradia. Tactile or chemical receptors are also scattered over the surface of the body, though they are most abundant in relation to the arms and suckers, where they subserve the investigation of the substratum over which the animal moves or of the objects which it handles.

The brain is presumably built on the same general pattern as in other molluscs, but has become so greatly enlarged that the homologies of its parts are not always clear. It lies around the oesophagus, surrounded by a protective skull of cartilage-like material. Part of it lies dorsal to the oesophagus and so may be regarded as the equivalent of the cerebral and pleural ganglia; part lies ventral to the gut and seems to correspond to the pedal and visceral ganglia of other molluscs. In addition to this central mass of nervous tissue there are outlying ganglionic masses related to special areas of the body; of these the buccal ganglia related to the buccal mass, the stellate ganglia lying in and controlling the mantle skirt and the gastric ganglion on the wall of the stomach may be mentioned.

The pedal and visceral areas of the brain form the lower motor centres. They work in relation to a part of the brain lying above the gut and forming the higher motor centres. Dorsal to these, and so the most dorsal part of the brain, is another series of lobes which are primarily sensory but are also the base of the complex learning, memory and other mental activities which distinguish the cephalopod from all other molluscs, and, indeed, from all other animals except the higher arthropods and vertebrates. Attached to the sides of this part of the brain are the optic lobes lying immediately alongside the eyes and together greater in volume and the number of nerve cells which they contain than all the rest of the brain put together, a fact which indicates the extent to which the cephalopods lean on the sense of sight as a means of catching food and learning about their environment.

With a brain of this degree of complexity it is not surprising that cephalopods have been proved capable of considerable feats of learning and memory, mainly in relation to visual, but also in relation to chemotactile stimuli, as shown by the great body of work carried out by Ten Cate, Young, Boycott and Wells. This store of learning is centred on the superior frontal and vertical lobes of the dorsal part of the brain and after their removal this kind of behaviour and the initiation of much spontaneous activity is lost, though all the animal's stereotyped responses still remain intact.

Minor groups of molluscs

The gastropods, bivalves and cephalopods constitute the immense majority of living molluscs, but there are a few small groups within the phylum which do not fall within their limits. One of these, the Scaphopoda or elephant tusk shells, is of so little importance as not to merit discussion here, whilst another, containing worm-like animals usually found sessile on hydroids, the Aplacophora, is perhaps not a true molluscan group at all. When these have been put aside, however, one group still remains as deserving of mention on the grounds of the abundance of its members, their adaptations to their mode of life and the thoughts which they raise in relation to the origin of the phylum: these are the Amphineura, the chitons and their relatives, about which something has already been said by way of introducing molluscan organization.

Since the head of a chiton is always sheltered from the environment by the overlying shell and mantle, sense organs are relatively valueless on it and there are neither tentacles nor eyes. These are functionally replaced by structures called aesthetes which lie within the substance of the shell. In British chitons these are small and simple in organization but in some tropical forms they are well developed, show the structure of a primitive eye and are light sensitive. In some chitons the shell becomes overgrown by mantle and so becomes internal (*Cryptochiton*): in this case pallial sense organs are the only ones which the animal has. The gills are always more numerous than the highest number (four in *Nautilus*) met elsewhere in the phylum. This has been explained as arising from the need to get an adequate respiratory surface within a cavity of such small dimensions.

The gut of a chiton is built on the standard molluscan pattern with salivary, oesophageal and mid-gut glands. It is marked by an extremely elongated intestine; some uptake of digested food has been recorded from this part, but otherwise its main function seems, as in the more primitive gastropods, to be the elaboration of faecal pellets. There is nothing remarkable about the other internal organs of chitons except perhaps the absence of ganglia in the central nervous system, nerve cells being scattered along the length of the nerves.

Related to chitons is a group of molluscs (Monoplacophora) containing a number of fossils (the Tryblideacea) and one living form, *Neopilina*, the first specimens of which were dredged off the Pacific coast of Central America in 1952. Inevitably the animals were dead after being brought up from 3 570 m, but their anatomy, as described by Lemche and Wingstrand, has aroused an enormous amount of interest and has possibly important bearings on the origin and relationships of the phylum Mollusca.

Neopilina has a conical shell up to 4 cm in length and said to show a trace of spiral winding at the apex. The shell is single, not made of several pieces as in the chitons, whence the name Monoplacophora. Under its shelter lies the head with a series of folds round the mouth apparently derived from the larval swimming lobes or velum, and a small, round foot so poorly provided with muscle as to suggest that the animal can neither travel much nor adhere strongly. The mantle cavity, as in chitons, is the space between the head and foot centrally and the shell and mantle skirt marginally. Into it discharge the gut, the excretory organs and the gonads; in it lie the ctenidia. The most important feature of *Neopilina* is the repetition, in an apparently metameric way, of a number of its organs – gills, branchial and pedal muscles nerves, excretory and genital organs. According to Lemche and Wingstrand this is a true metamerism and the molluscs are therefore regarded by those workers as close relatives of arthropods and annelids; according to other workers, such as Yonge, the metamerism is spurious and is perhaps to be explained in the same way as the large number of gills in chitons, as a multiplication of parts necessary to achieve a particular physiological end. In chitons two or four gills of the size which would fit the small mantle cavity would just not produce sufficient area for respiratory exchange; the number has therefore had to be increased, but this is not an indication of a metameric repetition of parts. At this stage of the argument it is not possible to come to any definite conclusion: the matter might be resolved if we had knowledge of the development of *Neopilina* and were able to see how the repetition of its parts came into existence, but this is something for which we shall have to wait a long time. In the meantime a cautious adoption of an intermediate position is perhaps a reasonable attitude to take; the molluscs may well show some signs of a short metameric body in their most primitive members but this metamerism has not been, or has not been used in any way to the animals' advantage. It has generally been lost, leaving the animals with a body form adaptable for a multiplicity of modes of life; indeed the further from metamerism the molluscs seem to go the more successful do they appear to have become. The monoplacophorans may then appear as archaic forms still retaining traces of the small-scale metamerism which their more progressive relatives have lost.

They are valuable pointers to some other features of molluscan evolution. They demonstrate the fact that the mantle cavity lies primarily all round the animal, as it still does in chitons, bivalves and scaphopods: it is only in gastropods and cephalopods that the cavity has enlarged posteriorly and diminished elsewhere. It is clear, too, that this has happened because of the gastropod's increasing ability to move in and out of the shelter of its shell, a thing which no chiton or bivalve does. The habit of extending the head from the shell accompanied the development of new feeding methods in which the radula became of greater and greater importance and the apparatus which worked it grew in complexity. At the same time a protrusible head became the natural place on which to site distance receptors such as eyes and chemoreceptive tentacles. Since *Neopilina* has a radula, but poorly developed tentacles and no eyes, it is arguable that the extensive development of a head was a gastropod–cephalopod innovation and that perhaps scaphopods and bivalves have never had one; in the latter case adaptation to a ciliary food-collecting way of life involved only the loss of the radula. It seems inevitable, too, to conclude from its occurrence in chitons, the most lowly gastropods, protobranchs, lamellibranchs, scaphopods and, presumably, *Neopilina*, that microphagy was the fundamental mode of feeding in the molluscs from which only gastropods and cephalopods have been able to break away.

Classification of molluscs

Mollusca
    [Aplacophora (= Caudofoveata)]
    Monoplacophora
    Polyplacophora
    Gastropoda
        Prosobranchia
            Diotocardia (= Archaeogastropoda)
                Zeugobranchia
                Patellacea
                Trochacea
            Neritacea
            Monotocardia
                Mesogastropoda – Platypoda and Heteropoda
                Neogastropoda
        Opisthobranchia
            Tectibranchia

      Bullomorpha (= Cephalaspidea)
      Pteropoda
      Aplysiomorpha (= Anaspidea)
      [Acochlidiacea]
      Sacoglossa
      Pleurobranchomorpha (= Notaspidea)
    Nudibranchia (= Acoela)
  Pulmonata
    Basommatophora
    Stylommatophora
[Scaphopoda]
Bivalvia
  Protobranchia
  Lamellibranchia
    Filibranchia
    Eulamellibranchia
    Septibranchia
Cephalopoda
  Nautiloidea
  [†Ammonoidea]
  Coleoidea
    Decapoda
    Octopoda
    Vampyromorpha

16

Supplementary reading

There are three text-books which present the results of the study of invertebrates in a comprehensive way, although two are incomplete at the time of writing. They give extensive accounts of the anatomy (not necessarily from a functional point of view), the taxonomy and of some aspects of the physiology and ecology of most phyla in the animal kingdom. They are:

Hyman, L. H. "The Invertebrates". McGraw-Hill, New York.
(1940). Vol. I. Protozoa through Ctenophora.
(1951). Vol. II. Platyhelminthes and Rhynchocoela.
(1951). Vol. III. Acanthocephala, Aschelminthes, and Entoprocta. (Includes nematodes and rotifers.)
(1955). Vol. IV. Echinodermata.
(1959). Vol. V. Smaller coelomate groups (lophophorates).
(1967). Vol. VI. Mollusca I. (Aplacophora, Polyplacophora, Monoplacophora and Gastropoda).

The series has so far been stopped at this point by the death of Dr Hyman.

Grassé, P-P. (ed.) "Traité de Zoologie". Masson et Cie, Paris.
(1952). Tome I. 1. Protozoaires (Généralités, Flagellés).
(1953). Tome I. 2. Protozaires (Rhizopodes, Actinopodes, Sporozoaires).
(1973). Tome III. 1. Spongiaires.
(1961). Tome IV. 1. Plathelminthes, Mésozoaires, Acanthocéphales, Némertiens.
(1965). Tome IV. 2. Némathelminthes (Nématodes).
(1965). Tome IV. 3. Némathelminthes, Rotifères, Gastrotriches, Kinorhynques.
(1959). Tome V. 1. Annélides, Myzostomides, Sipunculiens, Echiuriens, Priapuliens, Endoproctes, Phoronidiens.
(1960). Tome V. 2. Bryozoaires, Brachiopodes, Chétognathes, Pogonophores, Mollusques (Aplacophores, Polyplacophores, Monoplacophores, Bivalves).
(1968). Tome V. 3. Mollusques, Gastéropodes et Scaphopodes.
(1948). Tome XI. Echinodermes, Stomocordés, Procordés.

The volumes on ciliate protozoans, coelenterates and cephalopod molluscs have not yet been published.

Beklemischev, W. N. (1969). "Principles of Comparative Anatomy of Invertebrates" (Ed. Z. Kabata). Translated from the Russian by J. M. MacLennan. Oliver and Boyd, Edinburgh.

In addition to these, mention should be made of two other multi-volume works still incomplete, like two of the above. These are not primarily concerned with functional anatomy but contain introductory sections on the anatomy, physiology and ecology of the different invertebrate phyla which are extremely valuable, the more so as they are often written from a standpoint rather different from that of the ordinary zoologist. These works are:

R. C. Moore (ed.) "Treatise on Invertebrate Paleontology", Geological Society of America, Kansas University Press.

(1964). Part C. Protista 2 (Sarcodina chiefly "Thecamoebians" and Foraminiferida).

(1954). Part D. Protista 3 (Heliozoa, Radiolaria).

(1955). Part E. Archaeocyathida and Porifera.

(1956). Part F. Coelenterata.

(1953). Part G. Bryozoa.

(1965). Part H. Brachiopoda.

(1960). Part I. Mollusca 1 (General, Scaphopoda, Amphineura, Monoplacophora, Gastropoda in part).

(1964). Part K. Mollusca 3 (Cephalopoda: general, Nautiloidea).

(1959). Part L. Mollusca 4 (Ammonoidea).

(1969). Part N. Mollusca 6 (Bivalvia).

(1959), Part O. Arthropoda 1 (General).

(1955). Part P. Arthropoda 2 (Chelicerata).

(1961). Part Q. Arthropoda 3 (Crustacea Ostracoda).

(1969). Part R. Arthropoda 4 (Crustacea).

(1967). Part S. Echinodermata 1 (General, Crinozoa).

(1966). Part U. Echinodermata 3 (Echinacea, Asterozoans).

Florkin, M. and Scheer, B. T. (eds) "Chemical Zoology", Academic Press, New York and London.

(1967). Vol. I. Protozoa.

(1968). Vol. II. Porifera, Coelenterata, Platyhelmia.

(1968). Vol. III. Echinodermata, Nematoda and Acanthocephala.

(1969). Vol. IV. Annelida, Echiura, Sipuncula.

(1970). Vol. V. Arthropoda, part A.

(1971). Vol. VI. Arthropoda, part B.

(1972). Vol. VII. Mollusca.

A further series of books deals with specific activities in different invertebrate groups. These are given here rather than repeated under each group.

Bullock, T. H. and Horridge, G. A. (1965). "Structure and Function in the Nervous System of Invertebrates". 2 vols. W. H. Freeman, San Francisco and London.

Gray, J. (1968). "Animal locomotion". Weidenfeld and Nicolson, London.

Jennings, J. B. (1972). "Feeding, Digestion and Assimilation in Animals". Macmillan, Oxford.

Jones, J. D. (1972). "Comparative Physiology of Respiration". Arnold, London.

Jørgensen. C. B. (1966). "Biology of Suspension Feeding". Pergamon Press, Oxford.

Krogh, A. (1939). "Osmotic Regulation in Aquatic Animals". University Press, Cambridge.

Lockwood, A. P. M. (1971). "Animal Body Fluids and their Regulation". Heinemann Educational Books, London.

Potts, W. T. W. and Parry, G. (1964). "Osmotic and Ionic Regulation in Animals". Pergamon Press, Oxford.

Riegel, J. A. (1972). "Comparative Physiology of Renal Excretion". Oliver and Boyd, Edinburgh.

Trueman, E. R. (1975). "The Locomotion of Soft-Bodied Animals". Arnold, London.

Protozoans

There are many books dealing with

protozoans in a general way and others treating more particularly the protozoans which are parasitic in man and his domestic animals; of these we select the following four:

Mackinnon, D. L. and Hawes, R. S. J. (1961). "An Introduction to the Study of Protozoa". Clarendon Press, Oxford. (Primarily a practical approach to the study of a large variety of protozoans.)

Manwell, R. D. (1961). "Introduction to Protozoology." Arnold, London. (A more general and less practical treatment than the previous.)

Sleigh, M. A. (1973). "The Biology of Protozoa". Arnold, London. (Essays on a series of topics.)

Hoare, C. A. (1949). "Handbook of Medical Protozoology". Baillière, Tindall and Cox, London. (This deals predominantly with the anatomy, life history of parasitic forms and with their evolution from free-living ancestors.)

Slighter but interesting reviews of protozoans are to be found in:

Curtis, H. (1968). "The Marvelous Animals". American Museum of Natural History, New York.

Sandon, H. (1963). "Essays on Protozoology". Hutchinson Educational, London.

Books which deal in a general way, though often with an emphasis on taxonomy, with particular groups of protozoans:

Jeon, K. W. (ed.) (1973). "The Biology of Amoeba". Academic Press, New York and London.

Jepps, M. W. (1956). "The Protozoa, Sarcodina". Oliver and Boyd, Edinburgh.

Murray, J. W. (1973). "Distribution and Ecology of Living Benthic Foraminiferids". Heinemann, London.

Corliss, J. O. (1961). "The Ciliated Protozoa". Pergamon Press, Oxford.

Garnham, P. C. C. (1966). "Malaria Parasites and Other Haemosporidia". Blackwell, Oxford.

Hammond, D. M. and Long, P. L. (1973). "The Coccidia". Butterworths, London.

Books and papers on special aspects of protozoology:

FINE STRUCTURE

Grimstone, A. V. (1961). Fine structure and morphogenesis in Protozoa. *Biological Reviews*, **36**, 97–150.

Ehret, C. F. and Powers, E. L. (1959). The cell surface of *Paramecium*. *International Review of Cytology*, **8**, 97–133.

Pitelka, D. R. (1963). "Electron-microscopic Structure of Protozoa". Pergamon Press, Oxford.

Sleigh, M. A. (1974). "Cilia and Flagella". Academic Press, London and New York.

PHYSIOLOGY

Lwoff, A. (ed.) (1951). "Biochemistry and Physiology of Protozoa", Vol. I. Academic Press, London and New York.

Hutner, S. H. and Lwoff, A. (eds) (1955.) "Biochemistry and Physiology of Protozoa", Vol. II. Academic Press, London and New York.

Hutner, S. H. (ed.) (1964). "Biochemistry and Physiology of Protozoa", Vol. III. Academic Press, London and New York.

Allen, R. D. (1962). Amoeboid movement. *Scientific American*, **206**, 112–22.

Kitching, J. A. (1952). Contractile vacuoles. *Symposia of the Society for Experimental Biology*, **6**, 145–65.

Sonneborn, T. M. (1957). Breeding systems, reproductive methods and species problems in *Paramecium*. *In* "The Species Problem" (Ed. E. May). American Association for the Advancement of Science Publication, Washington D.C.

PARASITIC PROTOZOANS

Baer, J. G. (1971). "Animal Parasites". (Translated by K. Lyons). World University Library, McGraw-Hill, New York.

Taylor, A. E. R. and Muller, R. (eds) (1972). Functional aspects of parasite surfaces. *Symposia of the British Society for Parasitology*, **10**.

Taylor, A. E. R. (ed.) (1965). "Evolution of Parasites". Blackwell, Oxford.

IDENTIFICATION OF PROTOZOANS

Jahn, T. L. and Jahn, F. F. (1949). "How to know the Protozoa". W. C. Brown, Co., Dubuque, Iowa. (Since protozoans are cosmopolitan this book will serve to identify the common forms anywhere.)

For articles of all sorts on protozoans, including reviews and stimulating addresses the *Journal of Protozoology* may be consulted. It is published by the Society of Protozoologists, Lawrence, Kansas, USA, and is available in many libraries in this country.

Sponges

Apart from the general zoological texts mentioned at the beginning of this chapter there is no book on sponges. The following references are therefore limited to papers in scientific journals.

Fry, W. G. (ed.) (1970). The biology of the Porifera. *Symposia of the Zoological Society of London*, **25**. (This is the nearest approach to a book on sponges and contains a series of papers on a variety of topics.)

Parker, G. H. (1919). "The Elementary Nervous System". Lippincott, Philadelphia. (Though a little dated this contains the basic work on currents and pressures in sponges.)

Bidder, G. (1920). Notes on the physiology of sponges. *Journal of the Linnean Society*. **34**, 315–26.

Bidder, G. (1923). The relation of the form of a sponge to its currents. *Quarterly Journal of microscopical Science*, **67**, 293–323.

Jones, W. C. (1954). Spicule form in *Leucosolenia complicata*. *Quarterly Journal of microscopical Science*, **95**, 191–203.

Jones, W. C. (1958). The effect of reversing the internal water current on the spicule arrangement of *Leucosolenia variabilis* and *L. complicata*. *Quarterly Journal of microscopical Science*, **99**, 263–78.

Jones, W. C. (1961). Properties of the wall of *Leucosolenia variabilis*. I. The skeletal layer. *Quarterly Journal of microscopical Science*, **102**, 531–43; II. The choanoderm and the porocyte epithelium. *Quarterly Journal of microscopical Science*, **102**, 544–50.

Jones, W. C. (1966). The structure of the porocytes in the calcareous sponge *Leucosolenia complicata* (Montagu). *Journal of the Royal microscopical Society*, **85**, 53–62. (These papers by Jones present valuable work on the functional organization of the sponge body.)

Jones, W. C. (1962). Is there a nervous system in sponges? *Biological Reviews*, **37**, 1–50.

Burton, M. (1949). Observations on littoral sponges including the supposed swarming of larvae, movement and coalescence in mature individuals, longevity and death. *Proceedings of the Zoological Society of London*, **118**, 893–915.

Van Weel, P. B. (1948). On the physiology of the tropical freshwater sponge *Spongilla proliferans*: ingestion, digestion and excretion. *Physiologia comparata et Oecologia*, **1**, 110–26.

Coelenterates and ctenophores

References to these two groups are so intermingled that it is sensible to give a single list covering both. Much of our knowledge of the functional anatomy of sea-anemones comes from Pantin and his colleagues, who have laid special emphasis on the workings of the neuromuscular system. Their results are to be found in:

Pantin, C. F. A. (1935). The nerve net of the Actinozoa. I-IV. *Journal of experimental Biology*, **12**, 119–64, 389–96.

Pantin, C. F. A. (1950). Behaviour patterns in lower invertebrates. *Symposia of the Society for experimental Biology*, **4**, 175–95.

Pantin, C. F. A. (1952). The elementary nervous system. *Proceedings of the Royal Society* B, **140**, 147–68.

Batham, E. J. and Pantin, C. F. A. (1950). Muscular and hydrostatic action in the sea-anemone *Metridium senile* (L.) *Journal of experimental Biology*, **27**, 264–89.

Batham, E. J. and Pantin, C. F. A. (1951). The organization of the muscular system of *Metridium senile*. *Quarterly Journal of microscopical Science*, **92**, 27–54.

Batham, E. J., Pantin, C. F. A. and Robson, E. A. (1960). The nerve net of the sea-anemone *Metridium senile* (L.): the mesenteries and the column. *Quarterly Journal of microscopical Science*, **101**, 487–510.

A comparable investigation of the neuro-muscular system of jellyfish has been carried out by Horridge:

Horridge, G. A. (1955). The nerves and muscles of medusae. I-VI. *Journal of experimental Biology*, **31**, 594–600; **32**, 555–68, 636–41, 642–8; **33**, 366–83; **36**, 72–91.

Both Pantin and Horridge use the classical observations and experiments of Romanes and Parker as the starting point of their work. These are to be found in two interesting general accounts, still well worth reading despite their date.

Romanes, G. J. (1885). "Jellyfish, Starfish and Sea Urchins". Appleton, New York. International Science Series.

Parker, G. H. (1919). "The Elementary Nervous System". Lippincott, Philadelphia.

There are two collections of essays dealing with a variety of topics relating to coelenterates and ctenophores. They are:

Lenhoff, H. M. and Loomis, W. F. (eds) (1961). "The Biology of Hydra and some other Coelenterates". University of Miami Press.

Rees, W. J. (ed.) (1966). The Cnidaria and their evolution. *Symposia of the Zoological Society of London*, **16**.

A few other papers dealing with special topics may be mentioned:

Chapman, G. (1958). The hydrostatic skeleton in the invertebrates. *Biological Reviews*, **33**, 338–71.

Ewer, R. F. (1947). On the functions and mode of action of the nematocysts of Hydra. *Proceedings of the Zoological Society of London*, **117**, 365–76.

Goreau, T. F. and Goreau, N. I. (1956). The physiology of skeleton formation in corals. I-IV. *Biological Bulletin*, **116**, 59–75; **117**, 239–50; **118**, 419–29; **119**, 516–27.

Rees, W. J. (1957). Evolutionary trends in the classification of capitate hydroids and medusae. *Bulletin of the British Museum (Natural History), Zoology*, **4**, 456–534.

Skaer, R. J. and Picken, L. E. R. (1966). The structure of the nematocyst thread and the geometry of discharge in *Corynactis viridis* Allman. *Philosophical Transactions* B, **250**, 131—64.

Southward, A. J. (1955). Observations on the ciliary currents of the jelly-fish *Aurelia aurita* L. *Journal of the Marine biological Association of the U.K.*, **34**, 201–16.

Weill, R. (1934). Contribution à l'étude des Cnidaires et de leurs nematocystes. I-II. *Travaux du Station zoologique, Wimereaux*, **10**, 1–347; **11**, 349–701.

Yonge, C. M. (1964). The biology of coral reefs. *Advances in marine biology*, **1**, 209–60.

Yonge, C. M. (1968). Living corals. *Proceedings of the Royal Society* B, **169**, 329–44.

The interrelationships of invertebrate animals and their symbiotic algal cells are discussed in the following publications:

Keeble, F. (1910). "Plant Animals". Cambridge University Press, London. (This is a semipopular account of the earliest investigation of the symbiosis of an invertebrate, *Convoluta roscoffensis*, and an alga, carried out in the previous ten years by Keeble with Gamble.)

Smith, D., Muscatine, L. and Lewis, D. (1969). Carbohydrate movement from autotrophs to heterotrophs in parasitic and mutualistic symbiosis. *Biological Reviews*, **44**, 17–90.

Taylor, D. L. (1968). Chloroplasts as symbiotic organelles in the digestive gland of *Elysia viridis* (Gastropoda: Opisthobranchia). *Journal of the Marine biological Association of the U.K.*, **48**, 1–15.

Taylor, D. L. (1969). The nutritional relationship of *Anemonia sulcata* (Pennant) and its dinoflagellate symbiont. *Journal of Cell Science*, **4**, 751–62.

Taylor, D. L. (1973). The cellular interactions of algal-invertebrate symbiosis. *Advances in marine Biology*, **11**, 1–56.

Trench, R. K. (1971). The physiology and biochemistry of zooxanthellae symbiotic with marine coelenterates. I-III. *Proceedings of the Royal Society* B, **177**, 225–64.

Yonge, C. M. (1974). Coral reefs and molluscs. *Transactions of the Royal Society of Edinburgh*, **69**, 147–66.

Platyhelminths and nemertines

As might be expected the literature on the parasitic trematodes and cestodes is vast by comparison with that on free-living turbellarians and nemertines. It has, however, been better synthetized and summarized in books. The following list contains (1) some general books which are in part relevant, (2) some books dealing with one particular group and (3) papers, some of which are the basic material on which these books rest.

1. General

Dougherty, E. C. (ed.) (1963). "The Lower Metazoa, Comparative Biology and Phylogeny". University of California Press, Berkeley and Los Angeles.

Smyth, J. D. (1962). "Introduction to Animal Parasites". English Universities' Press, London.

Taylor, A. E. R. (ed.) (1965). "Evolution of Parasites". Blackwell, Oxford.

Wilmer, E. N. (1970). "Cytology and Evolution". Academic Press, London and New York. (Despite the title this contains much information presented from an individual point of view.)

2. Special

Erasmus, D. A. (1972). "The Biology of Trematodes". Arnold, London.

Gibson, R. (1972). "Nemerteans". Hutchinson University Library, London.

Pantelouris, E. M. (1965). "The Common Liver Fluke *Fasciola hepatica* L". Pergamon Press, Oxford.

Schell, S. C. (1970). "How to know the Trematodes". W. C. Brown Co., Dubuque, Iowa. (Keys and descriptions for the identification of flukes.)

Schmidt, G. D. (1970). "How to know the Tapeworms". W. C. Brown Co., Dubuque, Iowa. (Keys and descriptions for the identification of tapeworms.)

Smyth, J. A. (1966). "The Physiology of Trematodes". Oliver and Boyd, Edinburgh.

Smyth, J. D. (1969). "The Physiology of Cestodes". Oliver and Boyd, Edinburgh.

Wright, C. A. (1971). "Flukes and Snails". Allen and Unwin, London.

3. Papers

BODY FORM OF FREE-LIVING FORMS

Clark, R. B. and Cowey, J. B. (1958). Factors controlling the change of shape of certain nemertine and turbellarian worms. *Journal of experimental Biology*, **35**, 731–48.

Cowey, J. B. (1959). The structure and function of the basement membrane muscle system in *Amphiporus lactiflorus* (Nemertea). *Quarterly Journal of microscopical Science*, **93**, 1–15.

Skaer, R. J. (1965). The origin and continuous replacement of epidermal cells in the planarian *Polycelis tenuis* (Iijima). *Journal of Embryology and experimental Morphology*, **13**, 129–39.

LOCOMOTION

Pantin, C. F. A. (1950). Locomotion in British terrestrial nemertines and planarians: with a discussion on the identity of *Rhynchodemus bilineatus* (Mecznikow) in Britain, and on the name *Fasciola terrestris* O. F. Müller. *Proceedings of the Linnean Society of London*, **162**, 23–37.

DIGESTION

Jennings, J. B. (1957). Studies in feeding, digestion and food storage in free-living flatworms (Platyhelminthes: Turbellaria). *Biological Bulletin*, **112**, 63–84.

Jennings, J. B. (1959). Observations on the nutrition of the land planarian *Orthodemus terrestris* (O. F. Müller). *Biological Bulletin*, **117**, 119–24.

Jennings, J. B. (1959). Studies on digestion in the monogenetic trematode *Polystoma integerrima*. *Journal of Helminthology*, **33**, 197–204.

Jennings, J. B. (1960). Observations on the nutrition of the rhynchocoelan *Lineus ruber* (O. F. Müller). *Biological Bulletin*, **119**, 189–96.

Jennings, J. B. (1962). Further studies on feeding and digestion in triclad Turbellaria. *Biological Bulletin*, **123**, 571–81.

ATTACHMENT OF PARASITES

Llewellyn, J. (1956). The host specificity, microecology, adhesive attitudes and comparative morphology of some trematode gill parasites. *Journal of the Marine biological Association of the U.K.* **35**, 113–27.

Llewellyn, J. (1958). The adhesive mechanisms of monogenean trematodes: the attachment of species of the Diclidophoridae to the gills of gadoid fishes. *Journal of the Marine biological Association of the U.K.* **37**, 67–79.

Lyons, K. M. (1964). The chemical nature and evolutionary significance of monogenean attachment sclerites. *Parasitology*, **54**, 12P.

Rees, G. (1961). Studies on the functional morphology of the scolex and of the genitalia in *Echinobothrium brachysoma* Pintner and *E. affine* Diesing from *Raja clavata*. *Parasitology*, **51**, 193–226.

Rees, G. and Williams, H. H. (1965). The functional morphology of the scolex and the genitalia of *Acanthobothrium coronatum* (Rud.) (Cestoda: Tetraphyllidea). *Parasitology*, **55**, 617–51.

TEGUMENT

Beguin, F. (1966). Etude au microscope électronique de la cuticle et de ses structures associées chez quelques cestodes. Essai d'histologie comparée. *Zeitschrift für Zellforschung und mikroskopische Anatomie*, **72**, 30–46.

Hockley, D. J. (1973). Ultrastructure of the tegument in *Schistosoma*. *Advances in Parasitology*, **11**, 235–305.

Lyons, K. M. (1973). The epidermis and sense organs of the Monogenea and some related groups. *Advances in Parasitology*, **11**, 193–232.

Threadgold, L. T. (1962). An electron microscope study of the tegument and associated structures of *Dipylidium caninum*. *Quarterly Journal of microscopical Science*, **103**, 135–40.

Threadgold, L. T. (1963). The tegument and associated structures of *Fasciola hepatica*. *Quarterly Journal of microscopical Science*, **104**, 505–12.

PHYSIOLOGY AND IMMUNOLOGY

Stephenson, W. (1947). Physiology and histochemical observations on the adult liver fluke, *Fasciola hepatica* L. I.-IV. I. Survival *in vitro*. *Parasitology*, **38**, 116–22; II. Feeding. *Parasitology*, **38**, 123–7; III. Egg-shell formation. *Parasitology*, **38**, 128–39; IV. The excretory system. *Parasitology*, **38**, 140–4.

Taylor, A. E. R. (ed.) (1964). Host-parasite relationships in invertebrate hosts. *Symposia of the British Society for Parasitology*, **2**.

Taylor, A. E. R. (ed.) (1968). Immunity to parasites. *Symposia of the British Society for Parasitology*, **6**.

EVOLUTION

Jennings, J. B. (1971). Parasitism and commensalism in the Turbellaria. *Advances in Parasitology*, **9**, 1–32.

Llewellyn, J. (1963). Larvae and larval development of monogeneans. *Advances in Parasitology*, **1**, 287–326.

Llewellyn, J. (1965). The evolution of parasitic platyhelminths. *In* "Evolution of Parasites" (Ed. A. E. R. Taylor). Blackwell, Oxford.

Pearson, J. C. (1972). A phylogeny of life-cycle patterns of the Digenea. *Advances in Parasitology*, **10**, 153–89.

Valuable review articles are to be found in *Advances in Parasitology*, edited by B. Dawes, published annually by Academic Press, London and New York.

Nematodes

As with platyhelminths much of the general anatomy of nematodes and their life history has been summarized in textbooks of zoology or parasitology. Two smaller accounts dealing solely with nematodes are:

Lee, D. L. (1965). "The Physiology of Nematodes". Oliver and Boyd, Edinburgh.

Crofton, H. D. (1966). "Nematodes". Hutchinson University Library, London.

The functional anatomy of nematodes rests fundamentally on the facts recorded in:

Harris, J. E. and Crofton, H. D. (1957). Structure and function in the nematodes: internal pressure and cuticular structure in *Ascaris*. *Journal of experimental Biology*, **34**, 116–30.

Some other papers:

Bird, A. F. and Bird, J. (1969). Skeletal structures and integument of Acanthocephala and Nematoda. *In* "Chemical Zoology". (Eds M. Florkin and B. T. Scheer), **3**, 253–88.

Lee, D. L. (1972). The structure of the helminth cuticle. *Advances in Parasitology*, **10**, 347–79.

Hobson, A. D. (1948). The physiology and cultivation in artificial media of nematodes parasitic in the alimentary tract of animals. *Parasitology*, **38**, 183–227.

Annelids

Apart from the extended treatment given in Grassé's "Traité de Zoologie" there are several books dealing with annelids:

Dales, R. P. (1967). "Annelids". Hutchinson University Library, London. (This deals with the classes Polychaeta, Oligochaeta and Hirudinea but rather scantily with the last two.)

Stephenson, J. (1930). *The Oligochaeta.* Clarendon Press, Oxford. (One of the classical works on the group; though it has much taxonomic material there are introductory chapters on anatomy, but not functional anatomy.)

Laverack, M. S. (1963). "The Physiology of Earthworms". Pergamon Press, Oxford.

Edwards, C. A. and Lofty, J. R. (1972). "Biology of Earthworms". Chapman and Hall, London. (Contains a key for the identification of the common British worms.)

Mann, K. H. (1962). "Leeches (Hirudinea) their Structure, Physiology, Ecology and Embryology". Pergamon Press, Oxford. (Almost the only account in English. Contains keys for identification.)

The following papers either furnish the basic material on which the above books rest, or extend it by more recent investigation. They are grouped by topic rather than taxonomically.

LOCOMOTION AND BODY FORM

Gray, J. and Lissmann, H. W. (1938). Studies in animal locomotion VII. Locomotory reflexes in the earthworm. *Journal of experimental Biology*, **15**, 506–17.

Gray, J., Lissmann, H. W. and Pumphrey, R. J. (1938). The method of locomotion in the leech (*Hirudo medicinalis*). *Journal of experimental Biology*, **15**, 408–30.

Gray, J. (1939). Studies on animal locomotion VIII. The kinetics of locomotion in *Nereis diversicolor*. *Journal of experimental Biology*, **16**, 9–17.

Bhatia, M. L. (1941). *Hirudinaria* (the Indian cattle leech). *Indian Zoological Memoirs*, **8**, 1–85.

Wells, G. P. (1945). The mode of life of *Arenicola marina* L. *Journal of the Marine Biological Association of the U.K.* **26**, 170–207.

Wells, G. P. (1961). How Lugworms move. In "The Cell and the Organism" (Eds J. A. Ramsay and V. B. Wrigglesworth). University Press, Cambridge.

Wells, G. P. (1966). The lugworm (*Arenicola*)—a study in adaptation. *Netherlands Journal of Sea Research*, **3**, 294–313.

Newell, G. E. (1950). The role of the coelomic fluid in the movements of earthworms. *Journal of experimental Biology*, **27**, 110–21.

Clark, R. B. and Clark, M. E. (1960). The fine structure and histochemistry of the ligaments of *Nephtys*. *Quarterly Journal of microscopical Science*, **101**, 133–48. The ligamentary system and the segmental musculature of *Nephtys*. *Quarterly Journal of microscopical Science*, **101**, 149–76.

Clark, R. B. (1962). On the structure and functional significance of polychaete septa. *Proceedings of the Zoological Society of London*, **138**, 543–78.

Clark, R. B. (1964). "Dynamics in Metazoan Evolution. The Origin of the Coelom and Segments". Clarendon Press, Oxford.

Trueman, E. R. (1966). Observations on the borrowing of *Arenicola marina* (L.). *Journal of experimental Biology*, **44**, 93-118. The mechanism of burrowing in the polychaete worm, *Arenicola marina* (L.). *Biological Bulletin*, **131**, 369–77.

Mettam, C. (1967). Segmental musculature and parapodial movement of *Nereis diversicolor* and *Nephthys hombergi* (Annelida: Polychaeta). *Journal of Zoology, London*, **153**, 245–75.

Mettam, C. (1971). Functional design and evolution of the polychaete *Aphrodite aculeata*. *Journal of Zoology, London*, **163**, 489–514.

FEEDING AND DIGESTION

Orton, J. H. (1913). On ciliary mechanisms in brachiopods and some polychaetes. *Journal of the Marine biological Association of the U.K.* **10**, 283–311.

Watson, A. T. (1928). Observations on the habits and life-history of *Pectinaria (Lagis) koreni* Mgr., *Proceedings of the Liverpool Biological Society*, **42**, 25–60.

Nicol, E. A. T. (1930). The feeding mechanism, formation of the tube and physiology of digestion in *Sabella pavonina. Transactions of the Royal Society of Edinburgh*, **56**, 537–98.

MacGinitie, G. E. (1939). The method of feeding of *Chaetopterus. Biological Bulletin*, **77**, 115–18.

Harley, M. B. (1950). Occurrence of a filter-feeding mechanism in the polychaete *Nereis diversicolor. Nature, London*, **165**, 734–5.

Busing, K. H. (1951). *Pseudomonas hirudinis*, ein bakterieller Darmsymbiont des Blutegels (*Hirudo medicinalis*). *Zentralblatt für Bakteriologie* (I Abteilung, Originale), **157**, 478–84.

Busing, K. H., Doll, N. and Freytag, K. (1953). Die Bakterienflora der medizinischen Blutegel. *Archiv für Mikrobiologie*, **19**, 52–86.

Wells, G. P. (1952). The proboscis apparatus of *Arenicola. Journal of the Marine biological Association of the U.K.* **31**, 1–28.

Wells, G. P. (1954). The mechanics of proboscis movement in *Arenicola. Quarterly Journal of microscopical Science*, **95**, 251–70.

Kermack, D. M. (1953). The anatomy and physiology of the gut and vascular system of the polychaete *Arenicola marina* (L.). *Proceedings of the Zoological Society of London*, **125**, 347–81.

Jørgensen, C. B. (1955). Quantitative aspects of filter feeding. *Biological Reviews* **30**, 391–454.

Dales, R. P. (1955). Feeding and digestion in terebellid polychaetes. *Journal of the Marine biological Association of the U.K.* **34**, 55–79.

Dales, R. P. (1957). Some quantitative aspects of feeding in sabellid and serpulid fan worms. *Journal of the Marine biological Association of the U.K.* **36**, 309–16.

Sutton, M. F. (1957). The feeding mechanism, functional morphology and histology of the alimentary canal of *Terebella lapidaria* L. (Polychaeta). *Proceedings of the Zoological Society of London*, **129**, 487–523.

Arthur, D. R. (1965). Form and function in the interpretation of feeding in lumbricid worms. *Viewpoints in Biology*, **4**, 204–51.

RESPIRATION

Dam, L. Van, (1940). On the mechanism of ventilation in *Aphrodite aculeata. Journal of experimental Biology*, **17**, 1–7.

Wells, G. P. (1949). Respiratory movements of *Arenicola marina* (L.): intermittent irrigation of the tube, and intermittent aerial respiration. *Journal of the Marine biological Association of the U.K.* **28**, 447–64.

Wells, G. P. and Dales, R. P. (1951). Spontaneous activity patterns in animal behaviour: the irrigation of the burrow in the polychaetes *Chaetopterus variopedatus* Renier and *Nereis diversicolor* O. F. Müller. *Journal of the Marine biological Association of the U.K.* **29**, 661–80.

Jones, J. D. (1954). Observations on the respiratory physiology and on the haemoglobin of the polychaete genus *Nephthys*, with special reference to *N. hombergii* (Aud. et M.-Edw.). *Journal of experimental Biology*, **32**, 110–25.

Beadle, L. C. (1957). Respiration of the African swampworm *Alma emini* Mich. *Journal of experimental Biology*, **34**, 1–10.

VASCULAR SYSTEM

Ewer, D. W. (1941). The blood systems of

Sabella and *Spirographis. Quarterly Journal of microscopical Science,* **82**, 587–620.

Hanson, J. (1950–51). The blood-system in the Serpulimorpha (Annelida, Polychaeta). I–IV. *Quarterly Journal of microscopical Science,* **91**, 111–29; **91**, 369–78; **92**, 377–83.

Nicoll, P. A. (1956). The anatomy and behaviour of the vascular systems in *Nereis virens* and *Nereis limbata. Biological Bulletin,* **106**, 69–82.

See also Kermack's paper in § Feeding and digestion.

EXCRETION

The classical work of Goodrich on coelomoducts and nephridia is summarized and brought up to date in

Goodrich, E. S. (1946). The study of nephridia and genital ducts since 1895. *Quarterly Journal of microscopical Science,* **86**, 113–392.

Bhatia, M. L. (1938). On the structure of the nephridia and funnels of the Indian leech *Hirudinaria* with remarks on these organs in *Hirudo. Quarterly Journal of microscopical Science,* **81**, 27–80.

Bahl, K. N. (1947). Excretion in the Oligochaeta. *Biological Reviews,* **22**, 109–47.

Ramsay, J. A. (1949). The osmotic relations of the earthworm. *Journal of experimental Biology,* **26**, 46–56. The site of formation of hypotonic urine in the nephridium of *Lumbricus. Journal of experimental Biology,* **26**, 65–75.

Roots, B. I. (1960). Some observations on the chloragogenous tissue of earthworms. *Comparative Biochemistry and Physiology,* **1**, 218–26.

Robertson, J. D. (1936). The function of the calcareous glands in earthworms. *Journal of Experimental Biology,* **13**, 279–97.

NERVOUS SYSTEM

Bullock, T. H. (1945). Functional organization of the giant fiber system of *Lumbricus. Journal of Neurophysiology,* **8**, 55–71.

Nicol, J. A. C. (1948). The giant axons of annelids. *Quarterly Review of Biology,* **23**, 291–323.

Whitear, M. (1953). The stomatogastric nervous stem of *Arenicola. Quarterly Journal of microscopical Science,* **94**, 293–302.

Smith, J. E. (1957). The nervous anatomy of the body segments of nereid polychaetes. *Philosophical Transactions* B, **240**, 135–96.

Some other topics

REPRODUCTION, TUBE BUILDING, PHYLOGENY

Berrill, N. J. (1952). Regeneration and budding in worms. *Biological Reviews,* **27**, 401–38.

Hedley, R. H. (1958). Tube formation by *Pomatoceros triqueter* (Polychaeta). *Journal of the Marine biological Association of the U.K.* **37**, 315–22. See also papers by Watson (1928) and Nicol (1930) in § Feeding and digestion.

Durchon, M. (1960). L'endocrinologie ches lez annélides polychètes. *Bulletin de la Société Zoologique de France,* **85**, 275–301.

Dales, R. P. (1962). The polychaete stomodaeum and the inter-relationships of the families of Polychaeta. *Proceedings of the Zoological Society of London,* **139**, 389–428.

Crustaceans

The crustaceans are a sufficiently large group for two types of monograph to have been published – those which attempt a synthesis of knowledge of the whole group, and those that deal with only a component taxon. Much of the study of the evolution of the crustaceans, too, is part of that of the arthropods as a whole. The following list, therefore, contains references under these three headings as well as a series of others related to the study of functioning within the group.

BOOKS ON CRUSTACEA

Waterman, T. H. (ed.) (1960–61). "The physiology of Crustacea". **1**, 1960; **2**, 1961. Academic Press, New York and London.

Green, J. (1961). "A Biology of Crustacea". Witherby, London.

Lockwood, A. P. M. (1968). "Aspects of the Physiology of Crustacea". Oliver and Boyd, Edinburgh.

Schmitt, W. L. (1965). "Crustaceans". University of Michigan Press, Ann Arbor.

Moore, R. and McCormick, L. (1969). General features of Crustacea. *In* "Treatise of Invertebrate Paleontology". (Ed. R. C. Moore), vol. **R**, R57–R120.

BOOKS AND MONOGRAPHS ON SINGLE CRUSTACEAN GROUPS

Edney, E. B. (1953). The woodlice of Great Britain and Ireland – a concise systematic monograph. *Proceedings of the Linnean Society of London*, **164**, 49–98.

Siewing, R. (1956). Untersuchungen zur Morphologie der Malacostraca (Crustacea). *Zoologische Jahrbucher (Anatomie)*, **75**, 39–176.

Sanders, H. L. (1963). The Cephalocarida. Functional morphology, larval development, comparative external anatomy, *Memoirs of the Connecticut Academy of Arts and Science*, **15**, 1–80.

Mauchline, J. (1969). The biology of euphausiids. *Advances in marine Biology* **7**, 1–421.

Cloudsley-Thompson, J. L. (1972). "The Water and Temperature Relations of Woodlice (Isopoda: Oniscoidea)". Merrow Publishing Co., Watford.

Sutton, S. (1972). "Woodlice". Ginn & Co., London.

EVOLUTION OF ARTHROPODS AND CRUSTACEANS

The evolution of arthropods is dealt with in papers by Manton given in the next section.

Whittington, H. B. and Rolfe, W. D. I. (eds) (1963). "Phylogeny and Evolution of Crustacea". Museum of Comparative Zoology, Cambridge, Mass.

See also Siewing's article in previous §.

LOCOMOTION

There is a series of eleven papers by S. M. Manton dealing with the evolution of locomotory mechanisms in arthropods and all published in the *Journal of the Linnean Society (Zoology)* between 1950 and 1973. Two are concerned with Crustacea.

Manton, S. M. (1952). The evolution of arthropodan locomotory mechanisms. Part 2. General introduction to the locomotory mechanisms of the Arthropoda. *Journal of the Linnean Society (Zoology)*, **42**, 93–117; Part 8. Functional requirements and body design in Chilopoda, together with a comparative account of their skeleto-muscular systems and an Appendix on a comparison between burrowing forces of annelids and chilopods and its bearing upon the evolution of the arthropodan haemocoel. **46**, 251–484.

Some aspects of the work are summarized in

Manton, S. M. (1953). Locomotory habits and the evolution of the larger arthropodan groups. *Symposia of the Society for experimental Biology*, **7**, 339–76.

Other papers deal with particular crustaceans:

Garstang, W. G. (1896). Contributions to marine bionomics. I. The habits and respiratory mechanism of *Corystes cassivelaunus. Journal of the Marine biological Association of the U.K.* **4**, 223–32.

Hart, T. J. (1930). Preliminary notes on the bionomics of the amphipod *Corophium volutator* Pallas. *Journal of the Marine Biological Association of the U.K.* **16**, 761–89.

Dennell, R. (1933). The habits and feeding mechanism of the amphipod *Haustorius arenarius* Slabber. *Journal of the Linnean Society (Zoology).* **28**, 363–88.

Naylor, E. (1958). Spontaneous tidal and diurnal rhythms of locomotory activity in *Carcinus maenas* (L.). *Journal of experimental Biology,* **35**, 602–10.

Naylor, E. (1960). Locomotory rhythms in *Carcinus maenas* (L.) from non-tidal conditions. *Journal of experimental Biology,* **37**, 481–8.

Gauld, D. T. (1959). Swimming and feeding in crustacean larvae: the nauplius larva. *Proceedings of the Zoological Society of London,* **132**, 31–50.

Fryer, G. (1968). Evolution and adaptive radiation in the Chydoridae (Crustacea Cladocera): a study in comparative functional morphology and ecology. *Philosophical Transactions* B, **254**, 221–385.

Nicolaisen, W. and Kanneworff, E. (1969). On the burrowing and feeding habits of the amphipods *Bathyporeia pilosa* Lindström and *Bathyporeia sarsi* Watkin. *Ophelia,* **6**, 231–50.

FEEDING AND DIGESTION

The papers by Garstang, Hart, Dennell, Gauld, Fryer and Nicolaisen and Kanneworff, just quoted, are relevant.

The classical papers within this field are those by Cannon. These are given first followed by others.

Cannon, H. G. (1927). On the feeding mechanism of *Nebalia bipes. Transactions of the Royal Society of Edinburgh,* **55**, 355–69.

Cannon, H. G. (1928). On the feeding mechanism of the fairy shrimp *Chirocephalus diaphanus* Prevost. *Transactions of the Royal Society of Edinburgh,* **55**, 807–22.

Cannon, H. G. (1928). On the feeding mechanism of the copepods *Calanus finmarchicus* and *Diaptomus gracilis.* (British) *Journal of experimental Biology,* **6**, 131–44.

Cannon, H. G. (1933). On the feeding mechanism of the Branchiopoda. *Philosophical Transactions* B, **222**, 267–352.

Cannon, H. G. and Manton, S. M. (1927). On the feeding mechanism of a mysid crustacean, *Hemimysis lamornae. Transactions of the Royal Society of Edinburgh,* **55**, 219–53.

Cannon, H. G. and Manton, S. M. (1929). On the feeding mechanism of the syncarid Crustacea. *Transactions of the Royal Society of Edinburgh,* **56**, 175–89.

Herrick, F. H. (1911). Natural history of the American lobster. *Bulletin of the U.S. Bureau of Fisheries,* **29**, 149–408.

Yonge, C. M. (1924). The feeding mechanism, digestion and assimilation in *Nephrops norvegicus.* (British) *Journal of experimental Biology,* **1**, 343–89.

Orton, J. H. (1927). On the mode of feeding of the hermit-crab, *Eupagurus bernhardus,* and some other Decapoda. *Journal of the Marine biological Association of the U.K.* **14**, 909–21.

Nicholls, A. G. (1931). Studies on *Ligia oceanica.* Part II. The processes of feeding, digestion and absorption, with a description of the structure of the foregut. *Journal of the Marine biological Association of the U.K.* **17**, 675–706.

Nicol, E. A. T. (1932). The feeding habits of the Galatheida. *Journal of the Marine biological Association of the U.K.* **18**, 87–106.

Pearse, A. S. (1945). Ecology of *Upogebia affinis* (Say). *Ecology,* **26**, 303–5.

Fox, H. M. (1952). Anal and oral uptake of water by Crustacea. *Journal of experimental Biology,* **29**, 583–99.

Forster, G. R. (1953). Peritrophic membranes in the Caridea (Crustacea Decapoda). *Journal of the Marine biological Association of the U.K.*, **32**, 315–18.

Barnes, H. and Barnes, M. (1954). The general biology of *Balanus balanus* (L.) da Costa. *Oikos*, **5**, 63–76.

Schafer, N. (1954). Form und Funktion der Brachyuran-Schere. *Abhandlungen hrsg. von der Senckenbergischen naturforschenden Gesellschaft*, **489**, 1–65.

Naylor, E. (1955). The diet and feeding mechanism of *Idotea*. *Journal of the Marine biological Association of the U.K.* **34**, 347–55.

Southward, A. J. (1955). On the behaviour of barnacles. I-II. *Journal of the Marine biological Association of the U.K.*, **34** 403–22; 423–33.

Barnes, H. (1959). Stomach contents and microfeeding of some common cirripedes. *Canadian Journal of Zoology*, **37**, 131–6.

Crisp, D. J. and Southward, A. J. (1961). Different types of cirral activity in barnacles. *Philosophical Transactions* B, **243**, 271–308.

Berner, A. (1962). Feeding and respiration in the copepod *Temora longicornis* (Müller). *Journal of the Marine biological Association of the U.K.* **42**, 625–40.

Martin, A. L. (1964). The alimentary canal of *Marinogammarus obtusatus* (Crustacea, Amphipoda). *Proceedings of the Zoological Society of London*, **143**, 525–44.

Kanneworff, E. and Nicolaisen, W. (1969). The stomach (foregut) of the amphipod *Bathyporeia sarsi* Watkin. *Ophelia*, **6**, 211–29.

Schmitz, E. H. and Schulz, T. W. (1969). Digestive anatomy of terrestrial Isopoda: *Armadillidium vulgare* and *Armadillidium nasatum*. *American midland Naturalist*, **82**, 163–81.

Marshall, S. M. (1973). Respiration and feeding in copepods. *Advances in marine Biology*, **11**, 57–120.

Stone, R. L. and Barnes, H. (1973). The general biology of *Verruca stroemia* (O. F.

Müller). I. *Journal of experimental marine Biology and Ecology*, **12**, 167–85.

RESPIRATION

Some of the papers quoted above also deal with respiratory activities – Garstang, Hart, Dennell (§ Locomotion) and Pearse, Barnes and Barnes, and Marshall (§ Feeding and digestion).

Fox, H. M. and Johnson, M. L. (1934). The control of respiratory movements in Crustacea by oxygen and carbon dioxide. *Journal of experimental Biology*, **11**, 1–10.

Johnson, M. L. (1936). The control of respiratory movements in Crustacea by oxygen and carbon dioxide. II. *Journal of experimental Biology*, **13**, 467–75.

Fox, H. M. (1948). The haemoglobin of *Daphnia*. *Proceedings of the Royal Society* B, **135**, 195–212.

Fox, H. M. (1949). On Apus: its rediscovery in Britain, nomenclature and habits. *Proceedings of the Zoological Society of London*, **119**, 693–702.

Fox, H. M., Gilchrist, B. M. and Phear, E. A. (1951). Functions of haemoglobin in *Daphnia*. *Proceedings of the Royal Society* B, **138**, 514–28.

Gilchrist, B. M. (1954). Haemoglobin in *Artemia*. *Proceedings of the Royal Society* B, **143**, 136–46.

Fox, H. M. (1954). The effect of oxygen on the concentration of haem in invertebrates. *Proceedings of the Royal Society* B, **143**, 203–14.

Edney, E. B. and Spencer, J. O. (1955). Cutaneous respiration in woodlice. *Journal of experimental Biology*, **32**, 256–69.

Green, J. (1956). Variation in the haemoglobin content of *Daphnia*. *Proceedings of the Royal Society* B, **145**, 214–32.

Gray, I. E. (1957). A comparative study of the gill area of crabs. *Biological Bulletin*, **112**, 34–42.

VASCULAR SYSTEM

Drach, P. (1939). Mue et cycle d'intermue chez les Crustacés Decapodes. *Annales de l'Institut océanographique, Paris*, **19**, 103–391.

Krijgsman, B. J. (1952). Contraction and pacemaker mechanisms of the heart of arthropods. *Biological Reviews*, **27**, 320–46.

Burger, J. N. and Smythe, C.McC. (1953). The general form of circulation in the lobster, *Homarus. Journal of cellular and comparative Physiology*, **42**, 369–83.

Silen, L. (1954). On the circulatory system of the Isopoda Oniscoidea. *Acta Zoologica (Stockholm)*, **35**, 11–70.

George, C. J., Nair, K. K. and Muthe, P. T. (1955). The pericardial membrane and its role in crustacean circulation. *Journal of animal Morphology and Physiology*, **2**, 73–8.

EXCRETION AND OSMOREGULATION

Reference should be made to the relevant chapters in the books by Krogh, Potts and Parry and Riegel (p. 551) and to:

Edney, E. B. (1957). "The Water Relations of Terrestrial Arthropods". University Press, Cambridge.

Picken, L. E. R. (1936). The mechanism of urine formation in invertebrates. I. The excretory mechanism in certain Arthropoda. *Journal of experimental Biology*, **34**, 417–23.

Maluf, N. S. R. (1940). The uptake of inorganic electrolytes by the crayfish. *Journal of general Physiology*, **24**, 151–67.

Maluf, N. S. R. (1941). Experimental cytological evidence for an outward secretion of water by the nephridial tubule of the crayfish. *Biological Bulletin*, **81**, 127–33.

Spencer, J. O. and Edney, E. B. (1954). The absorption of water by woodlice. *Journal of experimental Biology*, **31**, 491–6.

Martin, A. W. (1957). Recent advances in knowledge of invertebrate renal func-tion. *In* "Recent Advances in Invertebrate Physiology" (Ed. B. T. Scheer). University of Oregon Press, Eugene.

Robertson, J. D. (1957). Osmoregulation and ionic regulation in aquatic inverte-brates. *In* "Recent Advances in Invertebrate Physiology" (Ed. B. T. Scheer). University of Oregon Press, Eugene.

Edney, E. B. (1968). Transition from water to land in isopod crustaceans. *American Zoologist*, **8**, 309–26.

ENDOCRINES

Carlisle, D. B. and Knowles, F. (1959). "Endocrine Control in Crustaceans". University Press, Cambridge.

Highnam, K. C. and Hill, L. (1969). "The Comparative Endocrinology of the Invertebrates". Arnold, London.

Charniaux-Cotton, H. (1956). Déterminisme hormonal de la différenciation sexuelle chez les Crustacés. *Année biologique*, **32**, 371–99.

Charniaux-Cotton, H. (1957). Croissance, régénération et déterminisme endocrinien des caractères sexuelles d'*Orchestia gammarella* (Pallas), Crustacé Amphipode. *Annales de Science naturelle, Zoologie et Biologie animale*, **19**, 411–599.

Kleinholf, L. H. (1957). Endocrinology of invertebrates, particularly crustaceans. *In* "Recent Advances in Invertebrate Physiology" (Ed. B. T. Scheer). University of Oregon Press, Eugene.

Scheer, B. T. (1957). The hormonal control of metabolism in decapod crusta-ceans. *In* "Recent Advances in Invertebrate Physiology" (Ed. B. T. Scheer). University of Oregon Press, Eugene.

NERVOUS SYSTEM

Johnson, G. E. (1924–26). Giant nerve fibres in crustaceans with special refer-ence to *Cambarus* and *Palaemonetes. Journal of comparative Neurology*, **37**, 291–307. Studies of the function of the giant nerve fibres of crustaceans with special reference to *Cambarus* and Palaemonetes.

Journal of comparative Neurology, **42**, 19–33.

Alexandrowicz, J. S. (1932–34). The innervation of the heart of the Crustacea. I–II. *Quarterly Journal of microscopical Science*, **75**, 181–249; **76**, 511–48.

Alexandrowicz, J. S. (1952). Innervation of the heart of *Ligia oceanica. Journal of the Marine biological Association of the U.K.* **31**, 85–96.

Alexandrowicz, J. S. (1954). Innervation of an amphipod heart. *Journal of the Marine biological Association of the U.K.* **33**, 709–19.

Alexandrowicz, J. S. (1955). Innervation of the heart of *Praunus flexuosus* (Mysidacea). *Journal of the Marine biological Association of the U.K.* **34**, 47–53.

Alexandrowicz, J. S. (1951). Muscle receptor organs in the abdomen of *Homarus vulgaris* and *Palinurus vulgaris. Quarterly Journal of microscopical Science*, **92**, 163–99.

Alexandrowicz, J. S. (1958). Further observations on proprioceptors in Crustacea and a hypothesis about their function, *Journal of the Marine biological Association of the U.K.* **37**, 379–96.

Wiersma, C. A. G. and Ripley, S. H. (1952). Innervation patterns of crustacean limbs. *Physiologica comparata et Ecologia*, **2**, 391–405.

Wiersma, C. A. G. (1957). Neuromuscular mechanisms. *In* "Recent Advances in Invertebrate Physiology" (Ed. B. T. Scheer). University of Oregon Press, Eugene.

Hughes, G. M. and Wiersma, C. A. G. (1960). Neuronal pathways and synaptic connections in the abdominal cord of the crayfish. *Journal of experimental Biology*, **37**, 291–307.

REPRODUCTION

Barnes, H. (1953). Orientation and aggregation in *Balanus balanus* (L.) da Costa. *Journal of Animal Ecology*, **22**, 141–8.

Knight Jones, E. W. (1953). Laboratory experiments on gregariousness during settling in *Balanus balanoides* and other barnacles. *Journal of experimental Biology*, **30**, 584–98.

Crisp, D. J. and Barnes, H. (1954). The orientation and distribution of barnacles at settlement with particular reference to surface contour. *Journal of Animal Ecology*, **23**, 142–62.

Yonge, C. M. (1955). Egg attachment in *Crangon vulgaris* and other Caridea. *Proceedings of the Royal Society of Edinburgh* B, **65**, 369–400.

Arthropods

The following list is limited to a small number of general references or others related to the topics discussed in the chapter.

The most important series of papers dealing with evolution of and within the arthropod grade is that by S. M. Manton dealing initially only with locomotion, but ultimately with the phylogeny of all the groups. It is published as a series of papers, all entitled "The evolution of arthropodan locomotory mechanisms" in the *Journal of the Linnean Society* (Zoology), between 1950 and 1973. The parts are:

(1950). 1. Locomotion of Peripatus. **41**, 529-70.

(1952). 2. General introduction. **42**, 93–117.

(1952). 3. Chilopoda and Pauropoda. **42**, 118–66.

(1954). 4. Diplopoda. **42**, 299–368.

(1956). 5. Pselaphognatha (Diplopoda). **43**, 153–87.

(1958). 6. Lysiopetaloida (Diplopoda); leg design. **43**, 487–556.

(1961). 7. Diplopod burrowing techniques. **44**, 383–461.

(1965). 8. Body design in Chilopoda. **45**, 251–484.

(1966). 9. Body design in Symphyla and Pauropoda. **46**, 103–41.

(1972). 10. Hexapod classes. **51**, 203–400.

(1973). 11. Uniramia and Arachnida. **53**, 257–375.

Summaries of Manton's views are given in

Manton, S. M. (1964). Mandibular mechanisms and the evolution of arthropods. *Philosophical Transactions* B, **247**, 1–183.

Manton, S. M. (1969). Evolution and affinities of Onychophora, Myriapoda, Hexapoda, and Crustacea. *In* "Treatise of Paleontology" (Ed. R. C. Moore), vol. **R**: R15–R56.

Manton, S. M. (1973). Arthropod phylogeny – a modern synthesis. *Journal of Zoology, London*, **171**, 111–30.

So far as insects are concerned we limit ourselves to mentioning only the following books, two dealing with morphology and habits, two with physiology. Many valuable reviews of knowledge on some topics are to be found in "Advances in Insect Physiology" (Eds J. W. L. Beament, J. E. Treherne, V. B. Wigglesworth) published annually from 1963 onwards by Academic Press, London and New York.

Imms, A. D. (1975). "A General Textbook of Entomology". 10th ed. Methuen, London.

Imms, A. D. (1971). "Insect Natural History". 3rd ed. Collins, London.

Wigglesworth, V. B. (1965). "Insect Physiology". 6th ed. Methuen, London.

Rockstein, M. (ed.) (1973–4). "The Physiology of Insecta". Academic Press, New York and London.

Echinoderms

The following books offer accounts – sometimes of unequal value – of various aspects of echinoderm anatomy and physiology.

Nichols, D. (1962). "Echinoderms". Hutchinson University Library, London.

Boolootian, R. A. (ed.) (1966). "Physiology of Echinodermata". Interscience Publishers, New York.

Millott, N. (ed.) (1967). Echinoderm biology. *Symposia of the Zoological Society of London*, **20**.

Binyon, J. (1972). "Physiology of Echinoderms". Pergamon Press, Oxford.

Two important reviews may be added to these:

Gislen, T. (1924). Echinoderm studies. *Zoologiska Bidrag fran Uppsala*, **9**, 1–316.

Nichols, D. (1964). Echinoderms: experimental and ecological. *Oceanography and Marine Biology, Annual Review*, **2**, 393–423.

The following papers deal, sometimes amongst other things, with aspects of functional anatomy:

LOCOMOTION, TUBE FEET, SKIN

Smith, J. E. (1937). The structure and function of the tube feet in certain echinoderms. *Journal of the Marine biological Association of the U.K.* **22**, 345–57.

Smith, J. E. (1946). The mechanisms and innervation of the starfish tube foot-ampullary system. *Philosophical Transactions* B, **232**, 279–310.

Smith, J. E. (1947). The activities of the tube feet in *Asterias rubens* L. I. The mechanisms of movement and posture. *Quarterly Journal of microscopical Science*, **88**, 1–14.

Nichols, D. (1959–61). The histology of the tube-feet and clavulae of *Echinocardium cordatum*. *Quarterly Journal of microscopical Science*. **100**, 73–87. The histology and activities of the tube-feet of *Echinocyamus pusillus*. **100**, 539–55. The histology and activities of the tube-feet of *Antedon bifida*. **101**, 105–17. A comparative histological study of the tube-feet of two regular echinoids. **102**, 157–80.

Buchanan, J. B. (1962). A re-examination of the glandular elements in the tube feet of some common British ophiuroids. *Proceedings of the Zoological Society of London*, **138**, 645–50.

Binyon, J. (1964). On the mode of functioning of the water vascular system of *Asterias rubens* L. *Journal of the Marine biological Association of the U.K.* **44**, 577–88.

Fontaine, A. R. (1964). The integumentary mucous secretions of the ophiuroid *Ophiocomina nigra*. *Journal of the Marine biological Association of the U.K.* **44**, 145–62.

FEEDING AND DIGESTION

Oomen, H. (1926). Verdauungsphysiologische Studien an Holothurien. *Publicazione della Stazione zoologica, Napoli* **7**, 215–97.

Anderson, J. M. (1953–60). Structure and function in the pyloric caeca of *Asterias forbesi*. *Biological Bulletin*, **105**, 47–61. Studies on the cardiac stomach of the starfish *Asterias forbesi*. **107**, 157–73. Studies on the cardiac stomach of a starfish *Patiria miniata* (Brandt). **117**, 185–201. Histological studies on the digestive system of a starfish, *Henricia*, with notes on Tiedemann's pouches in starfishes. **119**, 371–98.

Anderson, J. M. (1961). Structural peculiarities of the pyloric caeca in a particle-feeding sea-star, *Porania pulvillus*. *American Zoologist*, **1**, 338–9.

Lavoie, M. F. (1956). How sea stars open bivalves. *Biological Bulletin*, **111**, 114–22.

Christensen, A. M. (1957). The feeding behaviour of the seastar *Evasterias troschelii* Stimpson. *Limnology and Oceanography*, **2**, 180–97.

Nichols, D. (1959). Changes in the Chalk heart-urchin *Micraster* interpreted in relation to living forms. *Philosophical Transactions*, B, **242**, 347–437.

Goodbody, I. (1960). The feeding mechanism in the sand dollar *Mellita sexies-perforata* (Leske). *Biological Bulletin*, **119**, 80–6.

Farmanfarmaian, A. and Phillips, J. H. (1962). Digestion, storage and translocation of nutrients in the purple sea urchin, *Strongylocentrotus purpuratus*. *Biological Bulletin*, **123**, 105–20.

Goreau, T. F. (1964). On the predation of corals by the spiny starfish *Acanthaster planci* (L.) in the southern Red Sea. *Bulletin. Sea Fishery Research Station, Israel*, **35**, 23–36.

Lewis, J. B. (1964). Feeding and digestion in the tropical sea urchin *Diadema antillarum* Philippi. *Canadian Journal of Zoology*, **42**, 549–57.

Fontaine, A. R. (1965). The feeding mechanism of the ophiuroid *Ophiocomina nigra*. *Journal of the Marine biological Association of the U.K.* **45**, 373–85.

Warner, G. F. (1971). On the ecology of a dense bed of the brittle-star *Ophiothrix fragilis*. *Journal of the Marine biological Association of the U.K.* **51**, 267–82.

ASSOCIATED ACTIVITIES

Mines, G. R. (1912). Note on the mechanics of discharge of the Cuvierian organs of *Holothuria nigra*. *Quarterly Journal of microscopical Science*, **57**, 301–4.

Budington, R. A. (1942). The ciliary transport system of *Asterias forbesi*. *Biological Bulletin*, **83**, 438–50.

Endean, R. (1957). The Cuvierian tubules of *Holothuria leucospilota*. *Quarterly Journal of microscopical Science*, **98**, 455–72.

NERVOUS AND MUSCULAR SYSTEMS

Smith, J. E. (1937–50). On the nervous system of the starfish. *Marthasterias glacialis* (L.). *Philosophical Transactions* B, **227**, 111–73. The motor nervous system of the starfish, *Astropecten irregularis* (Pennant), with special reference to the innervation of the tube feet and ampullae. **234**, 521–8.

Smith, J. E. (1945). The role of the nervous

system in some activities of starfishes. *Biological Reviews*, **20**, 29–43.

Smith, J. E. (1950). Some observations on the nervous mechanisms underlying the behaviour of starfishes. *Symposia of the Society for experimental Biology*, **4**, 196–220.

Pople, W. and Ewer, D. W. (1954–8). Studies on the myoneural physiology of Echinodermata. I-III. *Journal of experimental Biology*, **31**, 114–26; **32**, 59–69; **35**, 712—30.

Boltt, R. E. and Ewer, D. W. (1963). Studies on the myoneural physiology of Echinodermata. IV-V. *Journal of experimental Biology*, **40**, 713–26; **40**, 727–33.

Cobb, J. L. S. and Laverack, M. S. (1966). The lantern of *Echinus esculentus* (L.). I-III. *Proceedings of the Royal Society* B, **164**, 624–40; 641–50; 651–8.

Bryozoans, brachiopods and pogonophorans

The following references supplement those in Hyman and Grassé, and (for bryozoans and brachiopods) those in the "Treatise of Invertebrate Paleontology" (Ed. R. C. Moore).

BRYOZOANS

Ryland, J. S. (1970). "Bryozoans". Hutchinson University Library, London. (A good general account.)

Borg, F. (1926). Studies on recent cyclostomatous Bryozoa. *Zoologiska Bidrag fran Uppsala*, **10**, 181–507.

Marcus, E. (1926). Beobachtungen und Versuche an lebenden Meeresbryozoen. *Zoologische Jahrbucher* (*Systematik der Tiere*), **52**, 1–102.

Harmer, S. F. (1930). Polyzoa. *Proceedings of the Linnean Society of London*, **141**, 68–118.

Harmer, S. F. (1931). Recent work on Polyzoa. *Proceedings of the Linnean Society of London*, **143**, 113–68.

Atkins, D. (1932). The ciliary feeding mechanism of the entoproct Polyzoa and a comparison with that of the ectoproct Polyzoa. *Quarterly Journal of microscopical Science*, **75**, 393–423.

Ryland, J. S. (1962). The association between Polyzoa and algal substrata. *Journal of Animal Ecology*, **31**, 331–8.

Bushnell, J. H. (1966). Environmental relations of Michigan Entoprocta and dynamics of natural populations of *Plumatella repens*. *Ecological Monographs*, **36**, 95–123.

Silen, L. (1966). On the fertilization problem in the gymnolaematous Bryozoa. *Ophelia*, **3**, 113–40.

Ryland, J. S. (1967). Polyzoa. *Oceanography and marine Biology, Annual Review*, **5**, 343–69.

BRACHIOPODS

Rudwick, M. J. S. (1970). "Living and Fossil Brachiopods". Hutchinson University Library, London. (The most up-to-date review.)

Bulman, O. M. B. (1939). Muscle systems in some inarticulate brachiopods. *Geological Magazine*, **76**, 434–44.

Chuang, S. H. (1956). The ciliary feeding mechanisms of *Lingula unguis* (L.) (Brachiopoda). *Proceedings of the Zoological Society of London*, **127**, 167–89.

Atkins, D. (1958). A new species and genus of Kraussinidae (Brachiopoda) with a note on feeding. *Proceedings of the Zoological Society of London*, **131**, 559–81.

Atkins, D. (1959). The growth stages of the lophophore of the brachiopods *Platidia davidsoni* (Eudes Deslongchamps) and *P. anomioides* (Philippi), with notes on the feeding mechanism. *Journal of the Marine biological Association of the U.K.* **38**, 103–32.

Atkins, D. (1959). The growth stages of the lophophore and loop of the brachiopod *Terebratalia transversa* (Sowerby). *Journal of Morphology*, **105**, 401–26.

Chuang, S. H. (1959). The structure and function of the alimentary canal in *Lingula unguis* (L.) (Brachiopoda).

Proceedings of the Zoological Society of London, **132**, 283–311.

Chuang, S. H. (1960). An anatomical, histological and histochemical study of the gut of the brachiopod *Crania anomala*. *Quarterly Journal of microscopical Science*, **101**, 9–18.

Atkins, D. (1960). The ciliary feeding mechanism of the Megathyridae (Brachiopoda), and the growth stages of the lophophore. *Journal of the Marine Biological Association of the U.K.* **39**, 459–79.

Rudwick, M. J. S. (1960). The feeding mechanisms of spire-bearing fossil brachiopods. *Geological Magazine*, **97**, 369–83.

Rudwick, M. J. S. (1961). "Quick" and "catch" adductor muscles in brachiopods. *Nature, London*, **191**, 1021.

Atkins, D. and Rudwick, M. J. S. (1962). The lophophore and ciliary feeding mechanism of the brachiopod *Crania anomala* (Müller). *Journal of the Marine biological Association of the U.K.* **42**, 469–80.

Chuang, S. H. (1964). The circulation of coelomic fluid in *Lingula unguis*. *Proceedings of the Zoological Society of London*, **143**, 221–37.

Fischer, A. G. (1964). Brackish oceans as the cause of the Permo-Triassic marine faunal crisis. *In* "Problems in Palaeoclimatology" (Ed. A. E. M. Nairn). Interscience Publishers, New York.

Rudwick, M. J. S. (1968). The feeding mechanisms and affinities of the Triassic brachiopods *Thecospira* Zugmayer and *Bactrynium* Emmrich. *Palaeontology*, **11**, 329–60.

Owen, G. and Williams, A. (1969). The caecum of articulate Brachiopoda. *Proceedings of the Royal Society* B, **172**, 187–201.

POGONOPHORANS

Ivanov, A. V. (1963). "Pogonophora". Academic Press, London and New York. (The main source of information on the group.)

Caullery, M. (1914). Sur les Siboglinidae, type nouveau d'invertébrés, recueilli par l'expédition du *Siboga*. *Compte rendu de l'Académie des Sciences Paris*, 2014–17. (The first account of the discovery of a pogonophoran.)

Jägersten, G. (1956). Investigations on *Siboglinum ekmani* n. sp. encountered in the Skagerrak. *Zoologiska Bidrag fran Uppsala*, **31**, 211–52.

Jägersten, G. (1957). On the larva of *Siboglinum* with some remarks on the nutrition problems of the Pogonophora. *Zoologiska Bidrag fran Uppsala*, **32**, 67–79.

Southward, A. J. and Southward, E. C. (1963). Notes on the biology of some Pogonophora. *Journal of the Marine biological Association of the U.K.* **43**, 57–64.

Southward, E. C. and Southward, A. J. (1966). A preliminary account of the general and enzyme histochemistry of *Siboglinum atlanticum* and other Pogonophora. *Journal of the Marine biological Association of the U.K.* **46**, 579–616.

Gupta, B. L., Little, C. and Phillip, A. M. (1966). Studies on Pogonophora. Fine structure of the tentacles. *Journal of the Marine biological Association of the U.K.* **46**, 351–72.

Gupta, B. L. and Little, C. (1969). Studies on Pogonophora. II. Ultrastructure of the tentacle crown of *Siphonobrachia*. *Journal of the Marine biological Association of the U.K.* **49**, 717–41.

Little, C. and Gupta, B. L. (1969). Studies on Pogonophora. III. Uptake of nutrients. *Journal of experimental Biology*, **51**, 759–73.

Gupta, B. L. and Little, C. (1970). Studies on Pogonophora. IV. Fine structure of the cuticle and epidermis. *Tissue and Cell*, **2**, 637–96.

Southward, E. C. (1971). Recent researches on the Pogonophora. *Oceanography and marine Biology, Annual Review*, **9**, 193–220.

The following papers deal with the uptake of organic material in solution in water:

Stephens, G. C. (1963). Uptake of organic material by aquatic intervertebrates. II. Accumulation of amino acids by the bamboo worm, *Clymenella torquata*. *Comparative Biochemistry and Physiology*, **10**, 191–202.

Ferguson, J. C. (1967). Utilization of dissolved exogenous nutrients by the starfishes *Asterias forbesi* and *Henricia sanguinolenta*. *Biological Bulletin*, **132**, 161–71.

Chapman, G. and Taylor, A. G. (1968). Uptake of organic solutes by *Nereis virens*. *Nature, London*, **217**, 763–4.

Stephens, G. C. (1968). Dissolved organic matter as a potential source of nutrition for higher organisms. *American Zoologist*, **8**, 95–106.

Taylor, A. G. (1969). The direct uptake of amino acids and other small molecules from seawater by *Nereis virens* Sars. *Comparative Biochemistry and Physiology*, **29**, 243–50.

Johannes, R. E., Coward, S. J. and Webb, K. L. (1969). Are dissolved amino acids an energy source for marine invertebrates? *Comparative Biochemistry and Physiology*, **29**, 283–8.

Molluscs

The documentation of this group is immense and only a small selection of references can be given. An extremely valuable account of the evolution and present range of the phylum is to be found in the account by C. R. Stasek in "Chemical Zoology". **7** (Eds M. Florkin and B. T. Scheer). Other books dealing with molluscs in general, or with special topics are:

Pelseneer, P. (1935). Essai d'éthologie zoologique d'après l'étude des mollusques. *Académie royale de Belgique. Classe des Sciences. Publications de la Fondation Agathon de Potter*, **1**, 1–662. (This contains an extraordinary assemblage of facts about molluscs but unfortunately gives poor references for further study.)

Lane, F. W. (1957). "Kingdom of the Octopus". Jarrolds, London.

Morton, J. E. (1958). "Molluscs". 4th ed. (1967) revised. Hutchinson University Library London.

Fretter, V. and Graham, A. (1962). "British Prosobranch Molluscs". Ray Society, London.

Wells, M. J. (1962). "Brain and Behaviour in Cephalopods". Heinemann, London.

Wilbur, K. M. and Yonge, C. M. (1964–6) "Physiology of Mollusca". 2 vols. Academic Press, New York and London.

Fretter, V. (ed.) (1968). Studies in the structure, physiology and ecology of molluscs. *Symposia of the Zoological Society of London*, **22**.

Purchon, R. D. (1968). "The Biology of the Mollusca". Pergamon Press, Oxford.

Runham, N. W. and Hunter, P. J. (1970). "Terrestrial Slugs". Hutchinson University Library, London.

Fretter, V. and Peake, J. F. (eds) (1975). "Pulmonates". Academic Press, London and New York.

A selection of papers dealing with malacological topics follows.

LOCOMOTION

Young, J. Z. (1938). The functioning of the giant nerve fibres of the squid. *Journal of experimental Biology*, **15**, 170–85.

Lissmann, H. W. (1945). The mechanism of locomotion of gastropod molluscs. I. II. *Journal of experimental Biology*, **21**, 58–69; **22**, 37–50.

Chapman, G. and Newell, G. E. (1956). The role of the body fluid in the movement of soft-bodied invertebrates. II. The extension of the siphons of *Mya arenaria* L. and *Scrobicularia plana* (da

Costa). *Proceedings of the Royal Society* B, **145**, 564–80.

Denton, E. J. and Gilpin-Brown, J. B. (1961). The buoyancy of the cuttlefish *Sepia officinalis* L. The distribution of gas and liquid within the cuttlebone. *Journal of the Marine biological Association of the U.K.* **41**, 319–42.

Denton, E. J. and Howarth, J. V. (1961). The osmotic mechanism of the cuttlebone. *Journal of the Marine biological Association* of the U.K. **41**, 351–64.

Denton, E. J. (1961). The buoyancy of fish and cephalopods. *Progress in Biophysics and biophysical Biochemistry.* **11**, 177–234.

Trueman, E. R., Brand, A. R. and Davis, P. (1966). The dynamics of burrowing of some common littoral bivalves. *Journal of experimental Biology*, **44**, 469–92.

Trueman, E. R. and Ansell, A. D. (1969). The mechanisms of burrowing in soft substrata by marine animals. *Oceanography and Marine Biology, an Annual Review*, **7**, 315–66.

Jones, H. D. and Trueman, E. R. (1970). Locomotion of the limpet, *Patella vulgata* L. *Journal of experimental Biology*, **52**, 201–16.

Jones, H. D. (1973). The mechanism of locomotion of *Agriolimax reticulatus* (Mollusca: Gastropoda). *Journal of Zoology, London*, **171**, 487–98.

Miller, S. L. (1974). Adaptive design of locomotion and foot form in prosobranch gastropods. *Journal of experimental Marine Biology and Ecology*, **14**, 99–156.

Miller, S. L. (1974). The classification, taxonomic distribution, and evolution of locomotor types among prosobranch gastropods. *Proceedings of the Malacological Society of London*, **41**, 233–72.

Feeding and digestion

GENERAL

Graham, A. (1949). The molluscan stomach. *Transactions of the Royal Society of Edinburgh*, **61**, 737–78.

Jørgensen, C. B. (1955). Quantitative aspects of filter feeding in invertebrates. *Biological Reviews*, **30**, 391–454.

Werner, B. (1959). Das Prinzip des endlosen Schleimfilters beim Nahrungserwerb wirbelloser Meerestiere. *Internationale Revue der gesamten Hydrobiologie und Hydrographie*, **44**, 181–216.

Morton, J. E. (1960). The functions of the gut in ciliary feeders. *Biological Reviews*, **35**, 92–140.

GASTROPODS

Orton, J. H. (1912). The mode of feeding of *Crepidula* etc. *Journal of the Marine biological Association of the U.K.* **9**, 444–78.

Orton, J. H. (1913). On ciliary mechanisms in brachiopods and some polychaetes . . . and an account of the endostyle of *Crepidula* and its allies. *Journal of the Marine biological Association of the U.K.* **10**, 283–311.

Yonge, C. M. (1926). Ciliary feeding mechanisms in the thecasomatous pteropods. *Journal of the Linnean Society (Zoology)*, **36**, 417–29.

Yonge, C. M. (1938). Evolution of ciliary feeding in the Prosobranchia, with an account of feeding in *Capulus ungaricus*. *Journal of the Marine biological Association of the U.K.* **22**, 453–68.

Ankel, W. E. (1938). Erwerb und Aufnahme der Nahrung bei den Gastropoden. *Verhandlungen der deutschen zoologischen Gesellschaft, Zoologischer Anzeiger* supplement, **10**, 77–86. (An important paper, well worth consulting even by those who do not read German for the value and beauty of its illustrations.)

Fretter, V. and Graham, A. (1949). The structure and mode of life of the Pyramidellidae, parasitic opisthobranchs. *Journal of the Marine biological Association of the U.K.* **28**, 493–532.

Forrest, J. E. (1953). Feeding habits and morphology and mode of functioning of the alimentary system in some littoral dorid nudibranchiate Mollusca. *Proceedings of the Linnean Society of London*, **164**, 225–35.

Werner, B. (1953). Über den Nahrungserwerb der Calyptraeidae (Gastropoda Prosobranchia). *Helgoländer wissentschaftliche Meeresuntersuchungen*, **4**, 260–315.

Ziegelmeier, E. (1954). Beobachtungen über den Nahrungserwerb bei der Naticide *Lunatia nitida* Donovan (Gastropoda Prosobranchia). *Helgoländer wissentschaftliche Meeresuntersuchungen*, **5**, 1–33.

Kohn, A. J. (1956). Piscivorous gastropods of the genus *Conus*. *Proceedings of the National Academy of Sciences*, **42**, 168–71.

Gascoigne, T. (1956). Feeding and reproduction in the Limapontiidae. *Transactions of the Royal Society of Edinburgh*, **63**, 129–51.

Miller, M. C. (1961). Distribution and food of the nudibranchiate Mollusca of the south of the Isle of Man. *Journal of Animal Ecology*, **30**, 95-116,

Graham, A. (1965). The buccal mass of ianthinid prosobranchs. *Proceedings of the Malacological Society of London*, **36**, 323–38.

Graham, A. (1973). The anatomical basis of function in the buccal mass of prosobranch and amphineuran molluscs. *Journal of Zoology, London*, **169**, 317–48.

Hurst, A. (1965). Studies on the structure and function of the feeding apparatus of *Philine aperta* with a comparative consideration of some other opisthobranchs. *Malacologia*, **2**, 281–347.

Carriker, M. R. (1969). Excavation of boreholes by the gastropod, *Urosalpinx*: an analysis by light and scanning electron microscopy. *American Zoologist*, **9**, 917–33.

BIVALVES

Yonge, C. M. (1925). The digestive diverticula in the lamellibranchs. *Transactions of the Royal Society of Edinburgh*, **54**, 703–18.

Yonge, C. M. (1926). Structure and physiology of the organs of feeding and digestion in *Ostrea edulis*. *Journal of the Marine biological Association of the U.K.* **14**, 295–386.

Yonge, C. M. (1927). Structure and function of the organs of feeding and digestion in the septibranchs, *Cuspidaria* and *Poromya*. *Philosophical Transactions* B, **216**, 221–63.

Yonge, C. M. (1939). The protobranchiate Mollusca: a functional interpretation of their structure and evolution. *Philosophical Transactions* B, **230**, 79–147.

Atkins, D. (1936–43). On the ciliary mechanisms and interrelationships of lamellibranchs. I-VII. *Quarterly Journal of microscopical Science*, **79**, 181–308, 339–73, 375–421; 423–45; **80**, 321–9; 331–44; 345–435.

Purchon, R. D. (1956–60). The stomach in the [lamellibranchs]. *Proceedings of the Zoological Society of London*, **127**, 511–22; **129**, 26–60; **131**, 487–525; **135**, 431–89.

Ballantine, D. and Morton, J. E. (1956). Filtering, feeding and digestion in the lamellibranch *Lasaea rubra*. *Journal of the Marine biological Association of the U.K.* **35**, 241–74.

Morton, J. E. (1956). The tidal rhythm and action of the digestive system of the lamellibranch *Lasaea rubra*. *Journal of the Marine biological Association of the U.K.* **35**, 563–86.

Purchon, R. D. (1971). Digestion in filter feeding bivalves – a new concept. *Proceedings of the Malacological Society of London*, **39**, 253–62.

OTHER GROUPS

Fretter, V. (1937). The structure and function of the alimentary canal of some species of Polyplacophora (Mollusca). *Transactions of the Royal Society of Edinburgh*, **59**, 119–64.

Bidder, A. M. (1950). The digestive mechanism of the squids *Loligo vulgaris, Loligo forbesii, Alloteuthis media* and *Alloteuthis subulata. Quarterly Journal of microscopical Science,* **91**, 1–43.

Dinamani, P. (1964). Feeding in *Dentalium conspicuum. Proceedings of the Malacological Society of London,* **36**, 1–5.

SKIN, RESPIRATION AND MANTLE CAVITY

Many of the papers dealing with bivalves in § Feeding and digestion inevitably cover this topic. Some others are:

Yonge, C. M. (1939). On the mantle cavity and its contained organs in the Loricata (Placophora). *Quarterly Journal of microscopical Science,* **81**, 367–90.

Yonge, C. M. (1947). The pallial organs of the aspidobranch Gastropoda and their evolution throughout the Mollusca. *Philosophical Transactions* B, **232**, 443–518.

Thompson, T. E. (1960). Defensive adaptations in opisthobranchs. *Journal of the Marine biological Association of the U.K.* **39**, 123–34.

Edmunds, M. (1966). Protective mechanisms in the Eolidacea (Mollusca Nudibranchia). *Journal of the Linnean Society (Zoology),* **46**, 27–71.

EXCRETION

Picken, L. E. R. (1937). The mechanism of urine formation in invertebrates II. The excretory mechanism in certain Mollusca. *Journal of experimental Biology,* **14**, 20–34.

Little, C. (1965). Osmotic and ionic regulation in the prosobranch gastropod mollusc, *Viviparus viviparus* Linn. *Journal of experimental Biology,* **43**, 23–37; The formation of urine by the prosobranch gastropod *Viviparus viviparus* Linn. *Journal of experimental Biology,* **43**, 39–54.

Potts, W. T. W. (1967). Excretion in the invertebrates. *Biological Reviews,* **42**, 1–41.

Andrews, E. B. and Little, C. (1972). Structure and function in the excretory systems of some terrestrial prosobranch snails (Cyclophoridae). *Journal of Zoology, London,* **168**, 395–422.

REPRODUCTION, DEVELOPMENT, EVOLUTION

Garstang, W. (1928). The origin and evolution of larval forms. Presidential address, Section D. *British Association Reports, Glasgow,* 77–98.

Crofts, D. R. (1937). The development of *Haliotis tuberculata,* with special reference to the organogenesis during torsion. *Philosophical Transactions,* B, **208**, 219–68.

Crofts, D. R. (1955). Muscle morphogenesis in primitive gastropods and its relation to torsion. *Proceedings of the Zoological Society of London,* **125**, 711–50.

Coe, W. R. (1943–4). Sexual differentiation in mollusks. I–II. *Quarterly Review of Biology,* **18**, 154–64; **19**, 85–97.

Fretter, V. (1941). The genital ducts of some British stenoglossan prosobranchs. *Journal of the Marine biological Association of the U.K.* **25**, 173–211.

Fretter, V. (1946). The genital ducts of *Theodoxus, Lamellaria* and *Trivia,* and a discussion on their evolution in the prosobranchs, *Journal of the Marine biological Association of the U.K.* **26**, 312–51.

Ankel, W. E. (1958). Beobachtungen und Überlegungen zur Morphogenese der atypische Spermien von *Scalaria clathrus* L. *Zoologischer Anzeiger,* **160**, 261–76.

Lemche, H. and Wingstrand, K. G. (1959). The anatomy of *Neopilina galatheae* Lemche, 1957 (Mollusca, Tryblidiacea). *Galathea Reports,* **4**, 9–72.

Galtsoff, P. S. (1961). Physiology of reproduction in molluscs. *American Zoologist,* **1**, 273–89.

Duncan, C. J. (1960). The evolution of the pulmonate genital system. *Proceedings of the Zoological Society of London,* **134**, 601–9. The genital systems of the freshwater Basommatophora. **135**, 339–56.

Thompson, T. E. (1961). The natural history, embryology, larval biology and post-larval development of *Adalaria proxima* (A. and H.) (Gastropoda Opisthobranchia). *Philosophical Transactions*, B, **242**, 1–58.

Thompson, T. E. (1962). The structure and mode of functioning of the reproductive organs of *Tritonia lineata* (Gastropoda Opisthobranchia). *Quarterly Journal of microscopical Science*, **102**, 1–14.

Ghiselin, M. T. (1966). The adaptive significance of gastropod torsion. *Evolution*, **20**, 337–48.

Streiff, W. (1966). Etude endocrinologique du déterminisme du cycle sexuel chez un mollusque hermaphrodite protandre, *Calyptraea sinensis* (L.) I-II. *Annales d'Endocrinologie*, **27**, 375–400; **28**, 461–72.

Index

Page numbers in *italics* refer to Figures